AUTOMOTIVE ENGINES
Systems, Diagnosis, and Service

AUTOMOTIVE ENGINES
Systems, Diagnosis, and Service

James G. Hughes
Rio Hondo College

HBJ

HARCOURT BRACE JOVANOVICH, PUBLISHERS
Technology Publications

San Diego New York Chicago Austin Washington, D.C.
London Sydney Tokyo Toronto

Cover and original design: Bill Agee
Production supervision: WordCrafters Editorial Services, Inc.

The information contained herein has been developed by James G. Hughes and Classic Examples, Inc., for publication by Harcourt Brace Jovanovich, Inc. Inquiries regarding the content of this work should be addressed to Classic Examples, Inc. and/or the publisher. The National Institute for AUTOMOTIVE SERVICE EXCELLENCE is neither the author nor the publisher of this work and has no responsibility for the material contained herein.

The information contained herein has been developed from the best available sources. Technology Publications, James G. Hughes, and Classic Examples, Inc. cannot assume responsibility for the accuracy of resource data or consequences of its application. Persons using this book are cautioned to refer to vehicle and equipment manufacturers' manuals for specific service information. Also be advised to observe all safety precautions herein and to refer to manufacturers' procedures for specific vehicle and personnel safety directions.

ISBN: 0-15-504366-8

Library of Congress Catalog Card Number: 87-82958

Printed in the United States of America

PREFACE

An essential part of any successful automotive training program is a solid foundation on which to develop the information to be presented. To save time and effort and to ensure that students receive the training they need to prepare them to enter the automotive service field, the author has built this book around the task list prepared by the National Institute for AUTOMOTIVE SERVICE EXCELLENCE (ASE).

This comprehensive list of job-related diagnostic and repair tasks was initially prepared by representatives of the automotive service industry. ASE updates each of the individual certification areas by assembling a group of industry specialists from domestic and import manufacturers, after-market suppliers, working technicians, and educators. This group reviews the task list, adding new and eliminating old technology. Each proposed change must be approved by consensus of the entire group. The result is a task list that represents the body of knowledge each working technician should possess.

Because this book was developed from the task list on which the ASE certification tests are based, using the book for initial or refresher training on automotive engine repair should help prepare individuals for the test.

Facts About ASE Certification

ASE is a nonprofit organization dedicated to improving the quality of automotive service and repair. It is governed by a board of directors representing all sectors of the automotive industry. ASE's voluntary certification program offers experienced technicians the opportunity to take any of the eight automotive certification tests during the May or November testing periods. Each test consists of either 40 or 80 questions on diagnosis and repair techniques. To be certified, the technician must pass one or more tests and be able to prove a minimum of two years' related education and/or work experience. Successfully completing all eight tests qualifies a technician as a Certified Master Automobile Technician (CMAT).

The eight ASE test areas are as follows:

1. Engine Repair (80 questions)
2. Automatic Transmission/Transaxle (40 questions)
3. Manual Drive Train and Axles (40 questions)
4. Suspension and Steering (40 questions)
5. Brakes (40 questions)
6. Electrical Systems (40 questions)
7. Heating and Air Conditioning (40 questions)
8. Engine Performance (80 questions)

Applicants must sign up in advance to take these tests. For complete information about ASE certification and test registration, contact:

National Institute for AUTOMOTIVE SERVICE EXCELLENCE
13505 Dulles Technology Drive
Herndon, VA 22071-3415
(703) 904-0100

Certification for Automobile Technician Training Programs

Future generations of automotive service professionals are in the hands of today's vocational educators. Automotive instructors have the opportunity not only to train technicians for entry-level jobs, but to encourage and foster their professionalism. To further this effort, ASE offers voluntary certification for automobile technician training programs upon the recommendation of the National Automotive Technicians Education Foundation (NATEF).

NATEF's goals are to develop and encourage automotive technical education. The foundation examines the structure and resources of automotive technician training programs and evaluates them in relation to certain nationally accepted standards. Among these standards is a task list nearly identical to the ASE task list.

To make effective use of this book, the training program should include this information and these procedures.

1. Appropriate theory, safety, and support instruction will be required to enable the trainee to perform each task. This instruction must include identification and use of tools and equipment required to accomplish the task. It must also include training in how to locate and use current reference materials from accepted industry publications that present the manufacturer's recommendations regarding the task (see Appendix A).

2. The training program should have performance standards for each task covered in the curriculum. For example: *Working alone or with one other person, you will remove and replace the cylinder heads on two instructor-assigned automobiles. The manufacturer's recommended procedure and specifications will be followed. Time limit will be twice the flat-rate time allowed in the Motors or Chilton flat-rate manual.* Individual progress should be monitored and evaluated against these performance standards.

3. A feedback system should be in place that informs the trainee as to which tasks have been completed successfully.

Special Features of this Book

1. True case history examples

2. Shop tips

3. Safety cautions

4. More than 200 questions similar to those on the actual ASE Engine Repair Test

5. Addresses for obtaining manufacturers' service manuals

6. Addresses for obtaining the most commonly used comprehensive shop manuals

7. Many useful reference tables and troubleshooting charts

ACKNOWLEDGMENTS

I am grateful for the comments and suggestions of the individuals who contributed to the development of this book. I extend special thanks to Bob Barkhouse (retired), Yuba College; William J. deKryger, Central Michigan University; Jay Ray, Rio Hondo College; Mike Slavich, Rio Hondo College; Bill J. Steen, Citrus College; and Dave Whitcomb, Rancho Santiago College.

Grateful acknowledgment is also made to the following companies, which provided reference literature used in the preparation of this book:

A–C Delco, General Motors Corporation; A-C Spark Plug Division, General Motors Corporation; American Honda Motor Company; American Motors Corporation; Automotive Engineering; Automotive Services Division/3M Company; Belban Manufacturing Corporation; Belden Corporation; Black and Decker Inc.; Brush Research Manufacturing Company; Buick Motor Division, General Motors Corporation; Cadillac Motor Division, General Motors Corporation; Champion Spark Plug Company; Chevrolet Motor Division, General Motors Corporation; Chrysler Corporation; Classic Examples Inc.; Clevite Industries Inc.; Cy-Lent Timing Gears Corporation; Dana Corporation; Dayco Corporation; Detroit Diesel Allison; Dorman Products; Easco/K-D Tools; Echlin Manufacturing Company; Engine Parts Inc.; Exide Corporation; Federal-Mogul Corporation; Fel-Pro Incorporated; Ford Parts and Service Division; Fram Corporation; Garrett Corporation; Gates Rubber Company; Hall-Toledo; Hastings Manufacturing Company; Hearst Corporation; Heli-Coil Products, Division of Mite Corporation; Hyundai Motor America; Jaguar Rover Triumph, Inc.; John Deere and Company; K-Line Industries; Kent-Moore Tool Company, Sealed Power Corporation; Kwick-Way Manufacturing Company; K & W Products; Lawson Products, Inc.; Lisle Corporation; Mac Tools, Inc.; Magnaflux Corporation; Marvel Oil Company; McKay Manufacturing Company; McQuay-Norris, Inc.; Mitsubishi Motor Sales of America, Inc.; Modine Auto-Cool Inc.; National Safety Council; Neward Enterprises, Inc.; Neway Manufacturing, Inc.; Nissan Motor Company, Ltd.; Oldsmobile Division, General Motors Corporation; Owatonna Tool Company (OTC); Peugeot Motors of America, Inc.; Pontiac Motor Division, General Motors Corporation; Quaker State Oil Refining Corporation; Safeguard Engine Parts, Inc.; Sealed Power Corporation; Silver Seal Products Company, Inc.; Sioux Tools Inc.; Snap-on Tools Corporation; Sun Electric Corporation; Sunnen Products Company; Thexton Manufacturing Company; T. Hoff Mfg. Company; Tobin Arp Manufacturing Company; Tomadur Engine Company; Toyota Motor Sales, U.S.A., Inc.; Trojan Battery Company; TRW Automotive Aftermarket Division; Vandervell Canada Limited; Winona Van Norman Machine Company; Zollner Corporation.

The author also gratefully acknowledges the National Institute for AUTOMOTIVE SERVICE EXCELLENCE for granting special permission to reproduce its Engine Repair task list.

ABOUT THE AUTHOR

James G. Hughes has been a full-time automotive training instructor at Rio Hondo College in Whittier, California, for twenty-two years, and was the national service training instructor for Toyota Motor Sales, U.S.A., Inc. He has written several textbooks and conducted numerous workshops for state and national audiences. Mr. Hughes is a Certified Master Automobile Technician and a member of the Society of Automotive Engineers. He was selected as the outstanding community college automotive instructor in California for 1985.

CONTENTS

Section E

AUTOMOTIVE ENGINES
Systems, Diagnosis, and Service

Section A

GENERAL ENGINE DIAGNOSIS

After studying this section, and with hands-on practice on a "live" vehicle, you should be able to perform the following tasks:

- Verify driver's complaint and/or road test vehicle; determine needed repairs.
- Inspect engine assembly for fuel, oil, coolant, and other leaks; determine needed repairs.
- Listen to engine noises; determine needed repairs.
- Diagnose the cause of excessive oil consumption, unusual engine exhaust color, odor, and sound; determine needed repairs.
- Perform engine vacuum tests; determine needed repairs.
- Perform cylinder power balance tests; determine needed repairs.
- Perform cylinder compression tests; determine needed repairs.
- Perform cylinder leakage tests; determine needed repairs.

ENGINE OPERATING PRINCIPLES

The process of converting heat energy into power in the gasoline engine is illustrated in Figure 1. The fuel mixture (air and gasoline) is pushed into the cylinder, compressed when the piston rises, and ignited (burned) to produce heat energy. This burning is rapid enough to be considered an explosion. At maximum rpm, it can happen in less than two-thousandths of a second (2 milliseconds). Inside the cylinder, high pressure and high temperature result as the fuel mixture gases expand. The piston, which moves up and down in the cylinder, is pushed down. As it moves downward, the piston moves the connecting rod which, in turn, rotates the crankshaft connected to it (Figures 2 and 3).

The action in the cylinder and related components from the intake of the air–fuel mixture to its compression, combustion, and the exhausting of the burned fuel mixture is called a *firing cycle* (Figure 4[a]).

The crankshaft turns two revolutions (720 degrees) for each firing cycle. The highest point of travel reached by the piston is called *top dead center (TDC),* and the lowest point of travel, *bottom dead center (BDC).*

In most automobile four-stroke-cycle engines, each cylinder is equipped with an intake valve and an exhaust valve. However, a number of passenger car engines are built with more than two valves per cylinder (Figure 4[b]). This helps to improve intake and exhaust flow efficiency, thereby increasing fuel economy.

DIAGNOSING ENGINE PROBLEMS

When diagnosing engine problems, become a "person of habit" and follow a set pattern. Follow the six-step VECTOR approach (outlined in the next paragraph). When you discover the reason for a problem, "file" it in your head for future reference. If you diagnose problems in this manner, you will increase the likelihood of fixing them correctly the first time.

The VECTOR approach has been structured for more efficient diagnosis of engine problems:

1. Verify the complaint by road testing. Can you actually witness the condition? Is the condition normal, or a legitimate complaint? Did the customer provide complete and accurate information? Analyze the symptoms and determine any related symptoms. See Figures 5 and 6.

2. Examine the systems involved. Could the problem be related to other systems on the vehicle? See Figure 7.

3. Check all related components both visually and mechanically. Isolate the problem and determine needed repairs (Figure 8).

4. Think about the information collected. Have you been systematic in your approach to the problem?

5. Organize yourself as well as your information, manuals, tools, and parts.

6. Repair the condition.

COOLING SYSTEM

If it were possible to build an internal combustion engine with 100% efficiency, there would be no need for a cooling system. Such efficiency, however, is not practical for passenger vehicles. As a result, a cooling system is needed.

Only part of the energy in the heat of combustion is converted to useful work. Some of this energy is lost as heat in the exhaust, some as heat radiated from the engine, and a large portion as heat transferred to the cooling system.

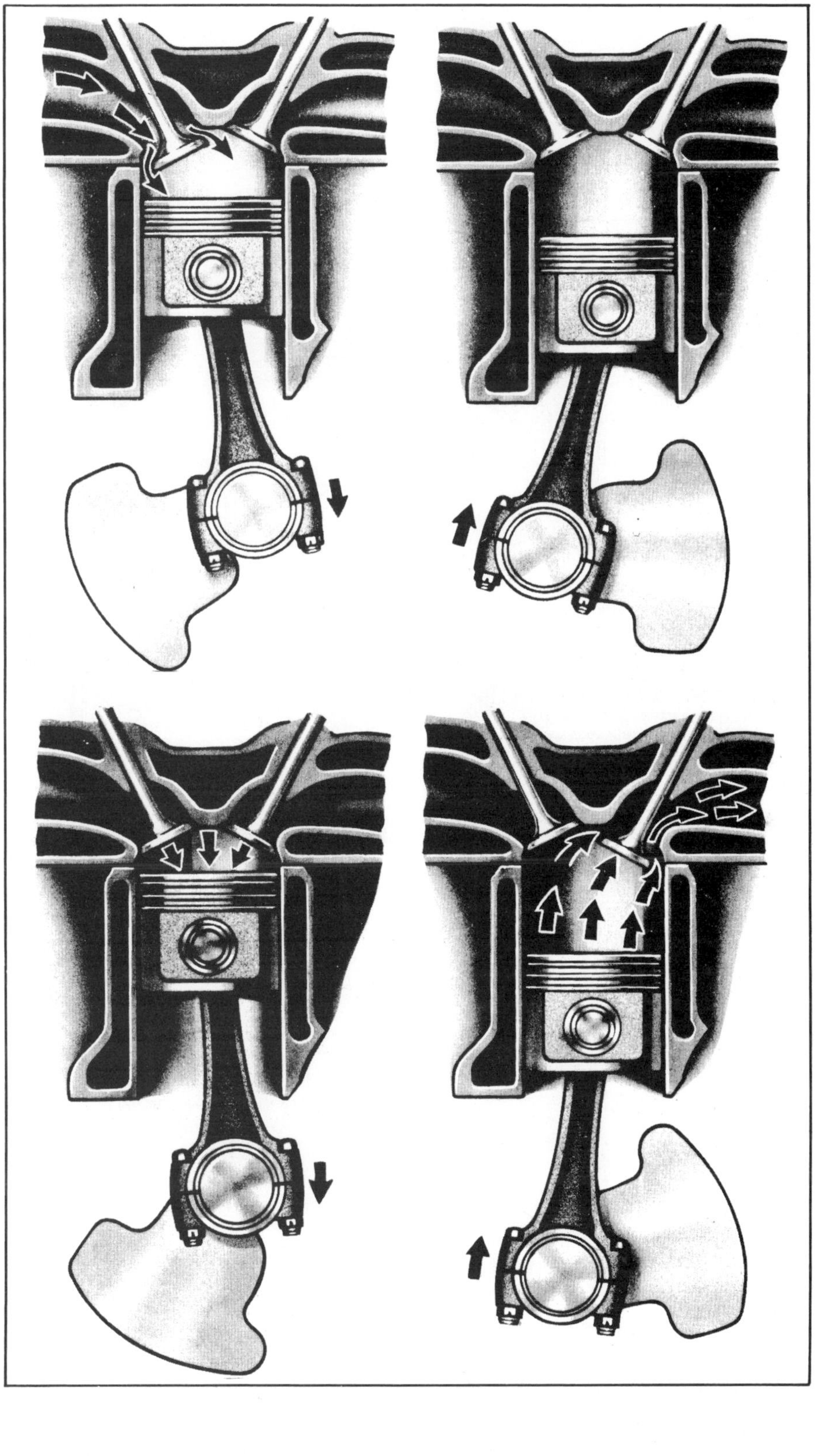

FIGURE 1

Four-stroke-cycle gasoline engine. (Courtesy of Chevrolet Motor Division, General Motors Corporation)

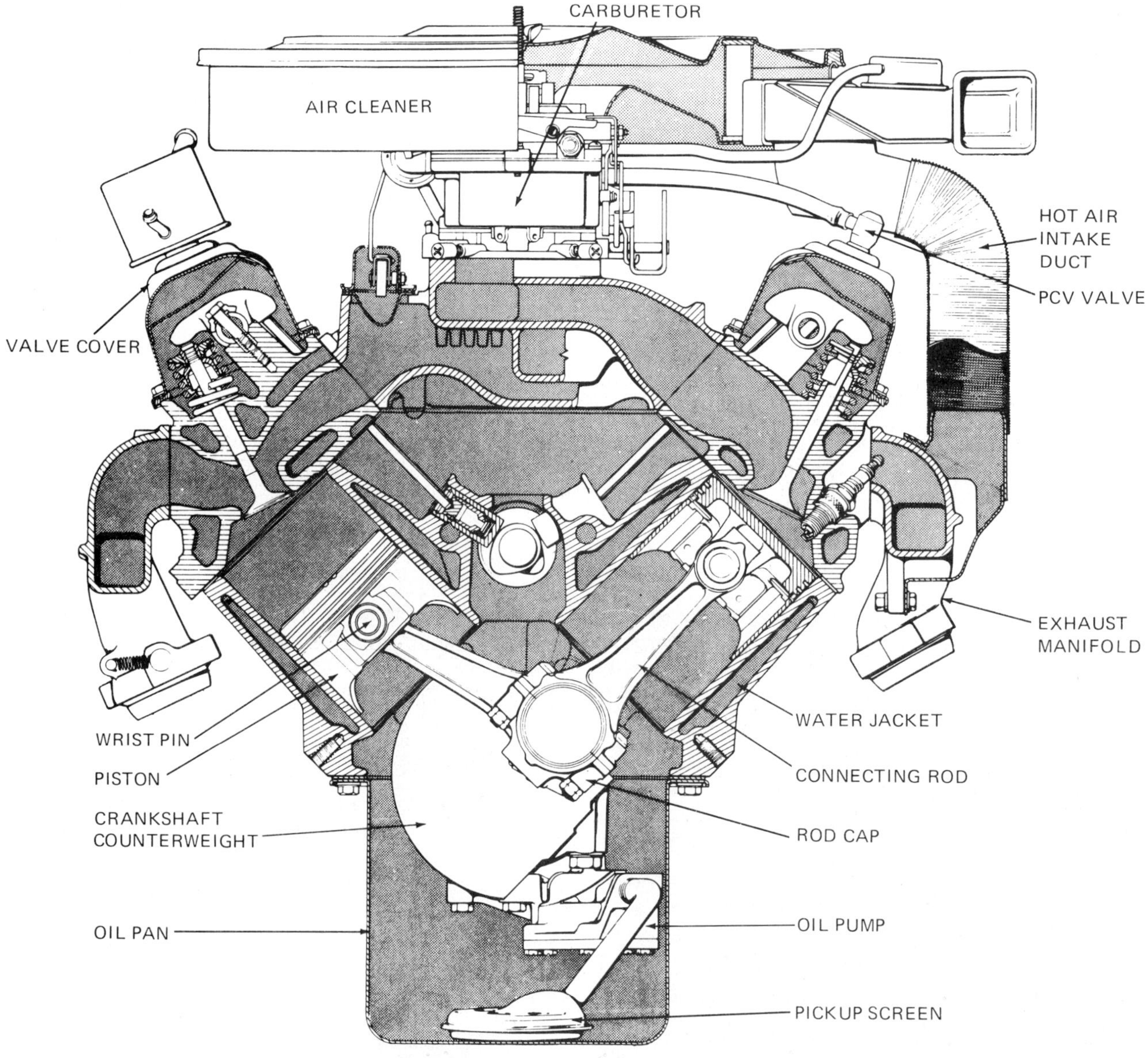

FIGURE 2
Cross-sectional view of an overhead valve (OHV) engine. The camshaft is not labeled. Can you locate it? (Courtesy of Chrysler Corporation)

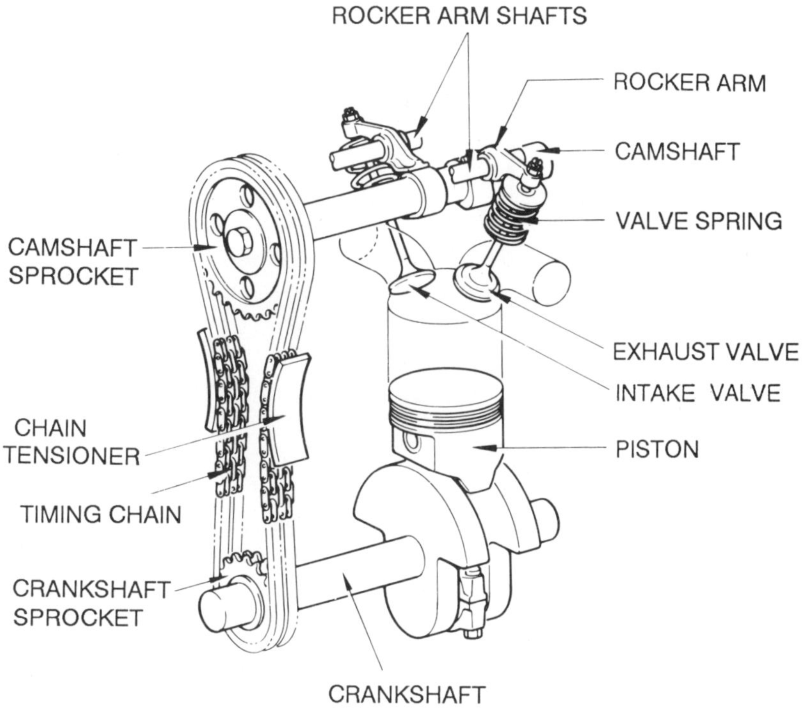

FIGURE 3
Piston and crankshaft mechanism. Notice that the camshaft is positioned overhead.

COOLING SYSTEM PROBLEMS

Engine overheating, overcooling, and coolant loss are the most common cooling system problems. The main causes of problems are internal and external leaks, the accumulation of contaminants in the water jackets, a defective thermostat, and the passing of combustion gases into the coolant (Figure 9). Other causes will be discussed later.

A defective head gasket or cracked head or block can allow exhaust gas into the coolant. If a combustion leak is left unrepaired, it will allow coolant to enter the crankcase. Rapid oil contamination, corrosion of engine parts, and formation of sludge will occur. There are several ways to test for internal combustion gas leakage into the coolant:

1. Start the engine and watch the gauge on a cooling system pressure tester gauge after operating temperature is reached. If the gauge needle moves up as the engine is revved up, the head gasket is probably faulty.

2. Use a combustion leak tester kit (Figure 10[a] and [b]).

3. Look into the thermostat housing and watch for bubbles (Figure 11).

4. Use an exhaust analyzer at the radiator filler neck.

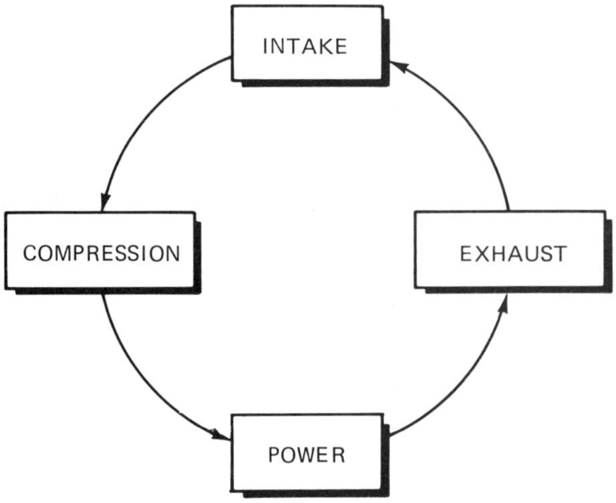

(a)

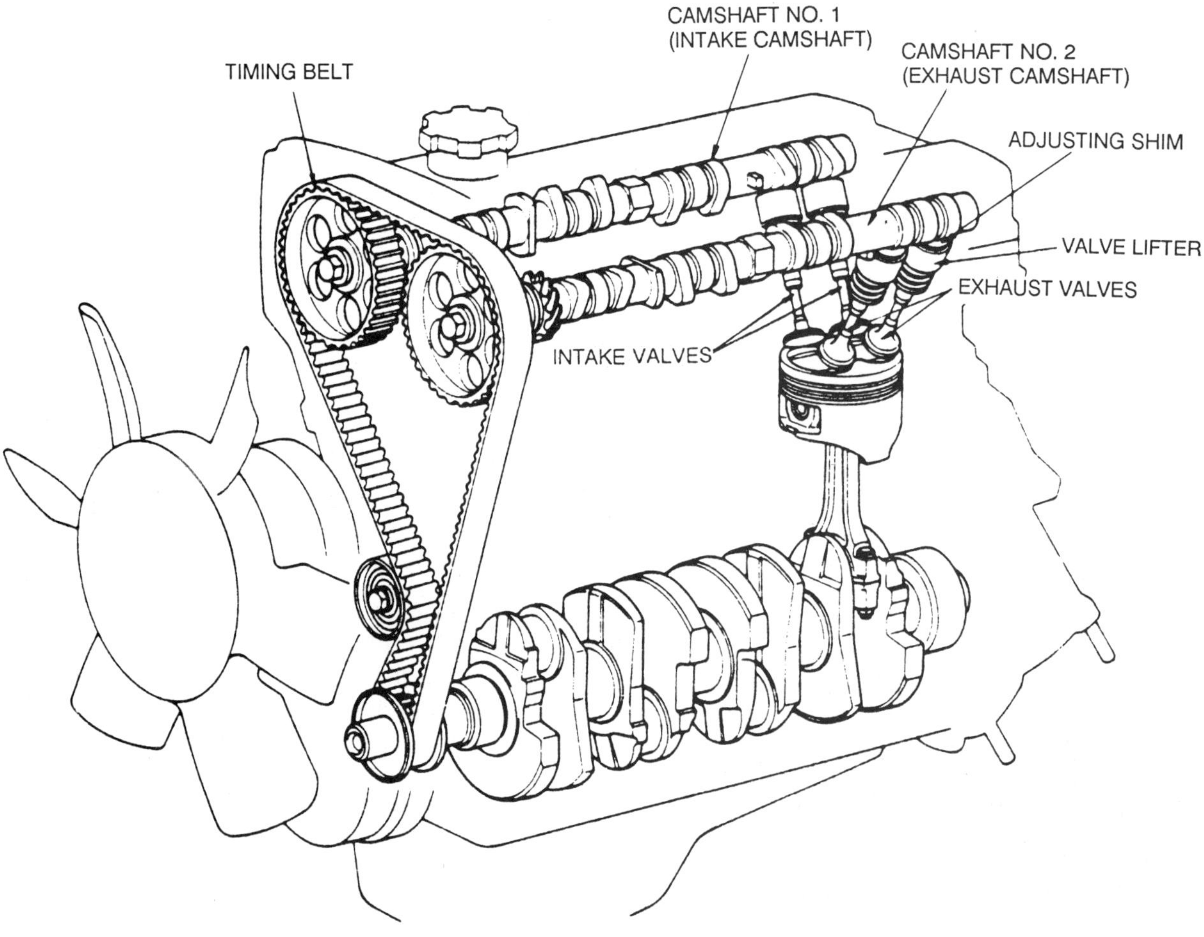

(b)

FIGURE 4

(a) One engine cycle produces power in four actions (strokes). (b) Phantom view of the Toyota DOHC 4A-GE engine used in the Corolla. (The same engine is used in the GM Nova.) In this engine, each cylinder is equipped with two intake valves and two exhaust valves so that the weight of each valve can be reduced. A valve mechanism designed in this manner will reduce inertia weight and friction loss. (Courtesy of Toyota Motor Sales, U.S.A., Inc.)

FIGURE 5
The diaphragm inside this automatic transmission modulator became torn and allowed automatic transmission fluid to be drawn into the intake manifold. This resulted in a badly smoking engine, rough idle, and poor shift quality.

FIGURE 7
Late-model Chevrolet small block 305-, 350-, and 400-CID engines often exhibit intake manifold gasket burn-through failure that can cause engine oil to be drawn into the cylinders. This problem is caused by high temperatures around the EGR valve. One solution to the problem is to use polyacrylic compound intake manifold gaskets. (Courtesy of Dana Corporation)

FIGURE 8
Metal head gaskets can corrode and allow combustion gas to enter the cooling system. In addition, small amounts of coolant can get into the cylinders. Sometimes, this problem can be isolated by pulling all the spark plugs (look for rust-colored firing ends) and pressurizing the cooling system. After waiting 10 minutes, crank the engine and check for "spitting" from the spark plug holes.

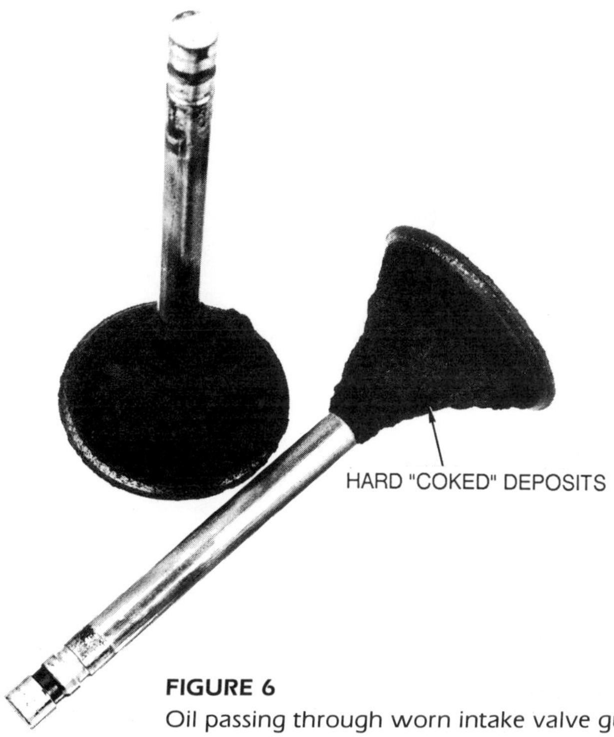

FIGURE 6
Oil passing through worn intake valve guides allows oil to be drawn into the combustion chamber and burned. In addition, hard black "coked" deposits form under the valve heads. This can upset air flow and cause rough idle.

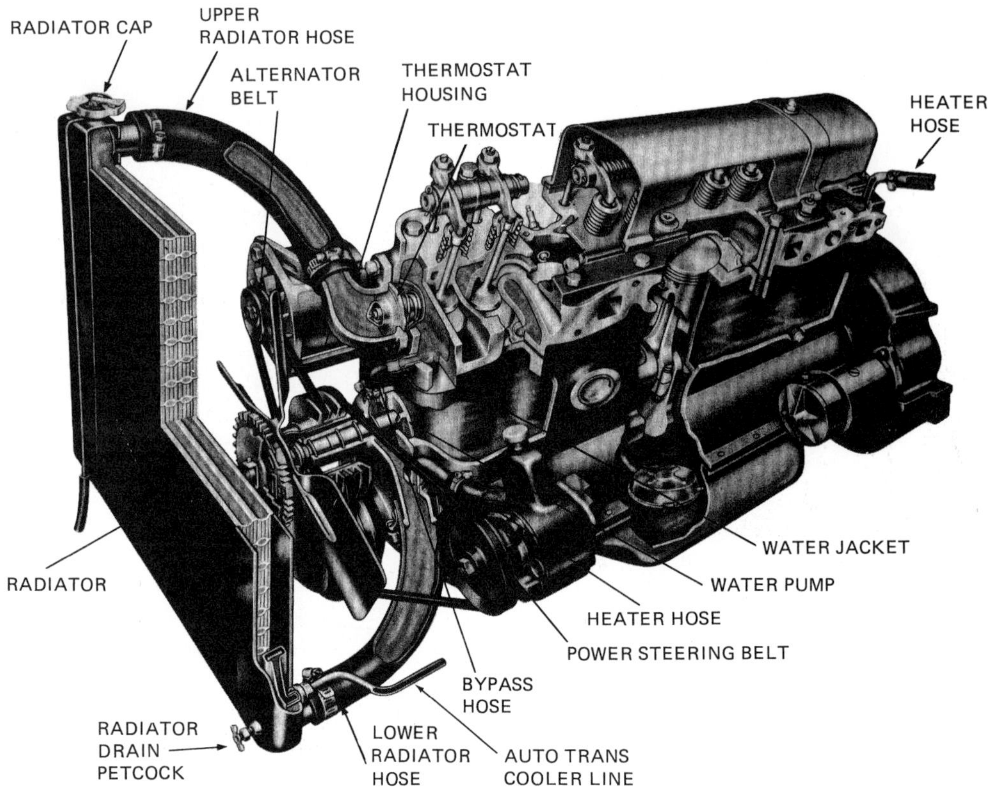

FIGURE 9

An engine cooling system. The water pump circulates the coolant in the engine block through water jackets which surround the cylinder walls, spark plugs, and valves. Hot coolant from the water jackets then returns to the radiator. (Courtesy of Ford Parts and Service Division)

FIGURE 10

(a) This test fluid will turn yellow in the presence of combustion gas. (b) This tube is filled with the test fluid and seated in the radiator filler opening. While the engine is running, air above the radiator coolant is drawn into the test fluid. A change in the test fluid color indicates a combustion gas leak into the engine coolant.

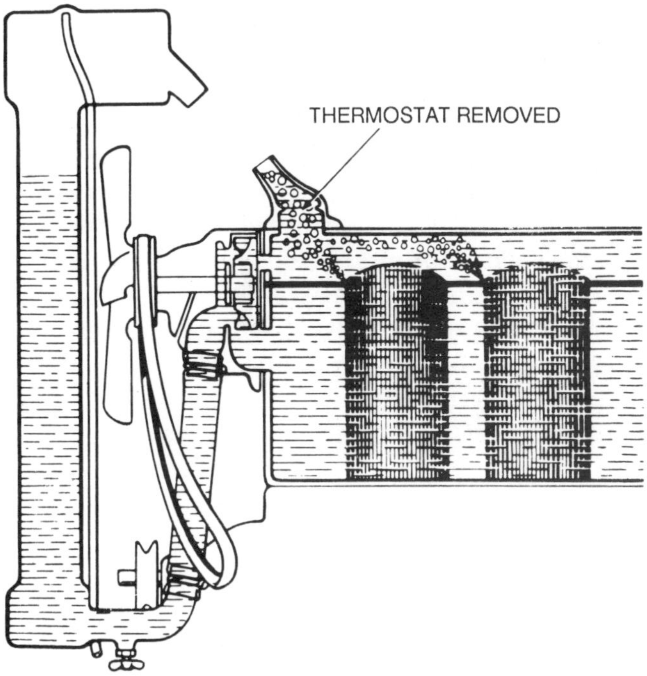

FIGURE 11
An old-fashioned diagnostic method that can still be used. After removing the upper radiator hose and thermostat, you can check for combustion leakage into the coolant. Start the engine and accelerate it six to eight times. Watch for bubbles or a sudden rise in liquid level while accelerating and also when rpm drops back to idle. The appearance of bubbles or a rise in liquid level indicates combustion leakage.

FIGURE 12
Scale and rust can cause an engine to overheat.

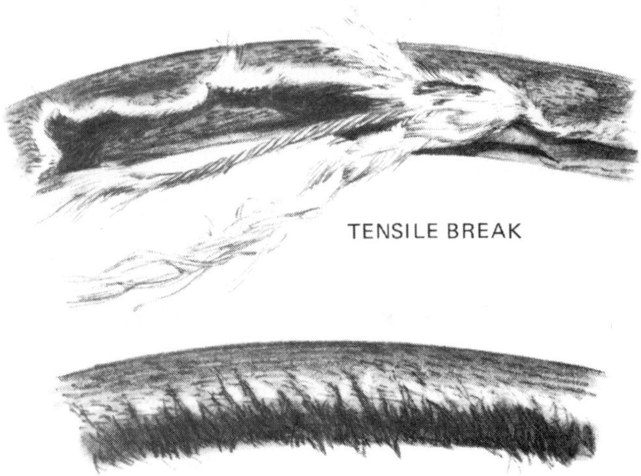

FIGURE 13
Belt tensile breaks can be caused by forcing the belt over the pulley during installation, high shock loads, or by a foreign object wedged between the bottom of the groove and the belt. Slip burns can be caused by loose belt tension or by excessively worn pulley grooves.

Oxygen from the air is the key ingredient in promoting rust. Aeration (mixing air with water) of the coolant can quickly produce rust. The principal causes for aeration are:

1. Turbulence in the upper radiator tank.
2. Small leaks at the water pump or hoses.

Scaling is a condition caused by certain minerals present in water. See Figure 12. These minerals deposit on the jacket walls, forming a white crusty coating that is a heat insulator.

Some other frequently found causes of engine overheating are:

1. Low coolant level.
2. Plugged, dirty, or blocked radiator fins.
3. Incorrect ignition timing.
4. Low engine oil level.
5. Broken or slipping fan belt (**Figure 13**).
6. Faulty radiator cap.
7. Defective water pump.
8. Faulty thermostat.

9. Faulty fan clutch (Figure 14).

10. Plugged exhaust system.

Overcooling

An engine that runs too cool can cause as much trouble as one that runs too hot. A too-cool engine causes excessive gas consumption and prevents the heater from warming up. Overcooling results in the engine not getting hot enough to burn away the water produced by combustion. This water (vapor) gets into the crankcase as blowby and produces sludge. Two possible causes of overcooling are:

1. Missing, malfunctioning, or improper (wrong) thermostat.

2. Faulty heater control valve that permits constant coolant flow through the heater core (Figure 15).

A too-cool running engine on a car equipped with a computer-controlled fuel system can result in the air–fuel mixture staying rich. The computer will keep the mixture rich because the coolant temperature sensor indicates the engine has not fully warmed up. This extra fuel causes the catalytic converter (often called a "cat") to overwork trying to burn up the rich mixture. However, the cat can handle only so much extra fuel. It will get hotter and hotter until it becomes red hot! This can set the underside of the car on fire. This author has seen several cars burned up this way.

Loss of Coolant

External loss of coolant can sometimes be detected by the rust stains or white spots that form around the leak area. Another way to check a cooling system for external leaks is to use a pressure tester (Figure 16). Look for external coolant leaks at these locations:

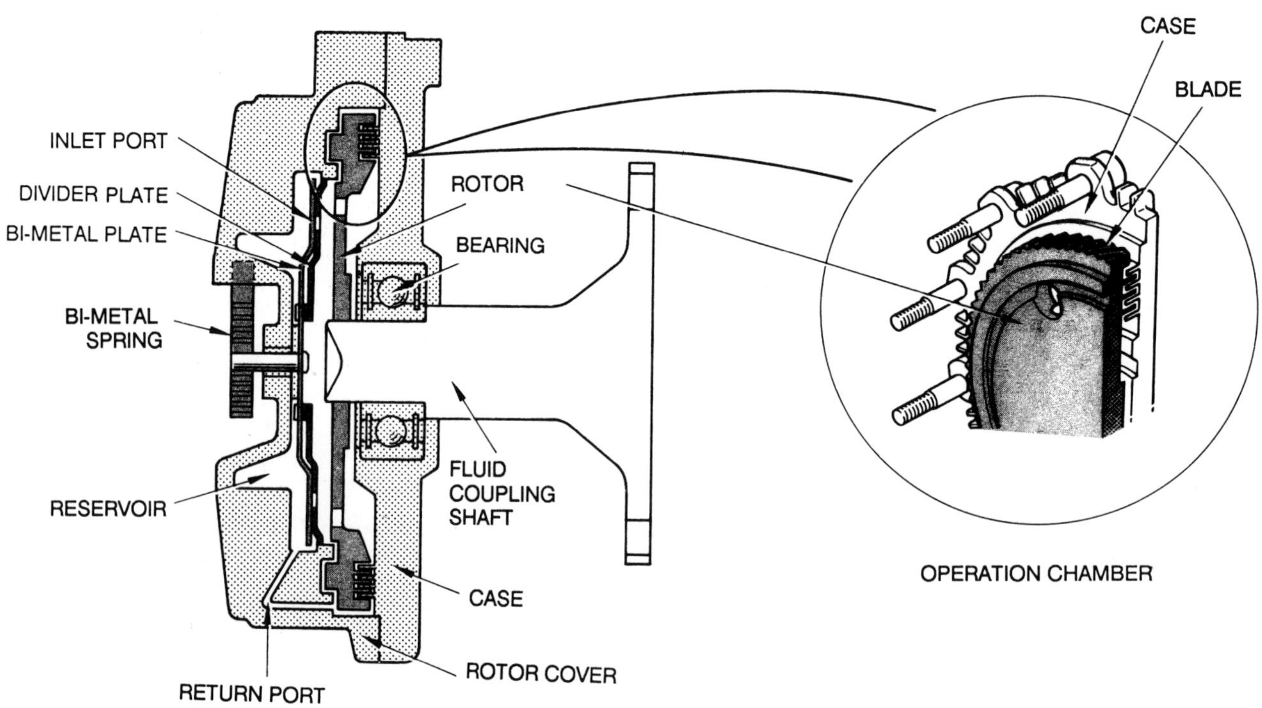

FIGURE 14
Temperature-controlled fluid coupling. Replace if you notice a noisy or rough feeling when you turn the assembly by hand.

FIGURE 15
A heater control valve that is stuck open will allow coolant flow through the heater core all the time. This can cause the engine to run cooler than it should.

1. Hoses and clamps.

2. Water pump shaft seal.

3. Radiator.

4. Heater control valve.

5. Water temperature sending unit.

6. Freeze plugs (check at back of block on some engines).

7. Thermostat housing.

8. Radiator petcock.

9. Water pump bolts.

10. Radiator cap.

11. Heater core (usually this leak will occur inside the passenger compartment).

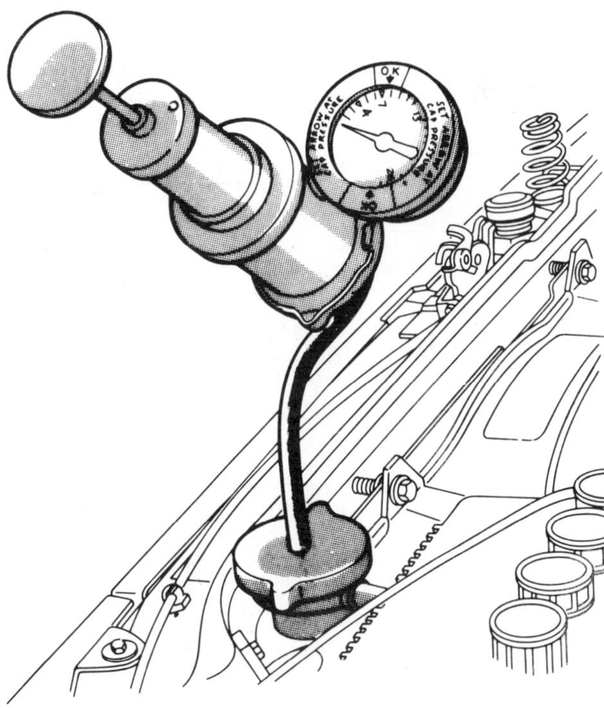

FIGURE 16
Pressure testing the cooling system. (Courtesy of Chrysler Corporation)

Loss of coolant through an internal crack or leak usually is detected by seeing water bubbles on the dipstick and/or noting chocolate colored oil. Internal leaks may occur at these locations:

1. A head gasket that is either improperly installed or defective.

2. Cracked head or block (Figure 17).

3. Head bolts that are loose or unevenly tightened.

4. Leaking transmission oil cooler.

FIGURE 17
The severe crack in this six-cylinder head allowed coolant to mix with oil. The inside of the valve cover was coated with frothy, yellowish deposits.

5. An intake manifold gasket that is either improperly installed or defective.

6. A front timing chain cover that is "pinholed" from electrolysis (Figure 18).

ENGINE ASSEMBLY FUEL LEAKS

A faulty mechanical fuel pump is sometimes responsible for oil consumption and poor fuel economy (Figure 19). A tear in the pump diaphragm can cause fuel to enter the crankcase. This dilutes the oil and causes it to be more easily consumed by the engine.

Fuel "pull-over" is an undesirable situation wherein fuel delivery (usually drippage) occurs at the carburetor main nozzle(s) during idle (Figure 20). This results in rough idle, surge, and black exhaust smoke from the tailpipe. Fuel "pull-over" is usually caused by one of these conditions:

1. High float level.

2. "Heavy" float.

3. Worn needle valve and/or seat.

4. Misaligned throttle valves.

5. Excessive fuel pump pressure.

6. Plugged air bleed(s).

7. Idle mixture screw tip damage.

SHOP TIP

A "pull-over" problem that is being caused by plugged air bleeds can often be corrected quickly. Run the engine at 2500 rpm and quickly choke and unchoke the carburetor several times. This will allow intake manifold vacuum to pull dirt from the bleeds.

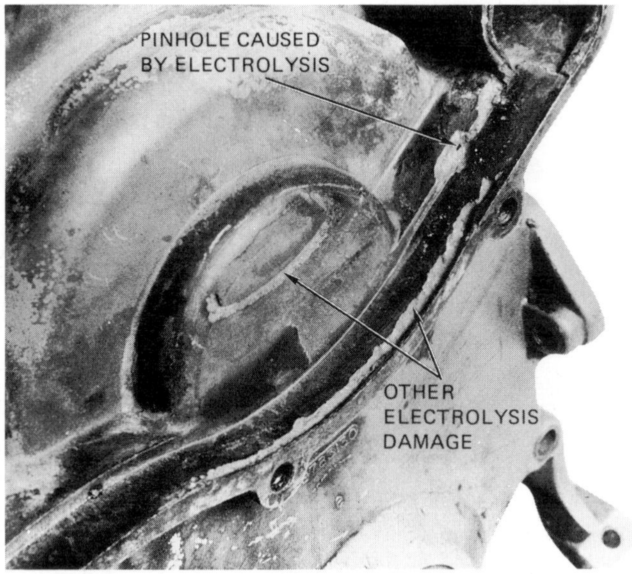

FIGURE 18
The water pump cavity in this aluminum front cover is "pinholed" from electrolysis. Coolant leaked into the timing chain area and into the crankcase.

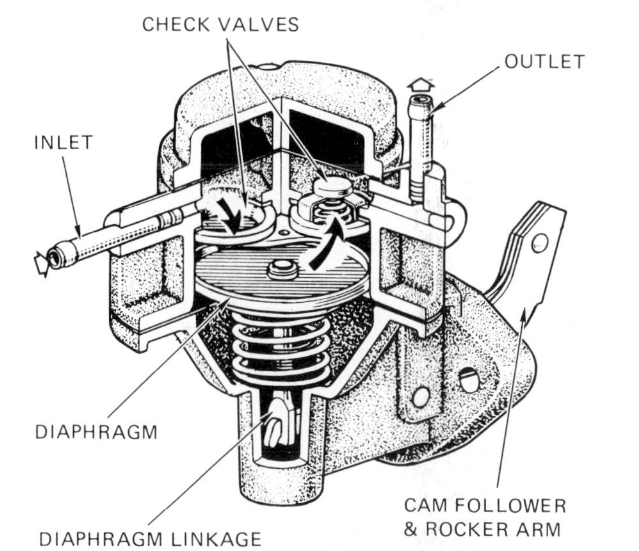

FIGURE 19
Mechanical fuel pump. The rocker arm is usually driven by an eccentric lobe on the camshaft or by a push rod.

FIGURE 20
There should be no fuel coming out the end of the carburetor main nozzle(s) during idle.

DIAGNOSING ENGINE NOISES

While it is very difficult to describe noises using words, in the following paragraphs we will attempt to do so. Experience is your best teacher. When you determine the cause of an abnormal noise condition, remember the situation for future reference.

Abnormal engine noises vary in sound intensity (loudness) and frequency (rate of occurrence). When diagnosing engine noises, make sure that an accessory or body part is not the cause. Also, noises can fool you by transmitting the sound to some other location.

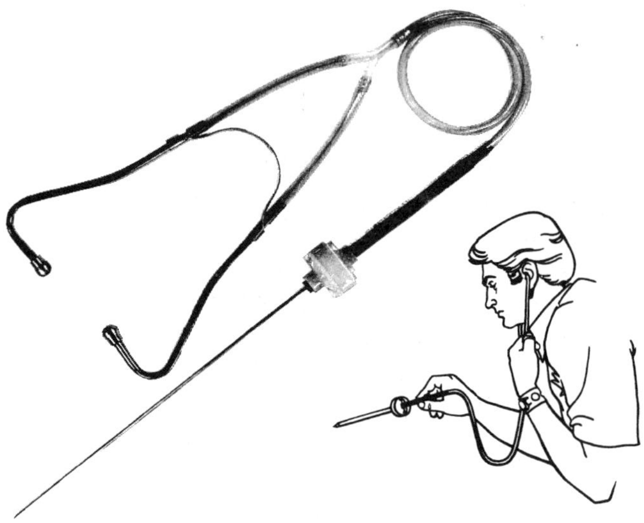

FIGURE 21
A stethoscope can be used to pinpoint the source of engine noise.

> **SHOP TIP**
> A very good tool for locating the source of engine noise is a stethoscope (Figure 21). In the absence of such a tool, you can use a length of heater hose and a funnel to carry sound directly to your ear.

Crankshaft Noises

Main bearing noises are usually *heavy metallic knocks.* These knocks increase in frequency as the speed and load on the engine is increased. Some possible causes of main bearing noise are:

1. Insufficient oil supply.
2. Low oil pressure.
3. Thin or diluted oil.
4. Excessive bearing clearance (Figure 22).
5. Out-of-round journal.

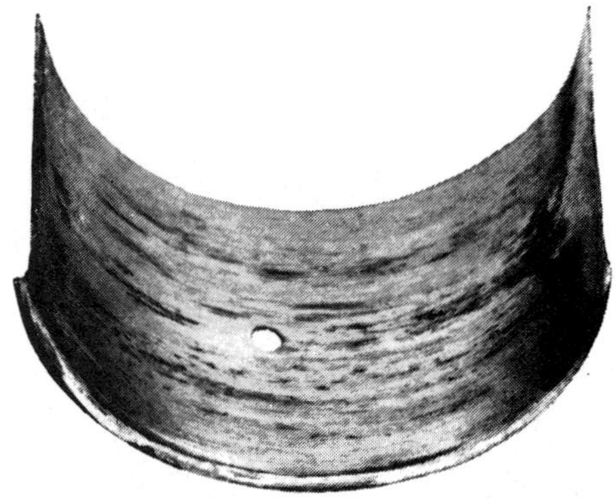

FIGURE 22
This bearing was making a knocking noise due to excessive clearance before it "spun."

Excessive crankshaft endplay can cause a *sharp rap* to occur at irregularly spaced intervals during idle (Figure 23).

The crankshaft vibration damper assembly can become separated (Figure 24). This will generally produce a *heavy rattling noise* that can be heard at low speed.

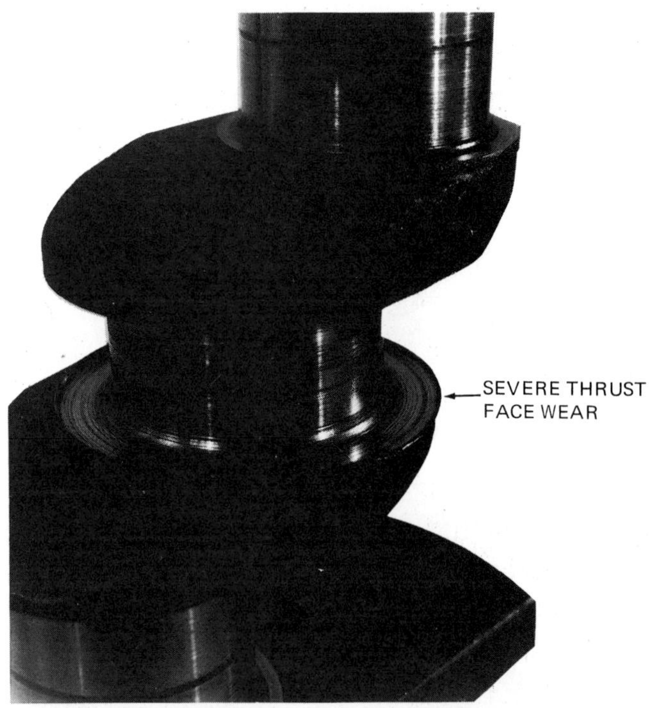

FIGURE 23
High lube oil pressure in the converter can cause excessive end thrust pressure on the crankshaft and severe thrust wear. This, in turn, can cause noise and cranking engagement problems.

FIGURE 24
This separated vibration damper produced a rattling noise heard only at idle and off-idle speed.

A sheared crankshaft snout key ruined this vibration damper hub and caused a rattling noise.

Sometimes a crankshaft will break off at the snout or at one of the bearing journals (Figure 25). A *heavy knocking sound* is then heard while the engine is running. Occasionally a crankshaft key will shear off, causing the vibration damper to rattle (Figure 26).

> **SHOP TIP**
> A loose flywheel or broken flex plate can produce a noise that sounds just like a rod knock (Figure 27).

FIGURE 25
Crank failure caused by a faulty vibration damper. This type of failure is almost always indicated by a 45° angle fracture across the bearing journal and through the oil hole.

FIGURE 27
Cracked automatic transmission flex plate. The noise produced while the engine was running sounded similar to a connecting rod knock.

Connecting Rod Noises

Connecting rod noises are usually *light raps or clatters* much less intense than main bearing knocks, and loudest when the engine is running under a light load. The noise becomes louder as engine rpm is increased. By individually grounding out each spark plug one at a time, you will be able to determine which cylinder the noise is coming from.

Connecting rod noise may be caused by:

1. Insufficient oil supply.
2. Low oil pressure.
3. Thin or diluted oil.
4. Excessive bearing clearance.
5. Out-of-round journal.
6. Misaligned rod.

Piston Noises

The most common piston noise is known as *slap.* This is a *hollow, bell-like sound* due to rocking of the piston from side to side in the cylinder. This condition is brought about when piston clearance becomes excessive or when a piston skirt collapses or breaks (Figure 28).

Piston slap is loudest when the engine is cold, and decreases or disappears as the engine reaches operating temperature.

Piston Pin Noise

Piston pin noise is generally caused by excessive pin clearance. It produces a *metallic, double-knock sound* most noticeable when the engine is idling.

FIGURE 28
The skirt on this piston had broken off and was found in the bottom of the oil pan. Severe slap noise resulted.

Valve Train Noise

Valve train noise has a characteristic *"clicking" sound.* The frequency of the noise is at one-half crankshaft speed. Some possible causes of valve train noise are:

1. High or low oil level in crankcase.
2. Thin or diluted oil.
3. Low oil pressure.
4. Faulty oil pump.
5. Worn or damaged lifter (Figure 29).
6. Excessive lifter leakdown (from wear, dirt, or varnish).
7. Improper valve adjustment (Figure 30).
8. Bent push rod (Figure 31).

9. Worn rocker arms (Figure 32).

10. Worn valve guides.

11. Broken valve spring.

12. Loose valve seat (Figure 33).

13. Worn cam lobes.

Other Engine Noises

Some other frequently encountered sources of engine noise are:

1. Belt "squish" or "squeal."

2. Manifold heat control valve rattle.

3. Air injection reaction (AIR) pump "squeak."

FIGURE 29
This cracked lifter was inoperative and produced an audible clicking sound.

FIGURE 30
Valve lash (clearance) on this Toyota engine is set by turning an adjustment screw. Excessive clearance can cause noise and poor performance. Not enough clearance can cause rough idle and valve burning.

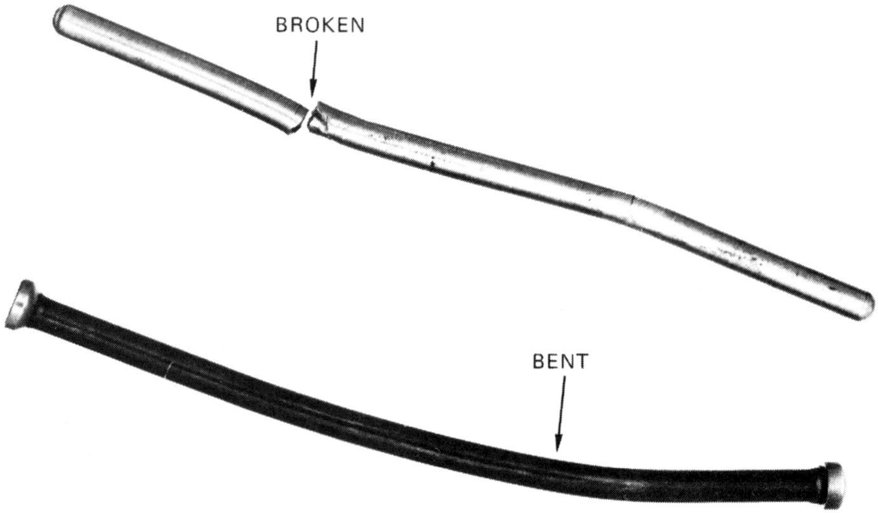

FIGURE 31
Damaged (bent) push rods.

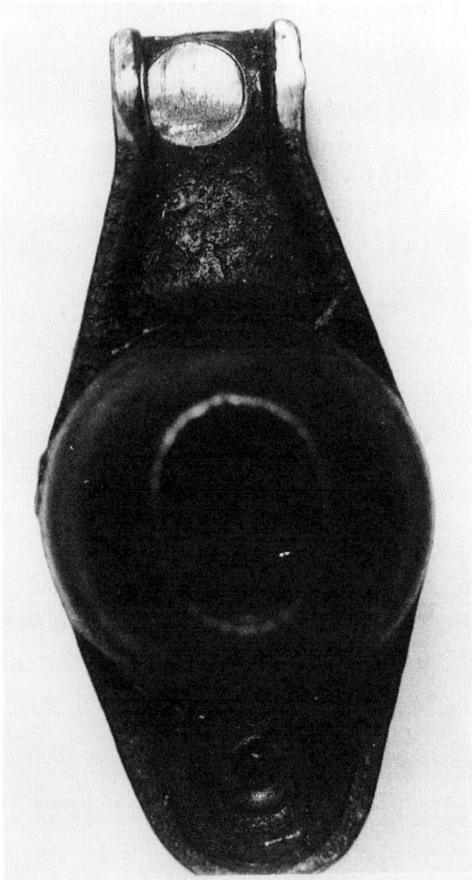

FIGURE 32
Worn rocker arm face (pad).

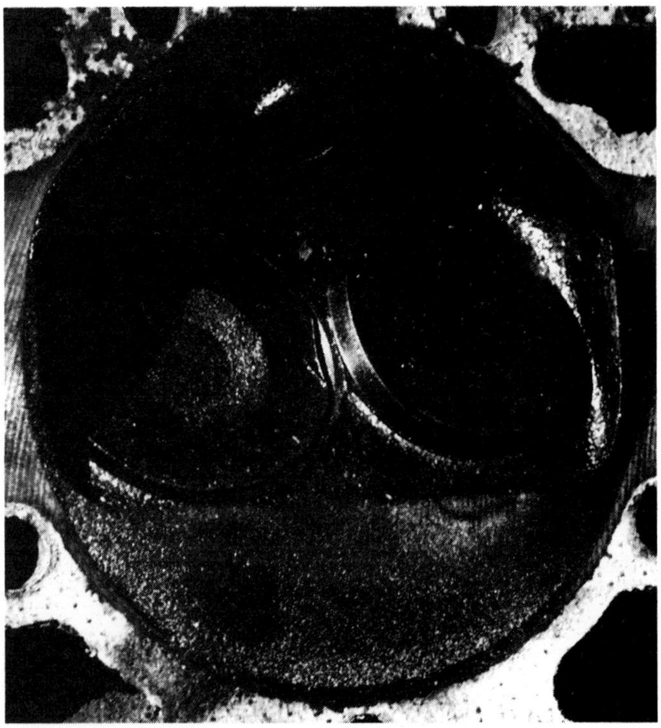

FIGURE 33
The exhaust valve seat ring in this head became loose and was wedged under the valve head as shown. This resulted in very rough idle, shake on take-off, and smoke from the tailpipe.

FIGURE 34
The grooves cut into this small-block Chevrolet front cover were caused by a loose timing chain. The noise sounded like a rod knock.

FIGURE 35
An accelerator pump ball check accidentally fell into the intake manifold during a carburetor repair. The picture tells the rest of the story.

4. Loose timing chain hitting the inside of the front cover (Figure 34).

5. "Pinging" from detonation or preignition.

6. Mechanical damage to the piston (Figure 35).

7. Connecting rod bolts hitting the inside of a dented oil pan.

8. Top of connecting rod bolts turned around and striking the camshaft.

9. Piston striking the edge of the head gasket.

10. Piston pin rubbing against the cylinder wall.

11. Torn engine (motor) mounts.

12. Defective fluid coupling on the fan.

13. Worn water pump.

14. Broken fuel pump rocker arm return spring (Figure 36).

15. Alternator bearing noise.

16. A/C compressor noise.

17. P/S pump bearing noise.

18. Popping noise from top of carburetor caused by "flat" exhaust lobe on camshaft.

CONDITIONS THAT CAN CAUSE EXCESSIVE OIL CONSUMPTION

A customer complaint that may lead to an engine overhaul is oil consumption. Engine oil consumption occurs in two ways:

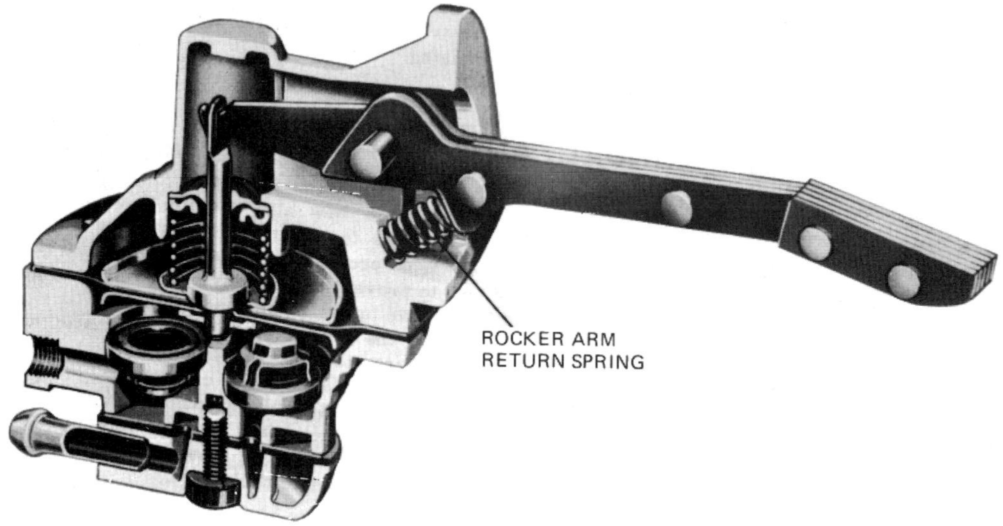

FIGURE 36
A broken or missing rocker arm return spring can cause the fuel pump to make a light metallic knock as it operates. (Courtesy of Ford Parts and Service Division)

External Leakage

1. External leakage.

2. Burning.

Begin by examining the engine exterior thoroughly. Some places where external leaks may occur are oil lines, crankcase drain plug(s), oil pan gasket(s), valve cover gasket(s), fuel pump gasket, timing gear cover gasket, front oil seal, rear oil seal, oil filter, camshaft expansion plug, and oil cooler and lines (Figures 37 and 38).

No possible source of leakage should be neglected. Even a very small leak will cause excessive oil consumption. As a case in point,

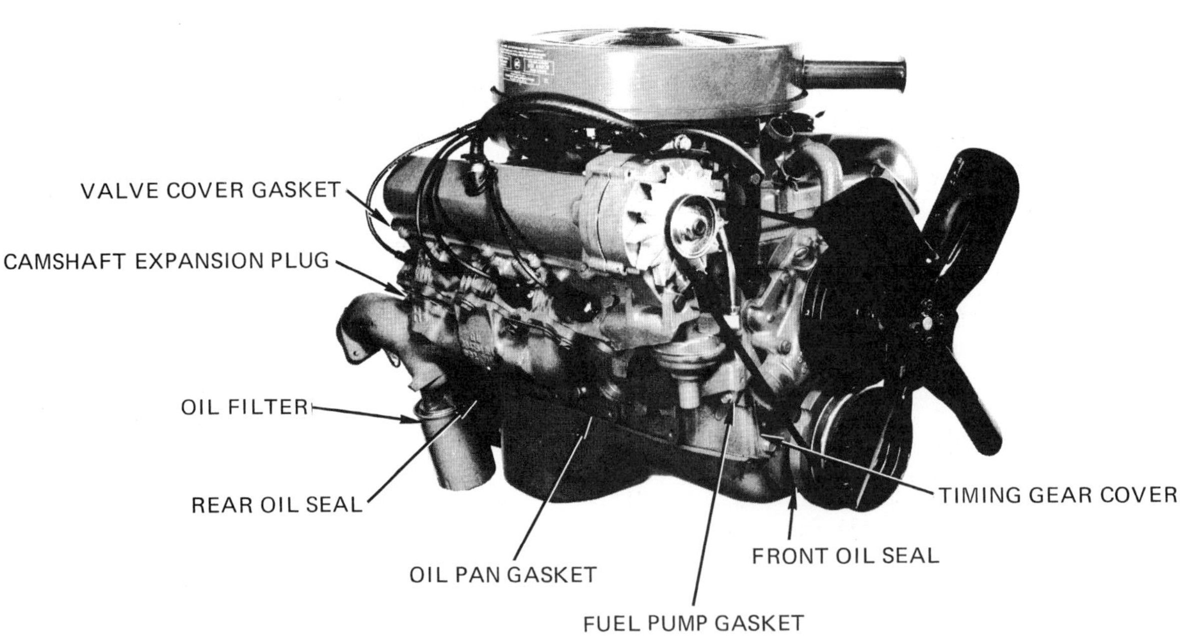

FIGURE 37
Some places where external oil leaks can occur. (Courtesy of Dana Corporation)

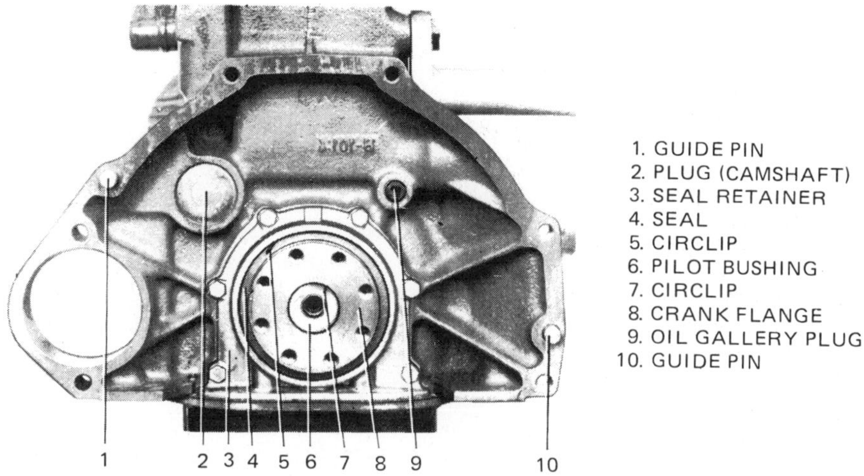

FIGURE 38
Parts 2, 3, 4, and 9 shown above can cause external oil leakage if installed incorrectly.

let's say that a vehicle is driven at 40 mph down the road. A leak of one drop of oil every 20 feet will add up to a loss of one quart of oil per 100 miles!

One quick check for external leaks is to raise the vehicle and run the engine (it may be necessary to steam clean the engine first). A special dye can be added to the oil to help locate the actual leak source (Figure 39).

> **SHOP TIP**
> A very high percentage of oil leaks from the bottom of the bell-housing are leaks in the oil pan gasket, rocker cover, side cover, and so on—not leaking rear main oil seals (Figure 40).

FIGURE 39
This bright red oil dye can be used to quickly pinpoint leaks in engines, transmissions, fuel lines, and power steering systems. (Courtesy of McKay Manufacturing Company)

Some additional causes of external oil leaks are:

1. Worn (grooved) vibration damper hub.

2. Stuck closed (clogged) positive crankcase ventilation (PCV) system. See Figures 41 and 42.

3. Excess oil in the crankcase.

4. Defective head gasket (Figure 43).

5. Leaking oil pressure sending unit (Figure 44).

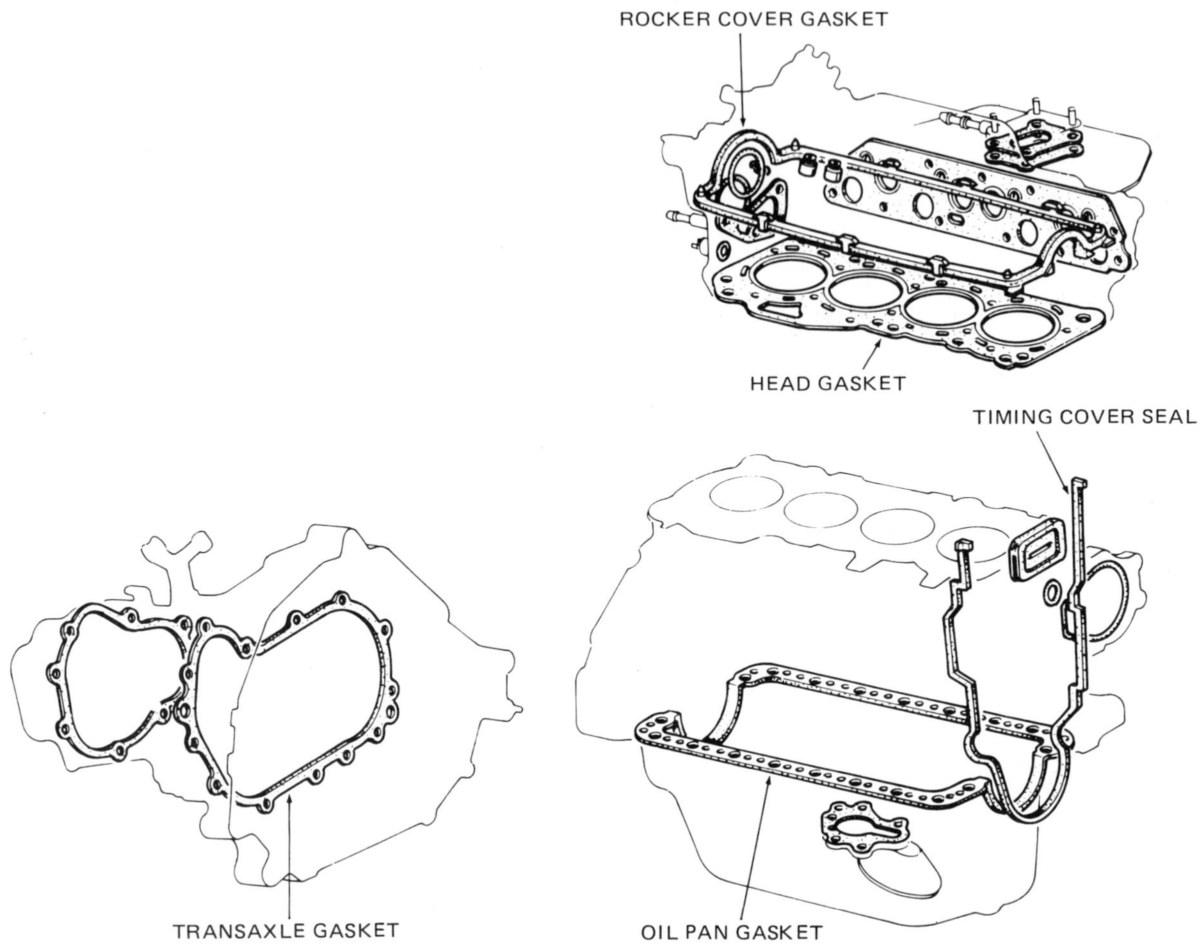

FIGURE 40
Typical locations of some important oil sealing gaskets in an engine. (Courtesy of American Honda Motor Company, Inc.)

6. Missing load spreaders on valve cover (Figure 45).

7. Distorted oil pan or valve cover flanges.

8. Damaged oil pan (Figure 46).

9. On many V-type engines, improperly installed intake manifold endstrip seals (Figure 47).

10. Failure to lubricate the rear main seal before installation.

Burning

Burning is the second way excessive oil consumption can occur. Burning results from oil entering the combustion chamber and being burned during the combustion process. Some of the many ways oil may enter into the combustion chamber are:

1. Clogged oil ring drain slots or holes.

2. Plugged cylinder head oil return passages (Figure 48).

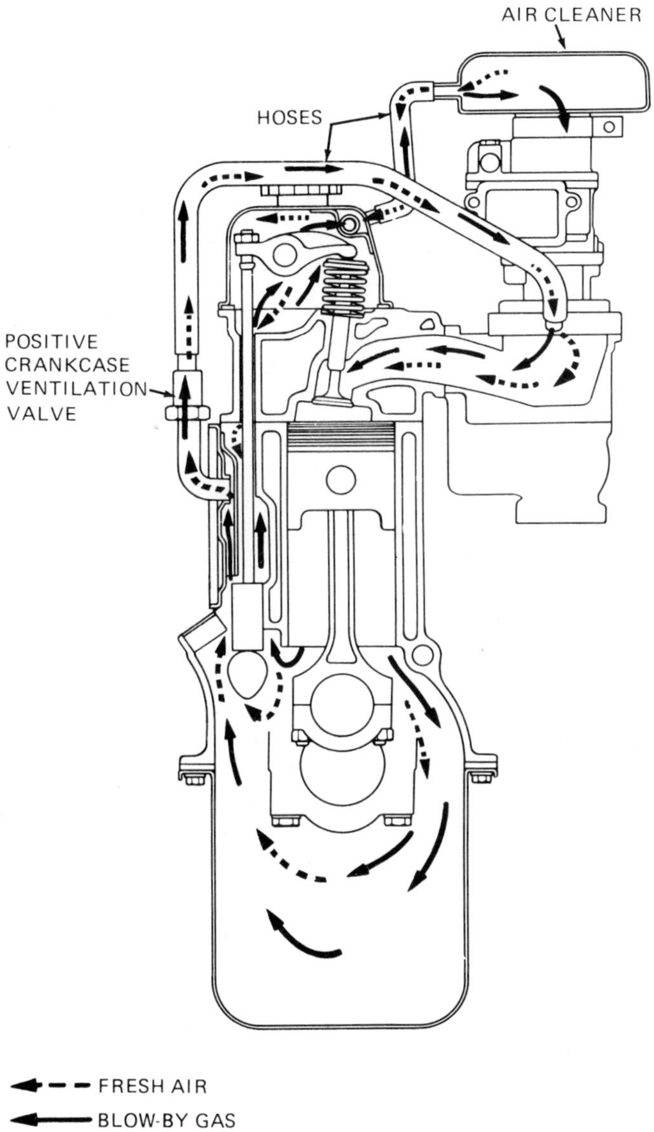

FIGURE 41
A PCV system closed with sludge and varnish deposits results in plugged oil rings, oil consumption, rapid engine wear, ruptured gaskets and seals, oil thrown out around the filler cap or inside the air cleaner assembly, carburetor problems, and consequent rough idle.

3. Piston rings stuck in grooves.

4. Worn or broken piston rings.

5. Cracked or broken ring lands.

6. Worn ring grooves (Figure 49).

FIGURE 42
Testing the PCV system for clogging with an air flow tester. If the test gauge indicates "danger," PCV system servicing is required.

7. Insufficient piston ring end clearance.

8. Wrong PCV valve.

9. Compression rings installed upside down or incorrect diameter rings.

10. Incorrectly installed oil control ring expander spacers (Figure 50).

11. Badly worn rings.

12. Improper ring face contact with the cylinder wall because of connecting rod misalignment.

13. Cylinder and/or piston wear (Figure 51).

14. Improperly fit wrist pins.

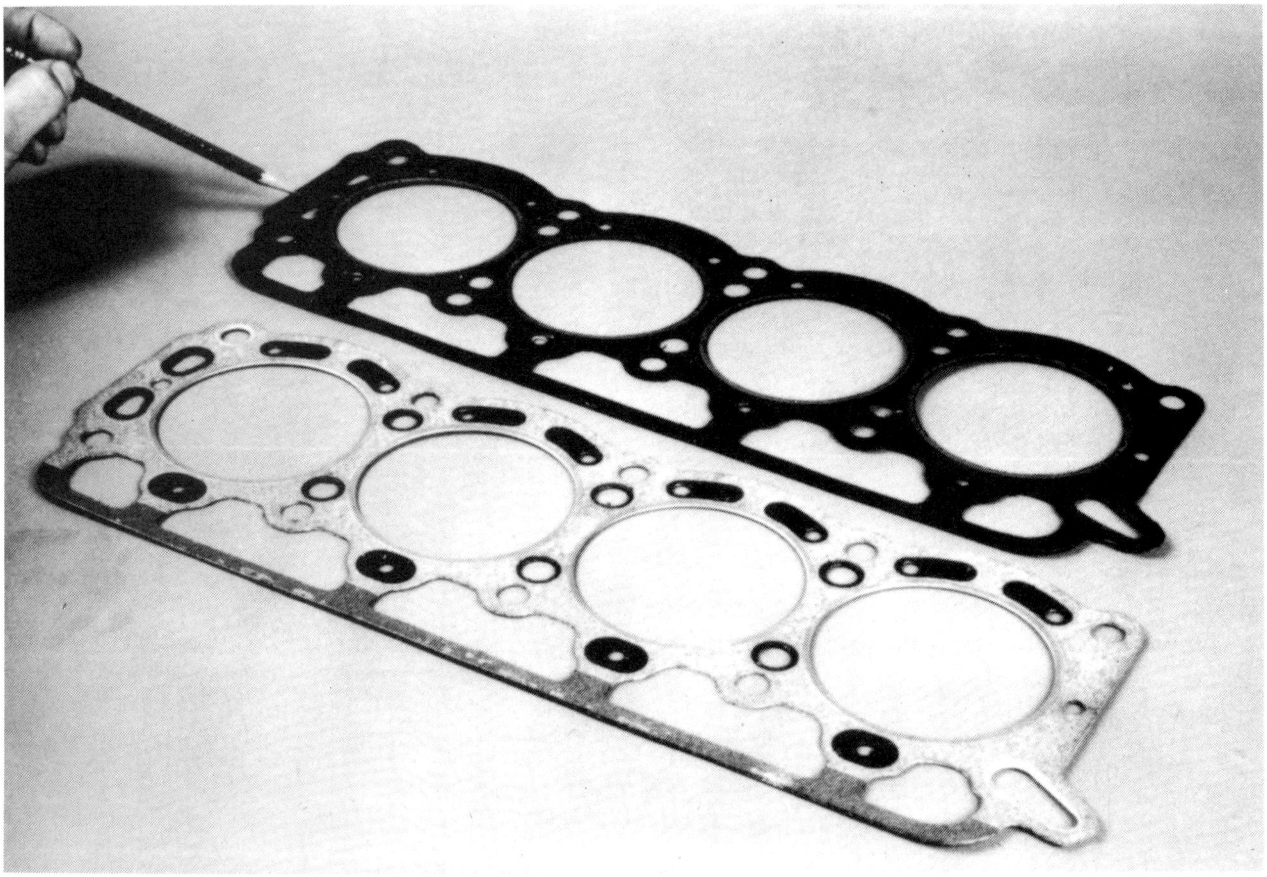

FIGURE 43
In some engines, the block has a drilled oil gallery that extends to the deck surface to carry oil to the rocker arms. If the head gasket does not seal this area properly, oil can escape into a cylinder or leak to the outside.

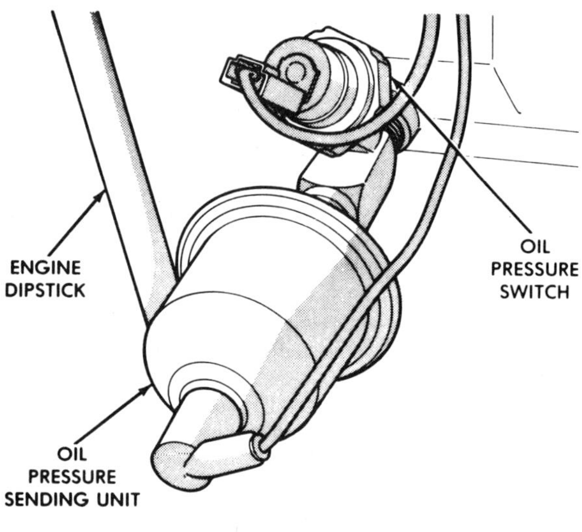

FIGURE 44
Leaks often develop at joints and crimps in oil pressure sending units.

15. Defective fuel pump.

16. Leaking intake manifold gasket (Figure 52).

17. Worn rocker arm pivot balls (Figure 53).

18. Worn or damaged crankshaft bearings (Figure 54).

19. Worn crankshaft journals.

20. Distorted cylinders resulting from unequal tightening of cylinder head bolts.

21. Worn valve stems and/or guides.

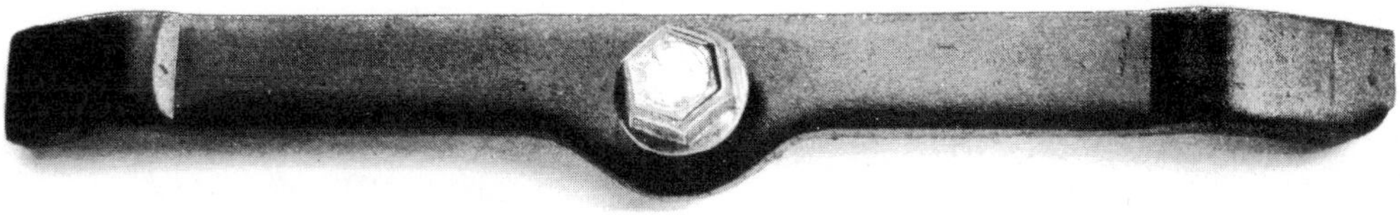

FIGURE 45
To ensure a positive valve cover seal, load spreader washers are often used under valve cover mounting bolts.

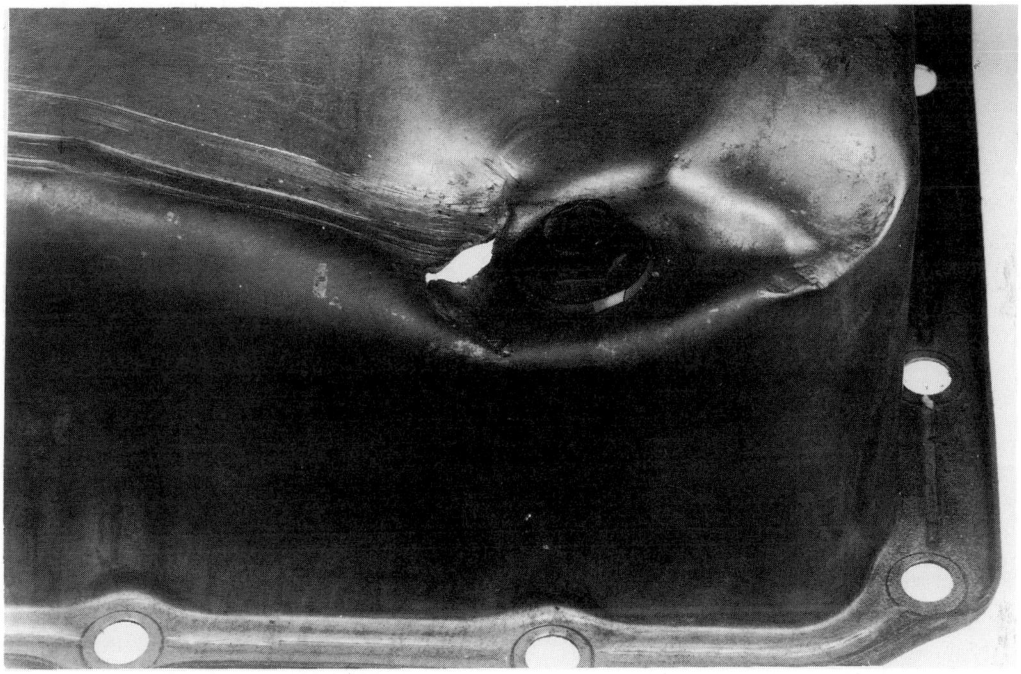

FIGURE 46
This oil pan was torn and all the engine oil was lost when the car "jumped" a curb.

22. Hardened valve stem seals.

23. Leaking turbocharger oil seals.

24. Plugged turbocharger oil drain passages.

ENGINE EXHAUST SMOKE

Determining the conditions under which the engine smokes and the color of the exhaust smoke can save considerable diagnosis time (Figure 55).

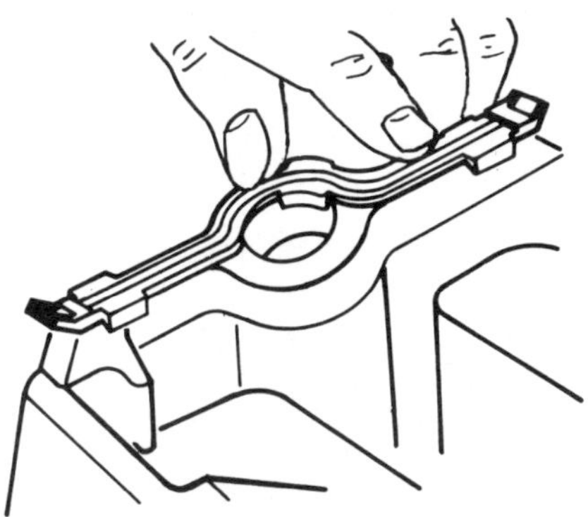

FIGURE 47
Installing endstrip seal. Always add a dab of RTV sealer at the intake manifold gasket and endstrip seal intersections to assure a positive seal. (Courtesy of Fel-Pro, Inc.)

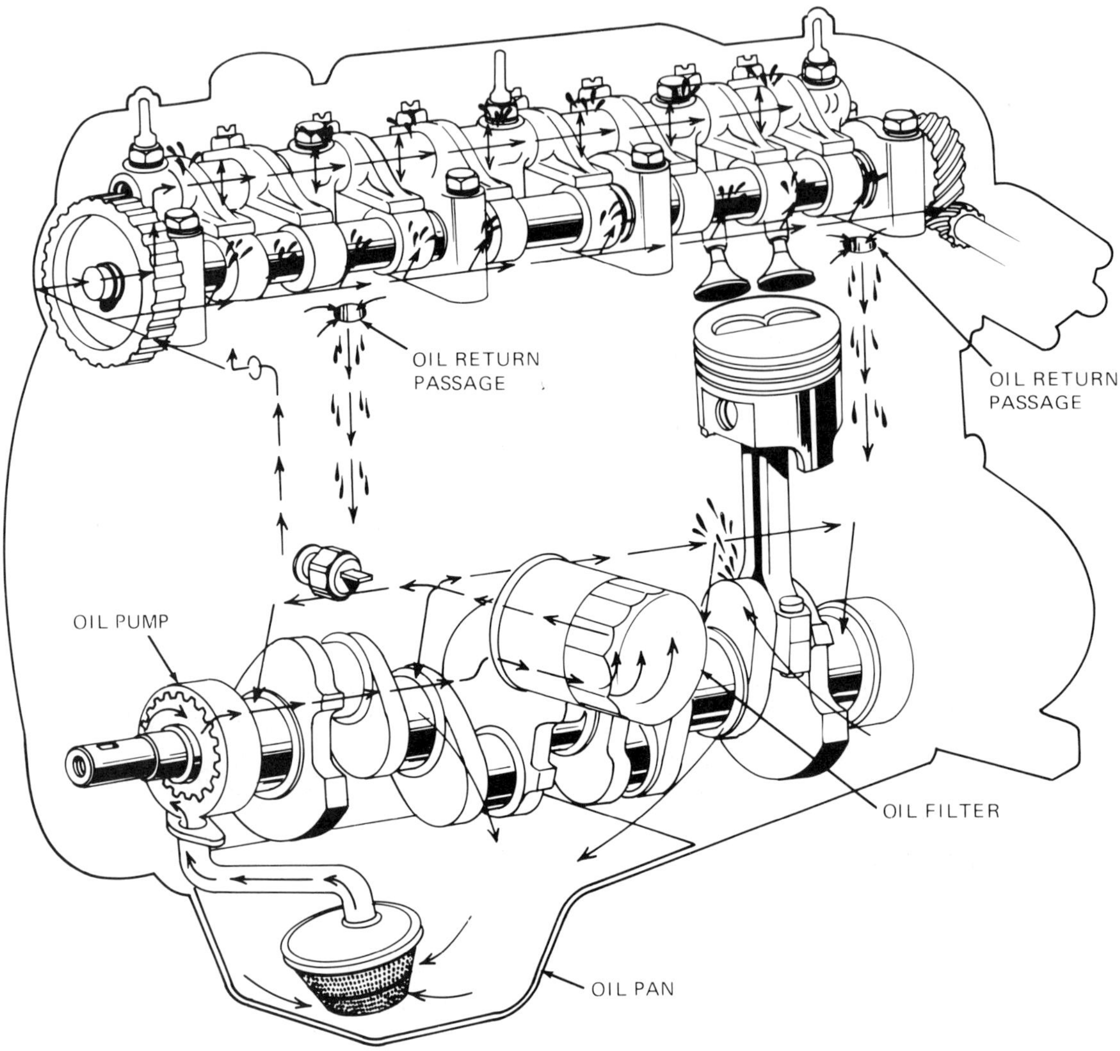

FIGURE 48
Cylinder head oil return passages must be free of foreign material and sludge deposits. If the holes are clogged, oil will flood the valve guides and result in high oil consumption.

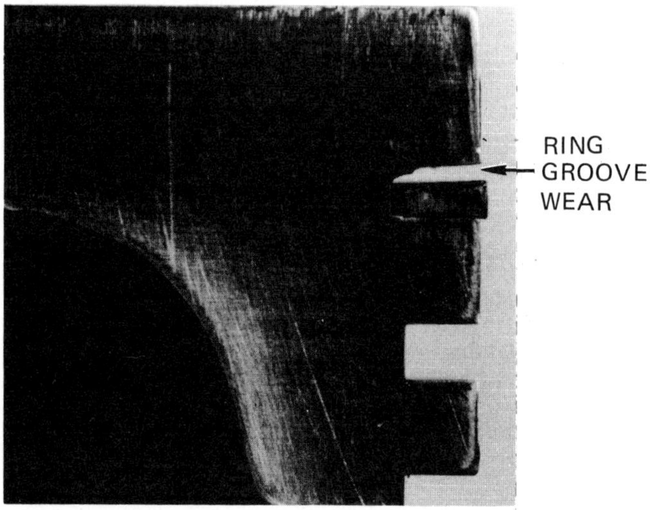

FIGURE 49
Excessive top groove and ring wear. The sides of the ring grooves must be true and flat (not flared or shouldered) for the rings to form a good seal. (Courtesy of Dana Corporation)

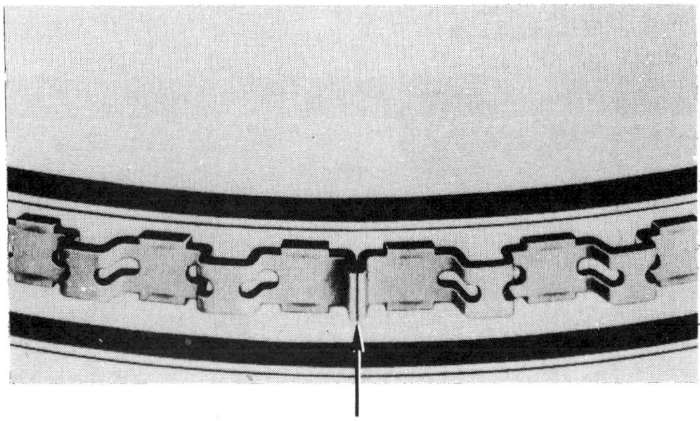

FIGURE 50
When installing a multiple-piece oil control ring, care must be taken to be sure that the expander spacer ends are butted together and not overlapped.

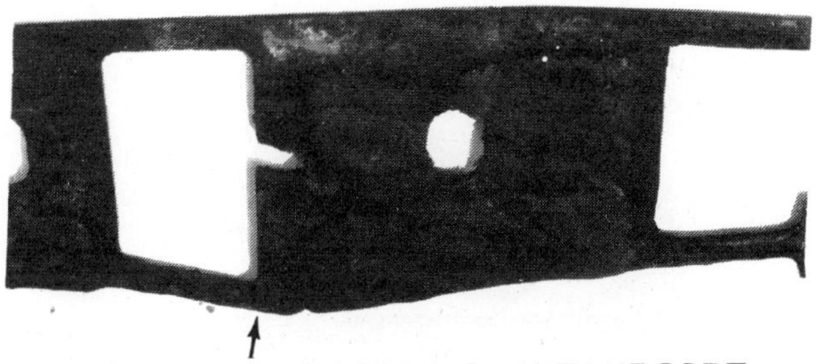

FIGURE 52
This defective intake manifold gasket came off a 302-CID Ford. Notice the pulled-out of position intake port. This caused a vacuum leak, which resulted in rough idle and excessive oil consumption.

FIGURE 51
All the pistons in an engine looked similar to this after severe overheating. Observe the skirt scoring and head damage.

FIGURE 53
Worn rocker arm pivot ball.

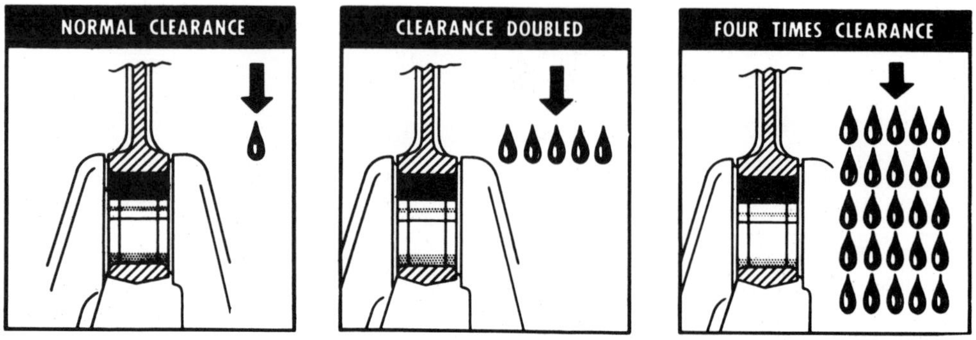

FIGURE 54
Normal clearance in automotive bearings is from .001″ to .003″. When normal clearance is doubled, the oil throw-off at higher speeds is multiplied approximately five times. When normal clearance is quadrupled, the oil throw-off is multiplied approximately 25 times. (Courtesy of Clevite Industries Inc.)

FIGURE 55
Exhaust smoke color can indicate certain conditions.

Blue-gray exhaust smoke during acceleration can be an indication of worn piston rings. In the early stages of ring wear, blue-gray smoke when accelerating after deceleration is similar to a worn valve guide indication. The cause of blue-gray smoke can usually be determined by a compression test and valve stem and guide clearance check (Figure 56). Also, check the valve stem seals (Figure 57).

Black exhaust smoke is the result of partially burned fuel. It can result from either too much fuel or incomplete burning of the normal fuel charge.

FIGURE 56
A quick guide clearance wear estimate can be made by side thrusting the spring retainer with a hammer handle when the valve has just started to open. You will be able to see and feel the relative movement between the stem and guide.

FIGURE 57
Valve stem seals lose flexibility and become hard with age. This makes it possible for oil to be drawn into the combustion chamber. These single-ring positive seals fasten to the guide chimney.

White vapor from the exhaust may be steam due to an internal water leak in the engine, a blown head gasket, or a cracked block or head. This condition usually shows up as a miss, and a wet or rust-colored spark plug is usually found.

ENGINE EXHAUST ODOR

A common complaint on vehicles with catalytic converters is a "rotten egg" or sulfur odor. This odor results from insufficient oxygen being present in the exhaust. An odor complaint can often be cured by properly adjusting the carburetor (idle speed and mixture) and making sure the air management system (smog pump and related hardware) is working properly.

ENGINE EXHAUST SOUND

Feel and listen to the tailpipe "pulse" at idle. It should be smooth and even. A burned or leaking valve or an ignition malfunction causes an irregular exhaust sound.

Quick Checks for a Burned Valve

1. Note the symptoms during engine idle. Rough idle and a "motorboating" exhaust sound at the tailpipe occur when a valve is burned.

2. Raise the engine speed to approximately 2500 rpm. With an ignition malfunction, the engine will continue to shake and run rough, but with a burned valve, the engine will smooth out.

3. Listen to the cranking rhythm. Disconnect the battery feed wire to the distributor or electronic ignition system and crank the engine over. A burned valve (which means that compression is low in one of the cylinders) will cause the cranking rpm to be uneven.

ENGINE VACUUM TESTS

An intake manifold vacuum test can be very useful in determining certain engine problems. The procedure is as follows:

1. Connect a vacuum gauge to a manifold vacuum source (Figure 58).

2. Disconnect the fuel vapor canister.

3. Run the engine until it reaches operating temperature.

4. Check the reading on the vacuum gauge at idle rpm. Note the position of the needle as well as any fluctuations.

5. You should get a steady reading of approximately 17 inches or more of Hg on most engines, without much flutter in the needle.

6. A low, steady reading indicates a loss of power in all cylinders. This could be caused by late ignition or valve timing, low compression, a loose timing chain, or a vacuum leak (Figure 59).

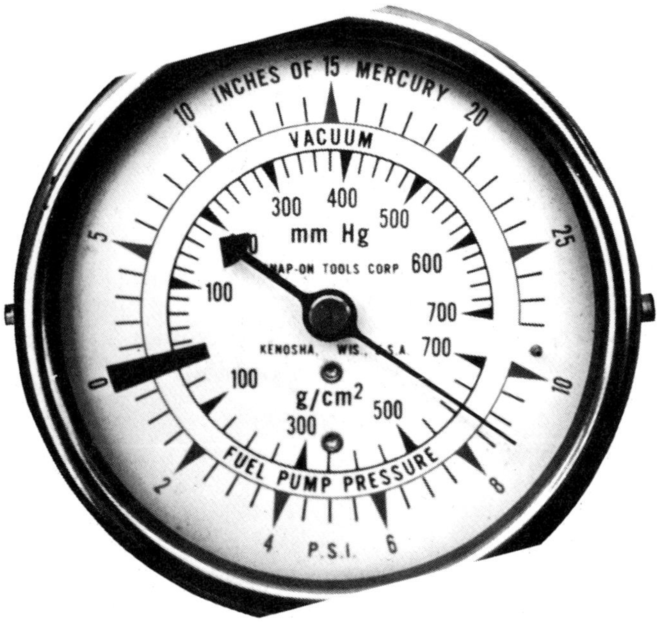

FIGURE 58
Vacuum and fuel pump gauge. What is the reading on this gauge?

7. A fluctuating needle means a partial or complete loss of power in one or more cylinders. Some probable causes are a burned valve, weak or broken valve spring, uneven compression, carburetor problem(s), or fouled spark plugs.

8. A needle that drops quickly with no apparent rhythm could mean a sticking valve (Figures 60 and 61).

9. Normal upon start-up, followed by a gradual drop of the needle means there is excessive back-pressure (a restriction) in the exhaust system.

NOTE: A restricted exhaust can also be detected by raising the engine speed quickly to 2500 rpm to produce a momentary low vacuum reading and then releasing the throttle quickly and observing the gauge. The needle should return smoothly with no hesitation. A slow, uneven return indicates a breathing restriction.

Figures 62 and 63 show examples of restricted exhaust systems.

FIGURE 59
Excessively worn (loose) timing chain. Look at the wear on the cam sprocket teeth after 91,000 miles.

CYLINDER POWER BALANCE TEST

The cylinder power balance test shows which cylinder (or cylinders) is not developing its share of power.

The cylinders are shorted out one at a time, and monitored with a tachometer and vacuum gauge. The tachometer shows how much engine rpm is lost when a cylinder is shorted out. The vacuum gauge shows how much the manifold vacuum drops.

If the power loss for each cylinder is about the same, the cylinders are all performing equally. However, if you find that one cylinder shows a smaller rpm drop than the others, there is an ignition, carburetion, or mechanical problem (Figure 64).

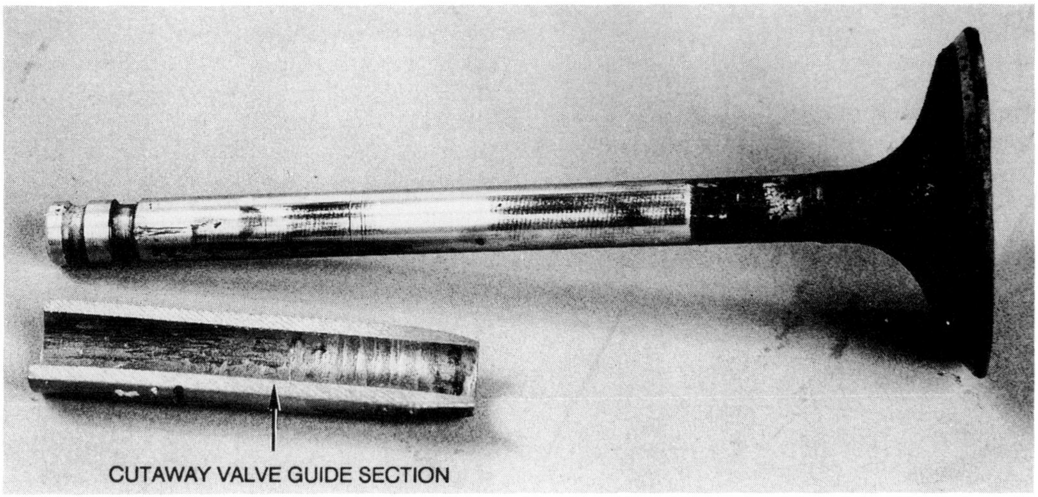

FIGURE 60
Both the valve stem and guide in this picture are badly scored. This resulted in a sticking valve that produced idle "hiccup."

Normal Test Procedure

1. Connect an oscilloscope to the engine.

NOTE: Disconnect and plug the EGR vacuum signal hose before power balance testing. Sometimes recirculation of the exhaust gas may cause incorrect readings in certain cylinders.

2. Start the engine and run it at 2000 rpm.
3. Short out the cylinders, one at a time.

CAUTION: Do not leave any cylinder shorted out for longer than 30 seconds; this can cause overheating and damage to the catalytic converter.

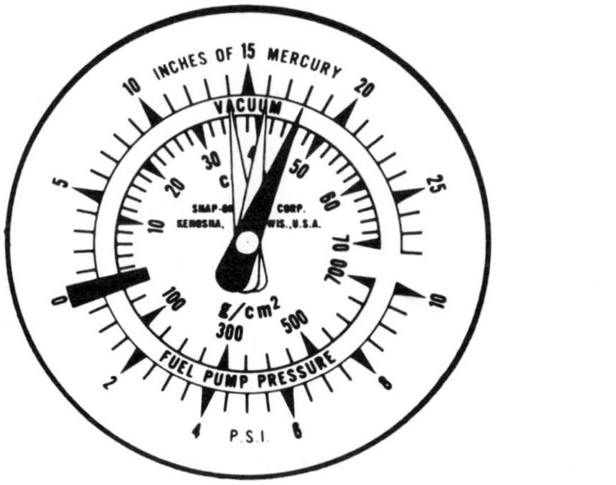

FIGURE 61
This vacuum gauge reading illustrates a sticking valve. With the engine idling, irregular drops below normal will occur. The drop is likely to be more pronounced with a sticking valve than if it were leaking as the result of improper valve seating.

FIGURE 62
The lead in gasoline will poison (coat) the inside of the "cat" so that it will no longer lower emission levels. As shown in this cutaway, leaded gasoline can cause converter plugging after extended use.

FIGURE 63
This tailpipe kink caused excessive back pressure in the exhaust system.

4. Record the rpm and vacuum readings for each shorted cylinder.

5. Compare the results for all cylinders. Results should be within 5% for each cylinder. If more than a 5% variation is noted, that cylinder (or cylinders) is low in power.

> **SHOP TIP**
> A low power reading combined with a high firing line usually indicates a lean mixture. A low power reading combined with a low firing line is often caused by low compression or an ignition malfunction.

Computer Control System Test Procedure

When a cylinder power balance test is performed on a computer-controlled engine, it is necessary to disconnect the computer's control of fuel, timing, and rpm. Otherwise, the computer will compensate for the dead cylinder.

FIGURE 64
Believe it or not, this engine was running with a broken camshaft when it was driven into the shop.

General Motors Port Fuel Injection (PFI) Test Procedure

1. Disconnect the O_2 sensor.

2. Disconnect the IAC motor.

3. Ground the ALDL diagnostic terminal and start the engine (field service mode).

Ford MCU, EEC I, and EEC II Systems (with 7200 vv carburetor) Test Procedure

1. Disconnect the O_2 sensor.

2. Disable the idle speed control.

3. Install a ⅝″ NF bolt in place of the stepper motor. Adjust to get approximately .5 V from the O_2 sensor.

General Motors C3 Carburetor Control Test Procedure

1. Ground the ALDL diagnostic terminal (put into self-test).

2. Disconnect the EST four-wire connector (found on only some applications).

3. Disconnect the O_2 sensor.

4. Unplug the ISC motor if testing is at idle.

General Motors Throttle Body Injection (TBI) Test Procedure

1. Disconnect the O_2 sensor.

2. Plug the idle air control (IAC) bleed hole in the throttle body.

3. Ground the ALDL diagnostic terminal and start the engine (field service mode).

Ford EEC III (with Central Fuel Injection) Test Procedure

1. Disable the idle speed control.

2. Disconnect the O_2 sensor.

NOTE: Timing may change with rpm and intake manifold vacuum changes.

Ford EEC IV (with Carburetor Control or Fuel Injection) Test Procedure

1. Disconnect the yellow wire at the distributor (brown or black hard-shell connector).

2. Disable the idle speed control.

3. Disconnect the O_2 sensor.

Chrysler Four-Cylinder Engines with Electronic Fuel Control (EFC) Test Procedure

1. Attach a jumper wire from the carburetor switch to ground.

2. Disconnect the vacuum hose to the vacuum transducer and plug hose.

Chrysler Six-Cylinder and Eight-Cylinder Engines (with EFC or EFI) Test Procedure

1. Disconnect and ground the O_2 sensor.

2. Wait 4 minutes.

3. Check the voltage at the duty-cycle solenoid for a fixed 7 to 9 V.

Chrysler Four-Cylinder Engines (with EFI and Turbo) Test Procedure

1. Start the engine.
2. Disconnect the coolant sensor.
3. The power loss light should be on.

After completing the cylinder power balance test, reconnect all components. Then clear the computer's memory to prevent a fault code from being stored.

ENGINE COMPRESSION TEST

A compression test reveals compression loss through the rings, valves, and head gasket. It can also show if there is excessive carbon buildup in the cylinders (higher than normal pressure readings will be obtained).

Engine cylinder pressures normally should not vary more than 25% from the highest to the lowest one. Variations beyond specifications will cause uneven idling, loss of power, and poor performance. Engine compression problems are pictured in Figures 65 and 66.

FIGURE 65
A hole had burned through this VW combustion chamber.

FIGURE 66
Here is what can happen if a vehicle continues to be driven with a blown head gasket.

A properly performed compression test is conducted in the following manner:

1. Run the engine at operating temperature.

2. Remove all the spark plugs.

3. Open the choke plate (if so equipped) and throttle valve(s) to a wide open position.

4. Disconnect the battery feed wire to the distributor or electronic ignition system.

5. Insert the compression gauge into a spark plug hole (Figure 67).

6. Crank the engine over so that each cylinder goes through five compression strokes.

7. Record the reading for each cylinder (Figure 68).

FIGURE 67

Compression tester connected to a cylinder. Remember that a compression test will identify the faulty cylinder, but cannot always identify the point of leakage.

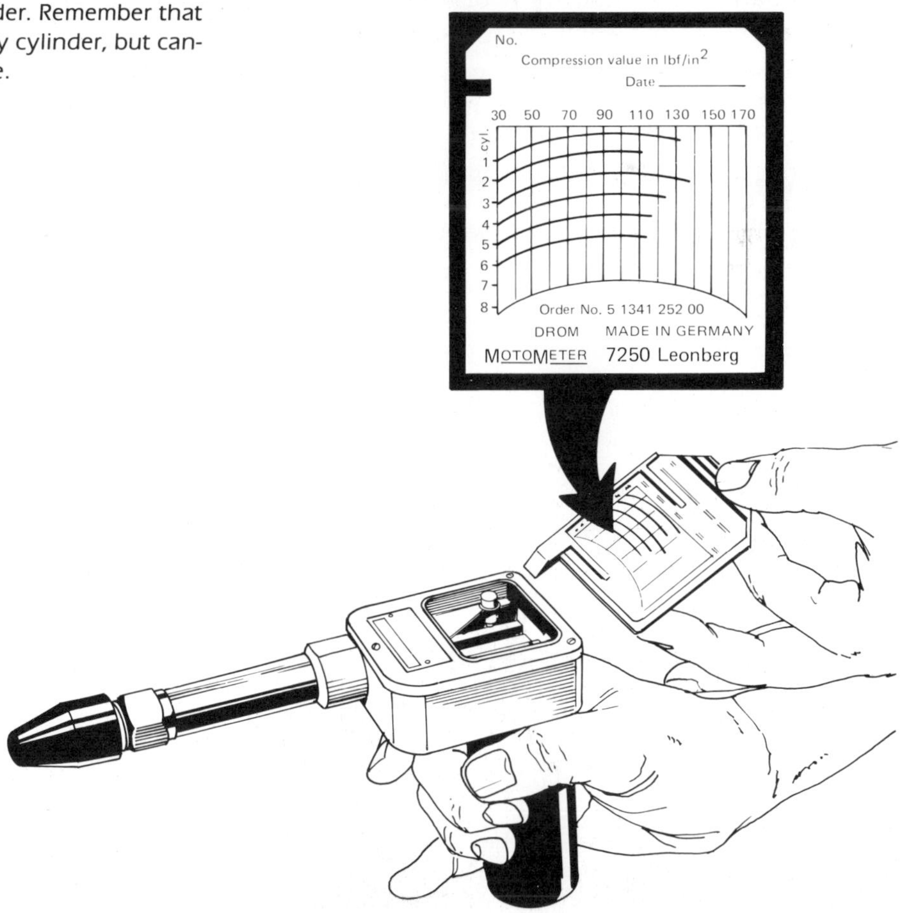

FIGURE 68

This compression tester has a recording chart to show the compression testing results.

NOTE: The battery should be fully charged during the test. An undercharged battery can cause low cranking rpm, and a lower pressure reading in the cylinders due to the increased time allowed for the compression stroke.

Wet Test

If any cylinders show a low reading, a wet test can be performed. Squirt about a teaspoon of motor oil into each low cylinder and recheck the compression. If the pressure reading is now normal, a piston ring problem is indicated.

CYLINDER LEAKAGE TEST

Use this test to pinpoint some of the causes of abnormal conditions seen in the compression test.

Test Procedure

1. Remove the spark plug from the suspect cylinder and install an air adapter in its place (Figure 69).

2. Position the piston at TDC on the compression stroke.

3. Calibrate the tester.

4. Direct compressed air into the cylinder, through the adapter.

5. Read the percentage of leakage on the cylinder leakage gauge (Figure 70). Leakage should read less than 20% on the face of the gauge.

Determining the Source of the Leak

To determine the exact source of the leak, listen at the following points:

1. **Air intake.** Air escaping through the carburetor means there is a leaking intake valve (Figure 71).

NOTE: Disconnect the PCV valve first.

2. **Oil filler.** Leakage at the oil filler tube indicates worn piston rings, or scored pistons or cylinder walls.

3. **Exhaust pipe.** Exhaust pipe leakage means burned or leaking exhaust valves.

COMMENT: A leak between the cylinder being tested and an adjoining cylinder indicates a blown head gasket between the two cylinders.

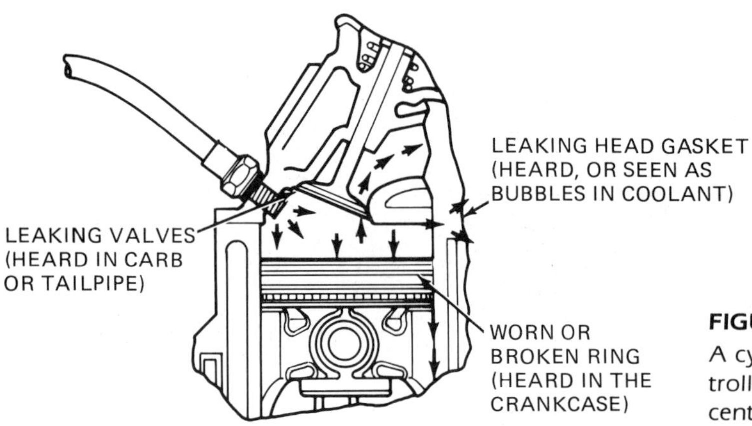

FIGURE 69

A cylinder leakage tester applies air to the cylinder at controlled volume and pressure, and then measures the percentage of leakage.

FIGURE 70

A commercially built cylinder leakage tester. This author's experience has determined on many occasions that a 40 to 50% leakage figure is equal to a failed compression tester reading.

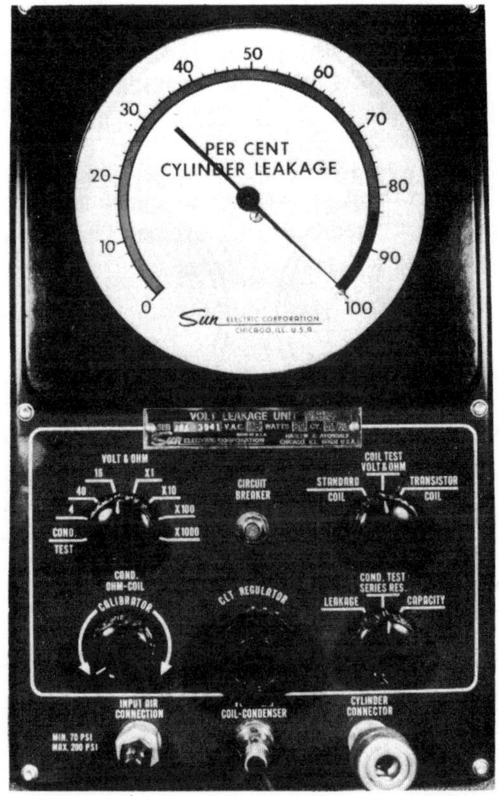

SHOP AIR

REGULATED AIR TO CYLINDER

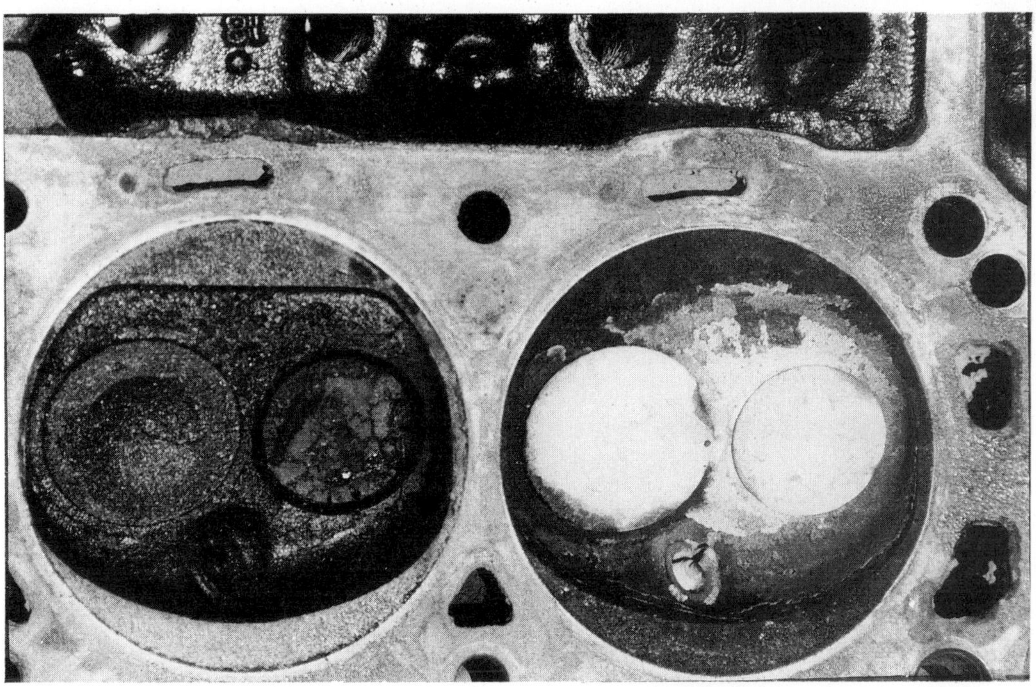

FIGURE 71

Do you see the burned intake valve? What else do you see?

After listening for leaks, look for air bubbles in the radiator or reservoir. Bubbles indicate a leaky head gasket, a cracked block, or a cracked cylinder head.

REMOVING AND REPLACING ENGINE (REAR WHEEL DRIVE)

1. Position fender covers on the fenders and grill cowling.

2. Disconnect the battery cables. *The ground cable should be disconnected first.* This eliminates the danger of causing a spark when disconnecting the positive cable.

3. Scribe a line to mark the position of the hood hinge attachment brackets (Figure 72). This will enable you to align the hood quickly when reinstalling it.

4. Remove the hood.

COMMENT: Disassembly organization is extremely important from this point on.

Considerable assembly time will be saved by keeping track of nuts and bolts, wire terminals, and other removed items. Place small parts in

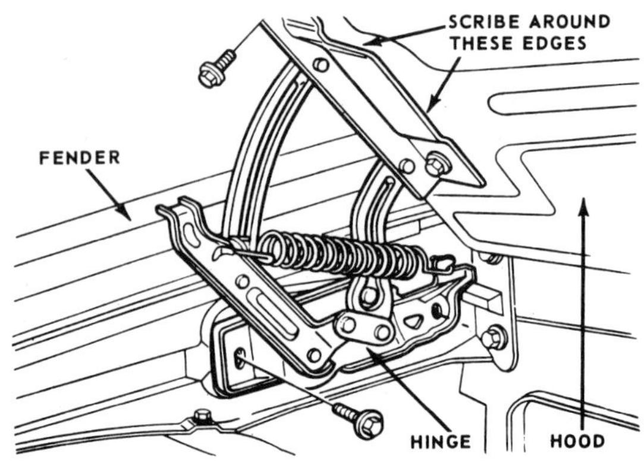

FIGURE 72
Scribing around the edges of the hood hinge attaching bracket will make hood alignment easy during reassembly. (Courtesy of Chevrolet Motor Division, General Motors Corporation)

marked cans. Place masking tape around disconnected wires and vacuum hoses, using a permanent ink pen to mark the location on the tape.

> **SHOP TIP**
> Individual telephone cable wires can be used to mark disconnected vacuum hoses (Figure 73). Put a piece of one wire on the end of the hose, and another piece of the same color on the connection where the hose was removed. This will make reassembly easier.

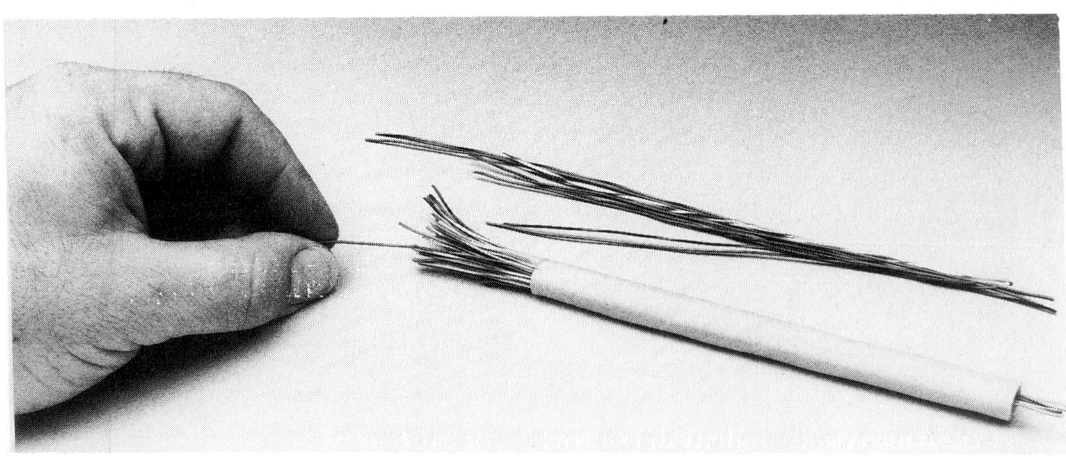

FIGURE 73
Telephone cable wires come in a variety of colors and can be used to mark vacuum hoses.

5. Drain the cooling system at the radiator and the cylinder block (Figure 74).

6. Remove the air cleaner and intake air duct assembly.

7. Disconnect the upper and lower radiator hoses.

8. Disconnect the heater hoses at the block fitting and water pump. Remove any engine grounding straps (Figure 75).

9. If the vehicle is equipped with an automatic transmission, disconnect and plug the transmission oil cooler lines from the radiator (Figure 76).

NOTE: Always use a flare nut wrench when removing tubing fittings (Figure 77). This will prevent rounding off the corners of the fittings.

10. Remove the fan shroud (if so equipped) and the radiator.

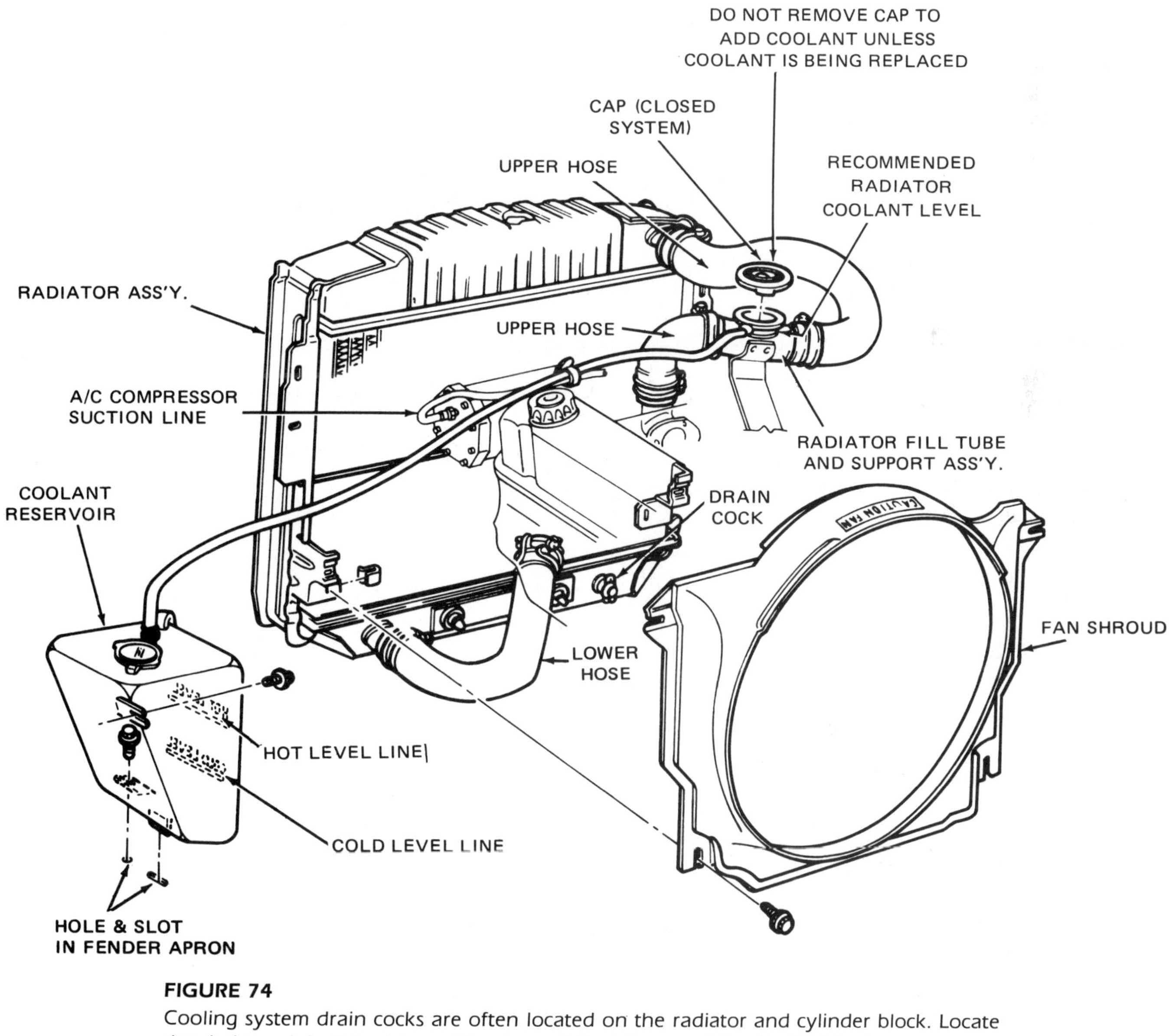

FIGURE 74

Cooling system drain cocks are often located on the radiator and cylinder block. Locate the drain cock in this picture. (Courtesy of Ford Parts and Service Division)

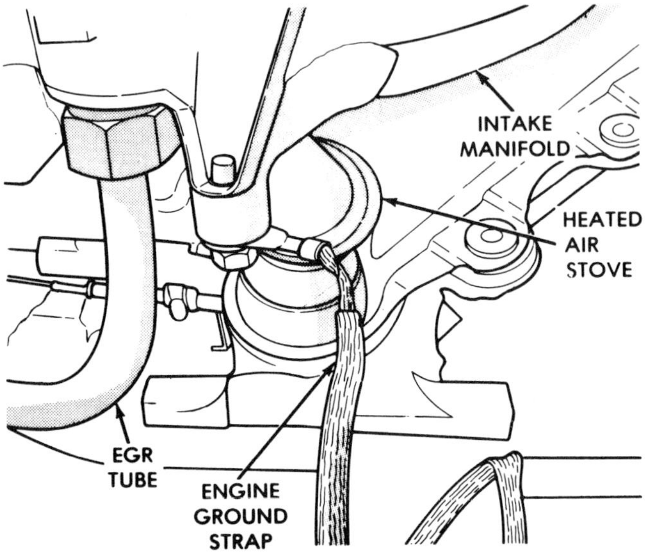

FIGURE 75
This engine has a ground strap attached to the intake manifold. Ground straps are important, and must be properly reconnected. (Courtesy of Chrysler Corporation)

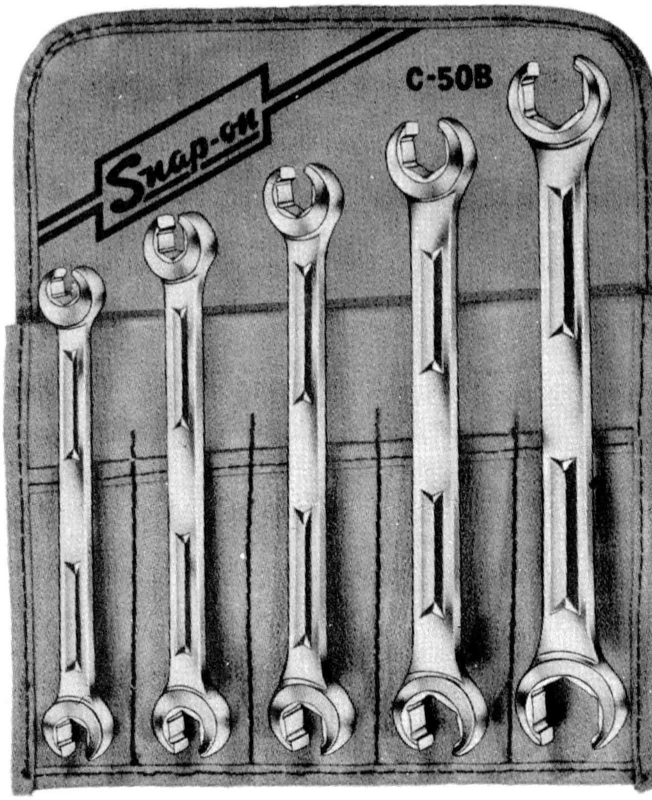

FIGURE 77
Flare nut wrenches. (Courtesy of Snap-On Tools Corporation)

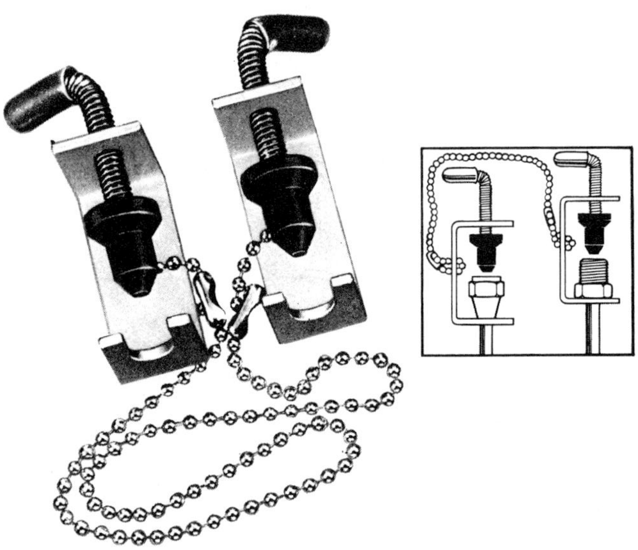

FIGURE 76
Disconnected transmission cooler lines should be plugged. These special plugs seal male or female fittings and allow the vehicle to be moved from the service stall with the radiator out. (Courtesy of Thexton Manufacturing Company, Inc.)

11. Remove the fan, spacer, and belt pulley.

12. On a vehicle equipped with air-conditioning, remove the compressor from the mounting bracket(s) and position it out of the way, leaving the refrigerant lines attached. *If refrigerant lines are disconnected, be sure to plug all openings immediately.* Moisture must not be allowed to enter the system.

NOTE: In some cases, it is easier and faster to remove the compressor by leaving it on the mounting bracket.

13. On a vehicle equipped with power steering, disconnect the power steering pump. If possible, leave the lines attached and position the pump so that fluid cannot leak out.

14. On a vehicle equipped with power brakes, disconnect the booster vacuum line from the intake manifold. Disconnect any other lines or hoses attached between the engine and chassis at this time.

15. Remove the alternator and mounting brackets. Move them out of the way, leaving the wires connected.

16. Disconnect the fuel inlet line at the fuel pump. Plug or cap the line to prevent fuel loss (Figure 78).

17. Disconnect the accelerator cable or linkage at the carburetor and intake manifold. Disconnect the transmission downshift linkage, if present.

18. On some vehicles it is necessary to remove the transmission fluid filler tube attaching bolt from the cylinder head.

19. Disconnect the engine wiring harness at the distributor, water temperature sending unit, and oil pressure sending unit. On engines equipped with computer-controlled systems, unhook the wire(s) to the various sensors and actuators (Figure 79).

FIGURE 78
These solid rubber plugs can be used to plug disconnected fuel lines. (Courtesy of Thexton Manufacturing Company, Inc.)

20. Remove the carburetor or throttle body injector assembly (if so equipped) from the engine to prevent damage from the lift sling as the engine is being removed. *Use two wrenches to avoid twisting the fuel line* (Figure 80).

21. It is recommended that you remove the distributor, distributor cap, and spark plug wires from the engine. This will prevent bulkhead (firewall) or lift sling interference damage as the engine is tilted during removal.

22. Raise the vehicle.

CAUTION: When you use a frame contact hoist, be sure the contact pads are in correct position (Figure 81). Cracked windshields, misaligned hoods, and incompletely closed doors can result when a vehicle is lifted incorrectly. This is especially true with unitized frame vehicles.

23. Drain the engine oil (Figure 82).

24. Disconnect the exhaust pipe(s).

25. Remove the nuts or bolts attaching the front motor mount rubber insulators to the crossmember (Figure 83).

26. Remove the starter (this is only necessary on some vehicles).

27. On a vehicle with a manual shift transmission, disconnect the clutch linkage.

28. Remove the clutch or converter housing bolt access cover.

29. On a vehicle with a manual shift transmission, disconnect the shift linkage, etc., and remove the transmission. *It is not necessary to remove the clutch assembly at this time.* It will come out with the engine.

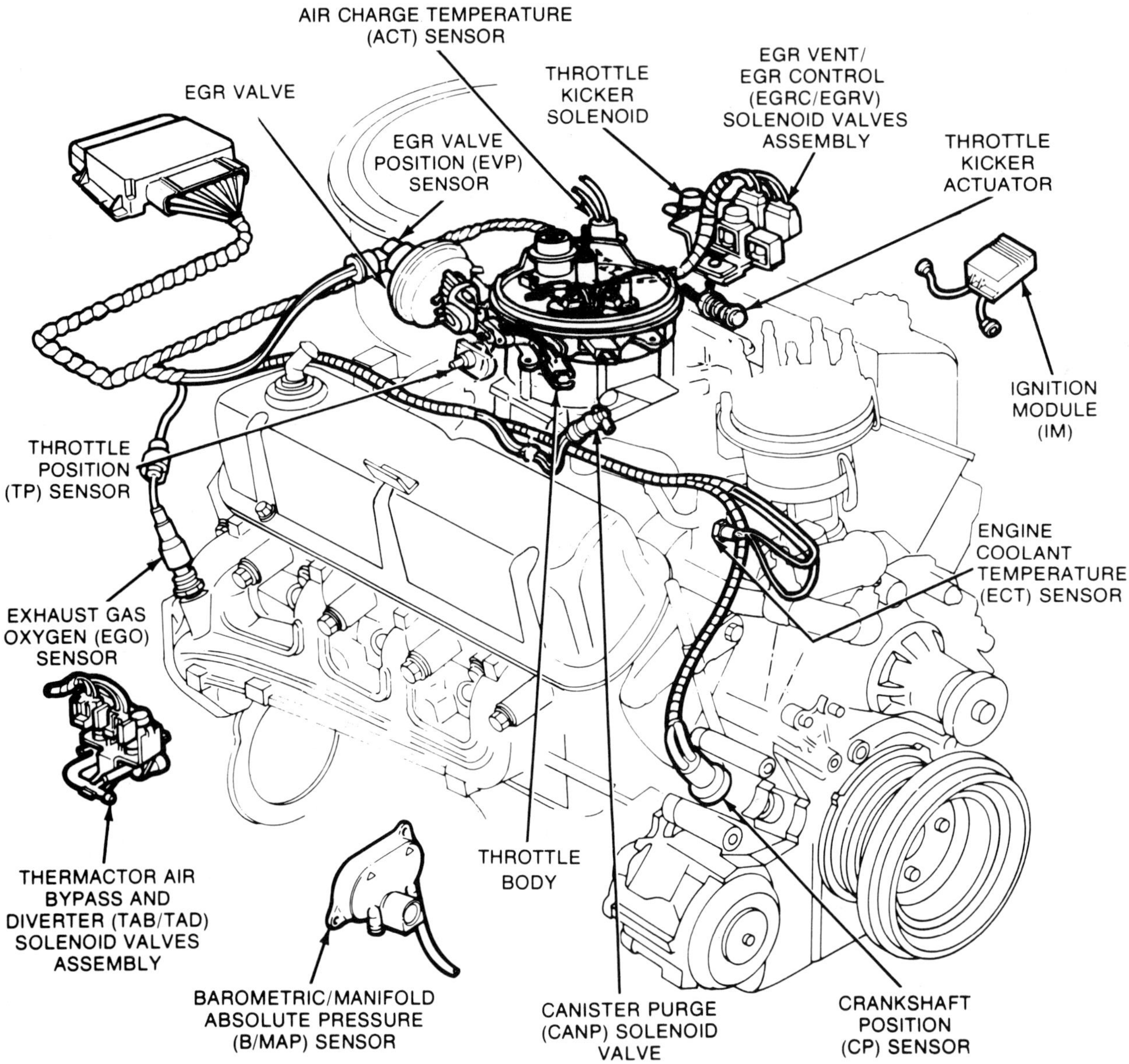

FIGURE 79

Ford EEC III system network. Similar sensor and actuator networks are found on many import and domestic automobiles.

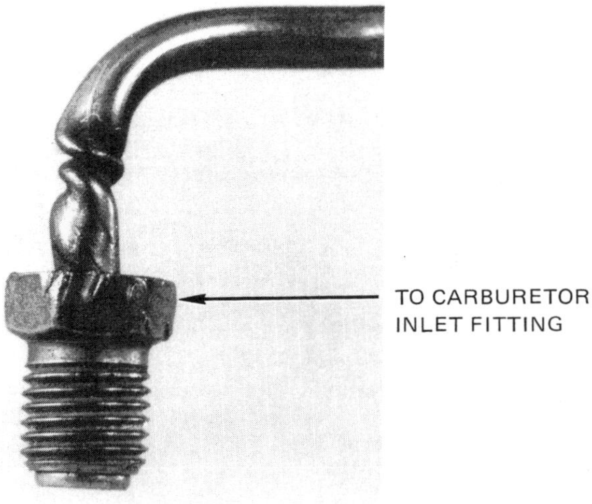

FIGURE 80
Twisted fuel line.

FIGURE 82
Does this oil pan drain plug tell you something?

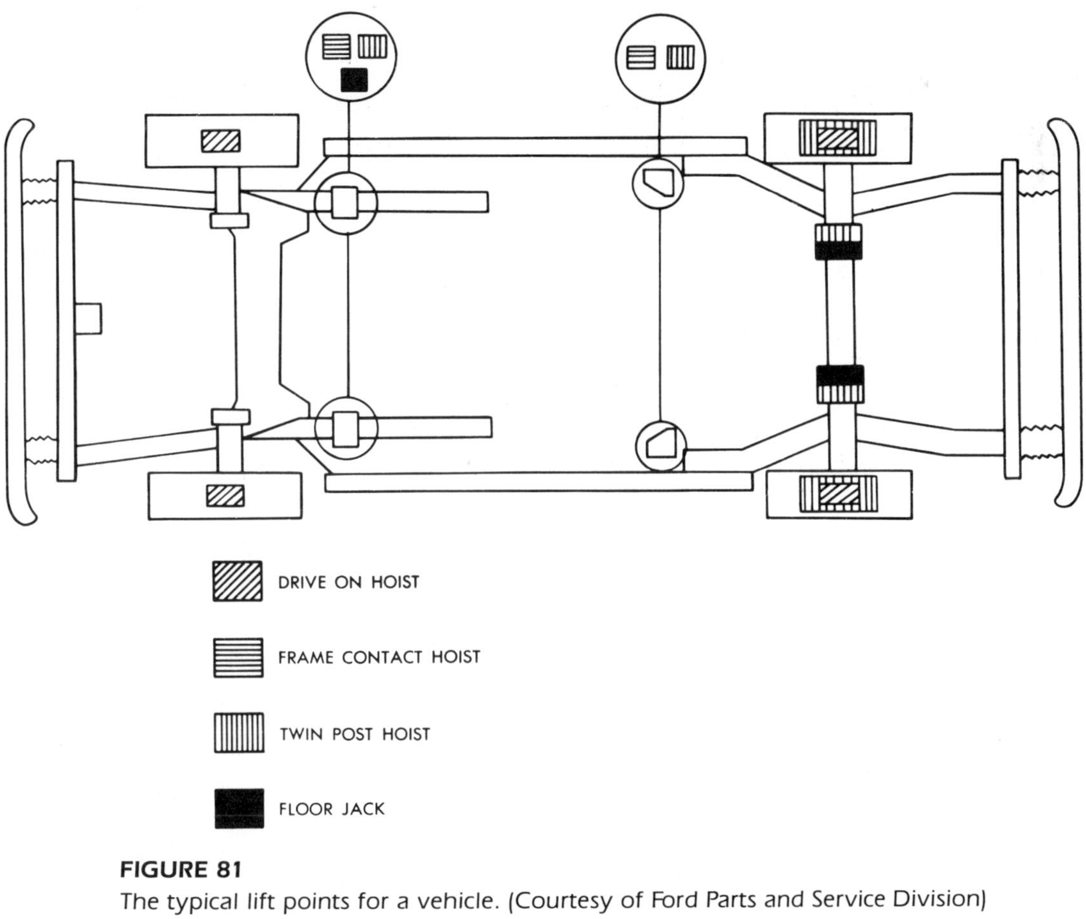

FIGURE 81
The typical lift points for a vehicle. (Courtesy of Ford Parts and Service Division)

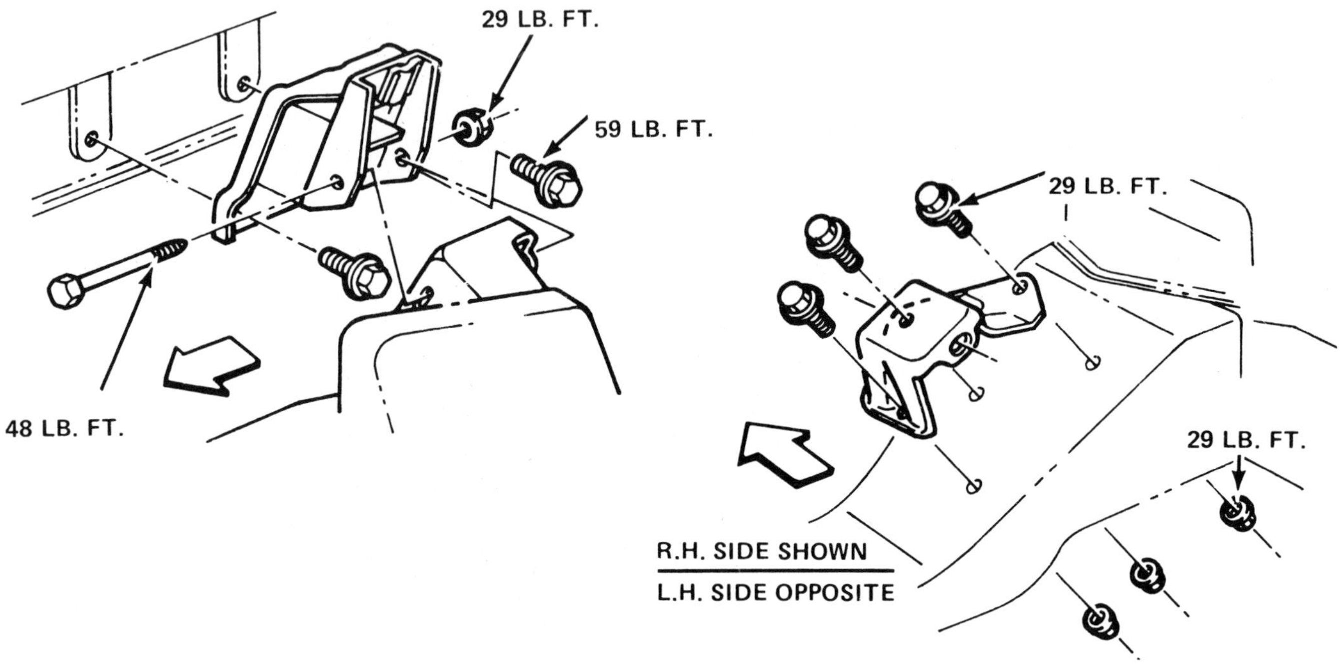

FIGURE 83
Front motor mount attachment details for a Buick 231-CID engine. (Courtesy of Buick Motor Division, General Motors Corporation)

30. On a vehicle equipped with an automatic transmission, disconnect the converter from the flexplate (Figure 84). Mark the converter relative to the flexplate position in order to maintain factory balance.

31. Remove the clutch or converter housing upper and lower attaching bolts.

NOTE: The upper bolts can be removed easily by lowering the rear of the transmission/engine assembly from 5° to 6° and using a long socket extension.

32. Lower the vehicle. Support the transmission with a jack or support bar.

Lifting Out the Engine

1. Attach an engine lifting sling (Figure 85).

CAUTION: The sling attachment bolts must be threaded into the block a minimum of 1 1/2 times the bolt diameter.

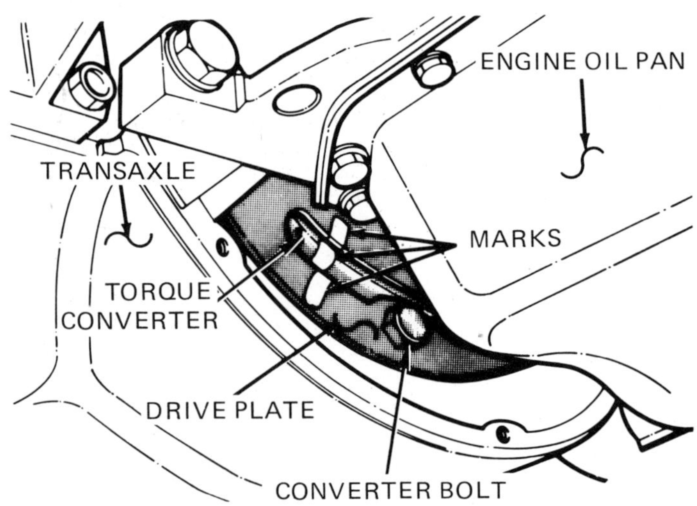

FIGURE 84
Removing the torque converter mounting bolts through a service hole at the rear of engine. (Courtesy of Chrysler Corporation)

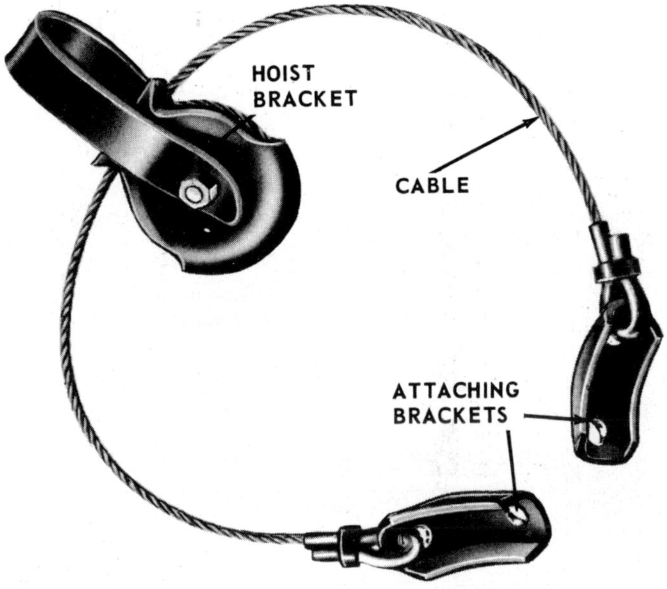

FIGURE 85
Cable-type engine pulling sling. The hoist bracket attaching point is adjustable.

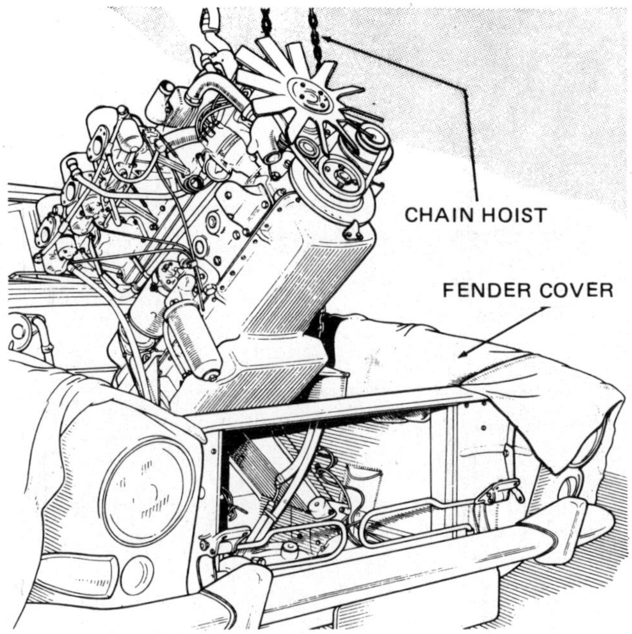

FIGURE 87
Pulling an engine with a chain hoist. Note the steep lifting angle required to provide engine compartment clearance. (Courtesy of Jaguar Rover Triumph, Inc.)

2. Use a chain hoist or a portable hydraulic crane (called a "cherry picker" in the trade) and raise the engine slightly. Carefully pull the engine forward from the transmission to disengage it from the converter (Figure 86). *Do not remove the converter from the transmission during engine removal.*

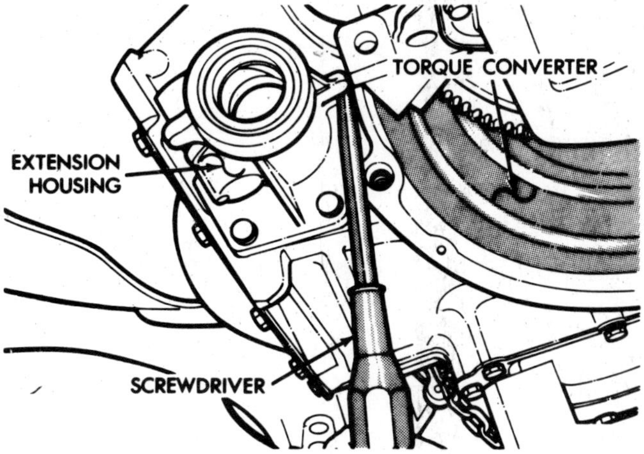

FIGURE 86
Disengaging the converter from the engine after the converter mounting bolts have been removed. (Courtesy of Chrysler Corporation)

3. Tilt the engine (it may be necessary to reposition the sling) and lift it out of the engine compartment. See Figure 87. Be careful not to bend or damage any components.

> **SHOP TIP**
> If working on a van, it is often easier to take the engine out from the bottom.

REMOVING AND REPLACING ENGINE (FRONT WHEEL DRIVE)

On transverse mounted engines with front wheel drive, it is generally recommended that the engine and transaxle assembly be removed as a unit (Figure 88).

When removing and replacing a front wheel drive engine, most of the procedures for removing and replacing a rear wheel drive en-

FIGURE 88
A transaxle assembly.

gine apply. The major exception is the drive axles. On front wheel drive vehicles, it is necessary to disconnect the drive axles from the differential gear splines (Figure 89). Also, disconnect the speedometer cable.

The engine and transaxle in some front wheel drive cars mount to a steel engine cradle (Figure 90). This cradle is secured to the vehicle itself, and can be dropped out the bottom with the engine and transaxle attached.

AFTER THE ENGINE HAS BEEN REMOVED FROM THE VEHICLE

Once the engine has been removed, visually inspect all external surfaces for any problems (Figure 91).

A careful evaluation of the automatic transmission condition should be made at this time. "Read" the contents in the transmission pan, and smell the ATF to determine if it's burned. Explain to the vehicle owner that, even though the automatic transmission was work-

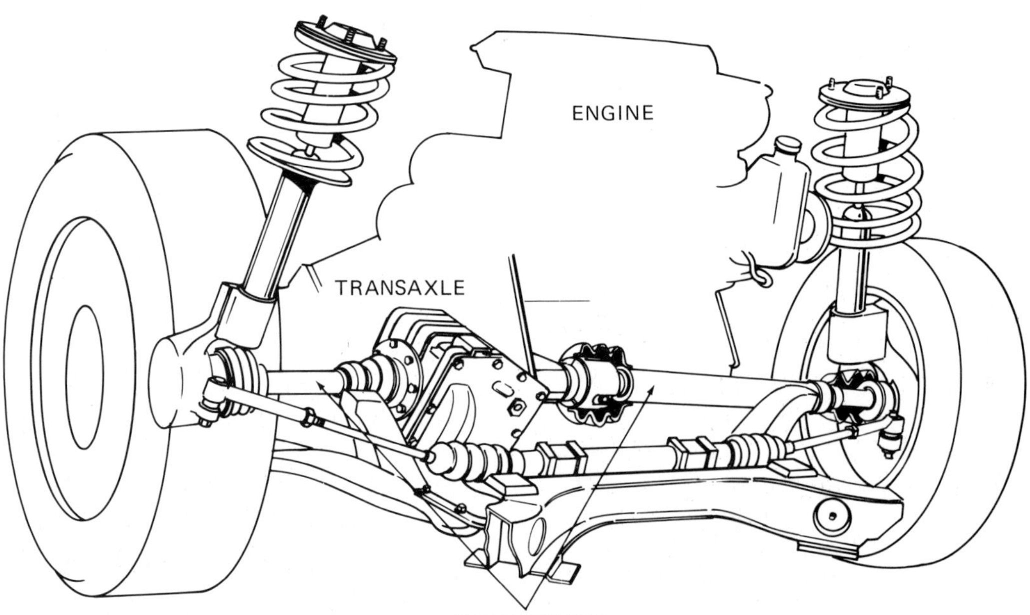

FIGURE 89
Front wheel drive axles (half shafts) are generally splined to the transaxle differential gears.

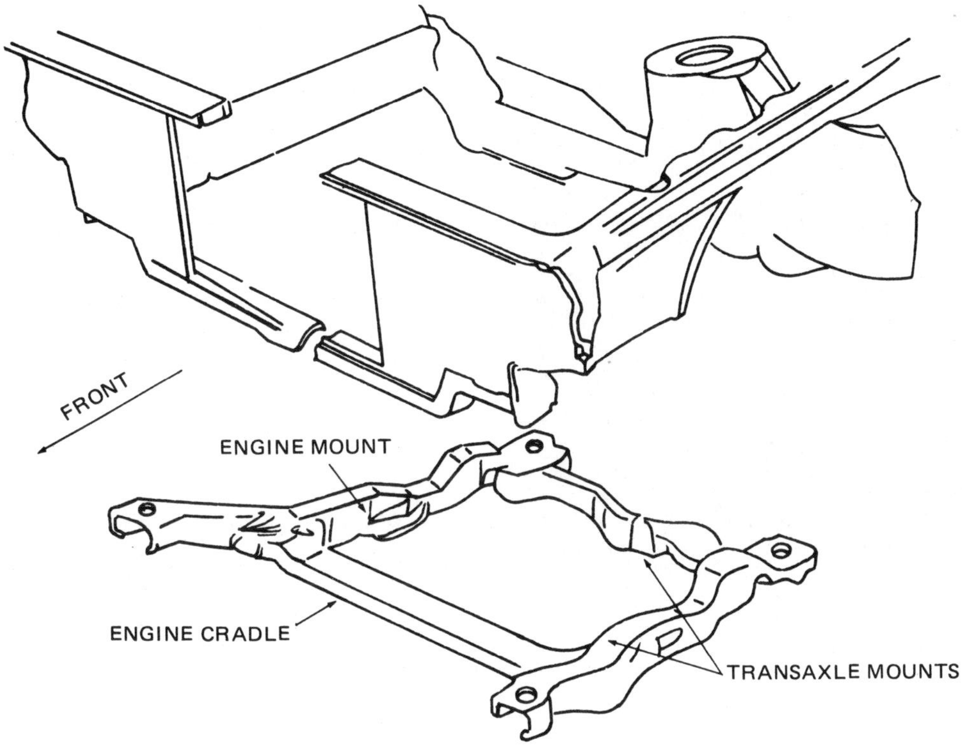

FIGURE 90
Engine cradle construction detail.

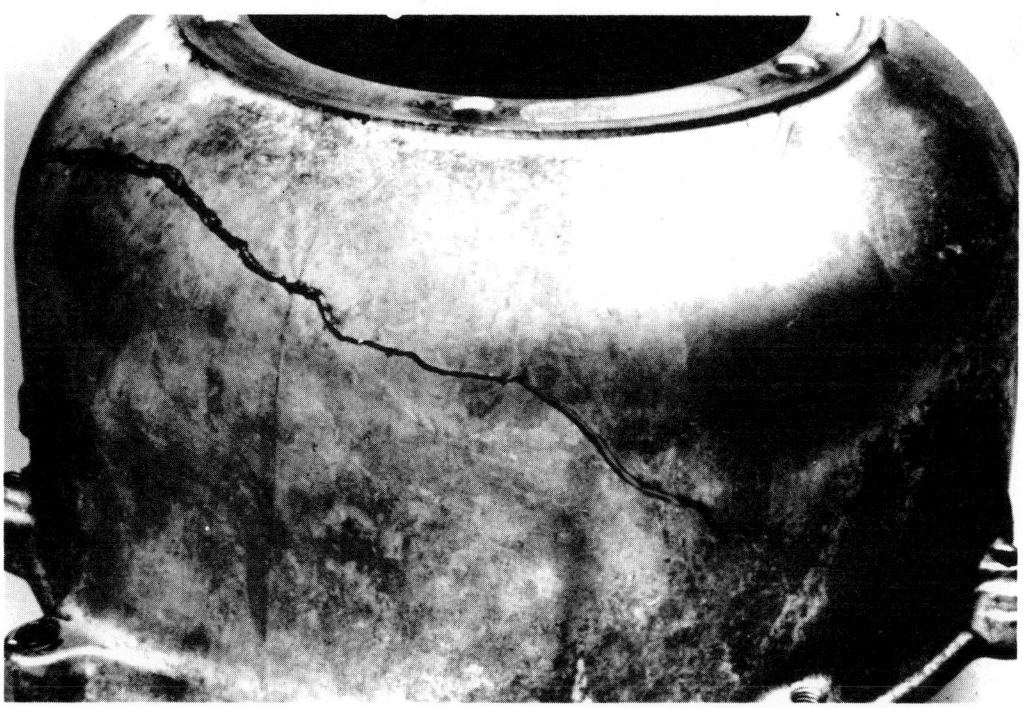

FIGURE 91
Badly cracked bell-housing.

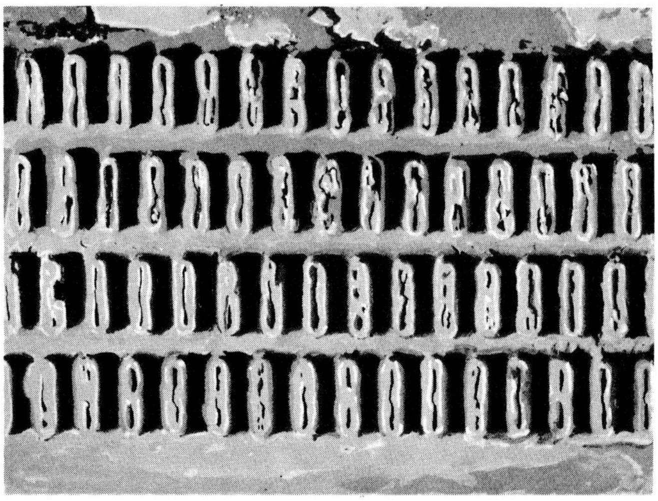

FIGURE 92
Rust-clogged radiator core. (Courtesy of Ford Parts and Service Division)

ing okay behind the old engine, a new engine (which develops more power) will sometimes "take out" a forward clutch assembly. This results in a no-drive condition and an unhappy customer. By correcting any transmission defects now, you will ensure that the vehicle is ready to perform with the new engine.

Have the radiator rodded out, not back flushed. Back flushing will not remove scale deposits from the tubes (Figure 92). In many cases, defective radiators have enough capacity to cool the old engine, since it is loose and has reduced internal friction. However, a new engine may run hotter than normal during the break-in period. The radiator will have to be in top condition to handle this extra heat.

REMOVING AND REPLACING ENGINE MOUNTS

Engine mounts are designed and installed to absorb shock and vibration (Figures 93, 94, and 95). A worn or broken mount (Figure 96) will permit vibrations to be transmitted over the entire vehicle. This can create damaging wear on other parts. Exhaust leaks can often be traced to faulty mounts. If the engine can rock on acceleration, it can cause the fan blades to cut into the radiator core or shroud. Mount failure can even cause the throttle linkage to bind.

Determining When to Replace Mounts

Look for separation of the rubber from the metal. Any separation is cause for replacement. If the rubber is cracked, or if the metal portion is cracked around a bolt hole, the mount should be replaced. Look for loose bolts or nuts. See that the stud on the mount is not forced to one side of the hole or slot. Any shiny spots on the studs indicate problems.

SHOP TIP

To check for a separated engine mount, place the transmission in both drive and reverse ranges while running the engine at a fast idle. KEEP THE VEHICLE BRAKED! A defective mount will cause the engine to lift.

FIGURE 93
Typical engine mounts. (Courtesy of Sealed Power Corporation)

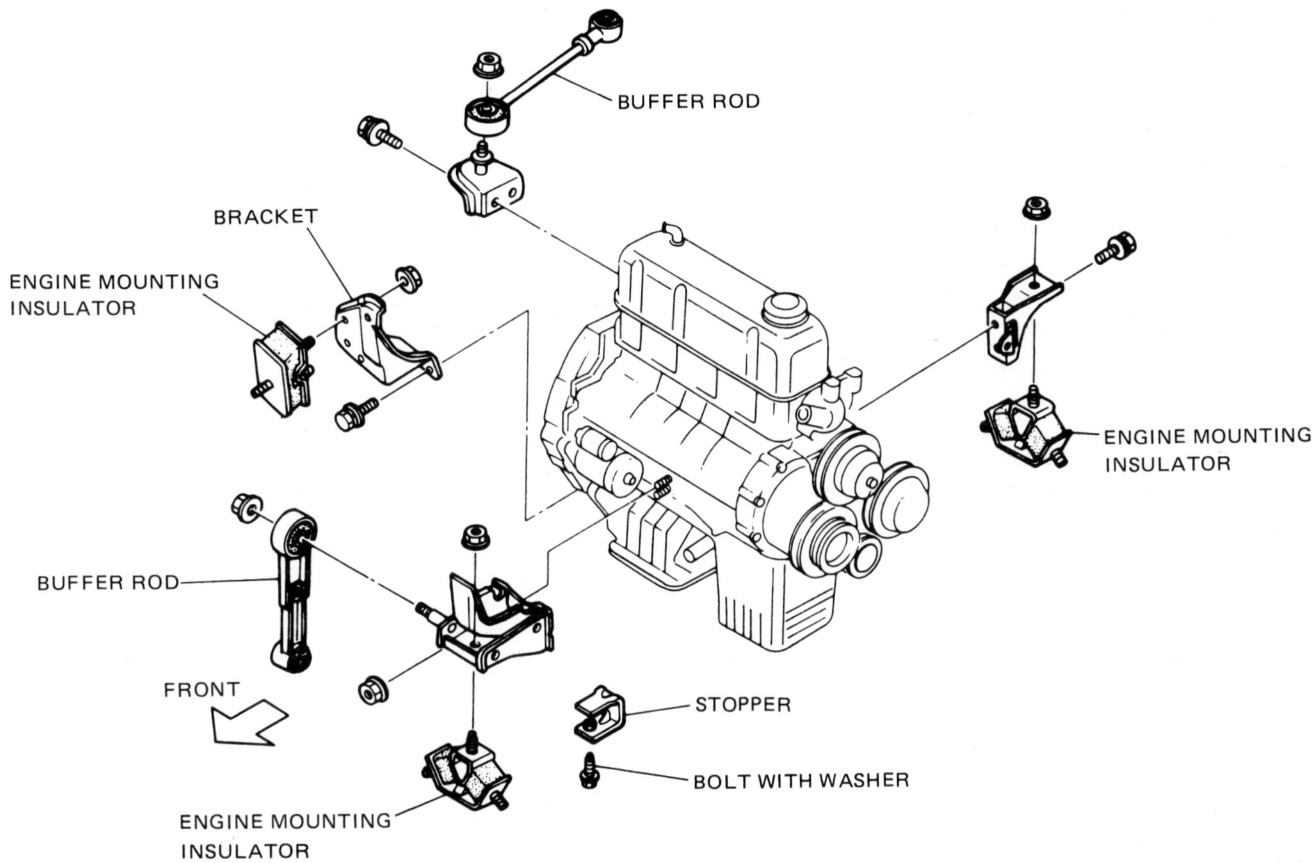

FIGURE 94
Engine mounting system for one popular import vehicle. (Courtesy of Nissan Motor Company, Ltd.)

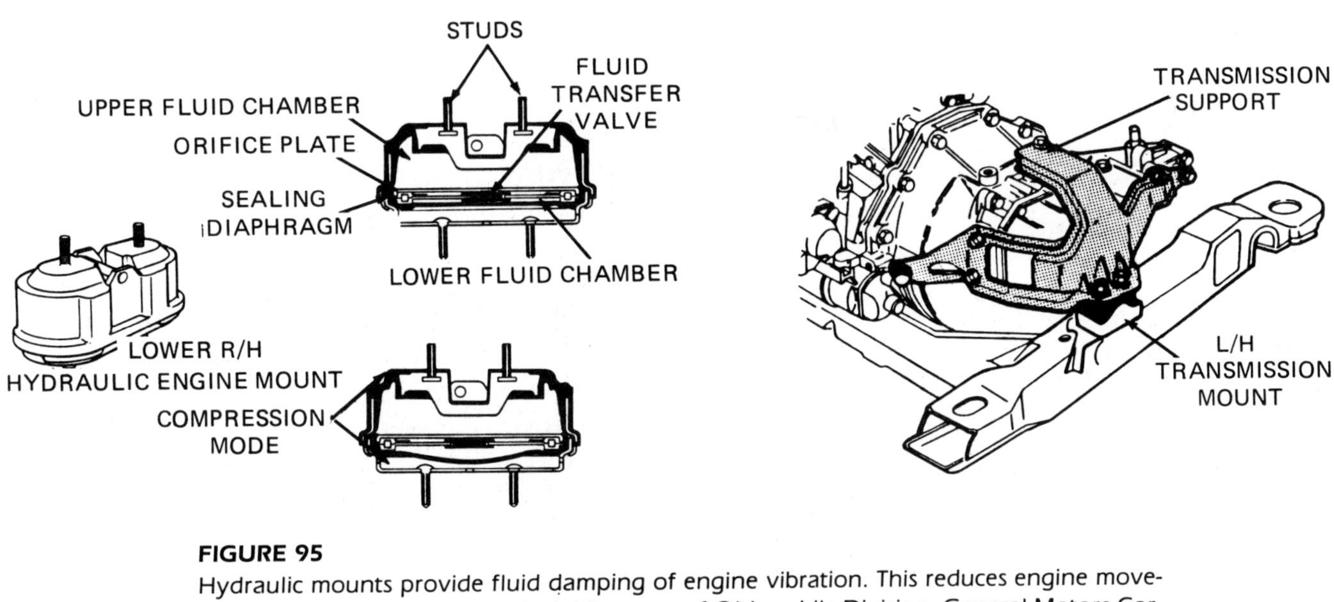

FIGURE 95
Hydraulic mounts provide fluid damping of engine vibration. This reduces engine movement and provides smoother feel. (Courtesy of Oldsmobile Division, General Motors Corporation)

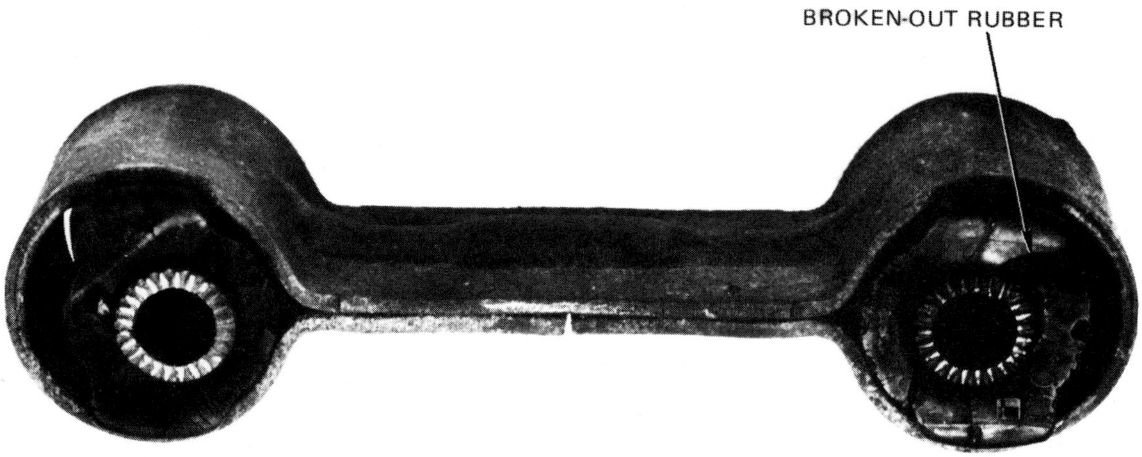

FIGURE 96
Broken "dog-bone" engine mount. Sometimes, this mount is called a strut or a buffer rod.

Typical Removal Procedure

1. Raise car.

2. Support weight of engine at forward edge of oil pan.

3. Remove mount to engine block bolts.

4. Raise engine slightly and remove mount-to-mount bracket bolt and nut.

5. Remove mount.

Typical Installation Procedure

1. Align mount and install. Do not tighten until all mounting bolts have been installed.

NOTE: Some front wheel drive cars have sliding engine mounts that allow for side-to-side positioning of the engine (Figure 97). If the bolts on these mounts have been loosened for any reason, or if the car has experienced front structural damage, half-shaft lengths must be checked and, if necessary corrected. A shorter-than-specified half-shaft length can result in noise. A longer-than-specified half-shaft length can cause transaxle damage.

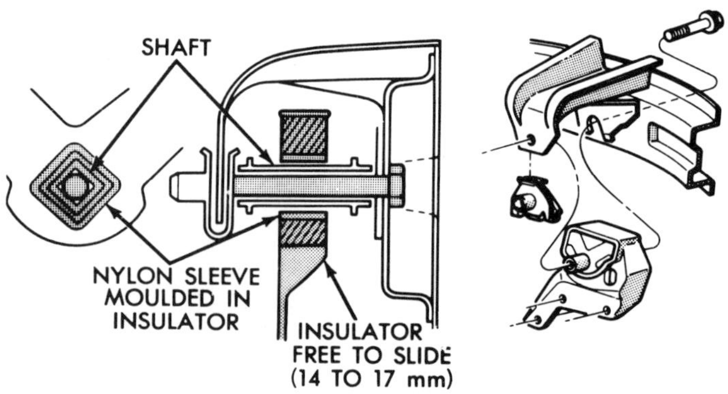

FIGURE 97
Adjustable engine mount. (Courtesy of Chrysler Corporation)

TRUE CASE HISTORY

Vehicle	Volkswagen Bus
Engine	Four-cylinder, 1679 cubic centimeters
Mileage	400 miles after an engine overhaul
Problem	1. Puffs of blue-gray exhaust smoke noticeable on deceleration. 2. Number three spark plug oil fouled.
Cause	One of the pin retainers for number three piston was accidentally left off when the engine was overhauled. This allowed the piston pin to move outward and cut a lengthwise groove in the cylinder wall (barrel). As a result, unwanted oil could pass into the combustion chamber. See Figure 98.

FIGURE 98
Grooved barrel.

TRUE CASE HISTORY

Vehicle	GMC K-15 truck equipped with 4-barrel carburetor
Engine	Eight-cylinder, 5.7 liter
Mileage	27,400
Problem	1. Heavy detonation noise during warm or hot engine operation. 2. Shop replaced EGR solenoid and thermac sensor, rebuilt carburetor, and tuned engine. None of this cured the problem.
Cause	Carbon build-up on pistons from oil getting past valve guides.
Cure	Removed heads and cleaned carbon from pistons. Installed reconditioned cylinder heads.

TRUE CASE HISTORY

Vehicle	Chevrolet C-10 pickup truck equipped with throttle body injection (TBI)
Engine	Eight-cylinder, 5.7 liter
Mileage	276 miles after purchasing vehicle new
Problem	1. Low power at wide open throttle (WOT) up to 3500 rpm. 2. Severe loss of power at 3500 rpm. 3. When engine reached 3500 rpm, the transmission would not shift unless the throttle was backed off.
Cause	Cam timing was advanced one tooth.

ASE STUDY QUESTIONS

1. A car owner complains of excessive exhaust smoke. During a road test the technician verifies the complaint and finds that the automatic transmission shifts in a rough manner. Which of these is the MOST LIKELY cause?

   a. Defective vacuum modulator
   b. Cracked flexplate
   c. Rich fuel mixture
   d. Leaking intercooler

2. Oil is dripping from the bottom of an engine as shown above.

   Technician A says that this could be from a leaky oil gallery plug.

   Technician B says that this could be from a leaky rear main seal.

   Who is right?

   a. Technician A
   b. Technician B
   c. Both A and B
   d. Neither A nor B

3. A light double knocking noise is heard from an idling engine.

   Technician A says that this could be from worn rod bearings.

   Technician B says that this could be from worn piston pins.

   Who is right?

   a. Technician A
   b. Technician B
   c. Both A and B
   d. Neither A nor B

4. **Technician A** says that oil can reach the combustion chamber through a torn intake manifold gasket.

   Technician B says that oil can reach the combustion chamber through worn valve guides.

   Who is right?

   a. Technician A
   b. Technician B
   c. Both A and B
   d. Neither A nor B

5. You are measuring vacuum on an engine that is in good condition. The MOST NORMAL reading (at idle) would be:

   a. 30''
   b. 17''
   c. 12''
   d. 3''

6. A power balance test is being performed on a non-computer-controlled engine.

   Technician A says to disconnect and plug the EGR vacuum signal hose.

   Technician B says to run the engine at 1000 rpm.

   Who is right?

   a. Technician A   c. Both A and B
   b. Technician B   d. Neither A nor B

7. A compression test has been made on the engine shown at the right. Cylinders 4 and 6 have readings of 60 psi. All other cylinders have readings between 130 and 135 psi.

   Technician A says that this could be caused by a "loose" timing chain.

   Technician B says that this could be caused by a blown head gasket.

   Who is right?

   a. Technician A   c. Both A and B
   b. Technician B   d. Neither A nor B

8. A cylinder leakage test shows all cylinders to be normal. Yet the engine just failed a compression test.

   Technician A says that slow cranking rpm could be the reason.

   Technician B says that an out-of-time camshaft could be the reason.

   Who is right?

   a. Technician A   c. Both A and B
   b. Technician B   d. Neither A nor B

9. A transverse mounted engine with front wheel drive must be removed from a car. All of the following are generally recommended EXCEPT:

   a. Remove the engine and transaxle as a unit
   b. Disconnect the half-shafts
   c. Disconnect the speedometer cable
   d. Remove the differential gears

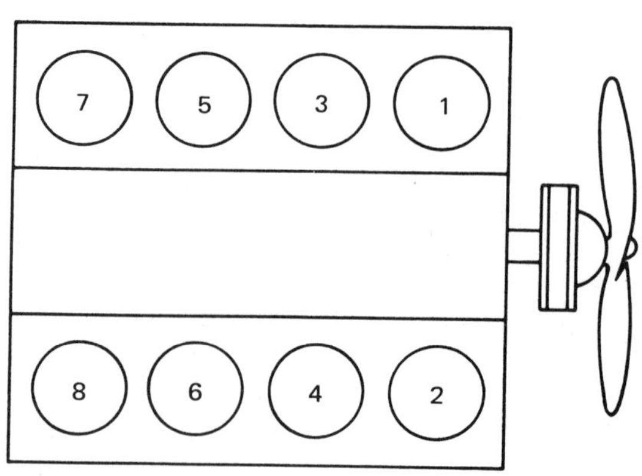

10. The engine shown above is going to be removed and replaced.

    Technician A says to mark the position of the hood hinges.

    Technician B says to drop the engine out the bottom.

    Who is right?

    a. Technician A **c.** Both A and B
    b. Technician B **d.** Neither A nor B

11. **Technician A** says to replace an engine mount if the rubber has separated from the metal.

    Technician B says to replace an engine mount if it has cracks around a bolt hole.

    Who is right?

    a. Technician A **c.** Both A and B
    b. Technician B **d.** Neither A nor B

SECTION REVIEW QUESTIONS

1. Describe the firing cycle that takes place in an engine.
2. What is the VECTOR diagnostic approach?
3. Why is a cooling system needed in an engine?
4. What are the three most common cooling system problems?
5. List four ways to test for internal combustion gas leakage into the coolant.
6. What are the principal causes of cooling system aeration?
7. A too-cool running engine on a car equipped with a computer-controlled fuel system can cause what problem(s)?
8. List eight possible sources of an external coolant leak.
9. List six possible sources of an internal coolant leak.
10. How can a faulty mechanical fuel pump be responsible for oil consumption?
11. What is fuel "pull-over"? List five causes.
12. A good tool for locating the source of engine noise is a ______________________.
13. List five possible causes of main bearing noise.
14. List six possible causes of connecting rod bearing noise.
15. What is piston slap? What does it sound like?
16. Piston pin noise produces what sound?
17. List ten possible causes of valve train noise.
18. A "flat" exhaust lobe on a camshaft generally produces what sound?
19. List ten places where external oil leakage may occur on an engine.
20. List fifteen ways that oil may enter into the combustion chamber.
21. How can knowing the color of exhaust smoke help to save diagnosis time?
22. How can a "rotten egg" exhaust odor often be cured?
23. How can you quickly diagnose a possible burned valve in an engine?
24. Explain the procedure for detecting a restricted exhaust with a vacuum gauge.

25. Describe a cylinder power balance test.

26. What is the procedure for making a cylinder power balance test on a General Motors C3 system?

27. Engine cylinder pressures normally should not vary more than ________ from the highest to the lowest one.

28. Outline the procedure for making a properly performed compression test.

29. What is a wet test?

30. What effect does low cranking rpm have on a compression test?

31. What test can help to pinpoint the source of compression leakage?

32. What methods can be used to mark disconnected wires and vacuum hoses when removing an engine?

33. Sketch a flare nut wrench.

34. Upper bell housing bolts can often be removed easily by following what procedure?

35. When attaching an engine lifting sling, the sling attachment bolts must be threaded in how far?

36. What is an engine cradle?

37. When installing a new engine, why is it important to make a careful evaluation of the automatic transmission condition?

38. How can you check for a separated engine mount?

39. Why do some front wheel drive cars have sliding engine mounts?

Section B

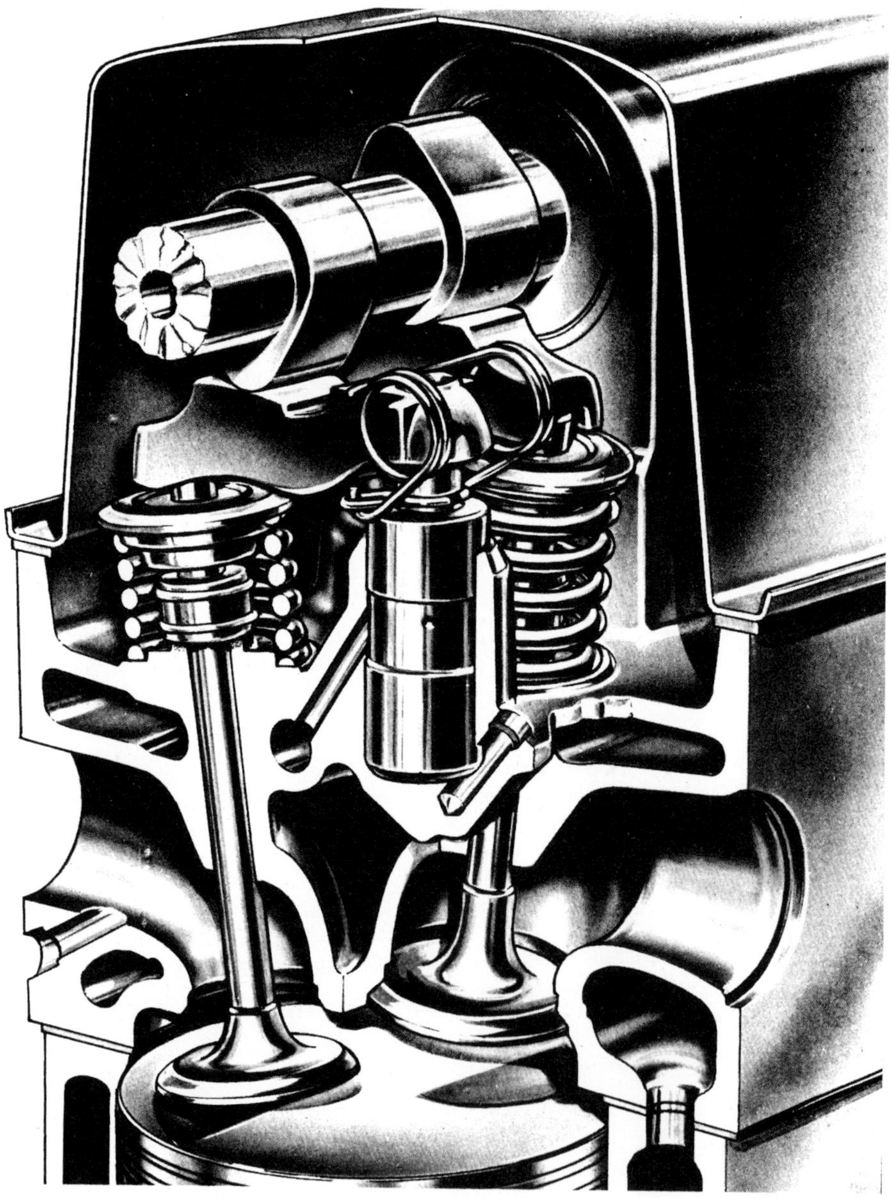

CYLINDER HEAD AND VALVE TRAIN DIAGNOSIS AND REPAIR

After studying this section, and with hands-on practice on a "live" vehicle, you should be able to perform the following tasks:

- Remove cylinder heads; visually inspect cylinder heads for cracks; gasket surface areas for warpage and leakage; check passage condition.
- Install cylinder heads and gaskets; tighten in accordance with manufacturers' specifications and procedures.
- Inspect and test valve springs for squareness, pressure, and free height comparison; replace as necessary.
- Inspect valve spring retainers, locks, and valve lock grooves.
- Replace valve stem seals.
- Inspect valve guides for wear; check valve guide height and stem-to-guide clearance; determine needed repairs (no questions about specific repair procedures).
- Inspect valves; resurface or replace.
- Inspect valve seats; resurface or repair/replace.
- Check valve face-to-seat contact and valve seat concentricity (runout); service seats and valves as necessary.
- Check valve spring assembled height and valve stem height; service valve and spring assemblies as necessary.
- Inspect push rods, rocker arms, rocker arm pivots, and shafts for wear, bending, cracks, looseness, and blocked oil passages; repair or replace.
- Inspect, test, and replace hydraulic or mechanical lifters (questions about leakdown tests limited to knowledge of purpose).
- Adjust valves on engines with mechanical and hydraulic lifters according to manufacturers' specifications and procedures.
- Inspect and replace camshaft drives. (Includes checking gear wear and backlash, sprocket and chain wear, overhead cam drive sprockets, drive belts, belt tension, and tensioners.)
- Inspect and measure camshaft journals and lobes.
- Inspect and measure camshaft bearing surfaces for damage, out-of-round, and alignment; repair or replace according to manufacturers' specifications.
- Measure camshaft timing according to manufacturers' specifications and procedures.

TOP END DISASSEMBLY

Certain procedures should be followed in the top end disassembly of an automobile engine (Figure 1). The following procedures may be considered typical. The steps may vary slightly depending on the make and model of the engine.

Cylinder Head Removal

1. Remove the intake manifold. If the manifold is stuck, jar the ends by hitting with a brass or "dead-blow" hammer.

NOTE: In some cases it will first be necessary to remove the valve covers.

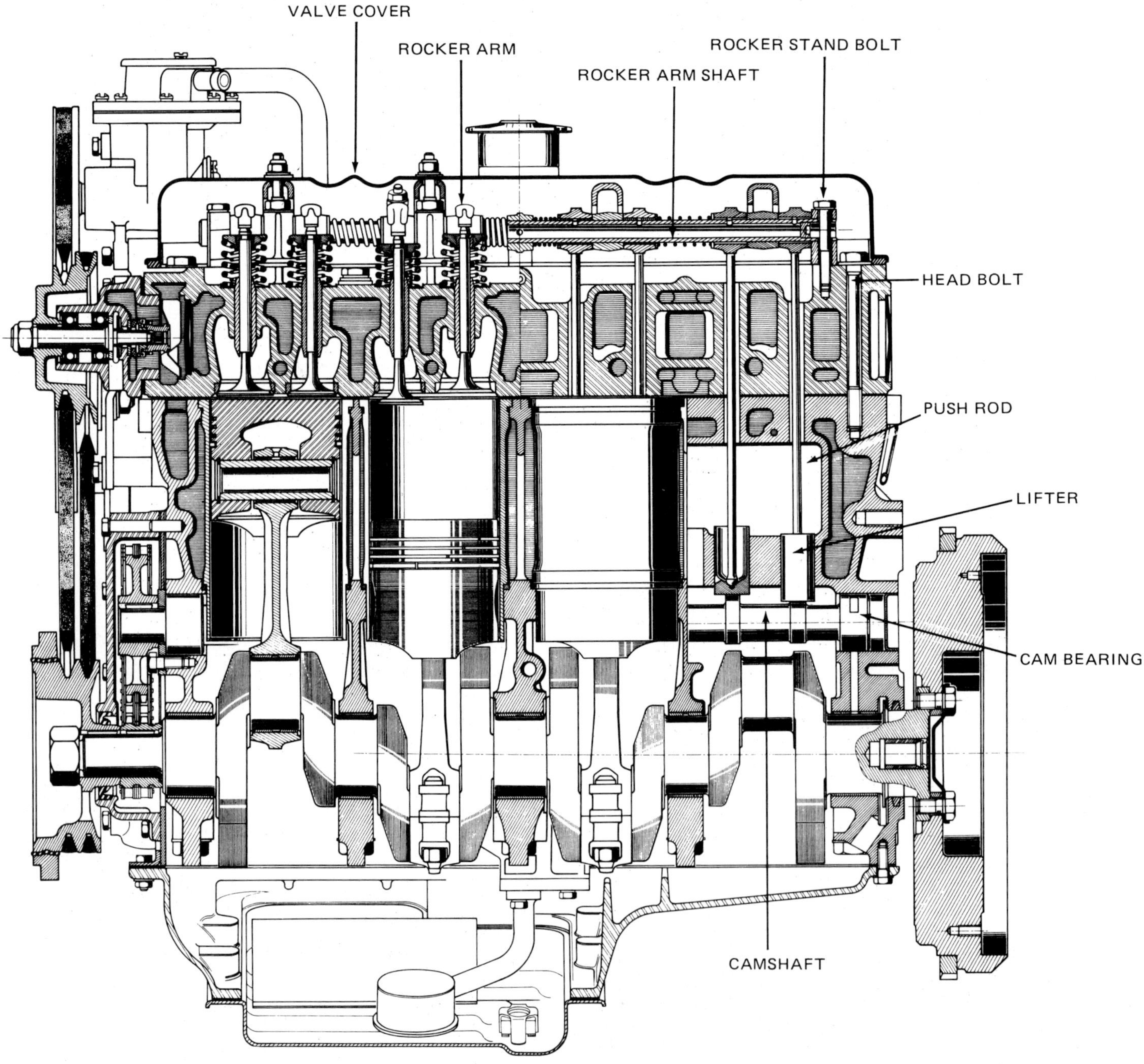

FIGURE 1
A four-cylinder diesel automobile engine. Study the labeled top end parts. (Courtesy of Peugeot Motors of America, Inc.)

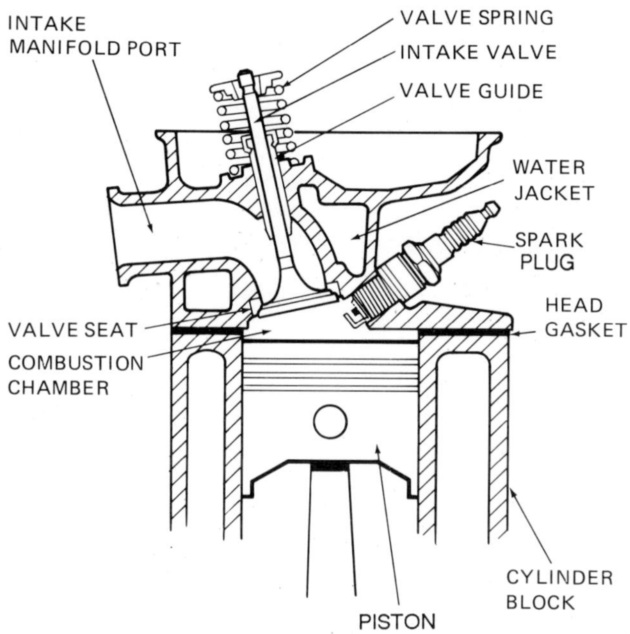

FIGURE 2
After removing the intake manifold, probe the cylinder head intake ports for oil. An oily condition indicates oil is being introduced from an outside source.

2. Inspect the intake manifold port passages by probing with a finger (Figure 2). If you detect oil in the passages and in the cylinder head intake ports, oil is being introduced from an outside source. A bad automatic transmission vacuum modulator, malfunctioning PCV system, or faulty fuel pump are common causes. Occasionally, an oil-bath type air cleaner (used mostly on older vehicles) is overfilled or has a light viscosity oil added. This results in oil being drawn into the engine by the incoming airstream.

FIGURE 3
Burned-through exhaust manifold gasket.

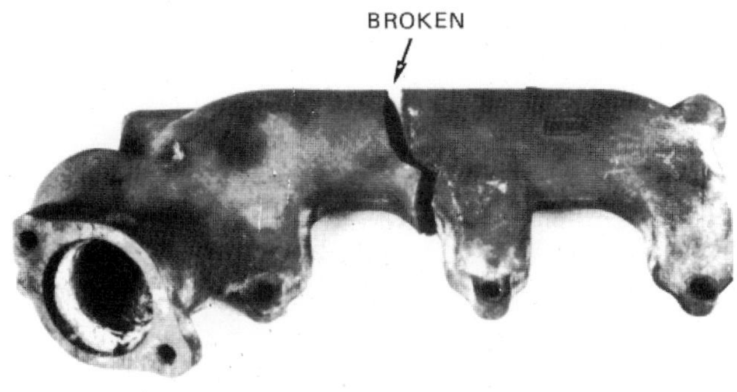

FIGURE 4
Broken exhaust manifold.

Shine a light down the cylinder head ports and, if possible, inspect the underside of the intake valve heads for hard carbon buildup. This indicates that oil is passing between the valve stem and guide.

3. Remove the exhaust manifold(s) or exhaust pipe(s). Inspect for broken gaskets and for any casting cracks (Figures 3 and 4).

4. Remove the valve cover(s). Look for sludge or any unusual deposits (Figure 5).

5. Remove any rocker arm shaft lubrication (oil delivery) pipes (Figure 6).

6. On shaft-mounted rocker arms, remove bolts in correct sequence in two or three steps (Figure 7). This will help prevent broken rocker arms and/or cracked stands.

FIGURE 5
This valve cover came off a cracked cylinder head. The sludge deposits inside are a "dirty mayonnaise" color. They resulted when the crack allowed water to mix with oil.

FIGURE 6
Removing union bolts and oil delivery pipes from a Toyota OHC engine.

FIGURE 7
Removing rocker shaft support (stand) bolts from the engine pictured in Figure 6. Always remove a rocker shaft assembly in correct sequence, in two or three steps.

7. Remove the push rods and examine them for visible defects (Figure 8).

8. On overhead camshaft (OHC) engines, remove the camshaft according to the manufacturer's specific procedure. See Figure 9.

9. Remove the cylinder head attaching bolts. It is extremely important that the head bolts be removed in correct sequence (Figure 10). Otherwise, head warpage or cracking can result.

CAUTION: Do not remove an aluminum head until it has cooled down completely; the head can become warped if this procedure is not followed.

10. Lift the cylinder head(s) from the block. Because dowel pins are installed on many blocks, never attempt to slide or push the head(s) off the block (Figure 11).

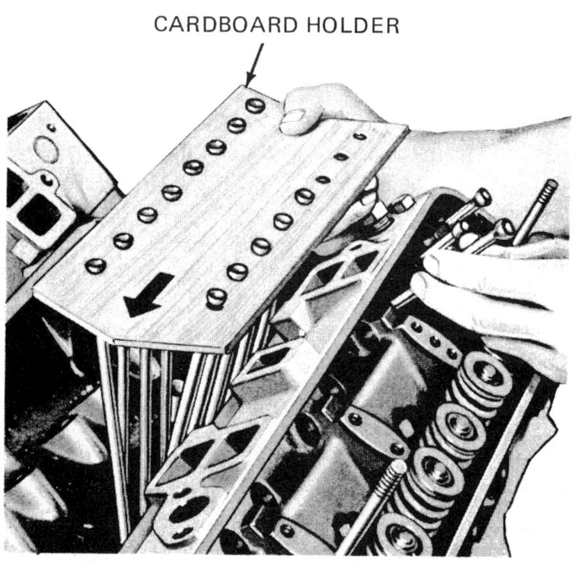

FIGURE 8
Removing push rods from an engine. Keeping them in a holder allows them to be replaced in their original order. (Courtesy of Ford Parts and Service Division)

SHOP TIP

Occasionally a cylinder head becomes stuck to the block and removal can be difficult. Here is a trick that can be used if the engine is still in the vehicle. First, check to be sure that every head bolt has been removed. Then loosely install one head bolt at each end of the head. Next, reinstall all the spark plugs (do not attach plug wires) and crank the engine. The compression pressure will break the bond between the block and head. The loosely installed head bolts at each end of the head will prevent the head from slipping off the block.

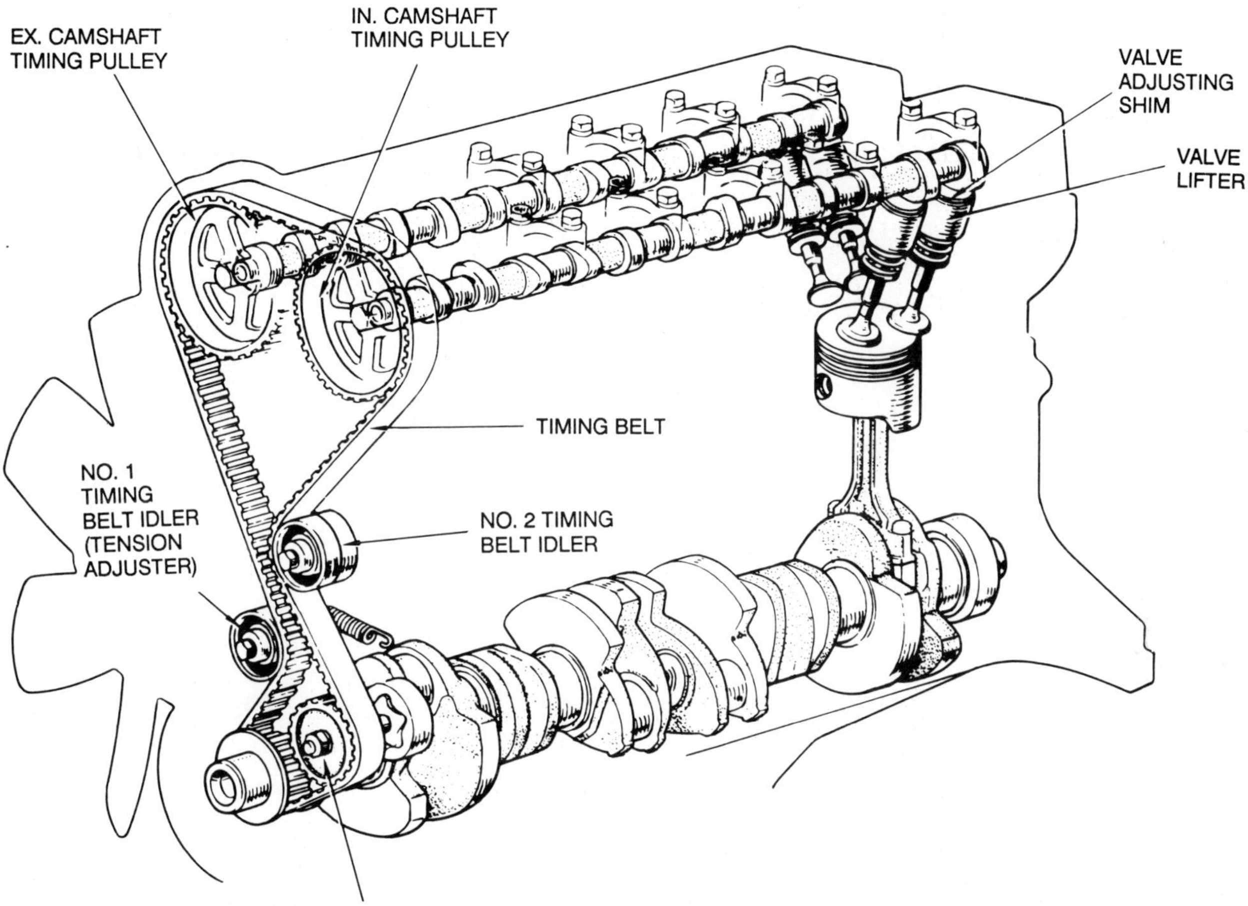

FIGURE 9
Double overhead camshaft (DOHC) engine. Bearing caps hold the camshafts. (Courtesy of Toyota Motor Sales, U.S.A., Inc.)

Cylinder Head Inspection

After removing the cylinder head(s), visually inspect the combustion chambers for cracks, burnthroughs, electrolysis damage, or other irregularities (Figures 12 and 13). Analyze combustion chamber and valve head coloring (Figure 14).

Clean all gasket surfaces of the head. Check the cylinder head gasket surface for warpage by using a precision straightedge and a feeler gauge (Figure 15). The surface should be checked in both diagonal directions, across the head, and lengthwise. *Maximum allowable warpage is .004" at any point.* If surfaces are warped beyond limits, surfacing will be required (Figure 16).

If surface *A* is on a V-type overhead valve (OHV) engine is cut more than .010″, it is recommended that metal should be removed from surface *B* (Figure 17). Otherwise, the holes in the intake manifold and cylinder head may not line up when reassembled. In addition, port mismatch will occur. The amount of corrective machining required depends on the make of the head and how much the head is surfaced. See the Appendix for corrective machining tables.

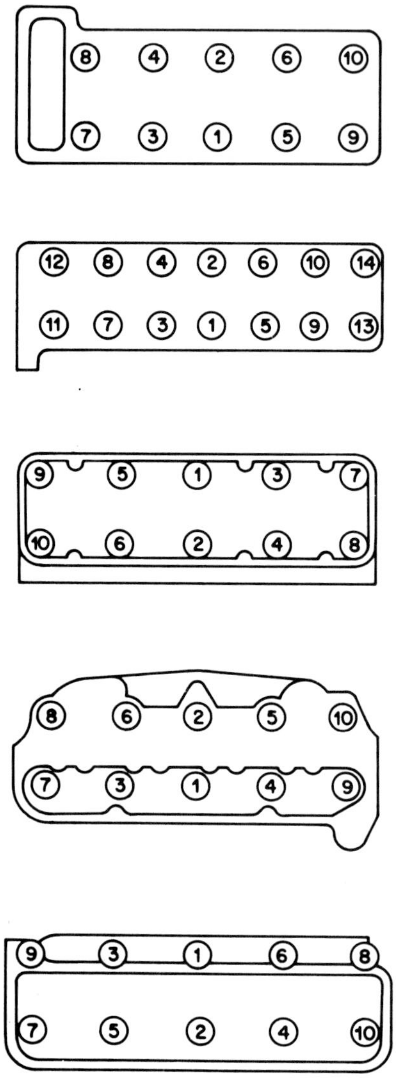

FIGURE 10
Head bolt installation sequence. Removal sequence is just the opposite—start at the ends and work inward. *Always refer to manufacturer torque specifications when installing head bolts. Some engines require a torque-to-yield method of tightening.* Refer to page 95. (Courtesy of Fel-Pro, Inc.)

Surfacing OHC Cylinder Heads

Often after surfacing an OHC head, the camshaft will not turn. That is because when this type of head warps, camshaft bearing alignment is also affected (Figure 18). Here are some approaches being used by shops to solve the problem.

1. If the head is aluminum, try to straighten it before surfacing. "Pull" the head casting into alignment by clamping it against a piece of flat steel plate. Then place the head in an oven for four hours at a temperature of 500°F. Shut the oven off and allow the head to cool slowly.

2. Heat the head in an oven to 600°F. Use a tempil stick to gauge the temperature. After the head has reached proper temperature, place it in a hydraulic press and press into straightness.

3. Place the warped head on a "dummy" block of the same description with no gasket and a .005″ strip of shim stock at each end between the block and head (Figure 19). Using only the two center head bolts, tighten them snugly. Next, heat the back side of the warped head with a rosebud torch. ***CAUTION:*** *Do not heat the warped head too hot, and heat it only in the middle area.* Torque the two center head bolts to factory specifications. Let the head cool for several minutes.

4. When the head is being set up in the surfacing machine, draw the camshaft bores into alignment. Then surface the head. When you take it off the surfacing machine, it will bow back. However, when torqued down on the block, it will be true again.

SHOP TIP

On certain OHC engines, the cylinder head overhangs the front timing cover. Always check alignment of the block surface to the top of the front timing cover surface. If the cover stands above the engine block surface, a cut should be taken to bring the timing cover surface back into alignment with the engine block (Figure 20).

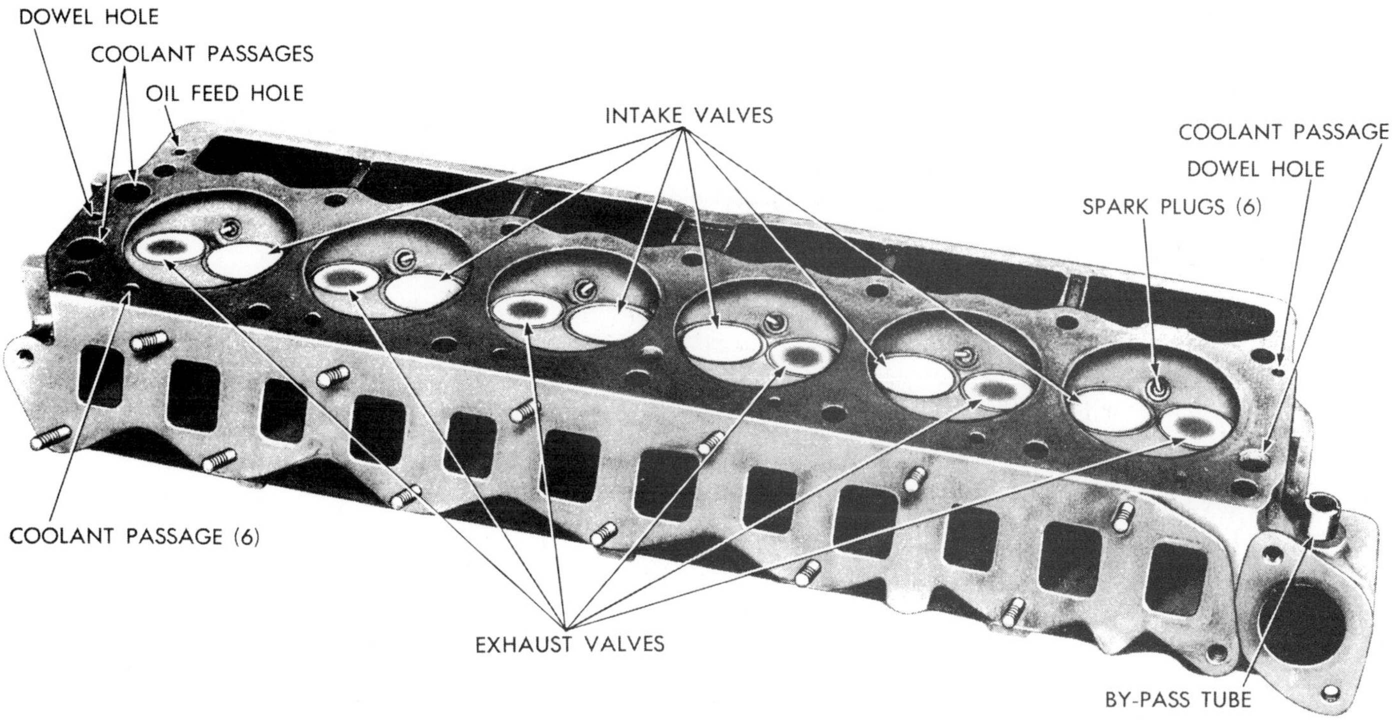

FIGURE 11
Cylinder head. Do you see the holes where the block dowel pins fit? (Courtesy of Chrysler Corporation)

FIGURE 12
Operating an engine beyond the normal rpm range can cause a valve impact failure.

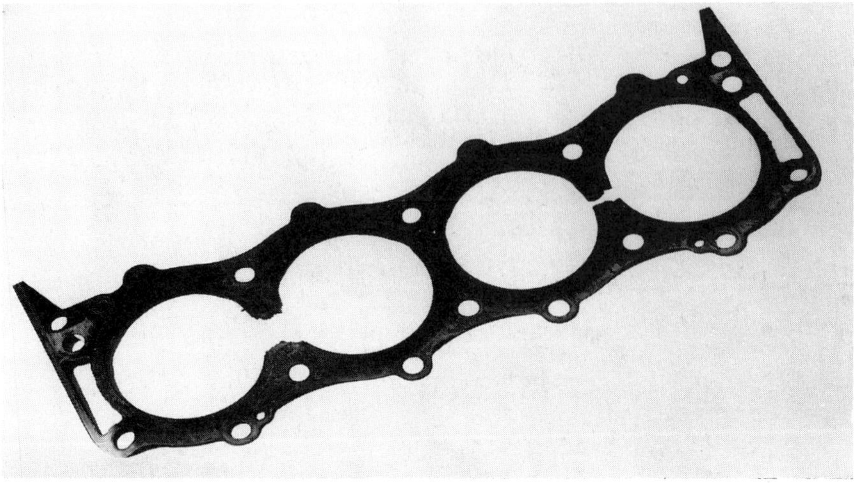

FIGURE 13
A head gasket blown out in two locations.

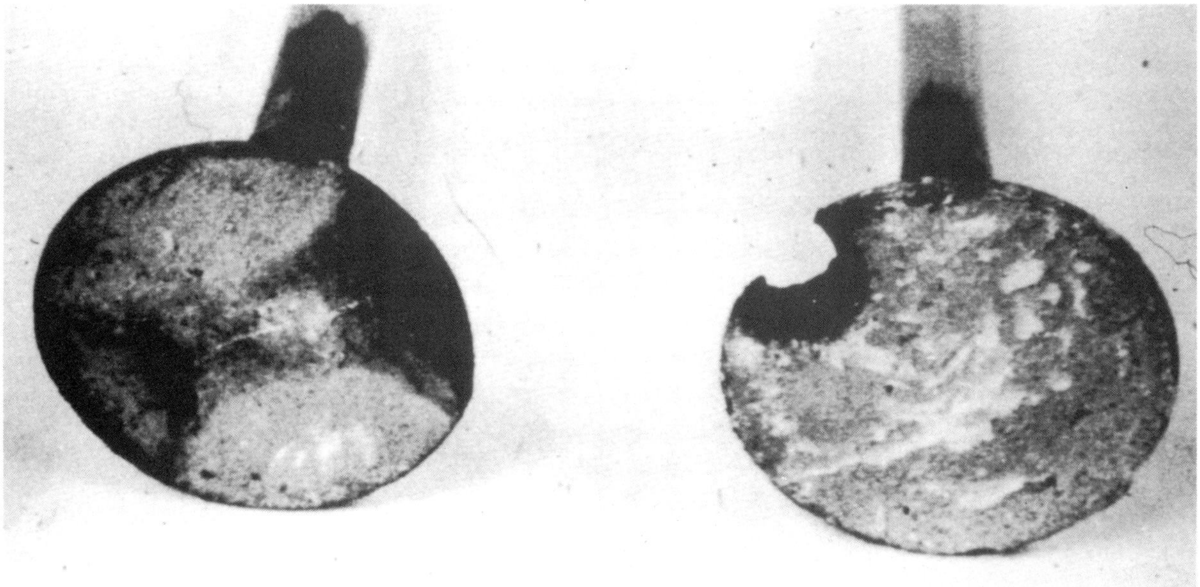

FIGURE 14
The valve on the left has a burned face, as indicated by the dark color zones on the valve head. The valve on the right shows what can happen when an air-pump-equipped engine is overheated.

5. True the camshaft bearing tunnel by boring (Figure 21).

6. Install semi-finished camshaft bearings and machine to size.

Locating Cracks

There are several trade-accepted methods (besides visual) of locating cracks:

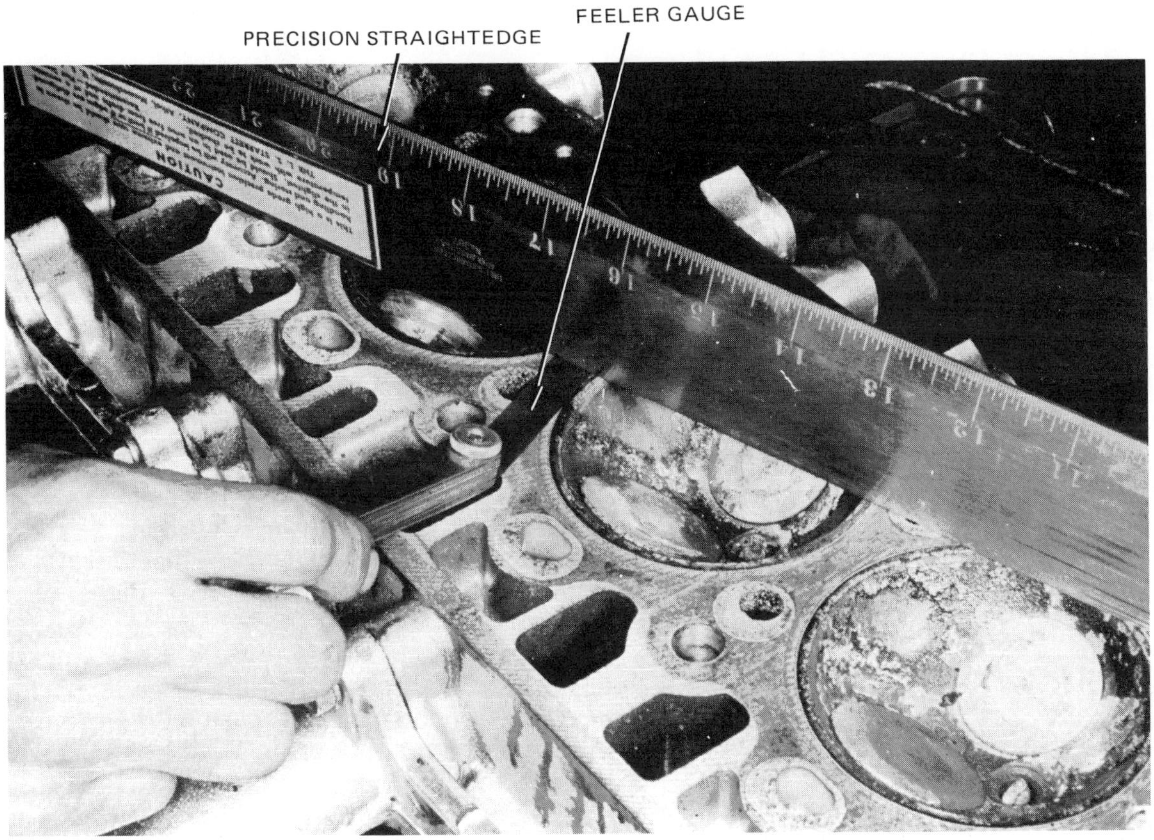

FIGURE 15
Checking the cylinder head for warpage.

FIGURE 16
Surfacing a cylinder head on a rotary broach. (Courtesy of Winona Van Norman Machine Company)

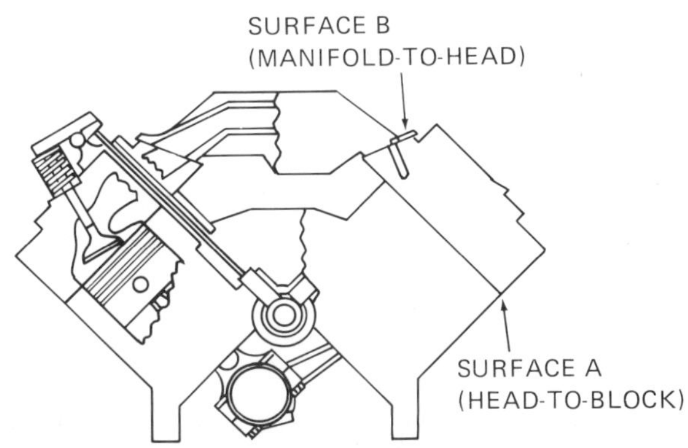

FIGURE 17
After machining the head at surface A, material should be machined from surface B for proper intake manifold alignment (fit).

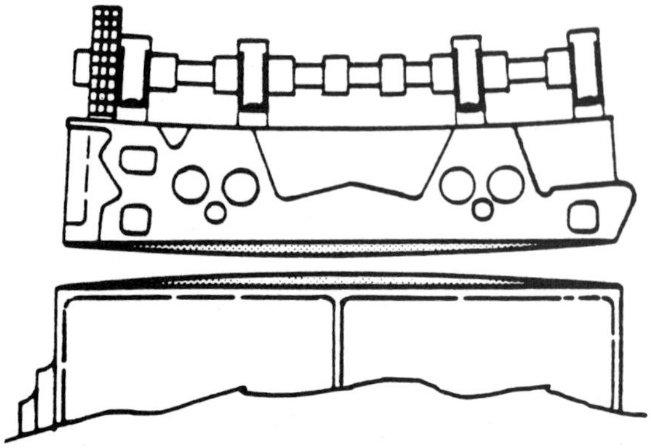

FIGURE 18
On OHC engines, the camshaft bearing tunnel will be affected when head warpage exists.

FIGURE 19
Getting ready to straighten a warped OHC aluminum head. Note the "dummy" block and the shim stock in place at each end. (Courtesy of Automotive Engineering)

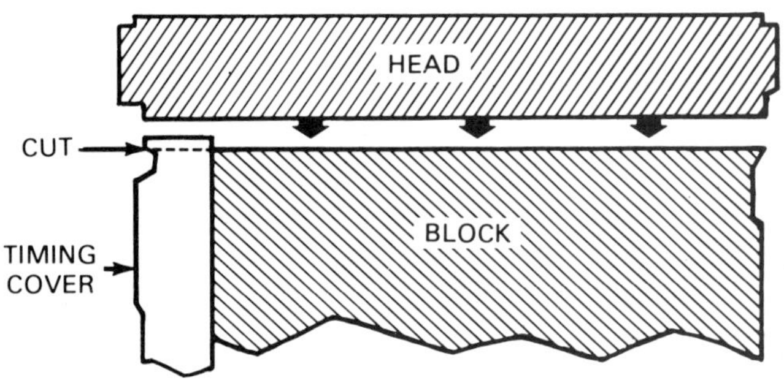

FIGURE 20
On certain OHC engines, the timing cover must be machined if the block deck is surfaced.

FIGURE 21
Truing the camshaft tunnel bore with a portable boring machine. (Courtesy of Belban Manufacturing Corporation)

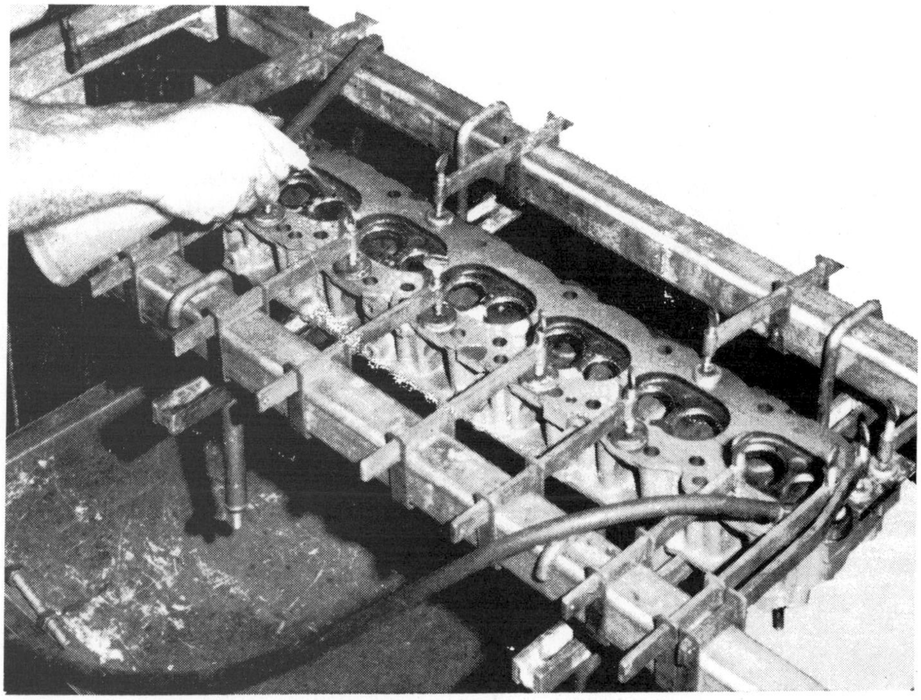

FIGURE 22
Cylinder head ready for pressure testing. Note the pads covering the water holes.

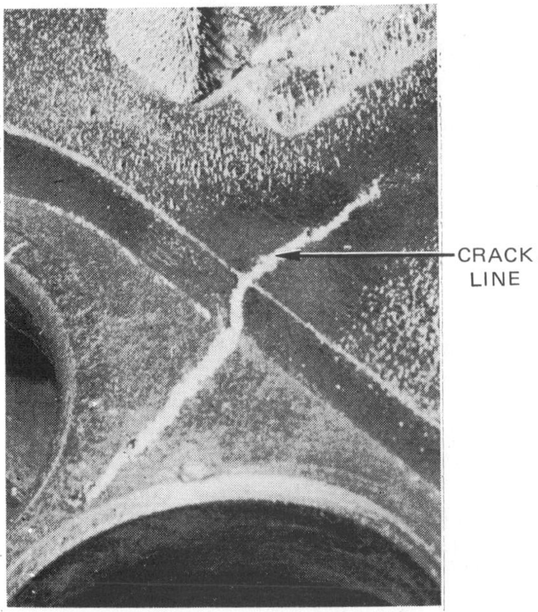

FIGURE 23
Magnetic particle method of inspecting ferrous parts. Cracks will show up as highly visible lines.

1. **Pressure testing** (Figure 22). Pressure testing is ideal for detecting leaks around rocker arm studs, valve guides, and injector tubes.

2. **Magnetic inspection.** When the part being inspected is magnetized, opposite and attracting poles will be set up at any crack. Fine colored (gray, white, or yellow) iron powder dusted on the surface will then be attracted. A definite colored outline of the crack can then be seen (Figure 23).

3. **Dye penetrant.** This inspection method can be used on magnetic and nonmagnetic materials. The process usually consists of a three-step spray system (Figure 24). Defects show up as bright red traces (Figure 25).

FIGURE 24
A dye penetrant crack detection kit.

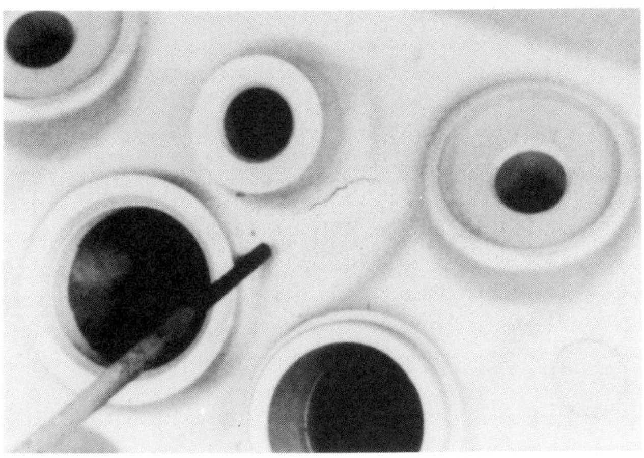

FIGURE 25
The crack in this aluminum head was located by dye penetrant inspection.

VALVE ASSEMBLY

The intake and the exhaust valves in an engine take a great deal of punishment. Valve heads are scorched by high temperatures and stressed by spring tension. Valve faces are exposed to corrosive combustion products and abrasives. As a result, the valves may become burned, pitted, worn, corroded, warped, grooved, and out-of-round (Figure 26).

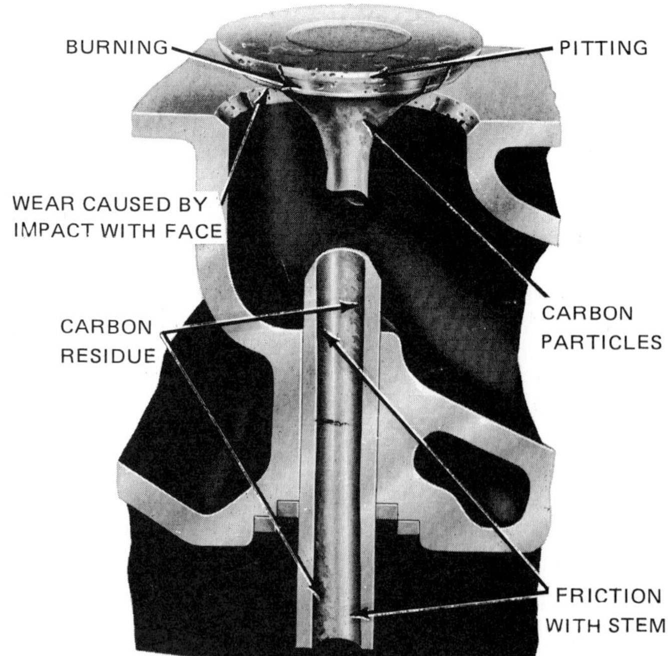

FIGURE 26
Valves, valve seats, and guides must be closely evaluated and reconditioned accordingly. (Courtesy of Sioux Tools Inc.)

FIGURE 27
Engine overheating caused this seat ring to come out.

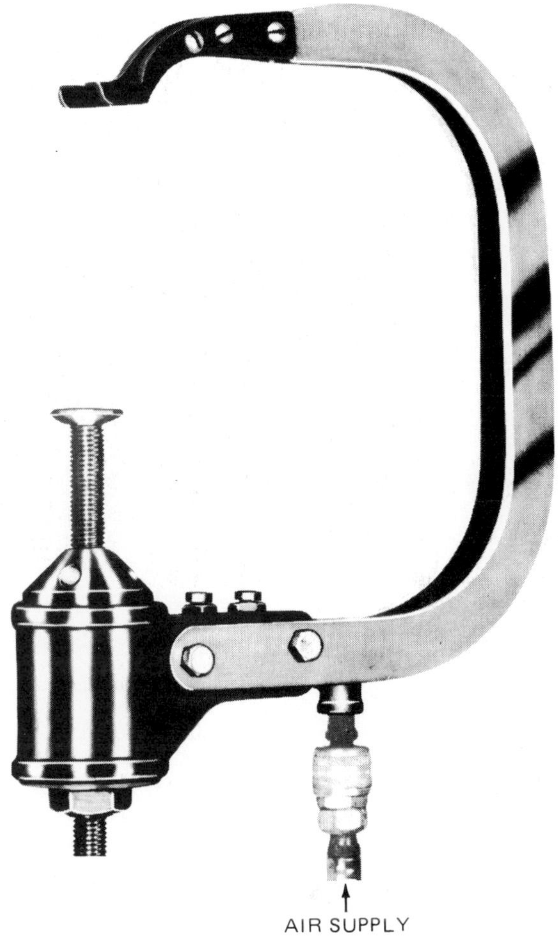

FIGURE 28
Air-operated valve spring compressor.

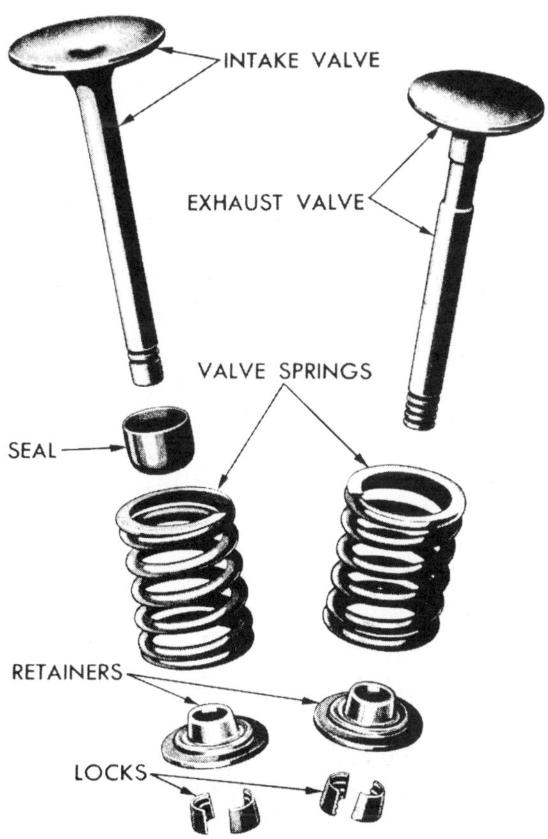

FIGURE 29
Details of valves and associated parts. (Courtesy of Chrysler Corporation)

Valve seats can distort, wear, pit, and recess. Seat inserts may become loose or crack due to thermal variations in the combustion chamber (Figure 27).

Valve guides wear in a pattern corresponding to valve stem wear. Carbon and/or varnish residues may form between the stem and guide and cause valve sticking.

Valve Removal

1. With the cylinder head removed, compress the valve springs using a spring compressor (Figure 28).

FIGURE 30
Valve stick.

SHOP TIP

Occasionally, a valve spring retainer and its keepers will not release and allow the spring to be compressed. In such cases, strike around the retainer with a hammer and a short length of pipe. This procedure will help break the varnish buildup so the spring can be compressed.

2. Remove the valve retaining locks (keepers), spring retainers, valve stem seals, and springs. See Figure 29.

3. Before removing valves, remove any burrs from the stem tips and keeper grooves with a mill-smooth file. This will prevent guide damage during valve removal.

4. As each valve is removed from the head, place it in a valve stick (Figure 30). This will ensure that the valves will go back into the same wear-mated guide and seat. If you know that the guides and seats are to be reconditioned, it is not necessary to use a valve stick at this time.

Valve Spring Measurement

The valve springs job is to ensure positive closing of the valves. On vehicles with high mileage, the springs become weak and lose tension. This may result in an engine power loss.

Valve springs can be measured for pressure by using a spring tester (Figure 31). *All springs should read within 90% of tension specifications.* Stand all springs on a flat surface and check free length height. *All spring heights should be within 1/16" of specifications.*

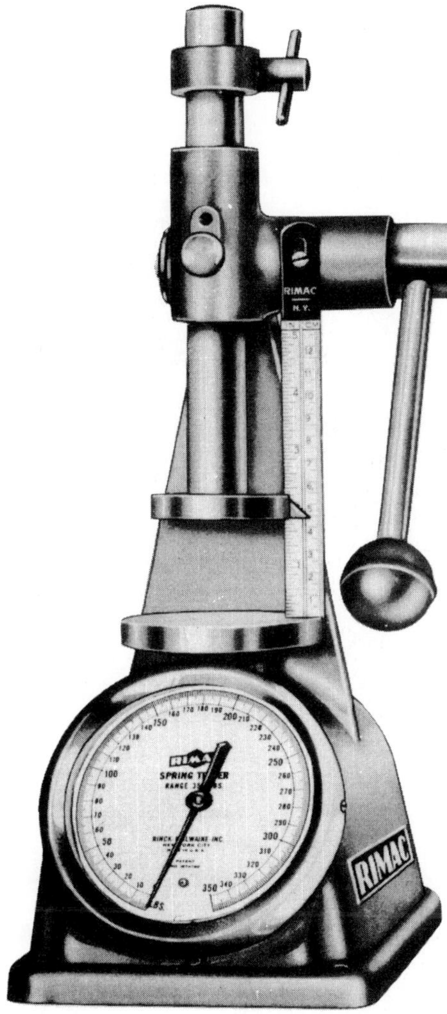

FIGURE 31
Direct-reading valve spring pressure tester. (Courtesy of Rimac)

Check the springs for squareness. Stand each spring up next to a square on a flat surface and turn it slowly (Figure 32). Measure the distance the top coil moves away from the square. The maximum allowable out-of-squareness limit is generally recognized as being 5/64″.

SHOP TIP

A broken valve spring (Figure 33) can be located by using a timing light. Strobe each spring and observe the movement. A broken spring will have an irregular action that is easily seen.

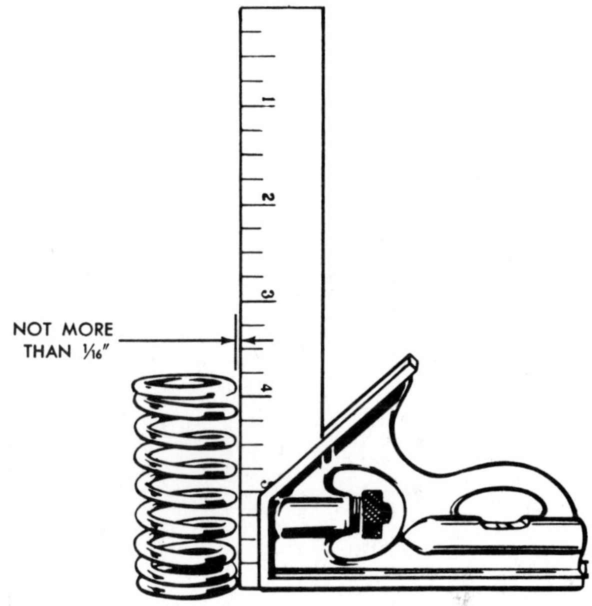

FIGURE 32
Determining whether or not a valve spring is square. (Courtesy of Chrysler Corporation)

Valve Spring Retainers

Retainers should be visually checked for cracks, wear, and scoring inside the hole (Figure 34).

Keepers

Split-type keepers (sometimes called locks or keys) are generally used in the automotive engine. Each keeper inner surface is generally provided with bead locks (Figure 35). The keepers are held in position against the valve stem by upward pressure from the spring retainer (Figure 36).

A visual inspection of the keepers should be made. Check for wear, metal shearing, and nicks. Discard all damaged keepers.

FIGURE 33
This broken valve spring was detected with a timing light.

FIGURE 34
The aprons (sides) on the rocker arm were hitting this spring retainer because the valve stems were cut too short. This can cause the valve keepers to pop out while the engine is running.

FIGURE 35
Valve keepers.

CAUTION: *New keepers must always be installed in pairs. If an unworn keeper is mated with a used keeper, the valve spring retainer may cock and break off the tip of the valve stem (Figure 37).*

Multigrooved Valve Stems

Certain engines use multigrooved exhaust stems in order to help achieve valve rotation (Figure 38). However, this type of valve stem is extremely subject to keeper groove wear.

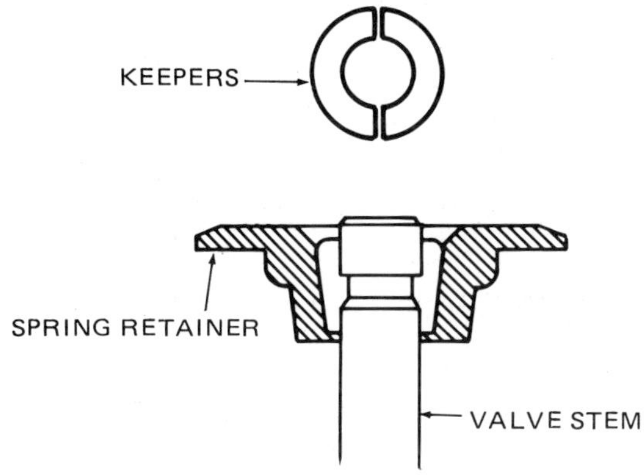

FIGURE 36
Keeper position when installed.

FIGURE 37
This damage was caused by faulty valve keeper seating. After the head has been assembled, tap the valve stem tips with a soft-faced hammer to ensure seating of the keepers.

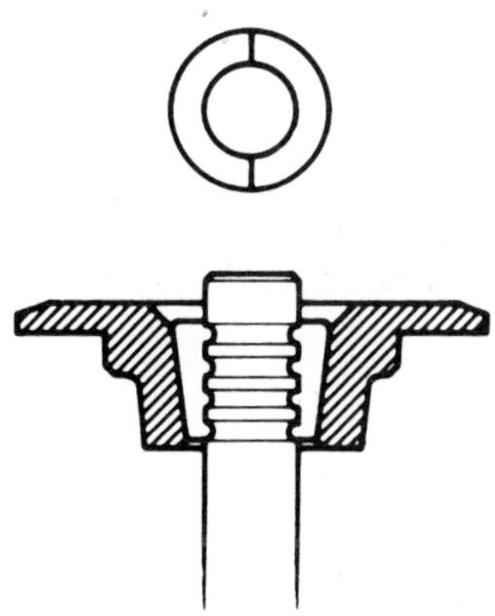

FIGURE 38
Four-bead keepers are used on certain engines to help rotate the valves.

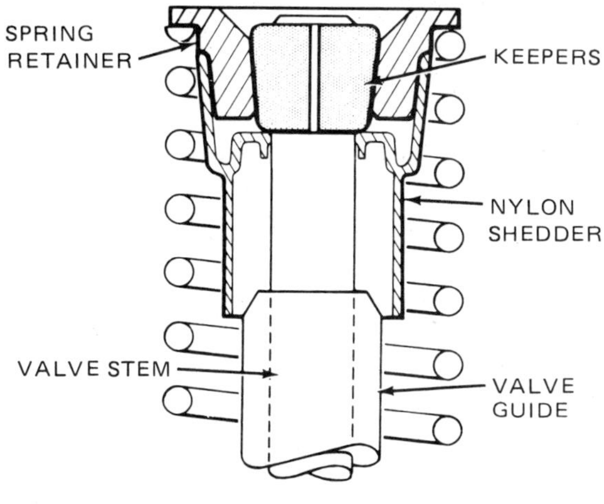

FIGURE 39
Metal shield (shedder) type of valve stem seal. Sometimes, nylon is used instead of metal. (Courtesy of Cadillac Motor Division, General Motors Corporation)

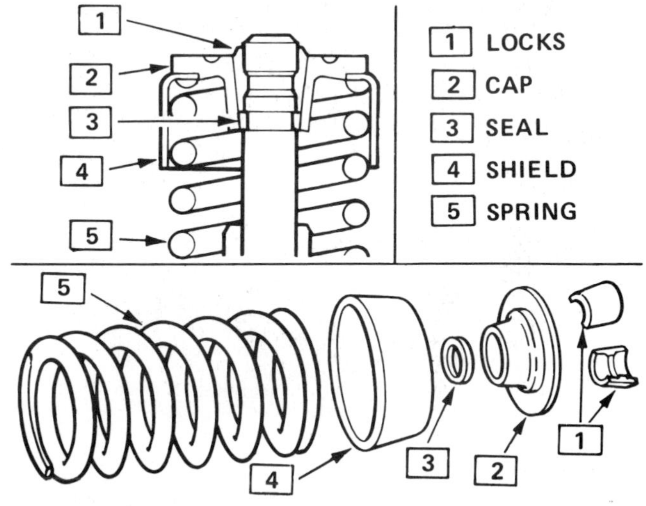

FIGURE 40
Square cut rubber "o" ring type valve stem seal. (Courtesy of Chevrolet Motor Division, General Motors Corporation)

Valve Stem Seals

A running engine generally has a large amount of oil under the valve covers. To prevent this extra oil from getting to the valve stem guide area, where it can be drawn into the engine and burned, seals of various types are used:

1. Rubber "umbrella" on the valve stem. This seal rides up and down with the valve stem and acts as a deflector. Refer back to Figure 29.

2. Metal shield that fits under the spring retainer (Figure 39).

3. Square cut rubber "o" ring seal in the keeper groove (Figure 40).

4. "Positive" insert type seal (Figure 41). This is generally considered the most effective guide seal. It attaches to the valve guide.

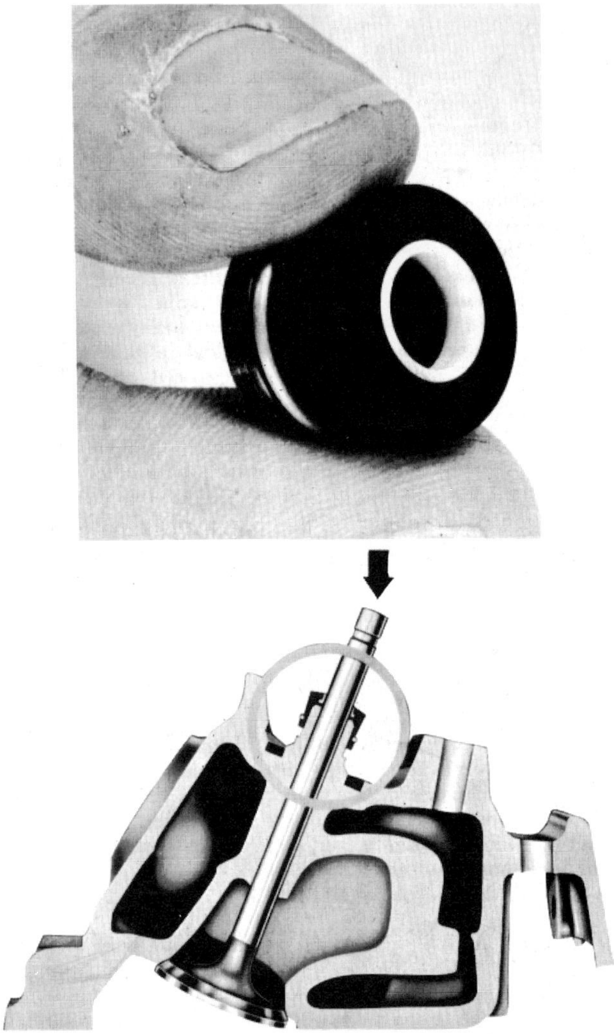

FIGURE 41
The "positive" insert type of valve stem seal is not much larger than a person's thumbnail. (Courtesy of Dana Corporation)

Installing Valve Stem "o" Ring Seals

Often an incorrect procedure is followed when installing valve stem "o" ring seals. *It is a mistake to install the "o" ring on the valve before the spring and the retainer are installed.* When this is done, the "o" ring is pushed out of place when the retainer is moved down to install the keepers. The correct procedure is as follows:

1. Insert the valve into the cylinder head valve guide. Position the valve spring and retainer assembly over the valve stem and compress with a spring compressor.

2. Install the "o" ring seal in the second valve stem groove.

3. Place the keepers in the first groove. Slowly release the spring compressor, keeping the hole in the spring retainer centered on the valve stem as it comes up. Otherwise, the "o" ring may be pinched, causing it to leak oil.

FIGURE 42
When a valve stem seal is used, it can be tested by placing a small suction cup over the retainer. Squeeze the bulb and press the cup against the retainer. When the bulb is released it should stay compressed, indicating an airtight seal.

Testing Valve Stem "o" Ring Seals for Leakage

After "o" rings are installed, test for leakage. This can be done in any one of several ways.

1. Fit a rubber bulb tightly over the spring retainer and squeeze (Figure 42). A bulb that does not hold pressure indicates that air is leaking past the seal. The seal and/or retainer must be replaced.

2. Fill the recess around the hole in the center of the spring retainer with oil. Allow the oil to stand for a moment, and observe for leakage down the valve stem.

Replacing Valve Stem Seals Without Removing the Cylinder Head

Defective valve stem seals, damaged retainers, or broken valve springs may be replaced without removing the cylinder head. The basic procedure is as follows:

1. Install an adapter with an air hose in the spark plug hole (Figure 43).

2. Position the piston at TDC on compression.

3. Turn the air supply on.

4. Compress the valve spring using one of the many tools that are available, and remove the keepers. See Figure 44.

> **SHOP TIP**
> Instead of using air to hold the valves, you can use a long piece of rope stuffed into the cylinder through the spark plug hole. Then gently bring the piston up against the rope.

FIGURE 43
Valve air holder set. Thread the proper size adapter into the spark plug opening and attach an air hose to hold valves in place. (Courtesy of Lisle Corporation)

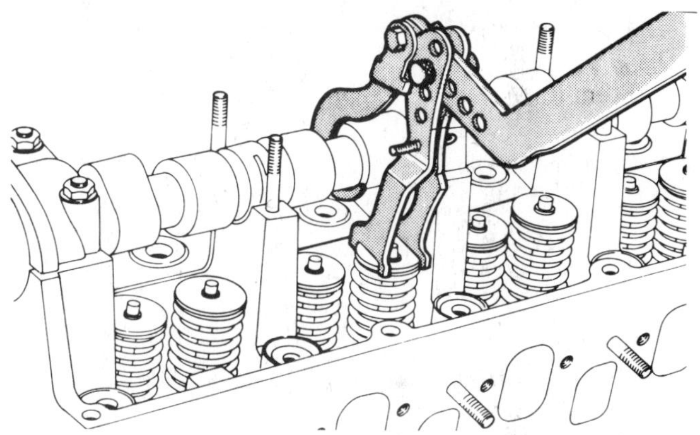

FIGURE 44
A valve spring compressor being used on an OHC engine. (Courtesy of Chrysler Corporation)

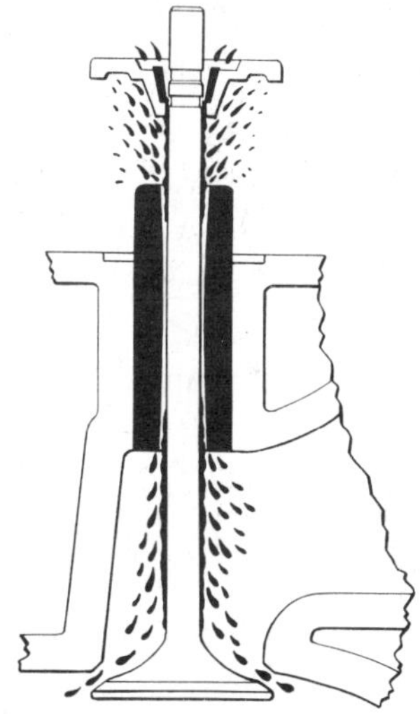

FIGURE 45
Worn valve guides cause excessive oil consumption. (Courtesy of Dana Corporation)

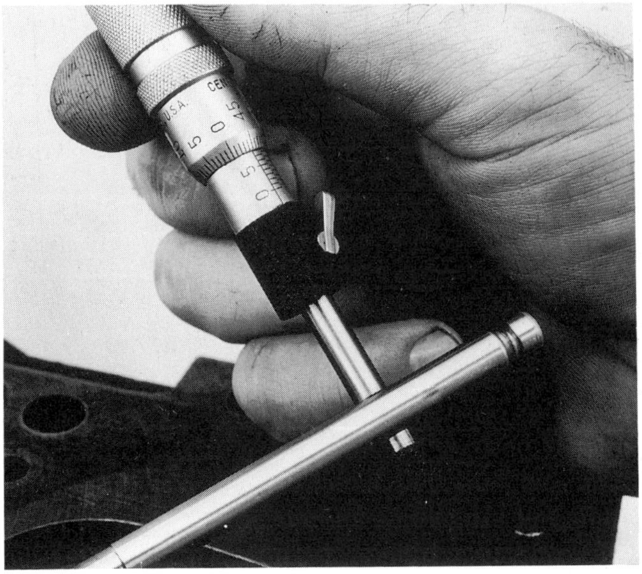

FIGURE 46
Miking a valve stem for wear.

FIGURE 47
Measuring valve guide wear.

VALVE GUIDE AND VALVE STEM MEASUREMENT

Valve stems and guides must be checked for wear. Too much stem-to-guide clearance will allow air and oil to be drawn into the combustion chamber. This can cause rough idle, intermittent high idle, increased oil consumption, foul spark plugs, carbon buildup and incorrect fuel mixture, and restricted engine breathing.

Valve guides generally show bell mouth wear at the extreme top and bottom (Figure 45). The direction of greatest wear is in line with rocker arm movement. Valve stem wear is greatest at the extreme top and bottom of the guide rub area (Figure 46).

On a computer-controlled engine, worn exhaust guides can cause a low oxygen sensor voltage signal. This happens because of the inducement of extra air into the exhaust stream.

There are several accepted methods for measuring valve guide wear.

1. A split-ball gauge is used to size the guide. Remove the gauge and transfer the size to an outside micrometer (Figure 47).

2. Use a valve guide dial bore gauge (Figure 48).

3. Insert a new valve into the guide. Position a dial indicator plunger against the valve stem in the direction of rocker arm thrust (Figure 49). Rock the valve back and forth against the indicator plunger.

SHOP TIP

Egg-shaped wear is often found on exhaust guides of cylinders next to a sticking EGR valve. When such wear is found, the corresponding valve spring(s) should be inspected. Hot exhaust gas can travel up the guide, overheat the spring, and cause tension loss.

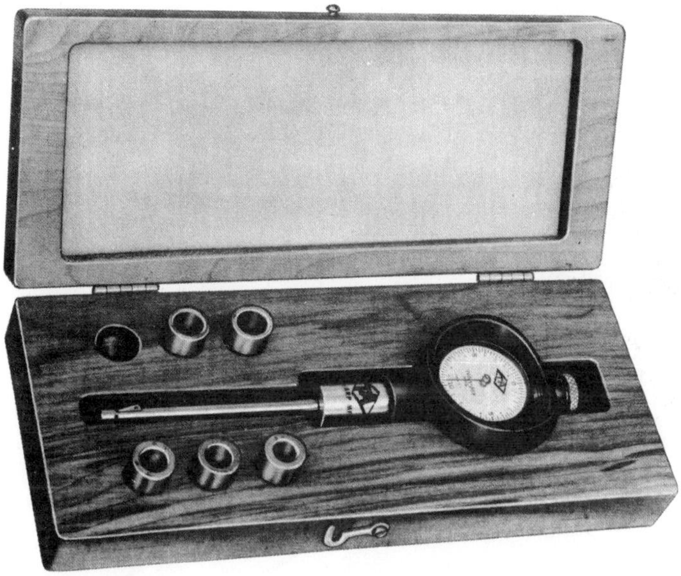

FIGURE 48
Valve guide dial bore gauge set. Included are five bushings for accurately setting the gauge. (Courtesy of Tobin-Arp Manufacturing Company)

FIGURE 49
Checking valve guide clearance with a dial indicator. This dial indicator is incorrectly positioned to measure guide wear in the thrust direction. Reposition it 90 degrees.

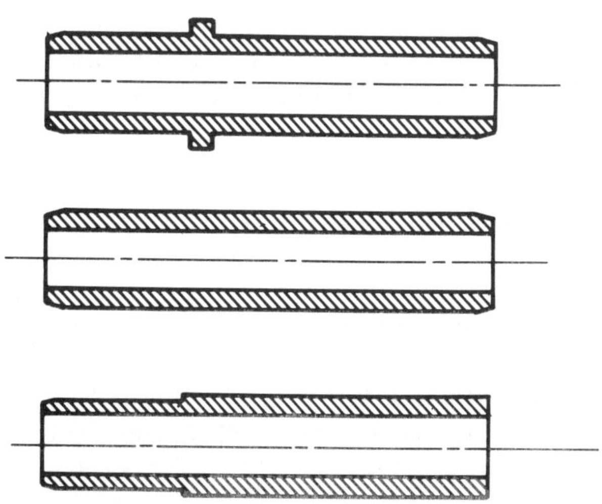

FIGURE 50
Various types of replaceable valve guides.

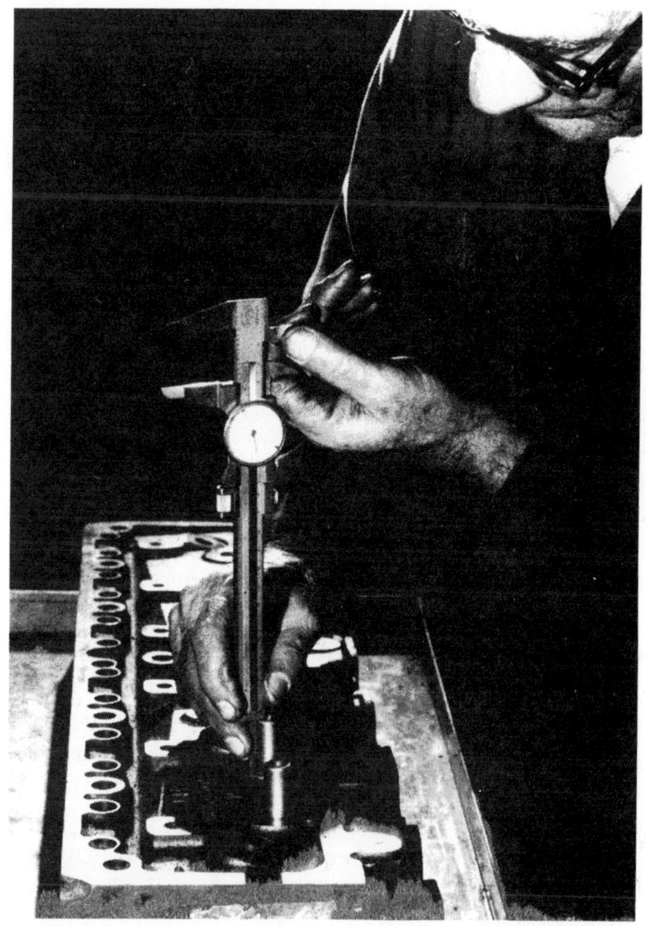

FIGURE 51
Measuring valve guide installed height with dial calipers.

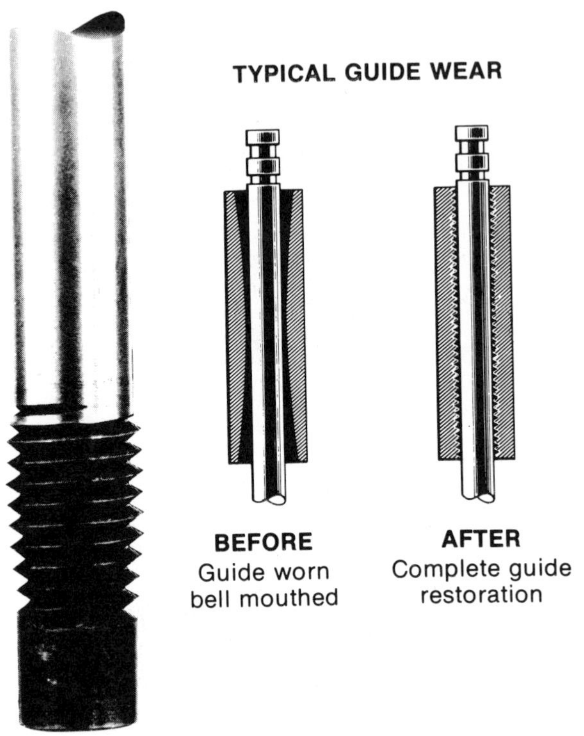

FIGURE 52
Worn guides can often be reconditioned by knurling. (Courtesy of Winona Van Norman Machine Company)

FIGURE 53
A very popular valve guide liner. (Courtesy of K-Line Industries, Inc.)

Valve Guide Replacement

A number of engines are equipped with replaceable guides (Figure 50). Worn guides can be pressed or driven out (by hand or with an air hammer), and new ones installed.

Measure the installed height of the guide before removing it (Figure 51). Drive the new guide into the head, keeping the original protrusion. If the guide is not positioned in far enough, the valve spring retainer can bottom out on top of the guide when the valve opens. If the guide is positioned in too far, accelerated valve stem wear will result and air flow in the port will be restricted.

When new replaceable guides are installed, distortion will occur in the bore inside diameter. Always make a pass with the correct size reamer even though the valve may appear to fit the guide.

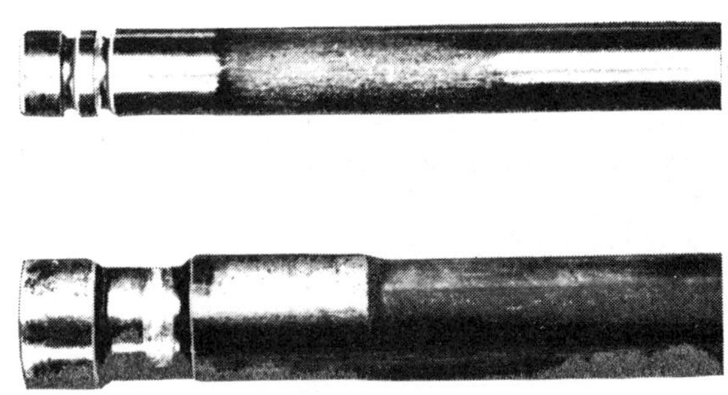

FIGURE 54
Scored and worn valve stems.

Valve Guide Reconditioning

Knurling is a guide machining operation used to recover clearances in worn integral guides (Figure 52). Many shops recommend core drilling worn integral guides and installing thin-wall inserts (Figure 53).

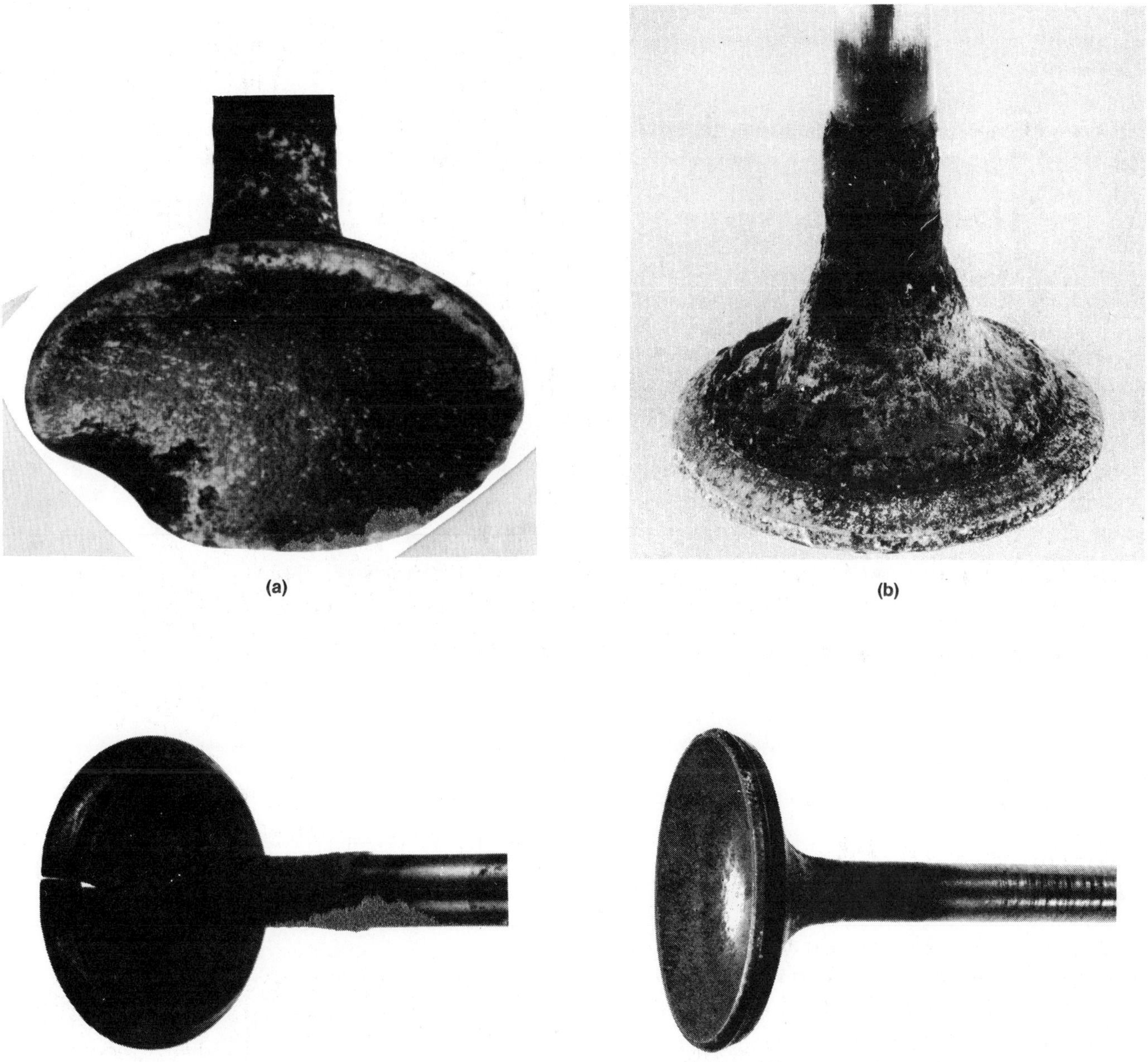

(a) (b) (c) (d)

FIGURE 55

(a) A cooling system deposit restriction caused this valve failure. (b) A badly pitted exhaust valve with heavy face deposits. (c) Hoop stress is a very common cause of valve failure. The condition starts because the center of a valve head cools more slowly than the outer edge. As the outer edge cools, it would normally contract. However, the center of the valve, still being hot, is expanded. Something has to give, and a crack in the outer edge of the valve face is the result. (d) Sometimes, excessive heat does not burn a valve. Instead, the valve head becomes cupped. Cupping makes the valve longer; taking up valve clearance and possibly holding the valve open.

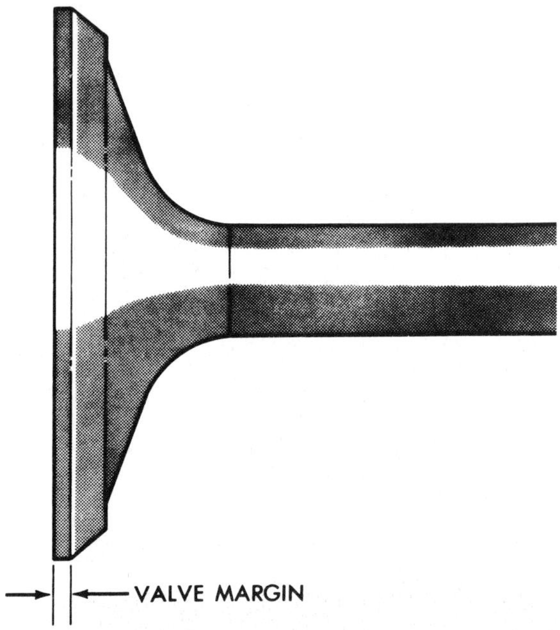

FIGURE 56
Valve margin.

FIGURE 57
Truing a valve stem end. (Courtesy of Sioux Tools Inc.)

VALVE INSPECTION

Inspect the valve stems for wear, nicks, score marks, corrosion, tip wear, and keeper groove wear (Figure 54). Maximum allowable valve stem wear is .001″. Never reuse valves that are badly worn, cupped, nicked, or bent (Figure 55[a], [b], [c], and [d]).

A valve must have margin (Figure 56). Margin helps the valve withstand pressure and control heat. As a general rule, do not grind the valve face if the margin will be reduced below .045″. The valve must be replaced.

VALVE RESURFACING (GRINDING)

Most kinds of valve grinding equipment operate in a very similar manner. The following instructions may be considered typical.

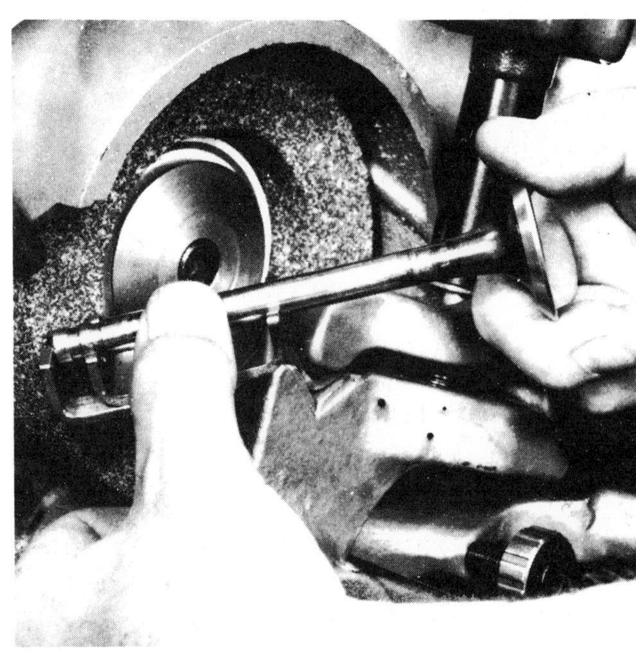

FIGURE 58
Chamfering a valve stem end. (Courtesy of Sioux Tools Inc.)

1. Dress the wheels. Have coolant flowing freely over the wheel to keep the diamond cool and wash away wheel grit.

2. True the valve stem ends (Figure 57).

3. Camfer the valve stem ends. The chamfer should be 1/32″ wide. Figure 58 shows renewing the chamfer using a chamfering vee.

4. Lock the chuck head at the exact angle you wish to refinish the valve.

5. Chuck the valve. Insert the valve so it is gripped just above the worn part of the stem.

6. Resurface the valves (Figure 59). When grinding, adjust the coolant to flow on the valve face. Keep the valve face moving back and forth on the stone during grinding. *Do not allow* the valve face to pass across the stone edge.

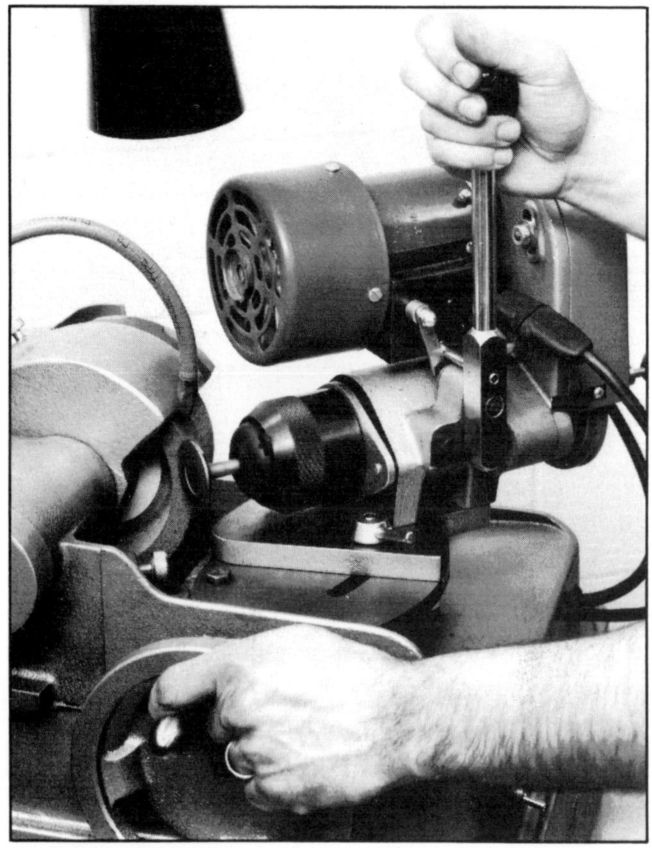

FIGURE 59
Grinding a valve face. (Courtesy of Tobin-Arp Manufacturing Company)

Valve Stem Height

Installed valve stem height is a critical dimension on OHV engines with nonadjustable rocker arms. On these engines, correcting the stem height is a must after a valve job. If it is not done, the lifters may bottom out and hold the valves open.

When correcting valve stem height, it is recommended that no more than .030″ be ground from the end of the valve stem because the heat-treating hardness may be removed.

Valve Stem Height Specifications

Specifications for this dimension are not generally available from the manufacturer because this is a machining measurement calculated during production (Chart 1).

Several different types of gauges are available for checking valve stem height (Figure 60[a] and [b]).

Interference Angles

When doing a valve job, the valve face is sometimes ground 1/2 degree to 1 degree less than the seat angle (Figure 61). Advocates of this practice claim that an interference angle helps compensate for heat expansion and provides a better seal during the engine break-in period.

VALVE SEAT INSPECTION

After a cylinder head is disassembled, always check insert-type seats for looseness. Hit the insert lightly with a brass hammer and feel for movement with your fingers.

On aluminum heads, visually inspect the combustion chamber for erosion. If erosion is excessive, or has exposed the side of any valve seat, the cylinder head must be repaired or replaced (Figure 62).

SEAT RESURFACING

Valve seats are resurfaced with either grinding stones or carbide cutters.

NOTE: Always check and/or recondition the guides before resurfacing the valve seats.

The actual steps and recommendations for grinding valve seats are as follows:

1. Select the proper size and grit stone. When grinding a seat, the proper size stone is approximately 1/8″ larger than the valve head diameter.

2. Set the dressing tool head at the desired angle and dress the face of the seat grinding stone (Figure 63).

3. Select the largest size grinding pilot that will anchor snugly on the entire length of the valve guide. Insert the pilot into the guide and lock it with a twisting motion (Figure 64).

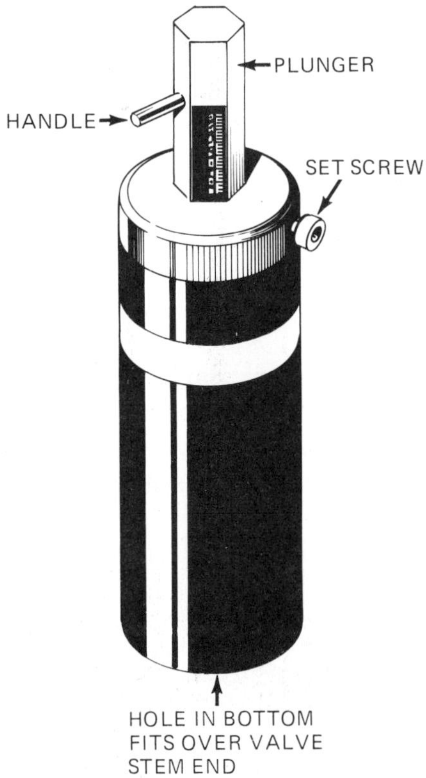

(a)

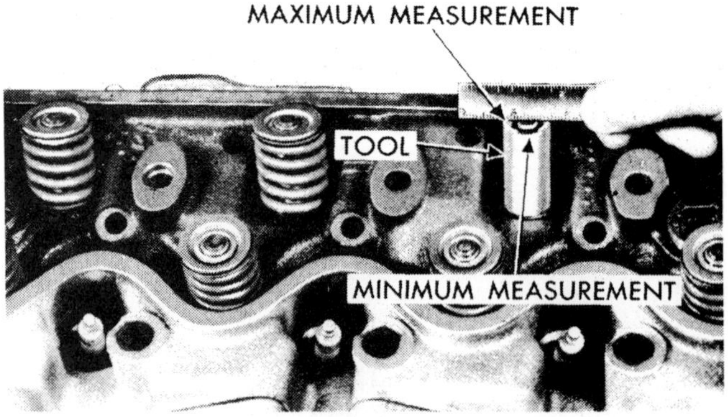

(b)

FIGURE 60
(a) Valve stem height gauge. This tool provides a fast, easy method for measuring valve stem height before and after servicing. (b) Installed valve stem height being checked with a special tool. (Courtesy of Chrysler Corporation)

CHART 1 VALVE STEM HEIGHT SPECIFICATION CHART

These dimensions have been established from numerous measurements made in the field and should be considered approximate. These may be used when manufacturer's specifications are not available. Measure stem heights from the valve spring seat to the top of the valve stem.

Engine	Displacement	Stem Height
AMC	192, 232, 237, 258	1.970″
	304	1.985″
	360, 390	1.965″
BUICK	231	1.955″
	350	1.950″
	430	2.115″
CADILLAC	368, 425, 472	2.060″
FORD	302	2.145″
	351C, 401	2.080″
	351W	2.200″
	351M	2.100″ intake 2.050″ exhaust
	360, 390	2.125″
	460	2.160″ intake 2.200″ exhaust
MOPAR	318, 360	1.920″
	383, 400, 440	2.130″
	413	2.080″ intake 2.125″ exhaust
	2.2, 2.5L	1.969″
OLDS	260, 455	2.050″
PONTIAC	250	1.975″
	350, 400	1.755″
	421	1.780″
VEGA	140	2.150″ intake 2.120″ exhaust

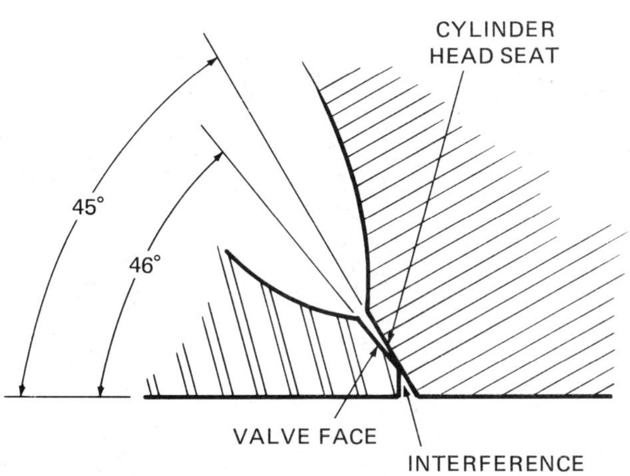

FIGURE 61
Interference angle between valve and seat.

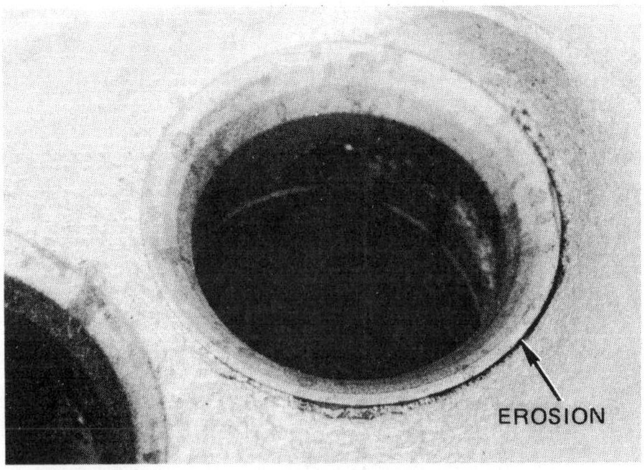

FIGURE 62
Unacceptable combustion chamber erosion.

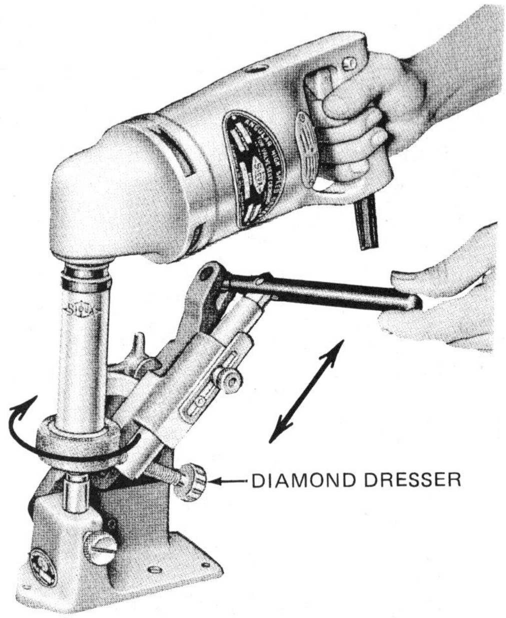

FIGURE 63
Dressing a valve seat grinding stone. Move the diamond steadily across the wheel, taking very light cuts. (Courtesy of Sioux Tools Inc.)

FIGURE 64
A self-centering expandable valve seat pilot. (Courtesy of Black and Decker Manufacturing Company)

FIGURE 65
A "bounce" (lifting) spring used under the valve grinding stone gives the operator better control of the stone against the seat. (Courtesy of Sioux Tools Inc.)

4. Place the stone and its holder on the pilot. Use a "bounce" spring under the grinding stone (Figure 65). This helps reduce stone breakage and provides the operator with better control of the stone against the seat.

5. Insert the driver motor into the stone holder (Figure 66). Grind with the dressed stone until the seat is just free of blemishes.

6. After the seat has been satisfactorily ground, determine its width and valve face contact location (Figure 67). A good seat will show a 360-degree contact pattern on the valve face, and will have a minimum of 1/32″ overhang. As a general rule, intake valves should have a contact width pattern of 1/16″ (.063″), and non-rotator exhaust valves, 3/32″ (.094″). Exhaust valves that use rotators should have the seat contact as wide as possible, while maintaining 1/32″ overhang at the top and bottom.

If a valve face contact pattern is too wide, correction angles must be ground (Figure 68). For instance, if the pattern is too wide with incorrect overhang, the seat must be "topped." To do this, use a 30-degree angle stone or cutter on 45-degree seats, and use a 15-degree angle stone or cutter on 30-degree seats.

If the valve face contact pattern is too wide and the overhang at the top is good, the seat must be narrowed by "throating." This is done by using a 60-degree angle stone or cutter.

7. Check valve seat concentricity (runout) with a dial indicator runout gauge (Figure 69).

COMMENT: *Some shops prefer cutting valve seats rather than using the grinding method (Figure 70[a], [b], and [c]).*

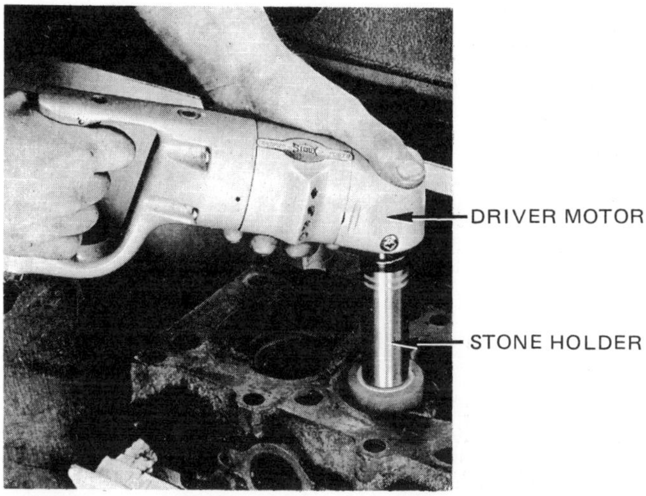

FIGURE 66
When grinding a seat, do not use heavy pressure. Use your lower hand to partially support the weight of the driver motor against the stone holder. (Courtesy of Sioux Tools Inc.)

FIGURE 67
Prussian blue paste can be used to determine valve face-to-seat contact location.

VALVE SEAT REPLACEMENT

When valve seats crack, burn, or become recessed, they must be replaced. There are two types of seats.

1. **Integral.** This seat is ground directly on the head casting material.

2. **Insert.** This seat is a metal ring that is press-fitted into a counterbore in the head.

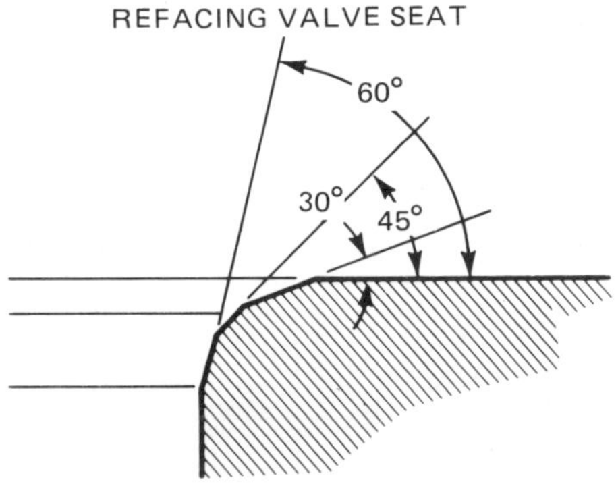

FIGURE 68
Valve seat correction angles.

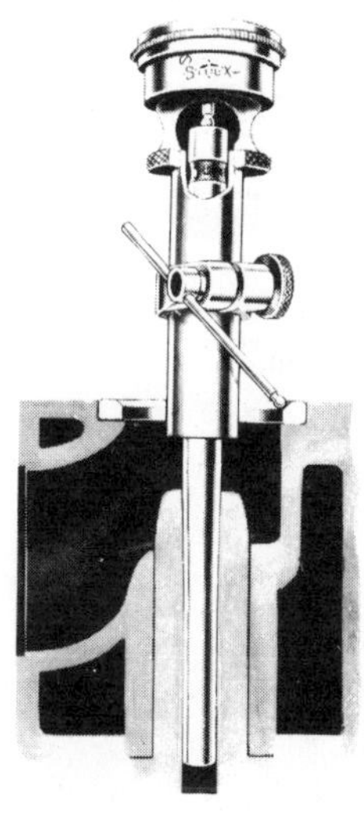

FIGURE 69
Dial indicator for checking valve seat runout. (Courtesy of Sioux Tools Inc.)

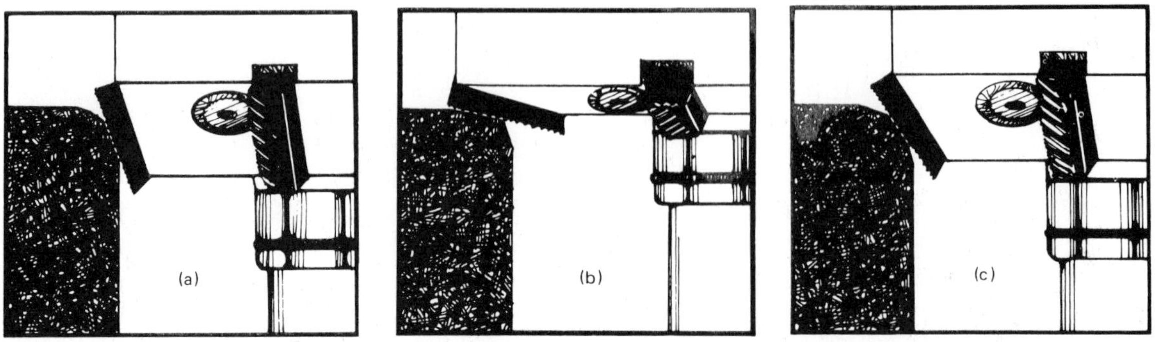

FIGURE 70
(a) Cutting the area below the valve seat. This is known as "throating." (Courtesy of Neway Manufacturing, Inc.) (b) Cutting the area above the valve seat. This is known as "topping." (Courtesy of Neway Manufacturing, Inc.) (c) Cutting the valve seat. (Courtesy of Neway Manufacturing, Inc.)

At times it becomes necessary to repair an integral seat. The repair is made by machining a counterbore into the old seat and installing a seat insert. The new insert is installed with a negative fit (generally about .003″). A seat and guide machine or a portable seat cutter is used (Figure 71).

FIGURE 71
Machining the valve port for a replacement seat ring insert. (Courtesy of Tobin-Arp Manufacturing Company)

CHECKING INSTALLED SPRING HEIGHT (OR ASSEMBLED HEIGHT)

The loss of metal from the valve face, the valve seat, or the valve spring seat has the effect of lengthening the valve stem (Figure 72). This will reduce valve spring tension because installed spring height is increased. Valve spring shims inserted under the valve spring can correct for this loss of tension (Figure 73). Correction shims are generally available in .015″, .030″, and .060″ thicknesses.

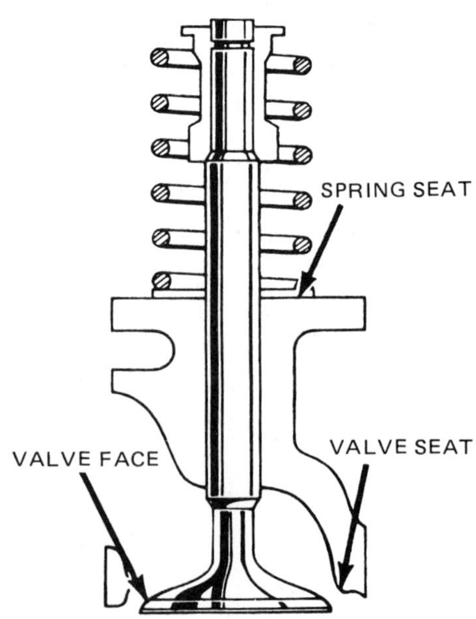

FIGURE 72
Loss of metal at any of these points will affect installed spring height.

FIGURE 73
Adding a spacer (shim) to correct valve spring assembled height. (Courtesy of Ford Parts and Service Division)

FIGURE 74
Measure the valve spring installed height from the machined surface of the head to the underside of the retainer. (Courtesy of Ford Parts and Service Division)

Proper installed valve spring height is usually determined by using one of these methods:

1. Using dividers, measure the height of the installed valve spring from the cylinder head spring seat surface to the underside of the valve spring retainer (Figure 74). Check the span of the dividers against a scale.

2. Using a caliper (Figure 75).

3. Position the valve in the cylinder head with the spring retainer and keepers installed. Holding the valve against the seat, measure the distance between the spring seat and the underside of the retainer with a telescoping gauge (Figure 76).

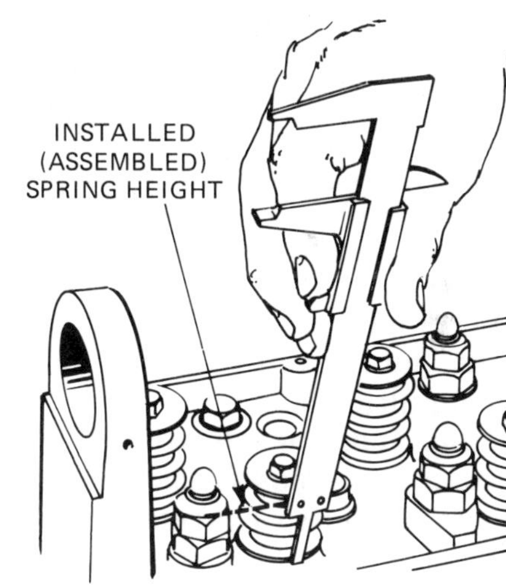

FIGURE 75
Valve spring installed height can be measured with a caliper.

Tips for Assembling Cylinder Heads

1. When installing variable pitch valve springs, the end with the close coils must be placed against the cylinder head (Figure 77).

2. After the head has been assembled, tap the valve stem tips with a soft-faced hammer to make sure that the valve keepers are seated (Figure 78).

3. Check each installed valve spring for looseness. The spring should not turn when attempting to rotate it with finger grip pressure.

1. Use new head bolts and washers (if so equipped).

2. Apply a light film of engine oil to the threads of the new head bolts.

3. Using an accurate torque wrench, tighten the head bolts in proper sequence to 44 ft.-lb. This will precompress the head gasket.

4. Back off two turns and retorque in proper sequence to 44 ft.-lb.

5. Turn each head bolt 90 degrees in proper sequence.

6. Turn each head bolt another 90 degrees in proper sequence.

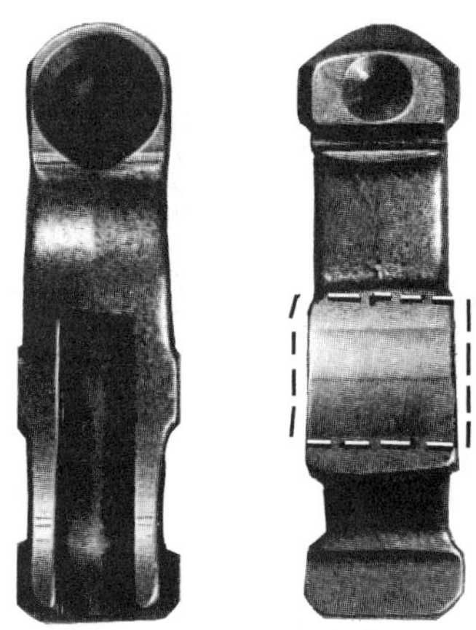

FIGURE 84
Worn OHC rocker arm. (Courtesy of Sealed Power Corporation)

ROCKER ARMS

The rocker arm-to-valve stem contact surface should be inspected for grooves, pits, scoring, and wear (Figure 82). In many cases, the contact surface can be refaced on a valve grinding machine (Figure 83).

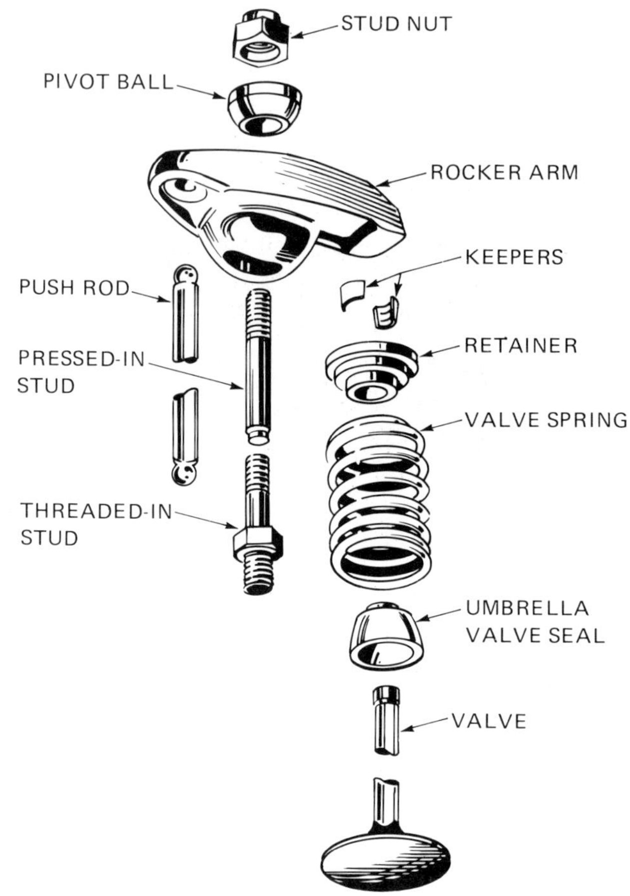

FIGURE 85
Rocker arm assembly as used on a number of different Ford engines. Locate the pivot ball in the picture.

Rocker arms for OHC engines are subjected to heavy operating loads. Inspect closely for scoring or wear on the rubbing surfaces (Figure 84).

CAUTION: OHC rocker arms will wear-mate to the camshaft lobes. If the rocker arms are being reused, it is very important to return them to original position. Failure to do so may cause lobe damage soon after installation.

Pivot Ball Inspection

Closely examine rocker arm pivot balls (Figure 85). If evidence of excessive wear or galling is found, replacement is necessary.

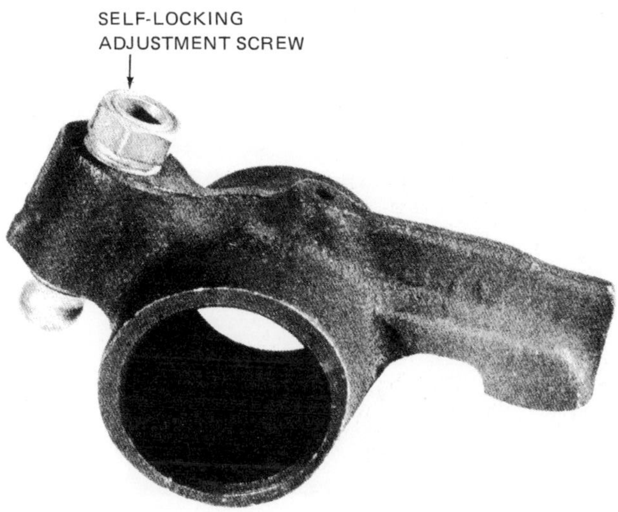

FIGURE 86
Adjustable cast iron rocker arm.

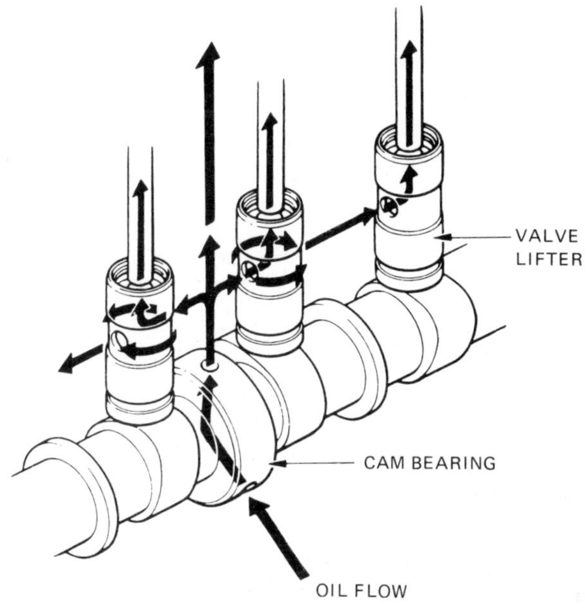

FIGURE 87
Oil supplied through hollow push rods.

Rocker Arm Adjusting Screw Inspection

If the rocker arm adjusting screw (Figure 86) breakaway torque is less than specifications, install a new screw and rocker arm.

Rocker Arm Lubrication

Oil flow to the rocker arm assembly usually occurs by one of the following means:

1. Through hollow push rods (Figure 87).
2. Through an oil hole in the rocker arm studs.
3. Through a drilled hole in the cylinder head (Figure 88).

NOTE: *When installing this type of rocker shaft assembly, be sure the rocker stand with the oil passage is positioned over this hole in the cylinder head.*

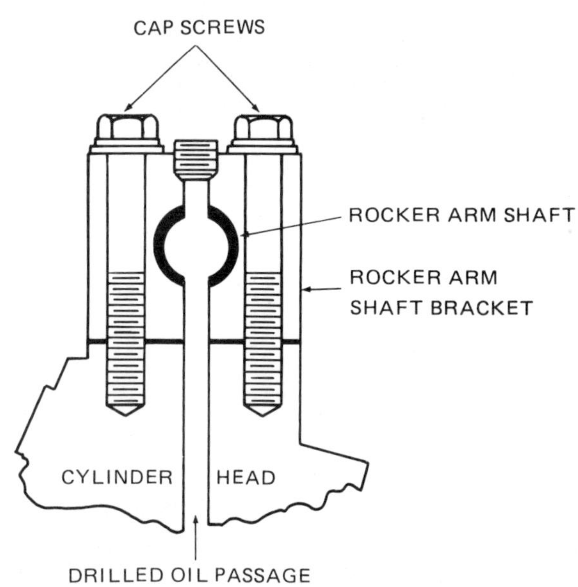

FIGURE 88
Oil supplied through a drilled hole in the cylinder head.

4. Through a hollow or undercut rocker stand screw that threads into an oil passage in the head.

5. Through an external oil supply tube (Figure 89).

NOTE: One side of a rocker arm shaft is usually drilled with lubrication holes (Figure 90). To provide proper lubrication, these holes must face the proper direction (generally down and towards the valves) when the rocker arm shaft is installed.

FIGURE 89
Oil supplied through an external line. (Courtesy of Dana Corporation)

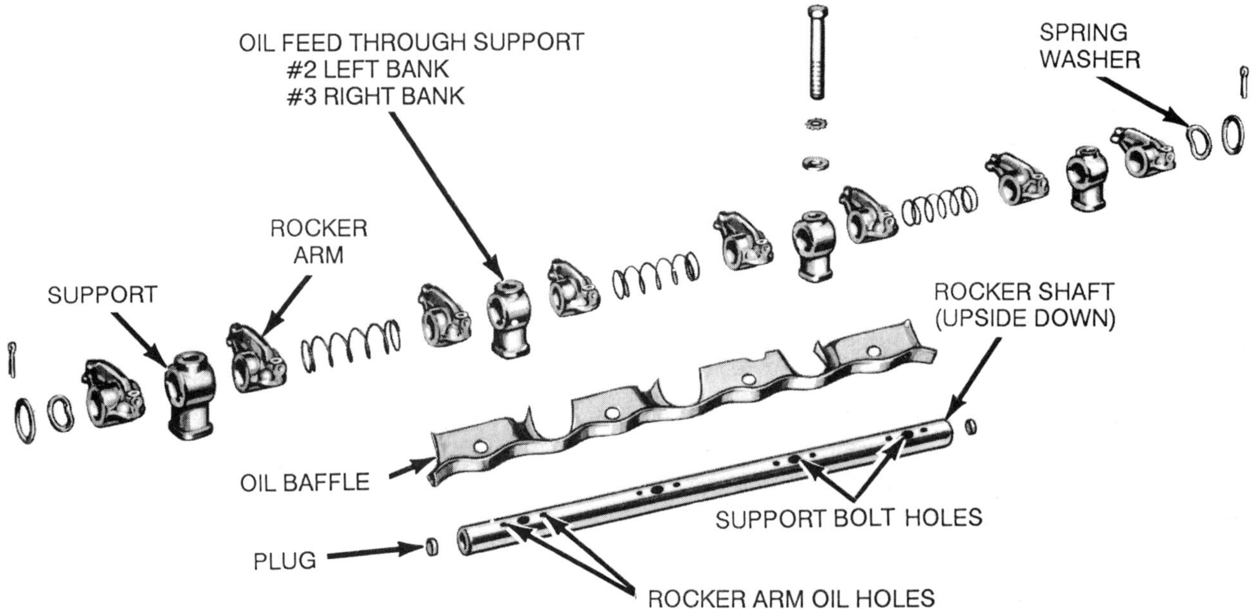

FIGURE 90
The rocker arm assembly used on Ford 330-, 352-, 361-, 390-, and 391-CID engines. Shaft installation is critical for proper lubrication of the assembly. (Courtesy of Ford Parts and Service Division)

FIGURE 91
Broken canoe-type rocker arm.

Rocker Arm Breakage

Broken rocker arms generally result from one of the following causes:

1. Excessive valve spring tension. If spring tension exceeds the design limit of the arm, cracking and breakage will occur (Figure 91).

2. Improper rocker shaft removal procedure. When removing the rocker shaft support bolts, loosen in sequence in two or three steps to prevent binding and rocker arm and/or shaft breakage.

3. Improper valve clearance. With excessive valve clearance, the impact force between the valve tip and the rocker arm will produce greater than normal wear, which can eventually lead to breakage.

4. Poor lubrication. As a rocker arm and shaft wear, an oil hole ridge starts to form, which may shut off oil flow. If so, rocker arm seizure and breakage will follow.

A shaft-mounted rocker arm will generally show the greatest wear on the bottom side of the pivot hole. The rocker shaft will also show the greatest wear on the bottom side surface (Figure 92).

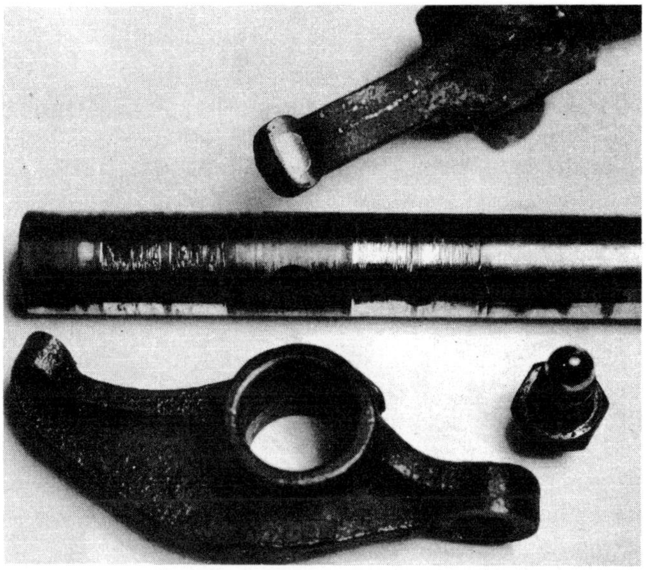

FIGURE 92
Worn rocker arm shaft.

PUSH ROD INSPECTION

Visually examine the push rod ends for ridges, chips, and "onioning" (Figure 93). Do not forget to inspect the push rod socket in the rocker arm for wear.

Roll the push rods on a flat surface and check for bend. Maximum allowable deflection is .010″. Replace any push rods that exceed this specification.

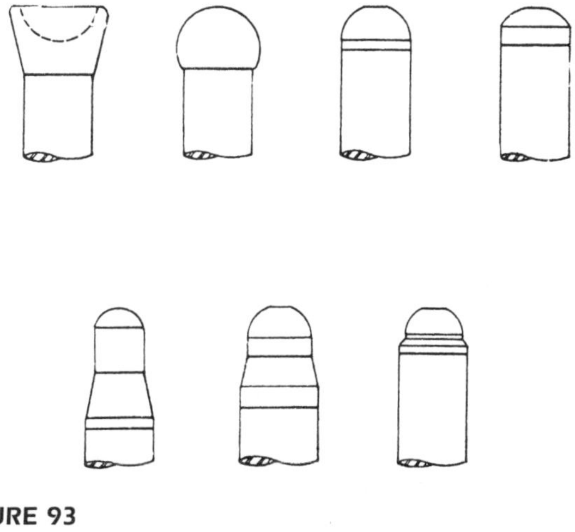

FIGURE 93
Typical push rod end shapes. Always inspect closely for wear.

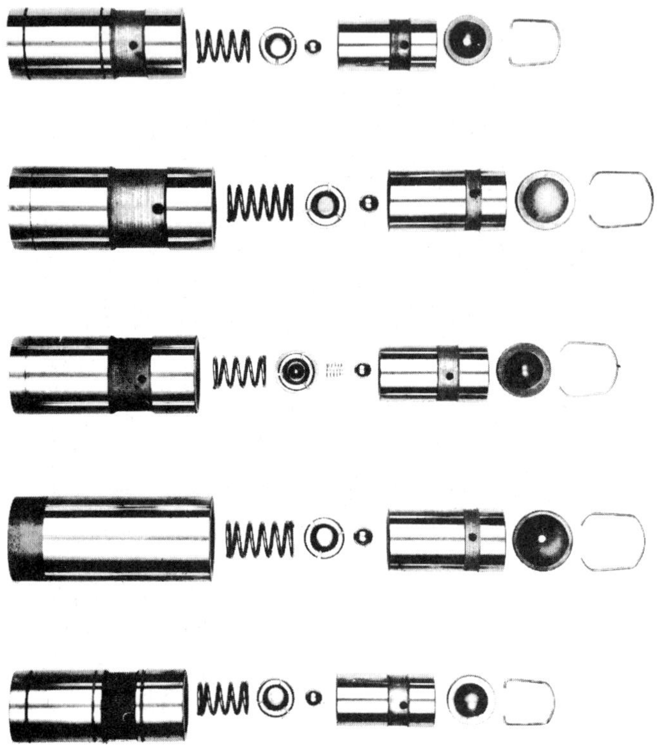

FIGURE 94
Construction comparison of various hydraulic valve lifters. (Courtesy of Dana Corporation)

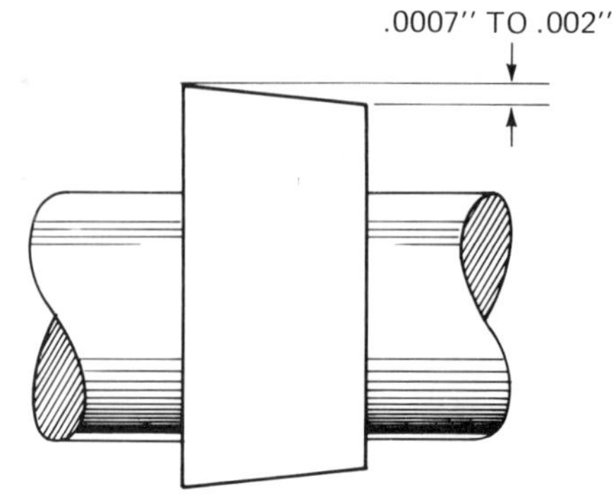

FIGURE 95
Camshaft lobes are generally ground to a slight taper.

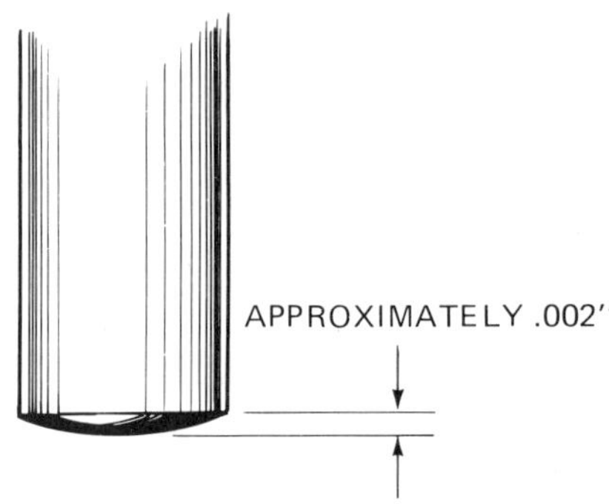

FIGURE 96
Lifter bottoms are radius ground.

VALVE LIFTERS

Lifters are either solid (mechanical) or hydraulic. A solid lifter transfers motion from the cam as a rigid mechanical piece. A hydraulic lifter functions like a solid lifter when a valve is opened. However, a hydraulic lifter is engineered to keep zero lash clearance automatically in the valve train at all times. Figure 94 shows hydraulic valve lifter construction.

CAUTION: Tightening rocker arm assemblies down too fast (especially when using an impact wrench) on engines equipped with hydraulic lifters does not give time for the lifters on the open valves to leak down. Valves may then contact pistons with sufficient force to bend the valves or push rods.

Hydraulic Valve Lifter "Pump-Up"

At high speed, valve train inertia may open the valves farther than they are supposed to open, resulting in additional valve train clearance. A hydraulic lifter "senses" this clearance, and the plunger may extend far enough to prevent the valve from closing. This condition can produce rough idle and valve-to-piston interference.

Hydraulic valve lifter "pump-up" can be caused by sticky valves, weak valve springs, or by excessive revving of the engine (especially during no-load operation).

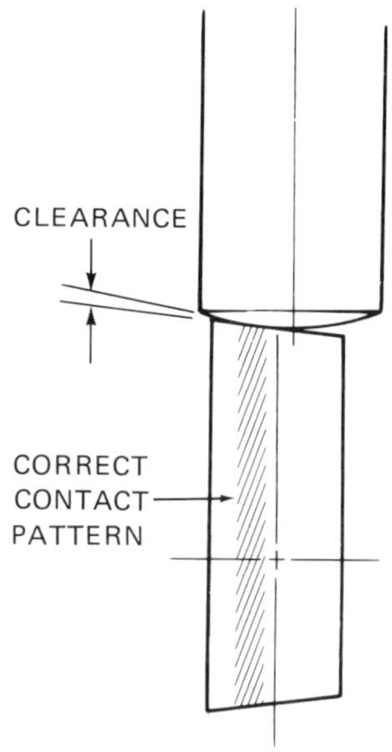

FIGURE 97
Correct mating path between a lifter and camshaft lobe.

FIGURE 98
Chipped and galled camshaft lobe.

Hydraulic Valve Lifter Leakdown

Leakdown is the leakage of oil from the lifter lower chamber through the clearance between the body and plunger during valve operation. Hydraulic valve lifters that leak down too slowly can affect performance and power output since the valves will fail to open and close properly. Lifters that leak down too quickly usually cause excessive noise.

> **SHOP TIP**
> An inoperative lifter can sometimes be found by stopping the engine and pushing down on each of the rocker arms (on the lifter side). A rocker arm that feels spongy usually indicates that the lifter is bad or not receiving oil from the engine.

Camshaft Lobe/Lifter Face Relationship

Most camshaft lobes are slightly tapered, though they appear to be flat (Figure 95). Lifter faces are ground spherically with a .002″ crown in the center (Figure 96). When the camshaft is in place and the lifters are installed, the lifters are offset slightly from the cam lobes. It is this offset, combined with the tapered lobes and the spherical lifter faces, that produces lifter rotation and prevents camshaft lobe edge loading.

Lifter/Camshaft Wear

The correct mating path between a valve lifter and its camshaft lobe should show a defined center track (Figure 97). The majority of lifter and camshaft wear difficulties occur when this mating track takes place on the edge.

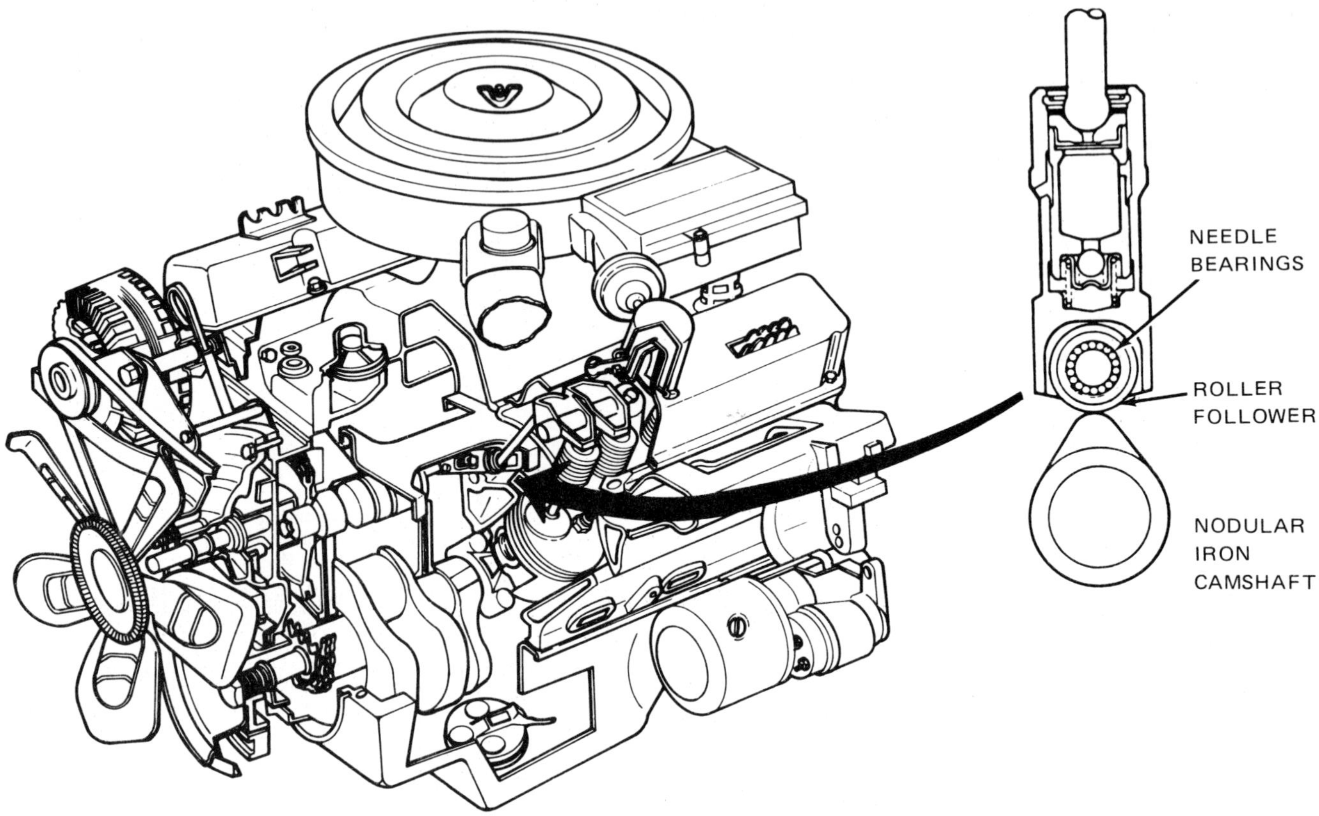

FIGURE 99
This engine is equipped with roller lifters. (Courtesy of Chrysler Corporation)

A camshaft cannot withstand the load it receives if the force is concentrated on the edges of the lobes. The lobes will first chip and gall, and soon will be ruined (Figure 98). The most frequent cause of edge contact on cam lobes is installing worn lifters with a new camshaft. Even if the lifters look good, the spherical shape may be worn flat. If so, the lifter faces will contact the edges of the cam lobes. Installing new lifters with a worn camshaft will also result in improper contact. *Always install new lifters and a new camshaft together.*

Roller Lifters

A number of car makers are using roller lifters as original equipment to reduce friction in the valve train, and thus improve fuel economy and performance (Figure 99).

Areas of a roller lifter that require careful inspection upon valve train disassembly are:

1. Roller axle and needles.

2. Surface of the roller.

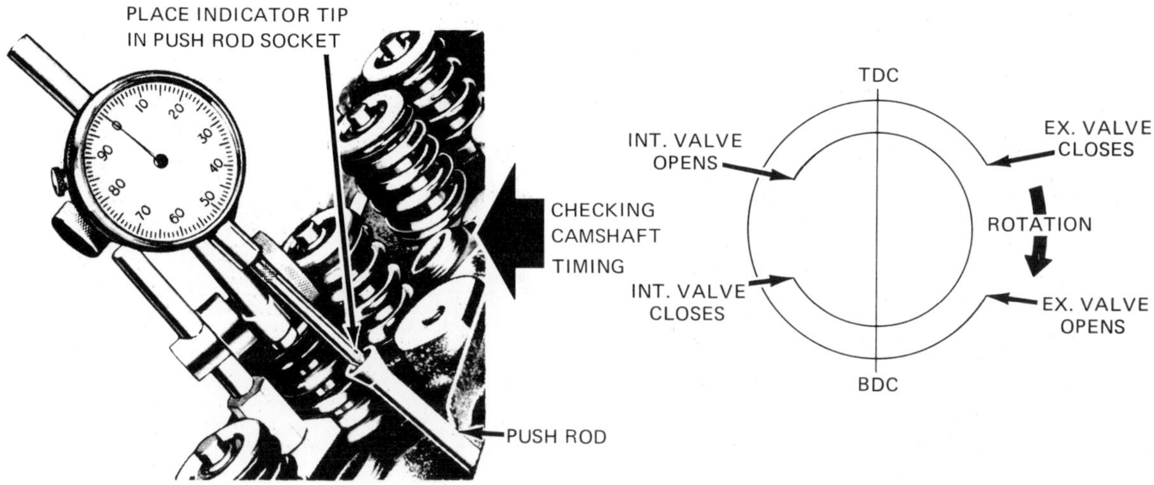

FIGURE 100
A dial indicator setup as shown here and a degree wheel attached to the vibration damper can be used to measure camshaft timing. A typical valve timing diagram is also pictured.

CAMSHAFT TIMING MEASUREMENT

In order to measure in-vehicle cam timing properly, the following items are needed:

1. Dial indicator with fixture for mounting it above a push rod (Figure 100).

2. 360-degree disc (degree wheel) or adhesive tape strip calibrated in one-degree increments.

The basic procedure for measuring camshaft timing is as follows:

1. Fasten the degree wheel or tape to the vibration damper, matching the zero mark on the damper to the zero mark on the degree wheel or tape.

2. Install the dial indicator with the tip resting on or in the end of the push rod of the cam lobe to be checked.

3. Rotate the crankshaft in the correct direction of rotation, noting the valve opening and closing points on the degree wheel. Specifications for cam or valve timing are given in crankshaft degrees before or after top or bottom dead center position.

Cam lobe lift can also be verified at this time by turning the crankshaft until a maximum dial indicator reading is observed.

Incorrect Camshaft Timing

Incorrect camshaft timing can be caused by any of the following:

1. Timing gear wear.

2. Timing chain and/or sprocket wear (Figure 101).

3. Incorrect location of the keyway, dowel pin, or bolt pattern on the camshaft end.

FIGURE 101
Badly worn camshaft sprocket. The timing chain on this engine slipped because the nylon coating on the sprocket teeth had worn off.

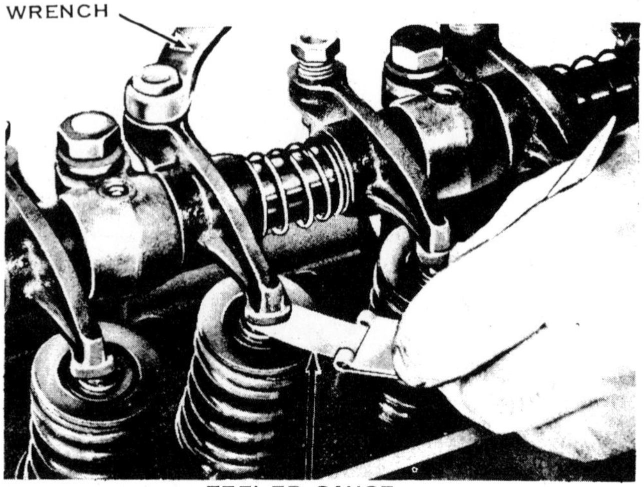

FIGURE 102
Making a valve lash adjustment. (Courtesy of Ford Parts and Service Division)

4. Incorrect position of the crank snout keyway, crank gear keyway, or crank gear timing mark.

5. Incorrect position of the camshaft gear or sprocket timing mark, keyway, dowel, pin hole, or bolt pattern location.

6. Incorrect valve clearance. If clearance is increased, valve duration and lift will be decreased.

7. Cam lobe wear.

ADJUSTING MECHANICAL LIFTER VALVE TRAINS

Tune-up specifications provide intake and exhaust valve clearances for mechanical (solid) lifters. Note whether the engine should be stopped or running and hot or cold (Chart 2).

CHART 2 AN EXAMPLE OF VALVE CLEARANCE SPECIFICATIONS

Intake—cold	.007″ (.18 mm)
hot	.008″ (.20 mm)
Exhaust—cold	.013″ (.33 mm)
hot	.014″ (.36 mm)

Hot means the engine has been running for at least 30 minutes; cold means it is not running at all. If a running adjustment is specified, start and idle the engine. Remove the rocker arm cover(s) and pass a feeler gauge of the correct thickness between the rocker arms and valve stem tips (Figure 102). For cold adjustment with engine stopped, the piston in each cylinder must be in firing position before the valve for that cylinder is adjusted. Adjust each valve until a slight drag is felt on the feeler gauge.

On certain OHC engines (Toyota, Datsun, and Pinto), valve clearance is set by adjusting the clearance between the cam lobes and the rocker arms. Check the factory manual if in doubt.

Some OHC engines (Audi, Fiat, Chrysler, Jaguar, Toyota, Volkswagen, and Volvo) use valve adjusting shims located on top of the valve lifters. A special tool is used to compress the valve springs in order to replace the shims (Figure 103).

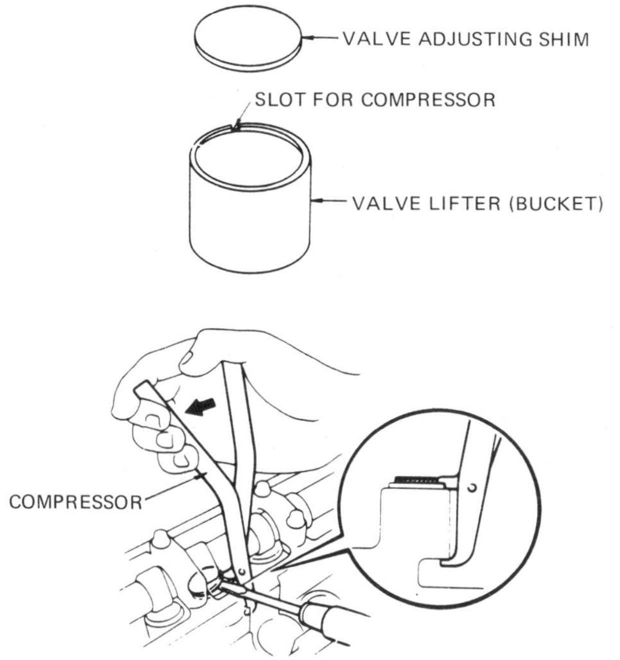

FIGURE 103
Some engines use valve adjusting shims located on top of the valve lifters (buckets). As shown, it is not necessary to remove the camshaft in order to replace shims when adjusting clearance.

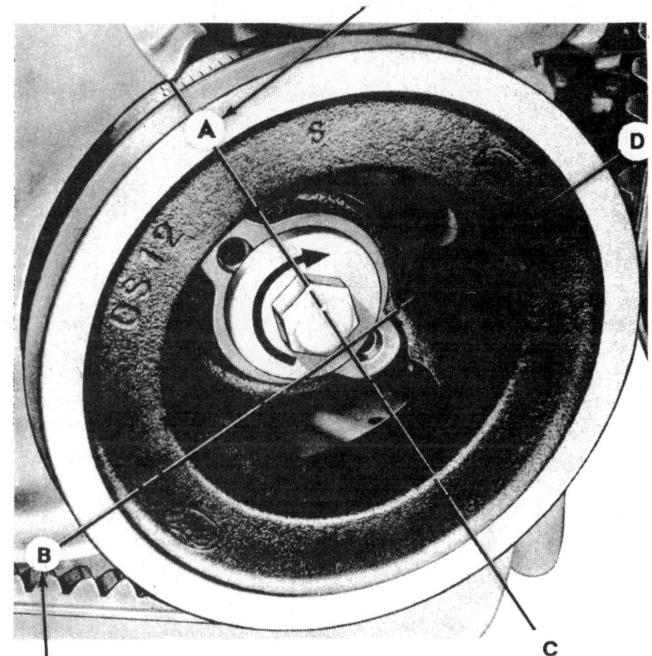

FIGURE 104
Details of the crankshaft damper division for a preliminary valve adjustment.

ADJUSTING HYDRAULIC LIFTER VALVE TRAINS

Correct hydraulic lifter adjustment positions the lifter plunger approximately at the midpoint of its travel. If the valve train has been disassembled, an initial valve setting may have to be made. The procedure that follows is for an eight-cylinder engine with adjustable hydraulic valve lifters:

1. Make three chalk marks 90 degrees on the crankshaft harmonic balancer (vibration damper) so that, with the factory TDC mark, the balancer is divided into four equal sections. See Figure 104.

2. Position the No. 1 piston at TDC on the end of the compression stroke. Now adjust intake and exhaust valve clearance for the No. 1 cylinder.

Begin by first loosening the rocker arm stud nut until there is end play between the push rod and rocker arm. Then tighten the nut to zero lash (clearance) position. This can be determined by rotating the push rod between your fingers (Figure 105).

Tighten the rocker stud nut slowly to the specified number of turns past zero (lash) (usually between one and two turns).

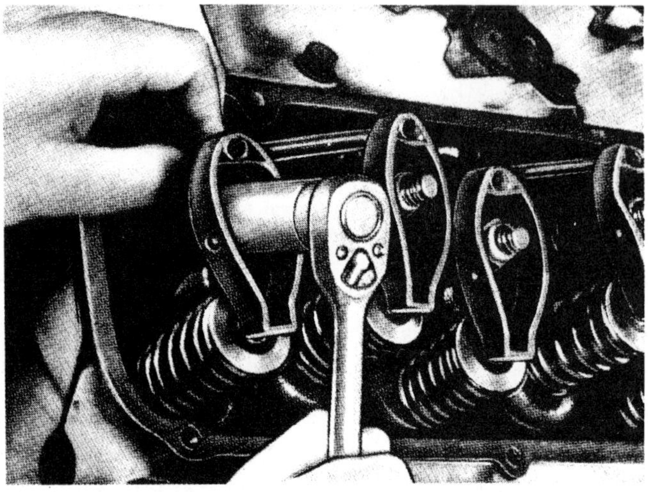

FIGURE 105
Determining zero lash position. (Courtesy of Ford Parts and Service Division)

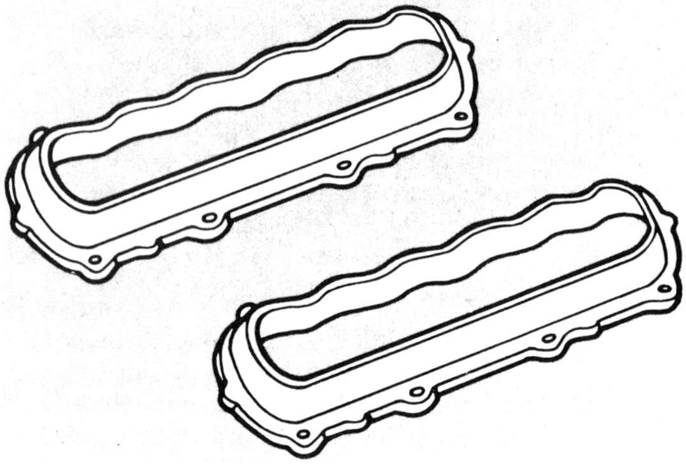

FIGURE 106
A "dummy" valve cover can be used to reduce oil splash when adjusting hydraulic valve lifters on small-block Ford and Chevrolet engines.

3. Turn the crankshaft 90 degrees (one-quarter turn) in the direction of engine rotation to the next mark on the harmonic balancer. Adjust the intake and exhaust valve clearance for next cylinder in the firing order.

4. Repeat this one-quarter turn procedure until all intake and exhaust valve sets are adjusted.

Adjustable hydraulic valve lifters can also be adjusted with the engine running. Loosen the rocker arm stud nut until the lifter starts to "clatter." Then turn the stud nut in slowly until the clatter disappears. This is the zero lash position. From this position, turn the stud nut in slowly the number of turns specified in the factory manual.

SHOP TIP

Adjusting hydraulic valve lifters with the engine running can be a messy job due to the oil flow coming off the rocker arms. Figure 106 shows one method that can be used to eliminate this problem.

Some hydraulic valve lifters are adjusted by bottoming the lifter (Figure 107). The clearance between the rocker tip and the valve stem end is then measured. If clearance is outside the specification range (.100″ to .200″, for example), correction may be made by using push rods that are a different length than stock. Many engines equipped with hydraulic valve lifters do not provide a method for valve adjustment. Correct adjustment is designed to occur as the rocker arm stands are tightened down. However, several conditions can disturb this adjustment and cause the lifter plunger to be bottomed. These conditions are:

1. Excessive cylinder head and/or block milling.

2. Excessive valve face and/or seat grinding.

Correction may be made by shimming the rocker arm stands. Some manufacturers make push rods of various lengths to accommodate these changes.

CAMSHAFT DRIVES

Correct valve timing is critical for proper engine operation. Valve action must be kept in a designated relationship to piston movement. This is done by means of timing gears, chain and sprockets, or belt and spockets (Figure 108).

Timing Gear Installation Tips

1. Locate the gear timing mark prior to pressing it on the camshaft. The mark must face outward.

2. Use Loctite® adhesive or an equivalent retaining compound to augment the press fit of a timing gear.

3. A pressed-on camshaft gear should be installed with the camshaft out of the engine. Trying to drive on the gear in some engines can knock out the cam plug at the rear of the block.

4. The timing mark on each gear must be aligned after gear installation (Figure 109). Check a factory service manual for proper alignment if there is any doubt.

5. Check camshaft thrust plate clearance (end play) after the gear is pressed on (Figure 110).

6. Check backlash between the mating gear teeth at several points with a feeler gauge or dial indicator (Figure 111).

Timing Gear Failure

Improper blacklash will cause timing gear failure. Excessive backlash can cause noise, high impact loads, and (often) breakage. Insufficient backlash can cause severe gear wear (Figure 112).

FIGURE 107
Checking valve clearance with the lifter bottomed out. Adjustment can be made by using a push rod that is a different length than stock. (Courtesy of Ford Parts and Service Division)

Timing Chain and Sprocket Installation Tips

1. Install the chain and sprocket assembly as an entire unit, keeping the timing marks aligned (Figure 113).

2. Visually inspect the rub contact surfaces on any dampers or tensioners before installing a new timing chain (Figure 114).

Timing Chain and Sprocket Wear

The following procedure may be used to determine when to replace a timing chain:

1. Turn the camshaft sprocket to get all the slack on one side of the chain.

2. Move the slack side in and out (Figure 115). As a rule of thumb, replace the chain if you have more than 1/2″ movement.

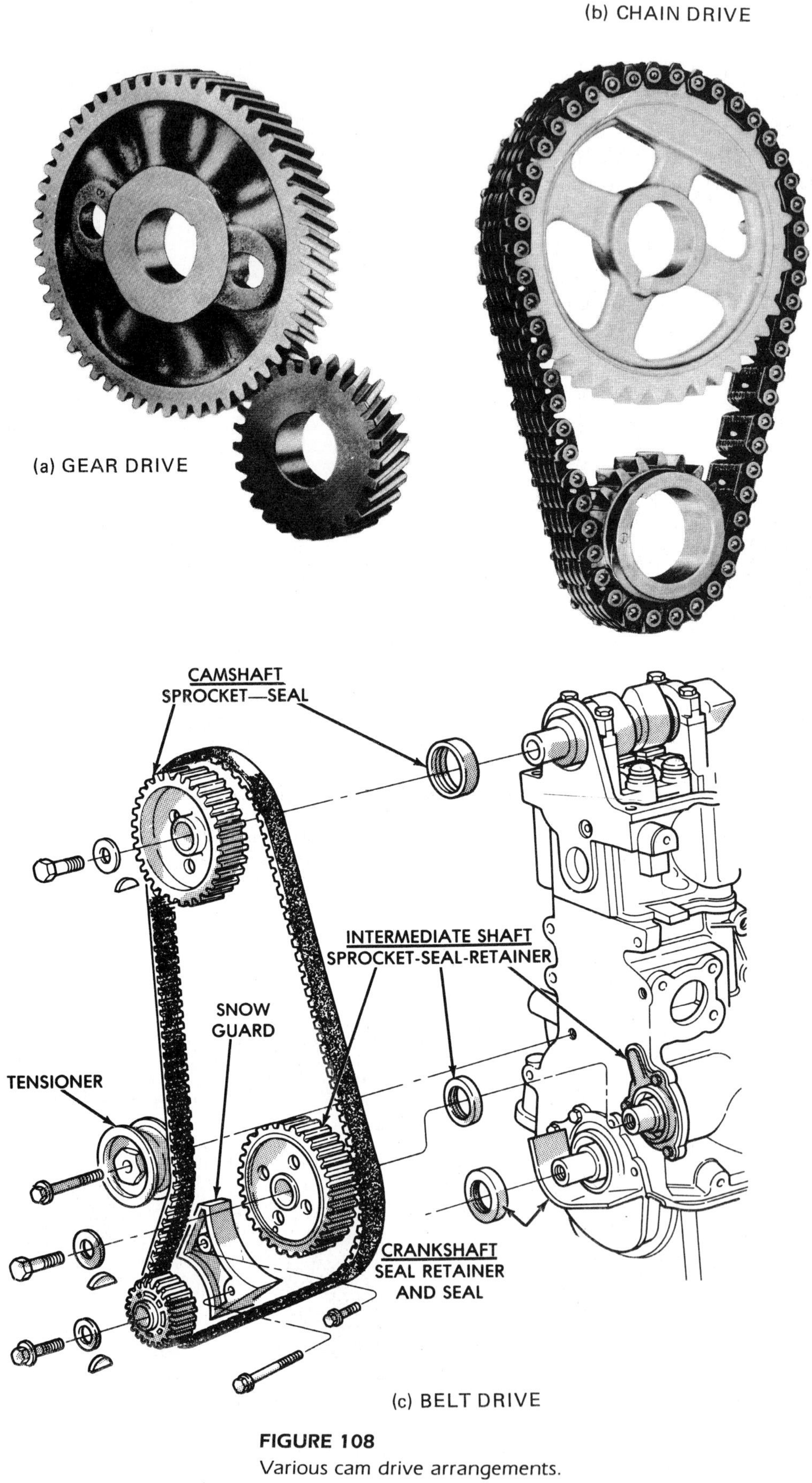

FIGURE 108
Various cam drive arrangements.

FIGURE 109
Typical timing gear mark alignment. (Courtesy of Ford Parts and Service Division)

FIGURE 110
Measuring camshaft thrust clearance.

FIGURE 111
Measuring timing gear backlash.

CAUTION: Before replacing the chain, visually inspect the teeth of both sprockets. Worn sprockets can cause premature chain wear.

Timing Chain and Sprocket Failure

Misalignment, overload, or hammering on the sprockets are some additional causes of chain and/or sprocket failure.

OHC Timing Belts

Cogged (toothed) timing belts are used on most OHC engines. See Figure 116.

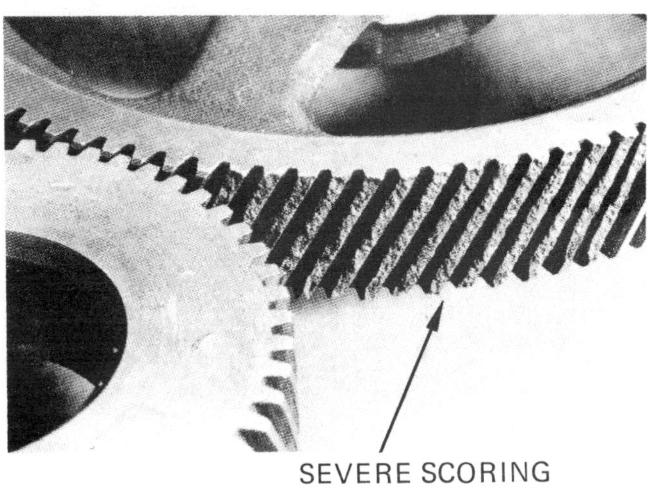

FIGURE 112
Badly worn timing gear. (Courtesy of TRW Automotive Aftermarket Division. Reprinted with permission)

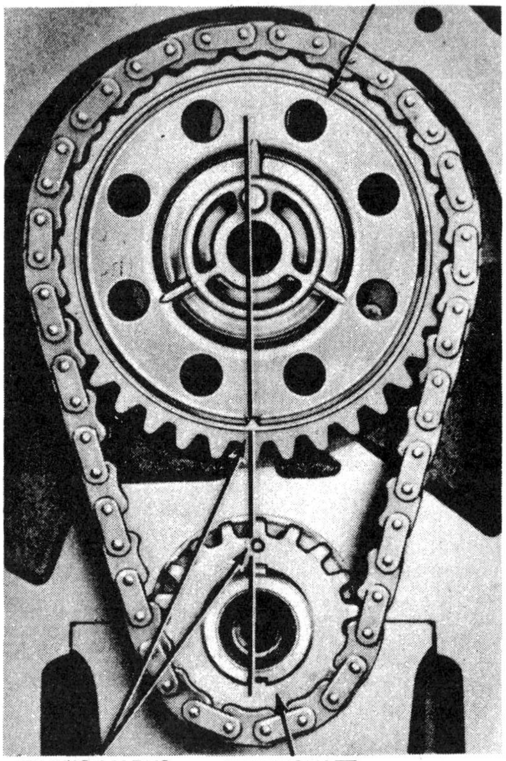

FIGURE 113
When installing a timing chain assembly, align the timing marks through the center of the gears, as shown. (Courtesy of Ford Parts and Service Division)

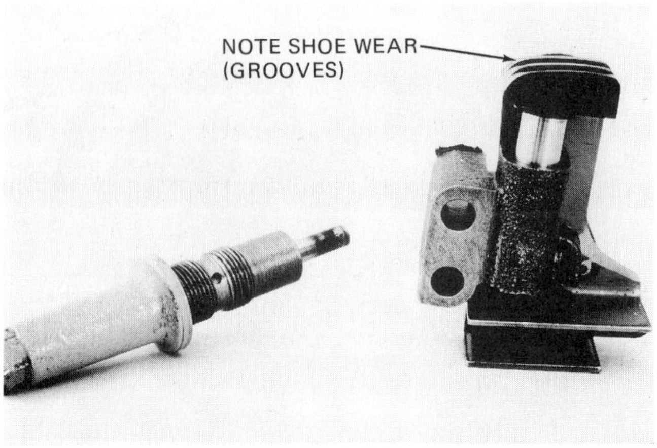

FIGURE 114
Timing chain tensioners.

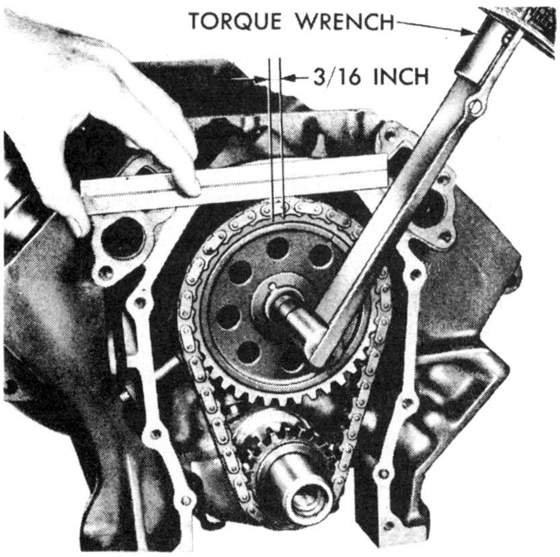

FIGURE 115
Measuring timing chain stretch. You can also check for timing chain stretch (looseness) by turning the crankshaft by hand and noting the degrees of movement before the distributor rotor moves. (Courtesy of Chrysler Corporation)

OHC Timing Belt Installation Tips

1. Handle belts carefully. Keep them free of oil, grease, and dirt.

2. Belt sprockets are relatively soft. *Do not turn them with a pipe wrench or chain wrench.* A gouge or nick on a sprocket can damage a belt.

3. Never turn the engine backwards with the belt installed. Doing so can cause the belt to jump one or more cogs and change valve timing.

4. Valve timing for most OHC engines is set by bringing the No. 1 piston to TDC and aligning the cam sprocket match marks.

5. Always follow factory recommendations when adjusting timing belt tension.

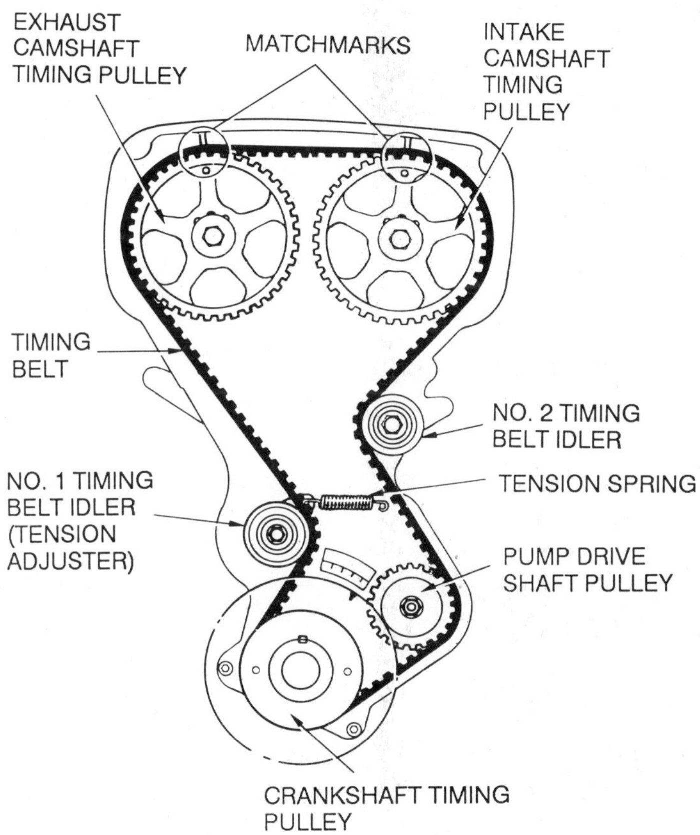

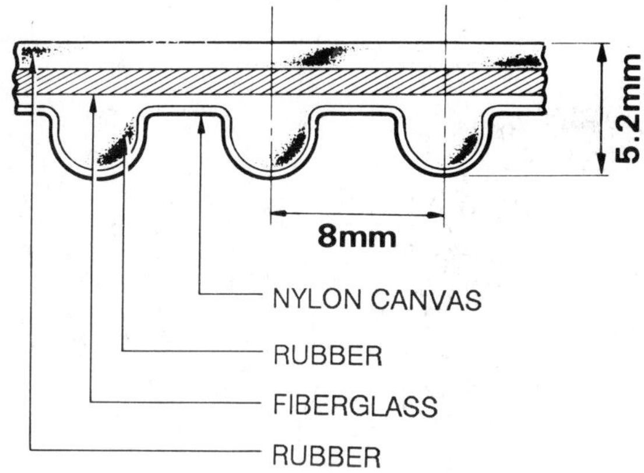

FIGURE 116
Timing belt used on the Toyota 5M-G DOHC engine. (Courtesy of Toyota Motor Sales, U.S.A., Inc.)

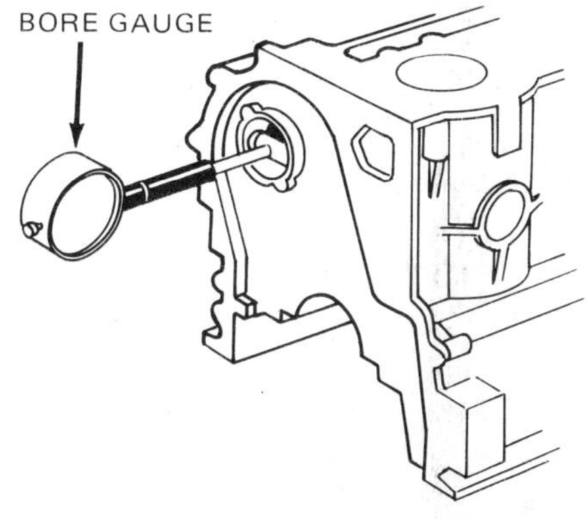

FIGURE 117
Measuring camshaft bearing out-of-roundness.

CAMSHAFT BEARINGS

Camshaft bearings will wear out-of-round (Figure 117). Cam bearing clearance greater than .003″ will result in more than normal oil leakage and a drop in engine oil pressure.

Clearance less than specifications can cause a cam bearing to spin in its bore. A dial bore gauge and an outside micrometer may be used to measure cam bearing clearance. See Section C for additional camshaft bearing information.

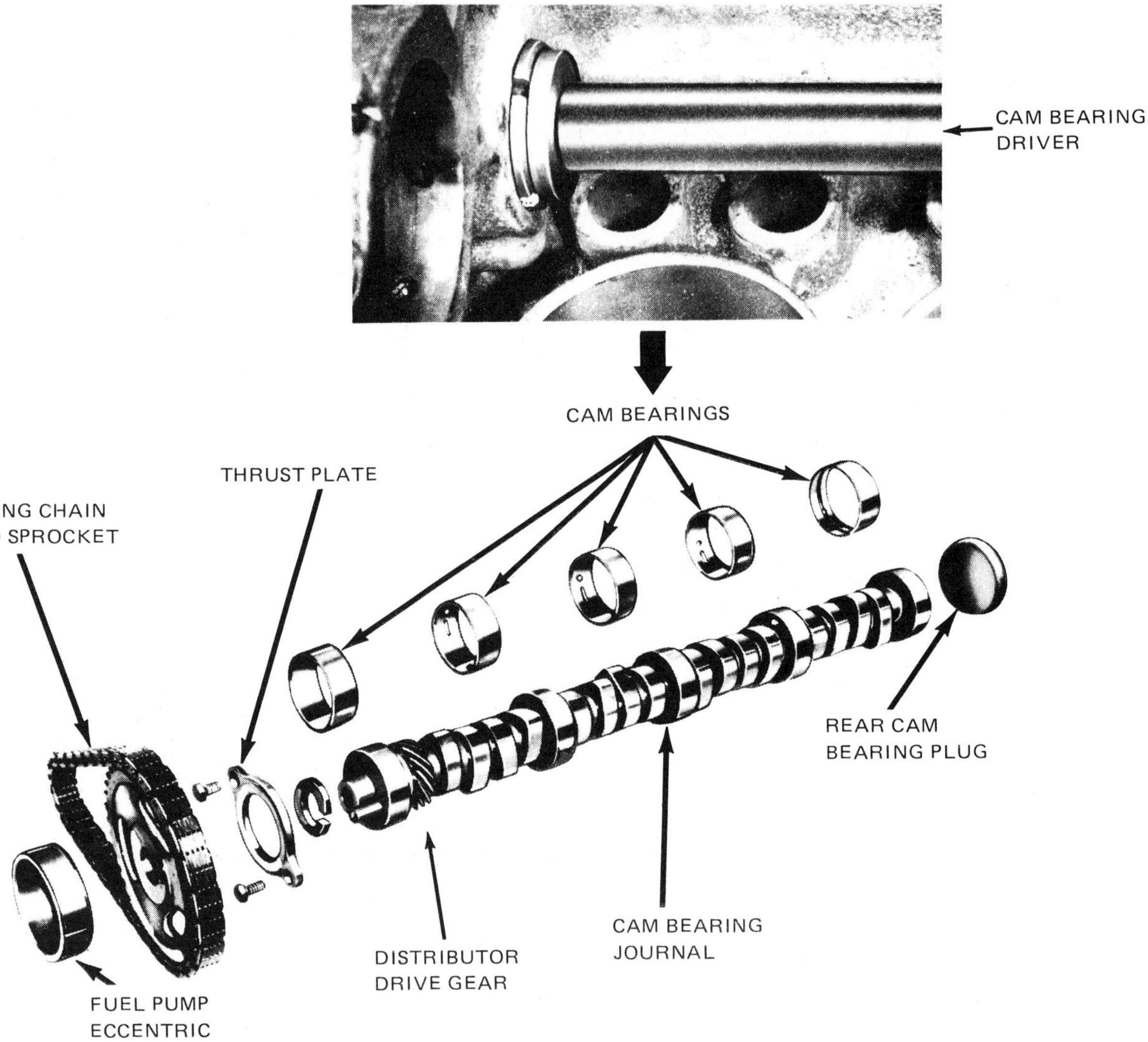

FIGURE 118
Installing cam bearings. (Courtesy of Ford Parts and Service Division)

Camshaft Bearing Installation

Camhaft bearing installation is a critical operation that requires attention to detail (Figure 118).

After the camshaft bearings are installed, check the alignment by installing the camshaft. If there is bind, check for a bent camshaft, an improperly installed cam bearing, or cam bearing bore misalignment.

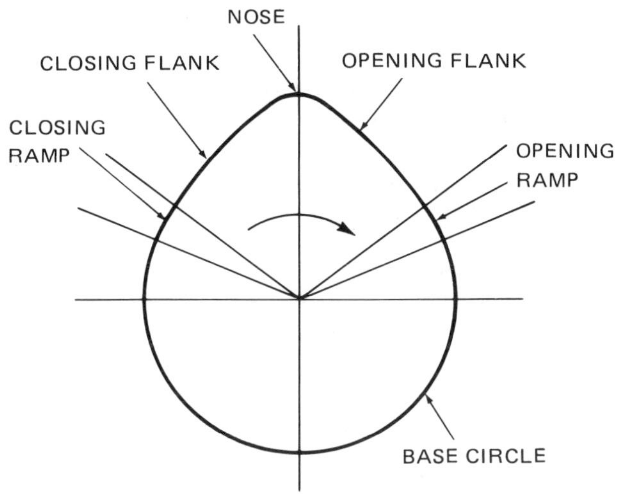

FIGURE 119
Measuring camshaft lobe wear is an important check often overlooked.

FIGURE 120
Measuring camshaft end play on an OHC engine with a feeler gauge. (Courtesy of Toyota Motor Sales, U.S.A., Inc.)

CAMSHAFT

The camshaft is designed to control the entry and exit of the air–fuel mixture at points that are timed to piston position.

Camshaft Inspection

Visually examine the camshaft. Replace it if there are rounded, chipped, pitted, worn, scored, or badly discolored lobes. Replace a camshaft that has worn journals or damaged distributor drive gear teeth.

Measuring Camshaft Bearing Journals

Measure the size of each bearing journal and compare to specifications. *Maximum wear limit on any journal is .001".*

Measuring Camshaft Lobe Lift

Lobe wear may be checked using a dial indicator setup, a caliper, or a micrometer (Figure 119).

Measuring Camshaft Base Circle Runout

Base circle runout may be checked with vee blocks and a dial indicator.

Checking Camshaft End Play

To check camshaft end play, mount a dial indicator against the timing sprocket or gear (Figure 120). If correction is necessary, install a new thrust plate or one of a different thickness.

TRUE CASE HISTORY

Vehicle	Cadillac Eldorado
Engine	Eight-cylinder, 425 CID
Mileage	37,000. Vehicle just purchased at an auto auction.
Problem	1. While driving, the engine would quit running at various times. After sitting for a period of time (usually overnight), the engine could be restarted. This random problem occurred about once a week. 2. The new vehicle owner had tried unsuccessfully for over six months to have this problem fixed.
Cause	A shop rag was found in the engine oil pan. Under the right driving conditions, the rag would be "sucked" against the oil pump pickup screen and would shut off oil flow. This would cause the hydraulic valve lifters to collapse, and the engine would stop running. Then, after sitting for a while, the rag would apparently float away from the pickup screen and the engine could be restarted.

TRUE CASE HISTORY

Vehicle	GMC S-10 truck equipped with TBI
Engine	Four-cylinder, 2.5 liter
Mileage	27,200
Problem	1. Idle surge and intermittent high idle. 2. Shop replaced IAC motor, ECM, EGR solenoid, EGR valve, and MAT sensor. None of this cured the problem.
Cause	1. Worn exhaust valve guides. 2. Base timing was incorrectly set at 20 degrees. Spec. is 8 degrees. 3. Minimum idle speed was incorrectly set at 450 rpm. Spec. is 550 rpm.
Cure	1. Reconditioned all the guides and replaced all exhaust valves, and reset base timing and minimum idle speed to factory specs.

ASE STUDY QUESTIONS

1. After each cylinder head on a V-8 OHV engine is surfaced .012″, the intake manifold mounting bolt holes will not line up.

   Technician A says to use thicker intake manifold gaskets and extra sealer.

   Technician B says to elongate the mounting bolt holes with a file.

   Who is right?

   a. A only **c.** Both A and B
   b. B only **d.** Neither A nor B

2. The cylinder heads are being installed on an engine. Some of the mounting bolt holes are "blind" and filled with oil.

   Technician A says that failure to clean these holes can result in an erroneous torque reading.

   Technician B says that failure to clean these holes can result in a cracked block.

   Who is right?

   a. A only **c.** Both A and B
   b. B only **d.** Neither A nor B

3. A valve job is going to be done on a cylinder head.

   Technician A says to resurface the seats before reconditioning the guides.

   Technician B says to check the valve springs for squareness.

   Who is right?

   a. A only **c.** Both A and B
   b. B only **d.** Neither A nor B

4. **Technician A** says that a used keeper should not be mated with a new keeper.

   Technician B says that keeper seating should be checked with a hammer.

   Who is right?

   a. A only **c.** Both A and B
   b. B only **d.** Neither A nor B

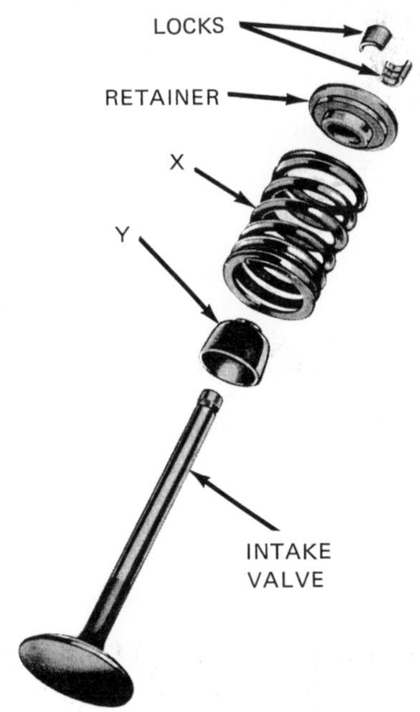

5. **Technician A** says that part X above can be replaced without removing the cylinder head.

   Technician B says that part Y above can be replaced without removing the cylinder head.

   Who is right?

   a. A only **c.** Both A and B
   b. B only **d.** Neither A nor B

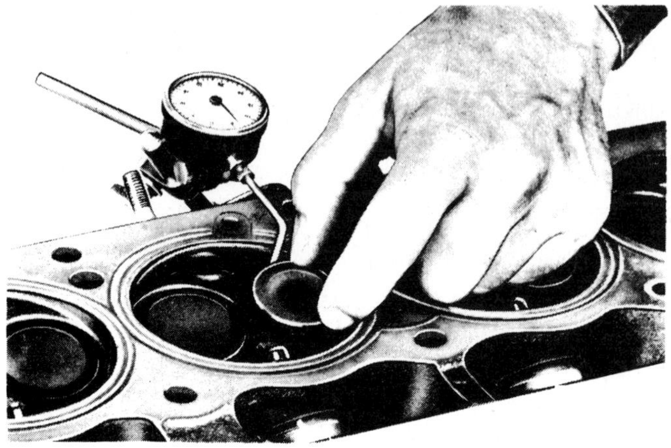

6. What is the technician measuring in the picture above?

   a. Valve face runout
   b. Valve lift
   c. Valve margin
   d. Valve guide wear

7. **Technician A** says that an interference angle must be used if the engine burns unleaded fuel.

   Technician B says that an interference angle makes a poor seal when the engine is first started after a valve job.

   Who is right?

   a. A only
   b. B only
   c. Both A and B
   d. Neither A nor B

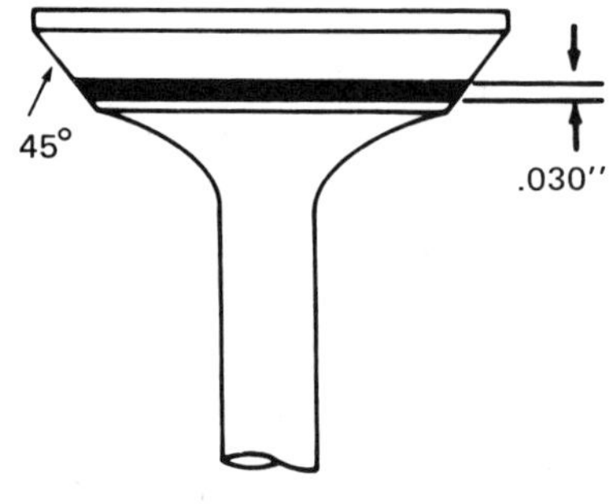

8. An exhaust valve has a seat contact pattern as shown above. What needs to be done?

   a. Grind the seat with a 45 stone, and then narrow by "topping"
   b. Grind the seat with a 45 stone, and then narrow by "throating"
   c. Widen the seat with a 45 stone
   d. Nothing

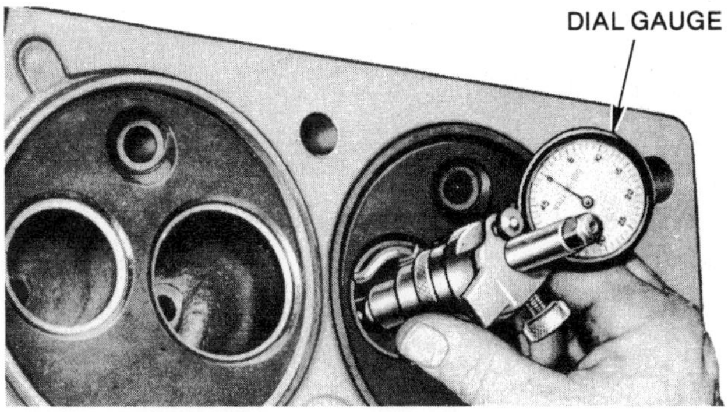

9. **Technician A** says that the tool shown above is measuring valve seat depth.

   Technician B says that the tool shown above is measuring valve seat concentricity.

   Who is right?

   a. A only
   b. B only
   c. Both A and B
   d. Neither A nor B

10. **Technician A** says that dividers can be used to measure valve spring assembled height.

    Technician B says that a telescoping gauge can be used to measure valve spring assembled height.

    Who is right?

    a. A only
    b. B only
    c. Both A and B
    d. Neither A nor B

11. A 6-cylinder OHV engine is being assembled. In what direction should the oil feed holes on the rocker arm shaft face?

    a. On the side opposite the push rods
    b. On the side toward the push rods
    c. Up
    d. Down

12. All the valve lifters in an engine are cupped as shown at the right.

    Technician A says to replace the camshaft.

    Technician B says to replace the lifters.

    Who is right?

    a. A only
    b. B only
    c. Both A and B
    d. Neither A nor B

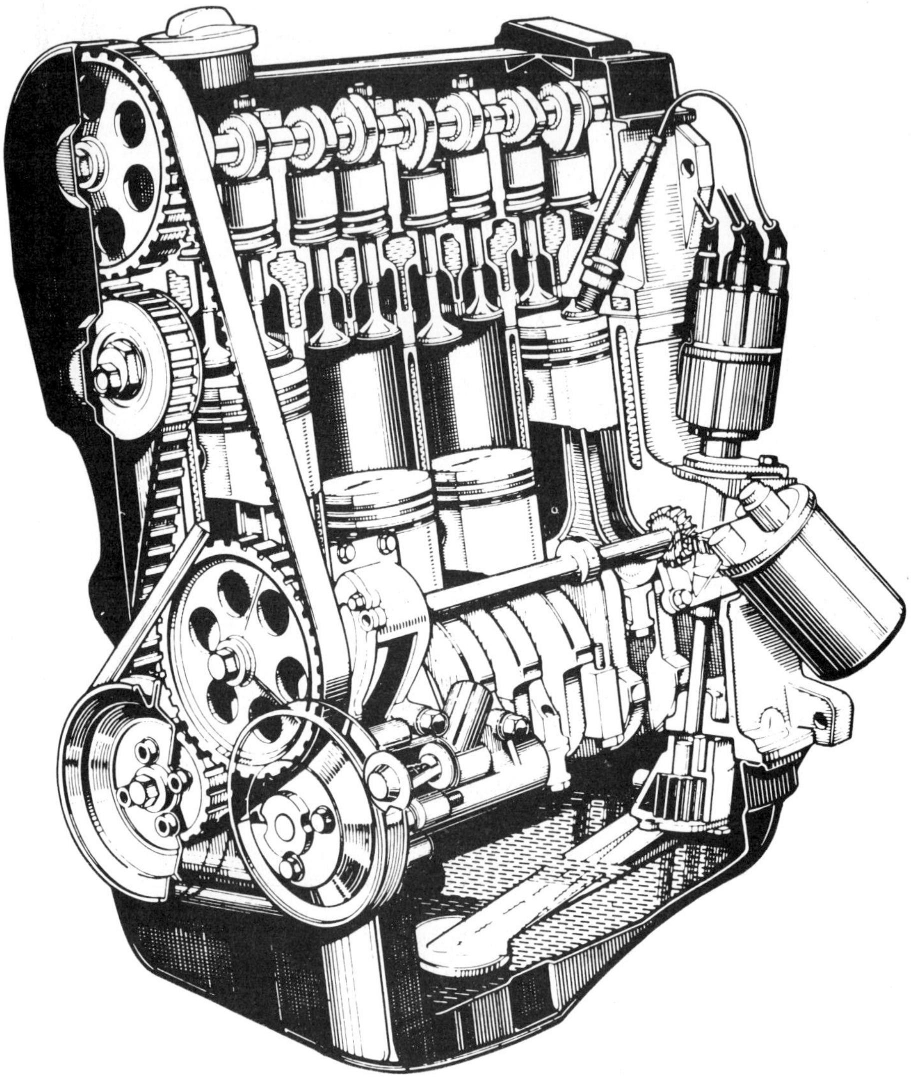

13. Valve clearance is not correct on the above OHC engine.

    Technician A says to grind the buckets.

    Technician B says to switch the buckets.

    Who is right?

    a. A only
    b. B only
    c. Both A and B
    d. Neither A nor B

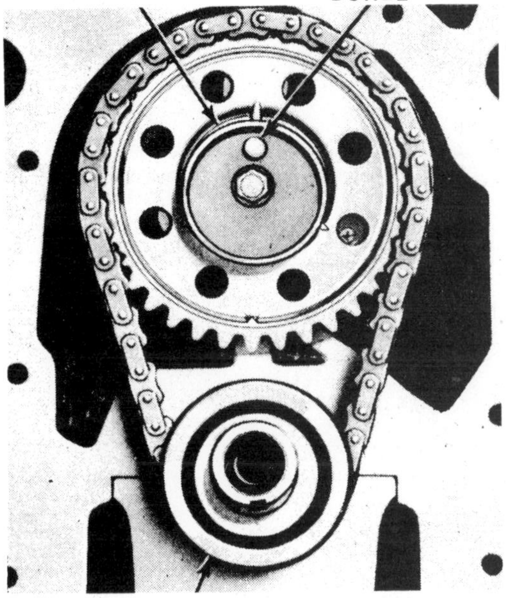

15. **Technician A** says that camshaft wear can cause retarded valve timing.

    Technician B says that camshaft wear can produce a "popping" noise out the top of the carburetor.

    Who is right?

    a. A only
    b. B only
    c. Both A and B
    d. Neither A nor B

14. Part X above is left out when replacing a timing chain and cam sprocket.

    Technician A says that fuel pump pressure will be zero.

    Technician B says that camshaft end play will be excessive.

    Who is right?

    a. A only
    b. B only
    c. Both A and B
    d. Neither A nor B

16. The bearings shown above are going to be installed in an engine.

    Technician A says that alignment must be checked.

    Technician B says that clearance must be checked.

    Who is right?

    a. A only
    b. B only
    c. Both A and B
    d. Neither A nor B

17. **Technician A** says that camshaft lobe lift is being checked in the above picture.

    Technician B says that valve timing is being checked in the above picture.

    Who is right?

    a. A only
    b. B only
    c. Both A and B
    d. Neither A nor B

SECTION REVIEW QUESTIONS

1. When removing a cylinder head, you detect oil in the intake manifold ports. What specific things could be causing this?
2. What trick can be used to remove a stuck cylinder head?
3. Why shouldn't an aluminum head be removed when hot?
4. After removing the cylinder head(s), you should visually look for what problems?
5. Maximum allowable cylinder head warpage is generally considered to be what amount?
6. What is meant by the term "corrective machining"?
7. Often after surfacing an OHC cylinder head, the camshaft will not turn. What are six approaches being used by shops to solve the problem?
8. What three methods (besides visual) can be used to locate cracks?
9. What procedure can be used to help release a stuck valve spring retainer and its keeper?
10. What is a valve stick?
11. How can a timing light be used to detect a broken valve spring?
12. What is the maximum out-of-squareness limit for a valve spring?
13. What is the maximum free length height limit for a valve spring?
14. Why must new keepers always be installed in pairs?
15. Why do some engines use multigrooved exhaust stems?
16. List four types of valve stem seals.
17. Why is it a mistake to install an "o" ring stem seal on the valve before the spring and the retainer are installed?
18. What two methods can be used to test valve stem "o" ring seals for leakage?
19. What two methods can be used to replace a broken valve spring without removing the cylinder head?
20. Too much valve stem-to-valve guide clearance can cause what engine problems?
21. Valve guides generally show what wear pattern?
22. What three methods can be used to measure valve guide wear?
23. Why is installed valve guide height important?

24. What is valve guide knurling?

25. What is valve margin?

26. What is a chamfering vee?

27. Why is valve stem height a critical dimension?

28. Why should no more than .030″ be ground from a valve stem end?

29. Describe an interference angle.

30. How can an insert-type seat be checked for looseness?

31. Outline the procedure for making a three-angle valve seat grind.

32. How is an integral valve seat usually repaired?

33. What is installed spring height? How is it corrected?

34. What methods can be used to determine installed valve spring height?

35. What needs to be observed when installing variable pitch valve springs?

36. Why should a head gasket never be reused?

37. Why is it important to clean head bolt threads prior to installation?

38. Should a sealer be used on a head gasket? Explain your answer.

39. What is a torque/angle meter?

40. Describe a pliant-type head bolt.

41. What various means are used to provide oil flow to the rocker arm assembly in an engine?

42. What things can cause a rocker arm to break?

43. After removing push rods, you should examine them for what problems?

44. What is hydraulic valve lifter "pump-up"? What can cause it?

45. Define leakdown.

46. What valve train design factors help produce lifter rotation?

47. Why shouldn't used lifters be installed on a new camshaft?

48. In order to measure in-vehicle cam timing, what tools are needed?

49. Name five conditions that can cause incorrect camshaft timing.

50. Outline the procedure for setting static valve clearance in an eight-cylinder engine with adjustable hydraulic valve lifters.

51. What is a "dummy" valve cover?

52. Improper timing gear backlash can cause what problems?

53. As a rule of thumb, replace a timing chain if it has more than ________″ of slack.

54. Why shouldn't an OHC engine be turned backwards if the timing belt is installed?

55. Why is cam bearing clearance so critical?

56. What measurements should be made when installing a new camshaft in an engine?

57. "Dirty mayonnaise" colored sludge deposits are found inside a valve cover. What does this indicate?

58. What is hoop stressing?

59. What is a "bounce" spring?

60. How can Prussian blue paste be used to determine valve face-to-seat contact location?

FIGURE 4
Straightening valve cover flanges. *Do not attempt to straighten cast aluminum or plastic valve covers in this manner.* (Courtesy of Fel-Pro, Inc.)

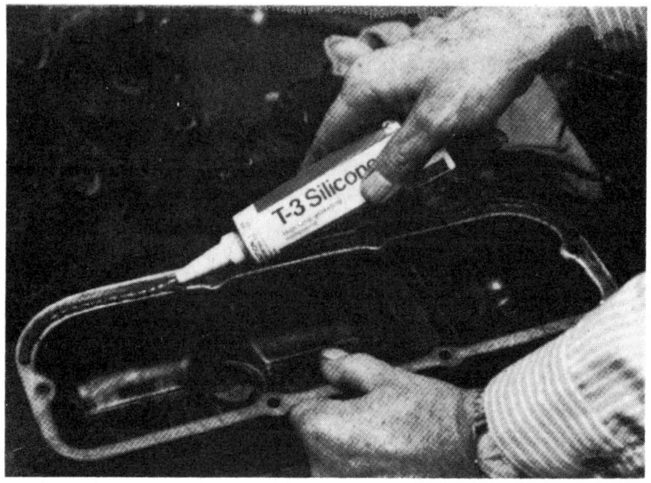

FIGURE 5
RTV (room temperature vulcanization) silicone rubber is sometimes used instead of a conventional gasket on valve covers. The formed-in-place gasket compound shown above is specially formulated to remain flexible at temperatures up to 600° F. In addition, it is noncorrosive and meets OEM specifications for use on oxygen-sensor-equipped engines. (Courtesy of Automotive Trades Division/3M)

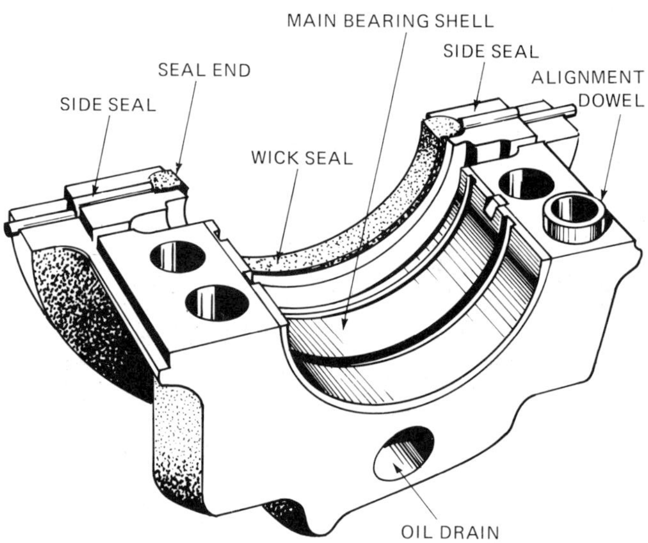

FIGURE 6
A wick-type (rope) rear main seal half installed in the bearing cap. The other half of the seal is installed in the block.

Valve Cover Installation

1. Clean gasket surfaces of foreign material.

2. Check the valve cover gasket flange for flatness by using a straightedge. Straighten with a hammer if required (Figure 4). *Do not attempt to hammer a cast aluminum or plastic valve cover;* it will break.

3. Make sure the cylinder head oil return holes are not plugged.

4. Attach gasket to the valve cover, or use RTV (Figure 5).

5. Install valve cover.

NOTE: *If load spreader washers are used, make sure they are in correct position.*

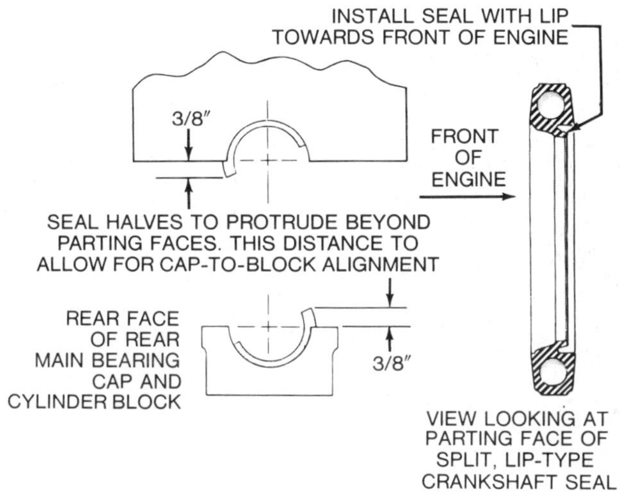

FIGURE 7
Split-lip crankshaft rear main oil seal. (Courtesy of Ford Parts and Service Division)

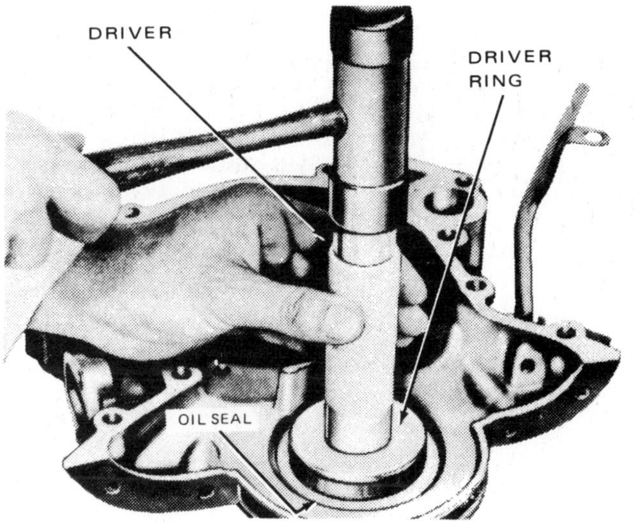

FIGURE 8
Installing a crankshaft front oil seal. (Courtesy of Ford Parts and Service Division)

Oil Seal Installation Tips

1. Always soak wick- (rope)-type rear main seals in engine oil for 10 minutes just prior to installation. Then shape the seal into position with a hammer handle or smooth piece of pipe and trim the seal ends (Figure 6).

2. When installing lip-type rear main seals, be sure the seal lip faces the lubricant supply (Figure 7).

3. When installing a crankshaft front oil seal, uniform pressure should be applied to drive the seal in its bore (Figure 8). Inspect the seal for proper seating by using a feeler gauge (Figure 9).

4. When installing a seal, it is generally recommended by the manufacturer that the seal lip be coated with the lubricant to be sealed.

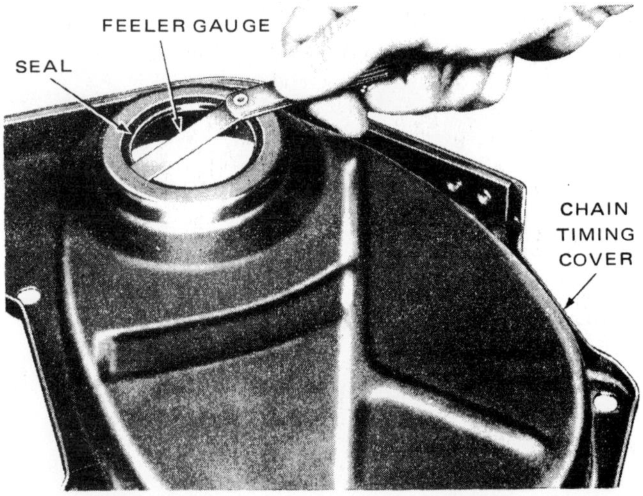

FIGURE 9
Inspecting a seal for proper seating. Attempt to insert a .0015″ feeler gauge between the seal body and the timing cover. If the seal is installed properly (fully bottomed), the feeler gauge cannot be inserted. (Courtesy of Chrysler Corporation)

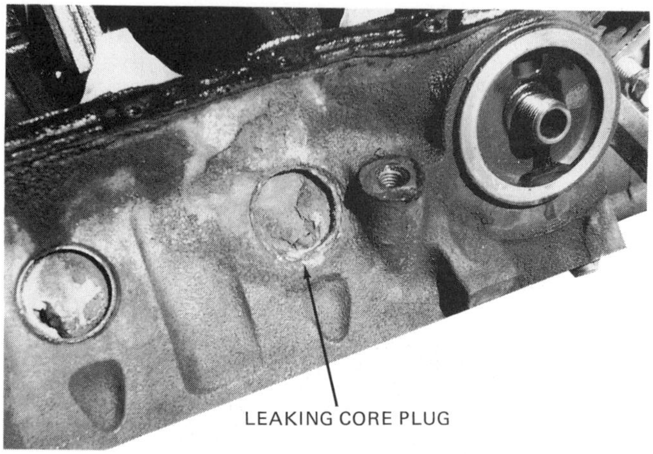

FIGURE 10
Rusted-through engine block core plug (often called freeze plug, expansion plug, or soft plug).

FIGURE 11
Oil and/or water leaks can occur at the back of this engine block.

VISUALLY INSPECTING THE ENGINE BLOCK

After the engine block has been cleaned, check for cracks, cylinder scoring, clogged oil passages, and leaking core and oil gallery plugs (Figures 10 and 11).

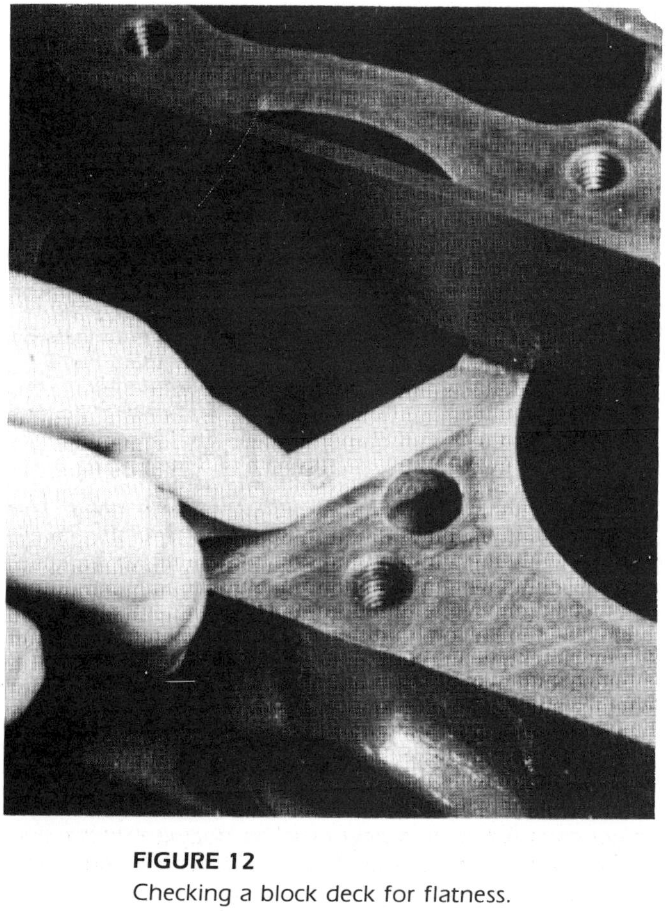

FIGURE 12
Checking a block deck for flatness.

Deck Flatness

The block deck surfaces must be flat to ensure proper head gasket sealing. Check for flatness with a straightedge and feeler gauge diagonally, vertically, and horizontally on the deck (Figure 12). A deck warped more than .004" should be resurfaced. Decks with corrosion around water passages or with deep scratches must be resurfaced (Figure 13).

> **SHOP TIP**
> Properly used, a "jitterbug" sander is an excellent tool for cleaning cylinder head and block deck surfaces (Figure 14).

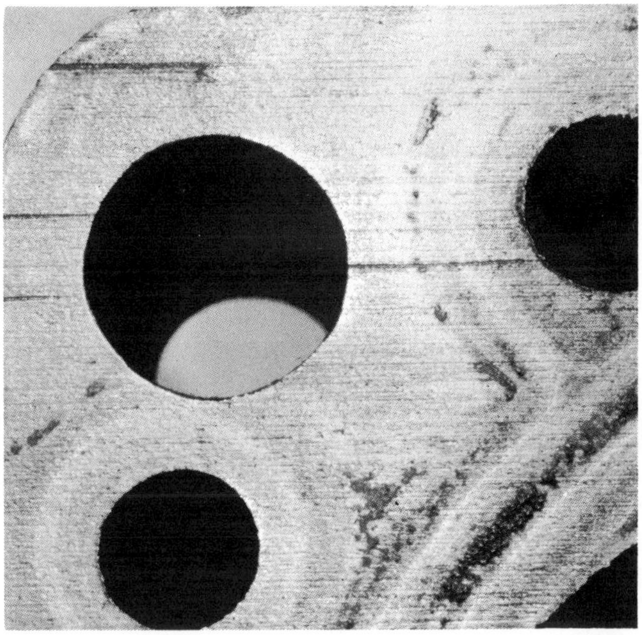

FIGURE 13
Deep scratches in this block deck allowed coolant to find its way up around a head bolt and leak into the valve cover area.

FIGURE 14
Cleaning a block deck.

Main Bearing Bore Alignment

After an engine has run thousands of miles, internal stresses can cause misalignment and/or stretch of the main bearing bores (Figure 15).

Alignment of the main bearing bores can be checked by placing a precision-ground arbor into the saddles (Figure 16[a]). It should be .001″ less in diameter than the low limit of the bore specification. After the arbor is in position, torque the main bearing caps. If the arbor will not turn, one or more of the bores is out of round, or the crankcase is warped. In either case, the block will have to be line-honed or line-bored.

If a precision-ground arbor is not available, bore alignment may be checked with a straightedge and a feeler gauge (Figure 16[b]).

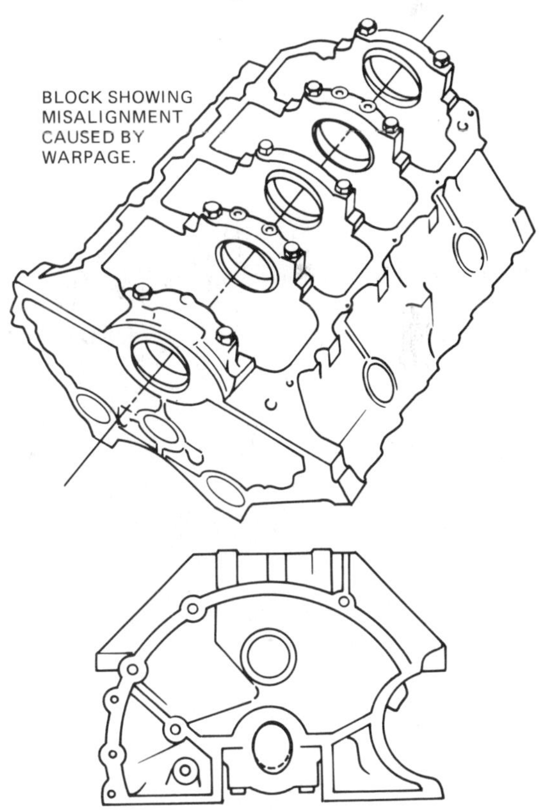

FIGURE 15
When an engine runs thousands of miles, stresses created by continual heating and cooling of the engine can cause block warpage and stretch. As this distortion takes place slowly over time, the original main bearings and crankshaft compensate through gradual wear. However, when a reground crankshaft and new main bearings are installed in the distorted block, this misalignment (even though slight) is all that is needed to make the crankshaft bind.

(a)

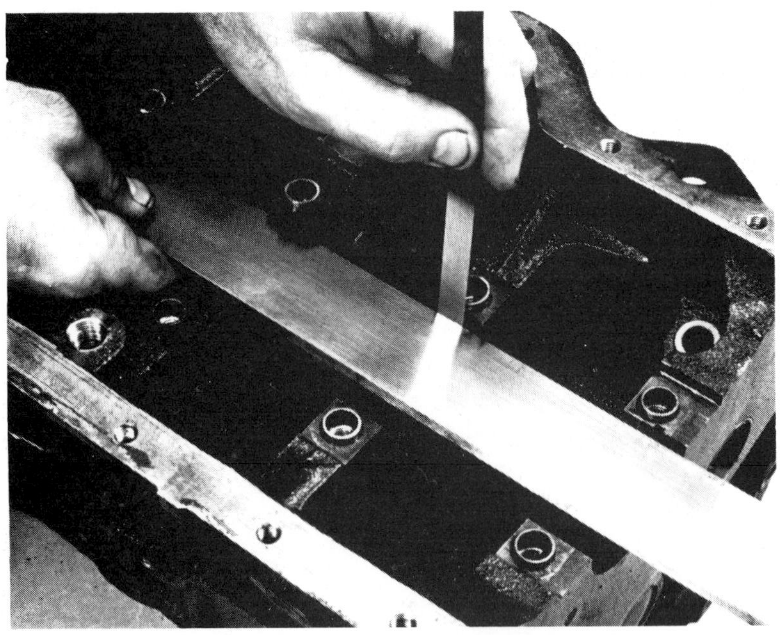

(b)

FIGURE 16
(a) Checking main bearing bore alignment with a precision ground arbor. (Courtesy of Clevite Industries Inc.) (b) Checking main bearing bore alignment with a straightedge and feeler gauge.

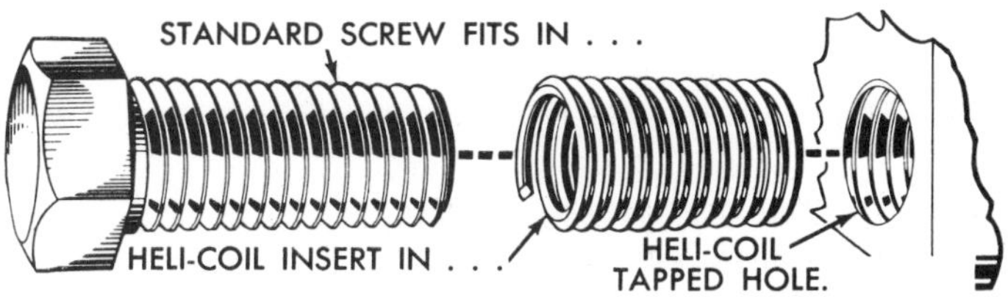

FIGURE 17
Heli-Coil® installation.

> **SHOP TIP**
> Bolts sometimes break, leaving the threaded portion stuck in the hole. Such a broken bolt can often be removed quickly by using a left-handed drill (Figure 18).

REPAIRING DAMAGED OR WORN THREADS

Damaged or worn threads in the engine can be repaired by the use of Heli-Coils® (Figure 17). This repair consists of drilling out worn or damaged threads, tapping the hole with a special Heli-Coil® tap, and installing a Heli-Coil® insert into the tapped hole. This brings the hole back to its original thread size.

FIGURE 18
A left-handed drill bit is an excellent tool for removing broken studs and bolts.

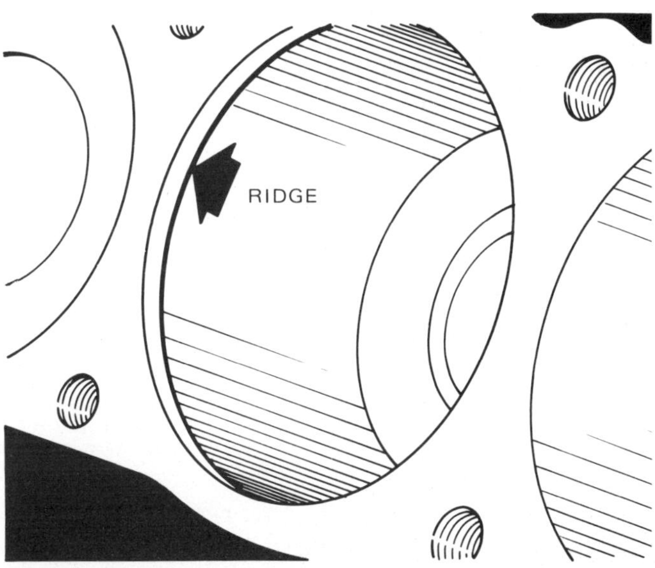

FIGURE 19
The ridge often found atop a cylinder. Always remove the ridge if the pistons are going to be reused. (Courtesy of Ford Parts and Service Division)

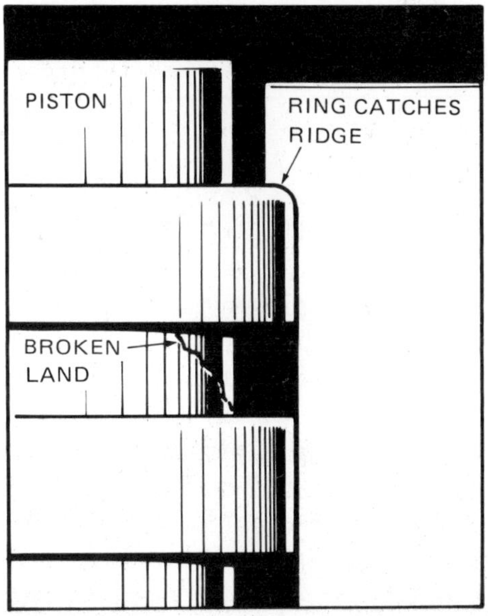

FIGURE 20
The cylinder ridge should be cut before a piston is removed. Failure to remove the ridge can cause broken rings or lands when pushing out the piston. Also, the corner of a new ring is unworn and square. Failure to remove the ridge can cause noise and possible damage after new rings are installed.

FIGURE 21
Lathe-type ridge reamer. This ridge reamer is positioned on the inside of the cylinder wall and cuts from the bottom up.

FIGURE 22
What could have caused the valve interference mark on this piston?

REMOVING CYLINDER WALL RIDGE

With the cylinder head and oil pan removed, turn the crankshaft to move the piston down in its bore. Check for a ridge at the top of the bore (Figure 19). On engines with high mileage, a ridge will form at the upper end of ring travel as the top compression ring wears into the cylinder. A ridge greater than .004″ should be removed before taking a piston out of its bore. Otherwise, the ridge will cause the piston rings to hit it and crack or crush the ring lands of the piston (Figure 20).

To cut the ridge away from the cylinder properly, follow these steps:

1. Stuff the cylinder with rags so no cuttings will wedge between the piston and the cylinder wall. Metal chips can scratch the skirts when the pistons are removed.

2. Attach a ridge reamer to the top of the cylinder to be cut (Figure 21). Adjust the cutter to take off just enough material to remove the ridge. The cut surface should blend smoothly into the cylinder wall.

> **SHOP TIP**
> When using a ridge reamer, use a spray lubricant on the cutter and "lead" the cutter by a few degrees as it is turned. This will minimize chatter and gouging.

3. When the ridge is cut, remove the protective rags and make sure there are no cuttings in the cylinder. Carefully remove the pistons. (If you need a review, refer to the information presented later in this chapter on removing and replacing piston assemblies.)

CHECKING CYLINDER BORES

After removing the pistons, inspect the bores for glazing, scuffing, unusual wear patterns, vertical scratches, and score marks. See Figure 22.

Glazing is an extremely shiny condition of the bore. Most piston ring manufacturers recommend that a glazed cylinder be lightly honed. This provides a surface that oil can cling to and helps the new piston rings break in.

If the cylinder is lightly scuffed or scored, the damage can usually be removed by honing. Deep scoring, however, usually requires boring the cylinder out to a larger diameter and fitting it with oversize pistons and rings. Boring and fitting an oversize piston is necessary also if cylinder taper or out-of-roundness exceed specified wear limits.

Measuring Cylinder Taper

Cylinder taper is the difference between the diameter of the cylinder at the top of the ring travel (in the direction of piston thrust) and the bottom of the cylinder. Subtracting the smaller diameter from the larger diameter equals the amount of taper. For example, if the diameter at the top is 3.687″ and the bottom diameter is 3.678″, the taper is .009″. *If maximum cylinder taper is .012″ or over, boring is recommended by most authorities.* A cylinder bore dial gauge, an inside micrometer, or a telescoping gauge can be used to obtain this measurement (Figures 23 and 24).

Measuring Cylinder Out-of-Roundness

Cylinder out-of-roundness is the difference between the left-to-right bore diameter and the front-to-rear bore diameter (Figure 25). Out-of-roundness must be checked at the top, center, and bottom of the bore. *If cylinder out-of-roundness is greater than .005″, boring is recommended by most authorities.*

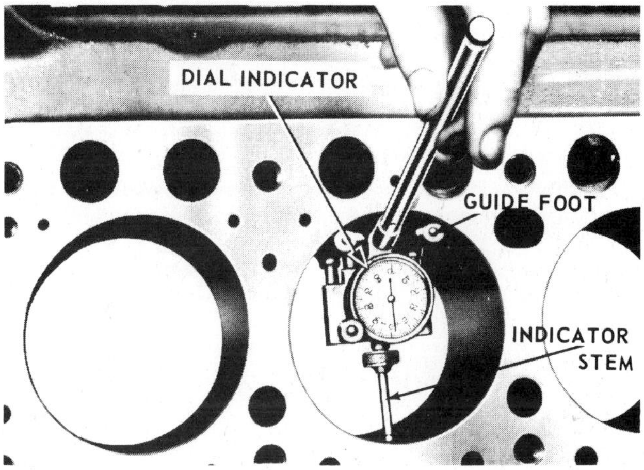

FIGURE 23
Checking cylinder taper and out-of-roundness with a dial bore gauge.

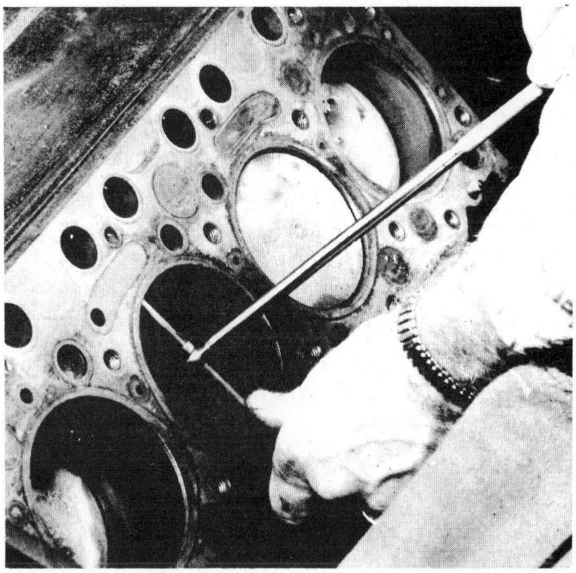

FIGURE 24
Measuring cylinder wear with a telescoping gauge. (Courtesy of Dana Corporation)

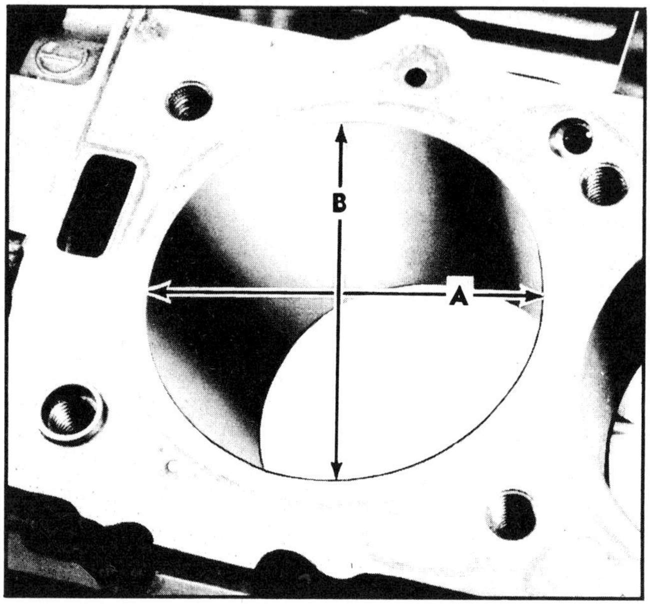

FIGURE 25
Cylinder bore out-of-roundness is determined by taking these two measurements and subtracting the difference. (Courtesy of Ford Parts and Service Division)

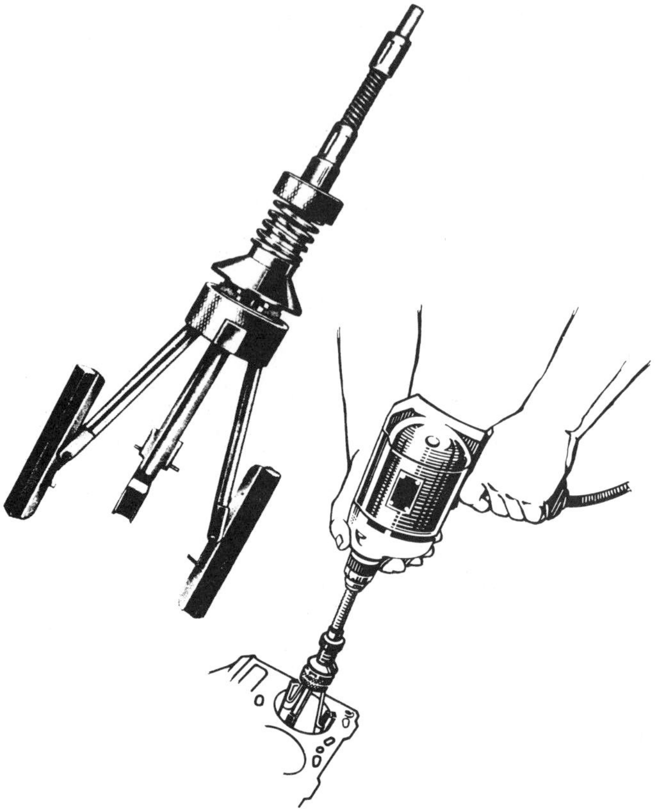

FIGURE 26
Spring-loaded hone. (Courtesy of Lisle Corporation)

HONING CYLINDER WALLS

Figure 26 shows a 220-grit spring-loaded hone being used to deglaze a cylinder wall. Operate the hone with a slow-moving drill motor (200 to 500 rpm). Do not use a high-speed drill. The recommended procedure is as follows:

1. Cover the crankshaft connecting rod journals with oil-soaked rags to prevent honing grit contamination.

2. Move the hone up and down in the cylinder while it is turning in order to obtain a 30- to 40-degree crosshatch pattern (Figure 27). Proper crosshatch pattern and wall roughness are essential to proper break-in of the new piston rings. Too smooth a pattern may cause the rings to scuff and score the cylinder. Too rough a pattern can make the rings wear quickly. Some wear patterns that are not acceptable are shown in Figure 28.

3. After each cylinder is deglazed, clean the stones.

FIGURE 27
Proper honing will leave a cross-hatch pattern as shown here. (Courtesy of Dana Corporation)

ITEM	PHOTOGRAPH	EFFECT ON ENGINE PERFORMANCE	COMMON MAJOR CAUSES
1. Cross-hatch grooves irregularly spaced	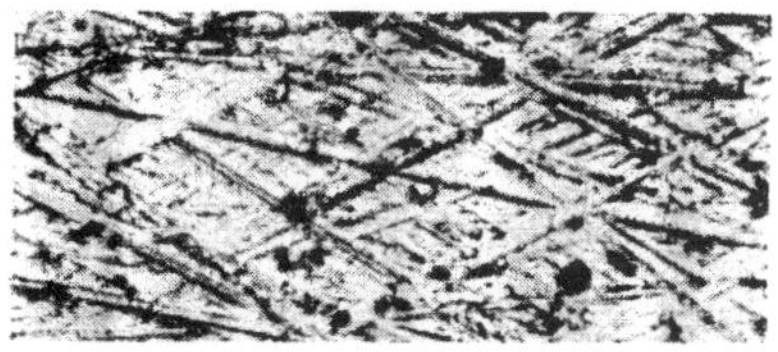	Poor oil distribution, erratic break-in, poor oil economy.	Varying reciprocation rate, poor stone breakdown with alternate loading and breakdown, stone grade too hard and/or grit too coarse.
2. Wide deep cross-hatch grooves		Causes abnormal wear, excessive oil consumption, poor and variable break-in period.	Stone grit too coarse, excessive stone pressure, poor stone breakdown, insufficient lubricant.
3. One directional cut	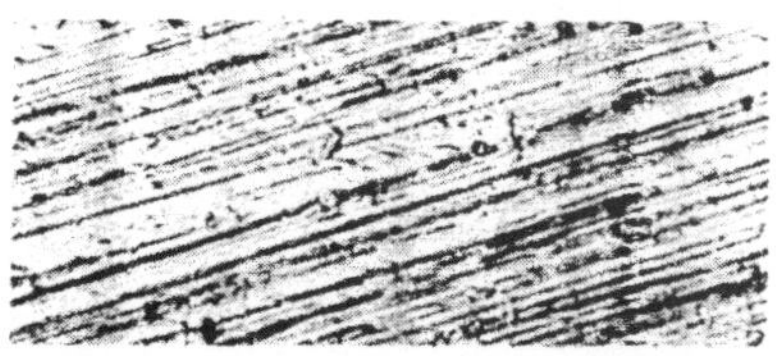	Causes rapid wear, poor seating, ring rotation, lowers oil control.	Excessive play in hone components.
4. Low cross-hatch angle		High impact forces causing excessive wear, poor oil distribution, slower break-in of rings.	Low reciprocation rate compared to rpm used.

FIGURE 28
Some common deviations from a good bore finish. (Courtesy of Muskegon Piston Ring Company)

If the cylinders are bored, the walls must be properly finished using a rigid hone (Figure 29[a]). At other times, boring cannot be justified economically, yet the cylinder walls have excessive taper and out-of-roundness. In such cases, the worn cylinders need to be trued up using a rigid hone, and the old pistons expanded by knurling.

Torque Plates

Torque plate honing is becoming a necessity with today's lightweight, thin-wall cast engines. By installing a plate (Figure 29[b]) that simulates the stresses the block will undergo with the heads installed, and then honing the cylinders, you will end up with cylinders that are round and free of "bulges" when the engine is assembled.

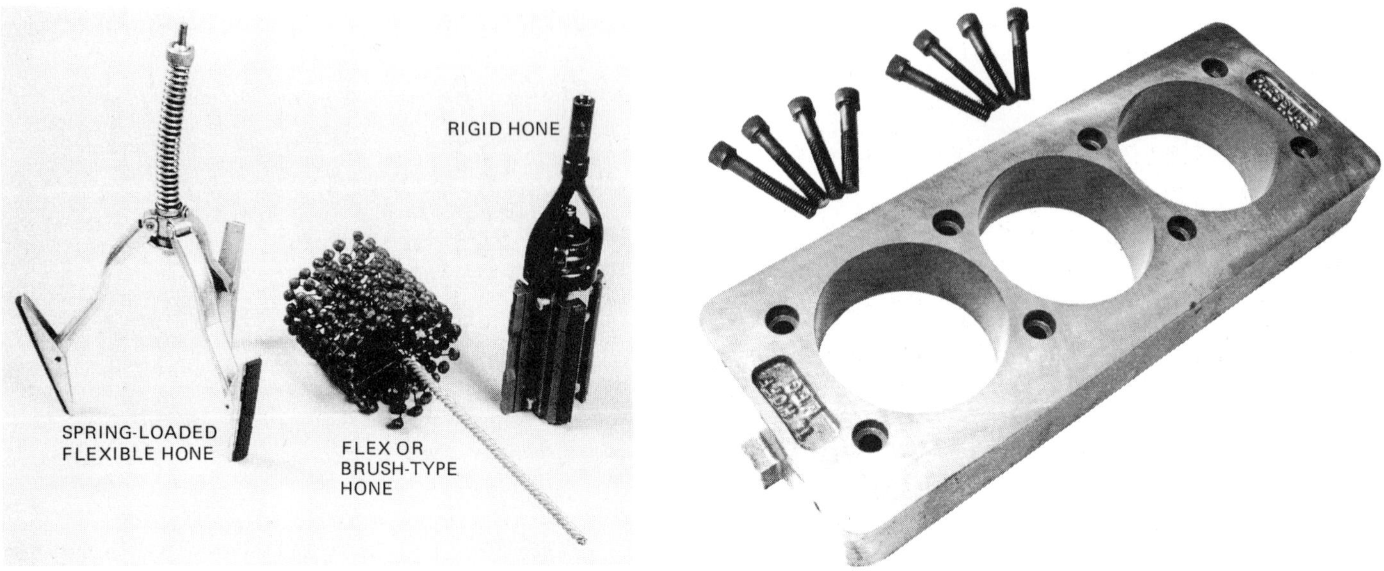

(a) (b)

FIGURE 29
(a) Various types of hones. (b) Torque plate and special attachment bolts. (Courtesy of T. Hoff Mfg. Company)

Cleaning the Cylinder Bores

After honing, *always* use soap and water to wash the cylinder bores (Figure 30). Continue washing until the soap and water no longer turns gray. This is the only positive way to remove the abrasive grit from the crosshatch crevices that result from honing.

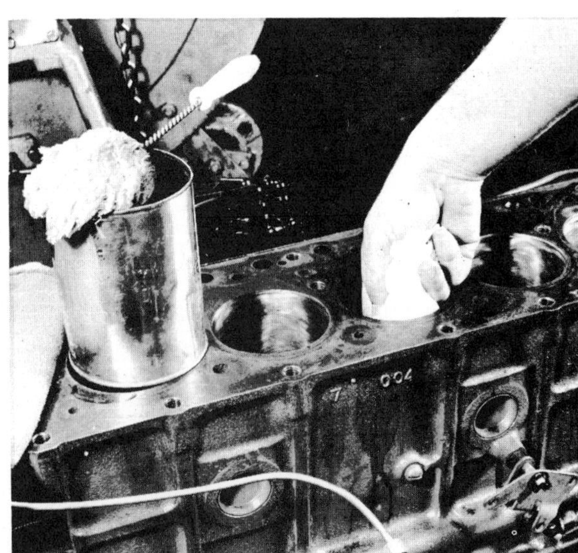

FIGURE 30
Hand washing cylinders with soap and water. (Courtesy of Dana Corporation)

CAMSHAFT BEARINGS

Camshaft bearings tend to wear out of round. Out-of-roundness may be determined by using a dial bore gauge (Figure 31).

***NOTE:** Cam bearing clearance of .003" or more will result in more than normal internal oil leakage and a drop in engine oil pressure.*

FIGURE 31
Using a dial bore indicator, check the diameter of each camshaft bearing bore. (Courtesy of Clevite Industries Inc.)

FIGURE 32
Installing a cam bearing. The bearing will be seated correctly when the driver mandrel is flush with the face of the block.

Camshaft bearings are generally press-fitted into housing bores in the block. The bearings are installed and removed with a special tool (Figure 32). The recommended basic procedure is as follows:

1. Clean the housing bores of rust, nicks, scratches, and scores.

2. Determine the correct bore location for each bearing.

3. Install the cam bearings dry (no lubricant). Drive in each bearing straight, and make sure that the oil holes are indexed (in line) with the oil holes in the block. See Figure 33.

4. After the cam bearings have been installed, check the alignment by installing the camshaft. If binding is encountered, check for a bent camshaft, an improperly installed cam bearing, or cam bearing bore misalignment.

Bearing high spots may be removed by carefully using a bearing scraper or a brush-type hone. In the case of severe cam bearing bore misalignment, the housing bores may be trued by boring and installing oversize outside diameter (OD) bearings, or by installing semi-finished bearings and machining to size.

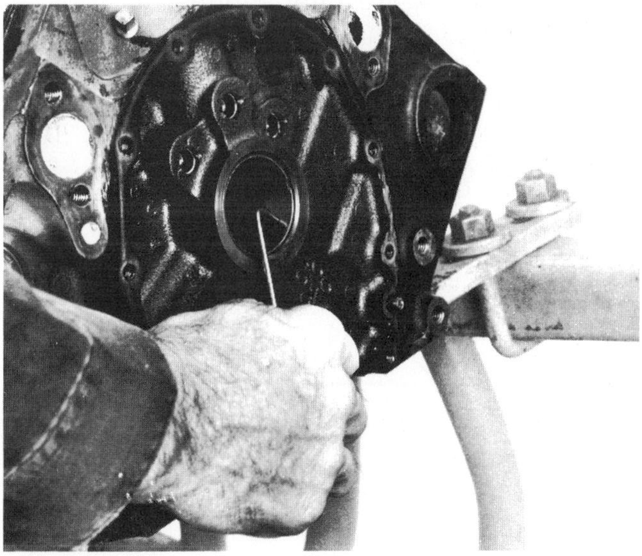

FIGURE 33
After installing new camshaft bearings, always check oil hole alignment. (Courtesy of Clevite Industries Inc.)

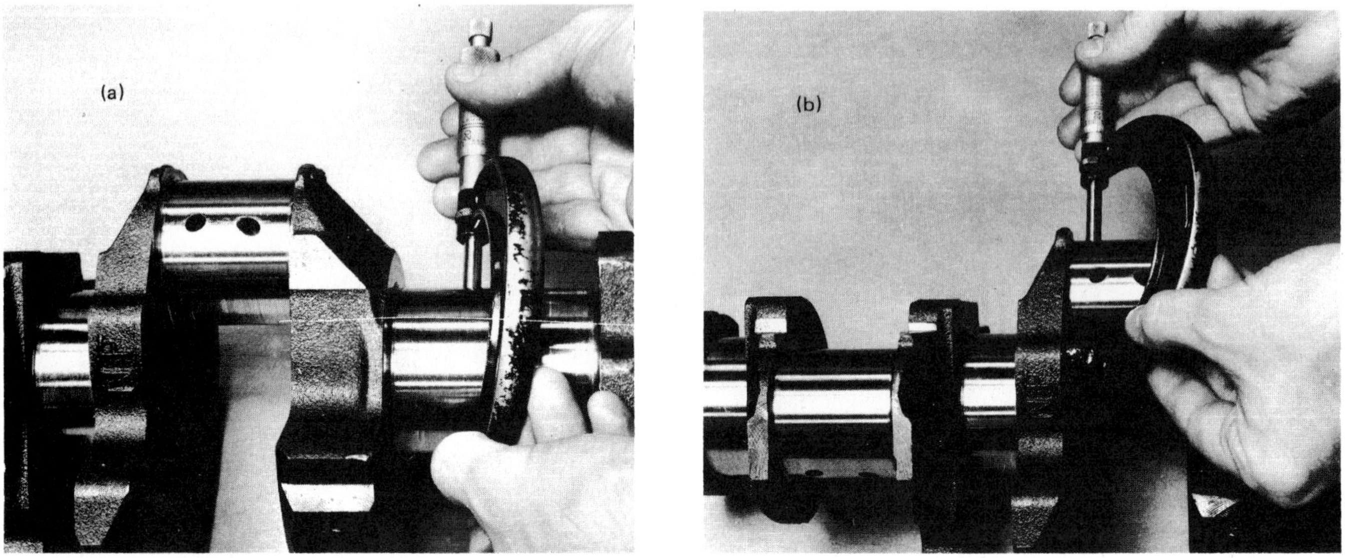

FIGURE 34
(a) Measuring a main bearing journal. (b) Measuring a connecting rod journal (crankpin).

CRANKSHAFT INSPECTION PROCEDURES

Each main bearing and connecting rod journal should be measured for taper, out-of-roundness, and diameter. Special gauges are available to measure the journals, or an outside micrometer may be used (Figure 34[a] and [b]).

Figure 35 shows which journal points to measure and demonstrates how to compare the measurements. If taper or out-of-roundness exceed specifications, or if the journal diameter is less than minimum, the journal must be refinished and an undersize bearing must be installed.

NOTE: An out-of-round or a tapered condition of .001" or more on any journal justifies regrinding.

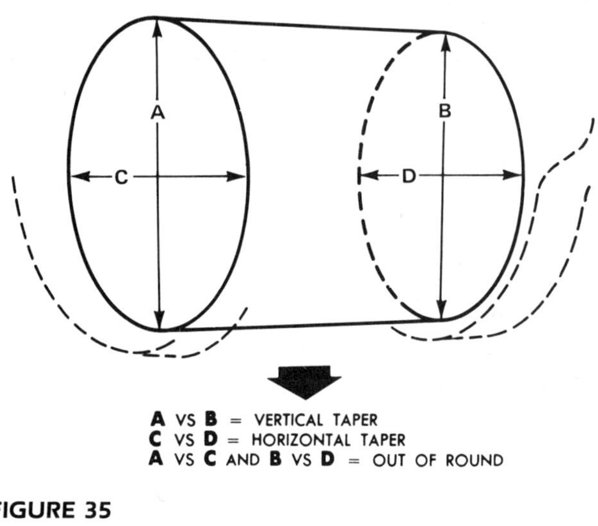

FIGURE 35
Crankshaft journal measurement points. Measure the diameter of each journal at four places to determine the wear, taper, and out-of-round condition.

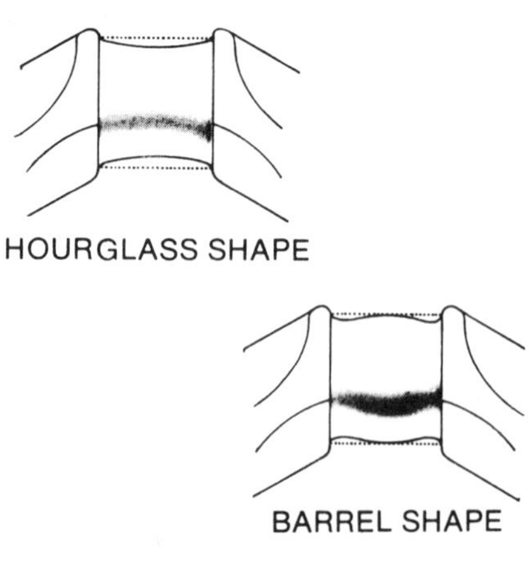

FIGURE 36
Two other crankshaft wear conditions that may be encountered. Out-of-shape journals will provide uneven oil clearance.

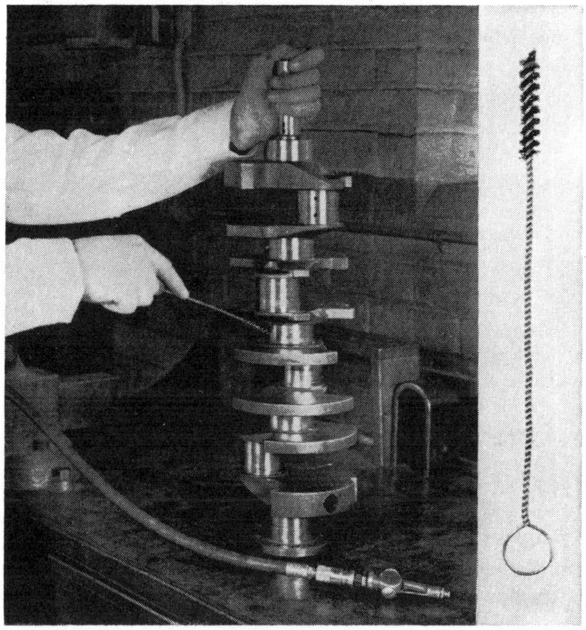

(a)

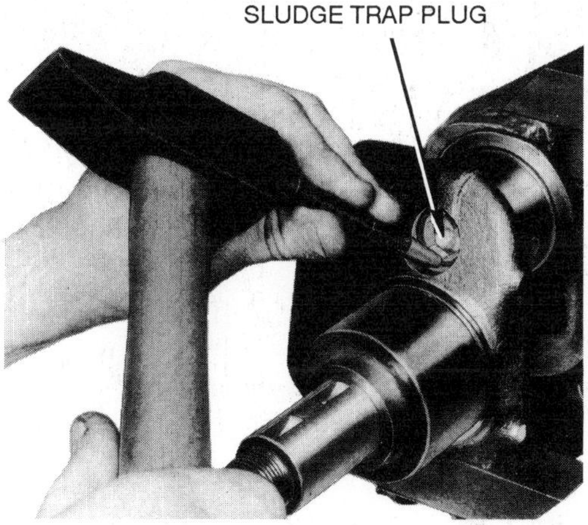

(b)

FIGURE 37
(a) Prior to installing a crankshaft, clean out all oil holes and check for restriction.
(b) Some crankshafts have sludge trap plugs. They can be removed in order to clean oil holes thoroughly. However, be sure to mark position when removing in order to maintain crankshaft balance.

Sometimes crankshaft journals show hourglass or barrel shaped wear (Figure 36). *Such wear should be limited to .001" maximum.*

The usual undersizes to which bearing journals are ground are .010″, .020″, or .030″. Bearings are commonly stocked in these undersizes to fit reground journals.

In addition to journal taper and out-of-roundness measurements, these areas of the crankshaft must be inspected:

1. **Oil passages.** All crankshaft oil passage holes must be clear. Hand brush with solvent and blow out with compressed air (Figure 37[a]).

NOTE: Some crankshafts have sludge trap plugs (Figure 37[b]). They can be removed in order to clean oil holes more thoroughly.

> **SHOP TIP**
> A rifle brush cleaning kit is an excellent tool for cleaning small oil passages.

2. **Crankshaft bolt end (snout) threads.** If there is damage, repair can be made by installing a Heli-Coil®.

3. **Snout keyway.** Torn keyways are usually the result of lower-than-specified vibration damper bolt torque.

NOTE: Always inspect the flat washer under the vibration damper bolt. If the washer is dished, replace it.

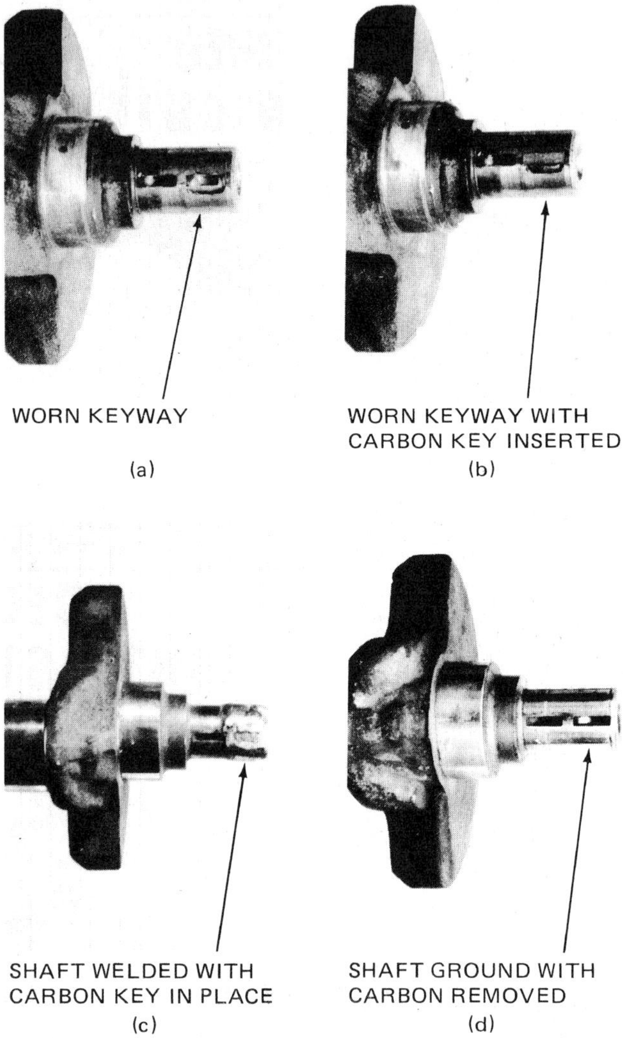

FIGURE 38
One method of reconditioning a crankshaft snout that has a worn keyway.

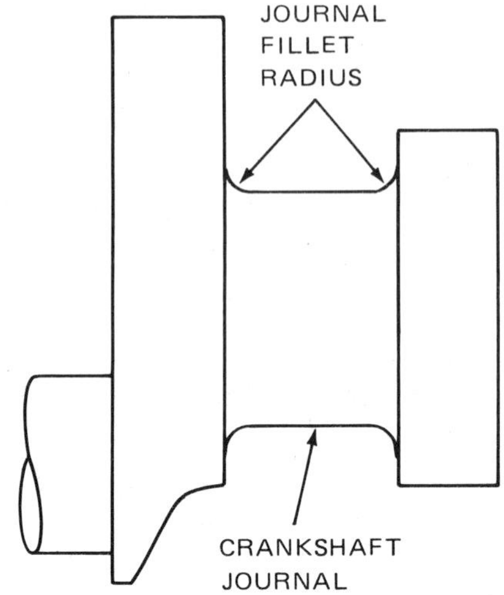

FIGURE 39
The fillets at the ends of a crankshaft journal must have a specified radius.

FIGURE 40
A magnetic fluorescent inspection booth was used to locate the cracks in the fillet area of this crankshaft. The white lines indicate cracking. (Courtesy of Magnaflux Corporation)

FIGURE 41
The nicks on this rod journal were caused by failure to use protective covers on the rod bolts when removing the piston assembly.

> **SHOP TIP**
> It is unnecessary to "junk" a good crankshaft due to a worn keyway, or because you are unable to have a new keyway cut. An "instant keyway" process can be performed (Figures 38[a], [b], [c], and [d]).

4. **Rear main seal surface.** To correct for wear, special sleeves or oversize seals are available for certain model engines.

5. **Fillet radii.** Check the fillet radii at the end of the rod and main bearing journals. Larger-than-specified fillet radii can cause bearing interference (Figure 39). Smaller-than-specified fillet radii tend to weaken the crankshaft. Check carefully for cracks in the fillet area also.

6. **Entire crankshaft surface.** Have the crankshaft magnetically inspected for cracks (Figure 40). Pay particular attention to the fillet radii, crank web, and crank cheek areas.

Visually inspect the journals. Nicks, dents, or pits in the journal OD must not exceed 1/8″ with no raised metal (Figure 41).

ROD AND MAIN BEARING SERVICE

The clearances between both main and rod bearings and their journals are critical (Figure 42). When the engine is disassembled, any bearings to be reused must be installed in the original location.

CHRYSLER CORP. (Continued) **ENGINE BEARINGS**

CATALOG INFORMATION				SHOP SPECIFICATIONS					
BORE & STROKE CUBIC DISP.	POSITION	NO. REQ'D.	STANDARD PART NO.	UNDERSIZES	STANDARD SHAFT SIZE	HOUSING BORE	OIL CLEARANCE	MAX. WALL THICKNESS	OVERALL LENGTH
L6				FORGED CRANKSHAFT 1976-60					
3.400x3.125	**ROD BRG**	6	**2375RA**	1-2-10-20-30-40-50-60	2.1865/2.1875	2.3125/2.3130	.0005/.0023	.0624	.990
170				CAST CRANKSHAFT 1980-76					
3.400x3.640	**ROD BRG**	6	**3695RA**	1-10-20-30	2.1865/2.1875	2.3125/2.3130	.0002/.0023	.0624	.720
198									
3.400x4.125				FORGED CRANKSHAFT 1976-60					
225	**MAIN SET**	1	**4092M**	1-2-10-20-30-40-50-60					
	1	1	2376RA	1-2-10-20-30-40-50-60	2.7495/2.7505	2.9425/2.9430	.0005/.0022	.0959	1.039
	3	1	▲2377RA	1-2-10-20-30-40-50-60	2.7495/2.7505	2.9425/2.9430	.0005/.0022	.0959	1.255
	2-4	2	2378RA	1-2-10-20-30-40-50-60	2.7495/2.7505	2.9425/2.9430	.0005/.0022	.0959	1.039
				CAST CRANKSHAFT 1980-76					
	MAIN SET	1	**5070M**	10-20-30					
	1-2-4	3	3696RA	10-20-30	2.7495/2.7505	2.9425/2.9430	.0002/.0023	.0959	.882
	3	1	▲3697RA	10-20-30	2.7495/2.7505	2.9425/2.9430	.0002/.0023	.0959	1.108
				NO CAMSHAFT BEARINGS USED IN ALUMINUM ENGINES					
18	**CAM SET**	1	**1454M**	STD-10, 1978-60					
	CAM SET	1	**1473M**	STD. ONLY 1980-79					

FIGURE 42
Sample section from a bearing catalog. Under SHOP SPECIFICATIONS, information is provided for the technician. This includes undersizes, standard shaft sizes, housing bores, oil clearance, maximum bearing wall thickness, and overall length. (Courtesy of Federal-Mogul Corporation)

FIGURE 43
A bearing worn and cut by excessive amounts of foreign particles in the oil.

Diagnosing Main and Connecting Rod Bearing Failure

The easiest way to determine if a main or connecting rod bearing has failed is to look at it. Whenever a bearing fails, the damage is generally caused by one of these conditions:

1. **Dirt.** Bearing material is soft enough that small dirt particles (metal, sand, plastic, and hard rubber) can become imbedded in it. The bearing material can "flow" slightly to maintain a normal clearance. However, if the dirt particle is too large, it will not imbed completely. Rather, it will be carried around by the crank journal and will scratch or score the bearing (Figure 43).

Always examine the bearing backs for correct fit and for possible dirt particle interference (Figure 44).

2. **Lack of oil.** A bearing overlay that is "wiped," indicates the bearing has been starved for oil (Figure 45).

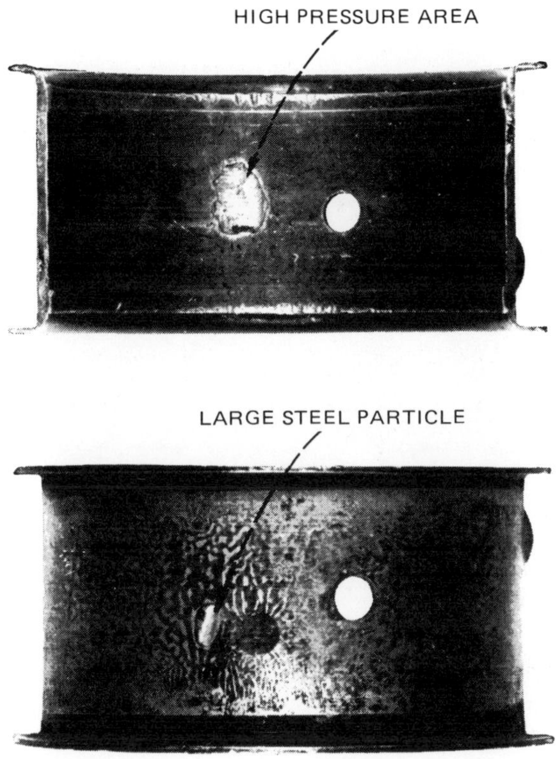

FIGURE 44
Bearing failure caused by a metallic particle on the bearing back. The high pressure area from the particle prevented the bearing from seating properly.

FIGURE 45
Oil supply failure.

Wiping is the result of metal-to-metal scrubbing action between a crank journal and its bearing. It can be caused by clogged oil lines or passages, a defective oil pump or pressure regulator valve, excessive heat, or insufficient oil clearance.

"Dry starts" are a prevalent cause of wiping. A dry start occurs when the lubrication system has not been primed with oil before starting the engine for the first time after an overhaul.

SHOP TIP

Internal parts on many engines can be prelubricated by turning the oil pump with a fabricated drive rod attachment (Figure 46). Seat the drive rod in the oil pump, and use an air ratchet or a 3/8″ or larger drill to rotate the drive rod in the proper direction.

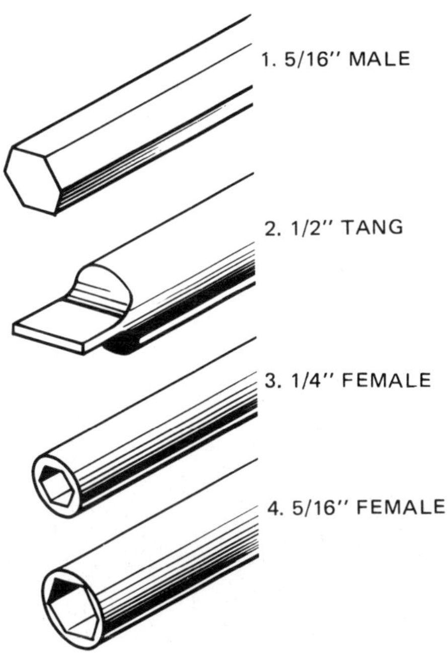

FIGURE 46
These oil pump drive rod attachments can be fabricated easily.

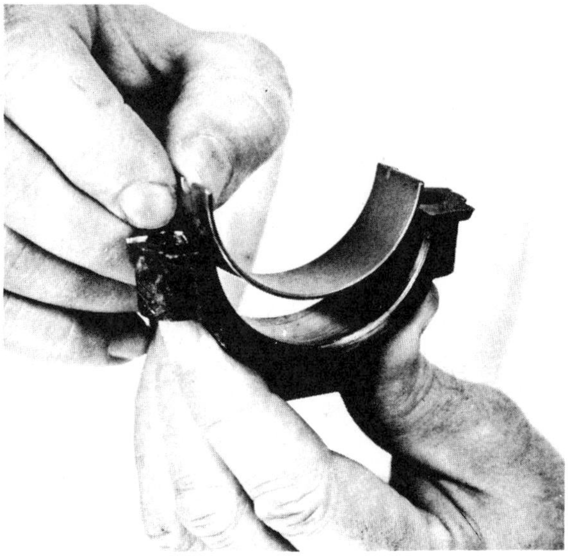

FIGURE 47
A bearing shell must have a positive fit against the total housing bore.

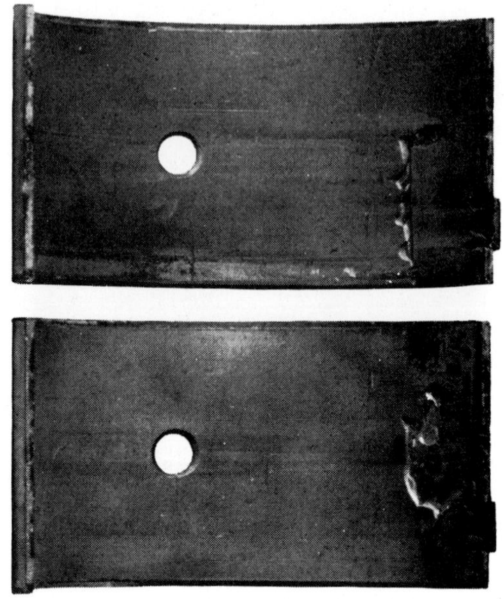

FIGURE 48
Excessive crush caused these bearing walls to buckle and bulge near the parting faces, resulting in the failure shown.

3. **Improper seating.** If the bearing is not seated properly in its bore (Figure 47), the result is localized "high spots." Clearance with the shaft is then inadequate at these high spots. These areas become brightly polished, while the rest of the bearing does not wear at all. High spots are caused by dirt under the bearing shell or by nicks in the block or connecting rod bores that were not cleaned up.

Crush prevents bearing movement in rod and main bearing housing bores. Excessive crush can cause bearing failure (Figure 48).

4. **Tapered journal.** Wear on one side of the bearing with none on the other side indicates the journal has too much taper (Figure 49).

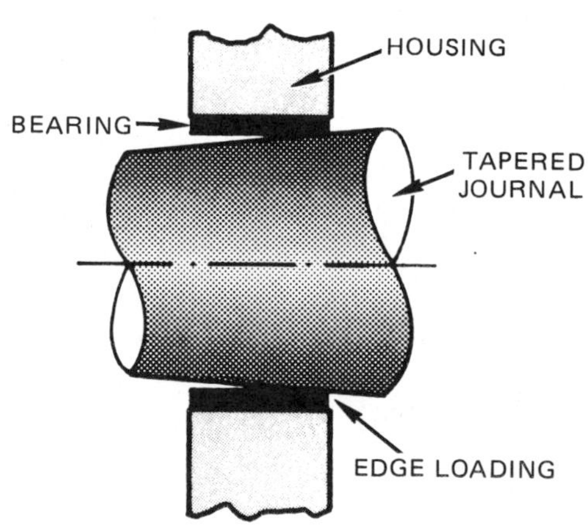

FIGURE 49
Bearing distress can be caused by a tapered journal.

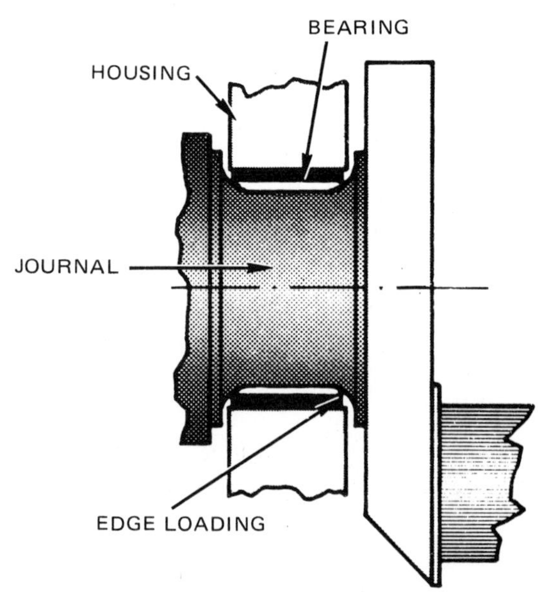

FIGURE 50
Bearing distress can be caused by radius ride.

FIGURE 51
Fatigue failure.

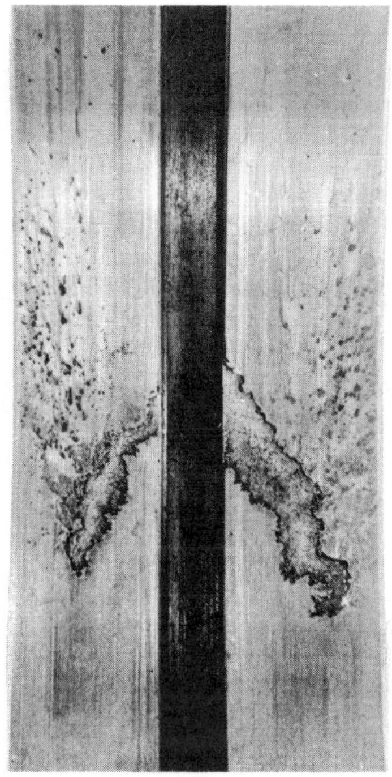

FIGURE 52
Severe discharge cavitation erosion damage to a lower main bearing.

COMMENT: A bent connecting rod causes a similar type of bearing failure, except that it occurs on opposite sides of the bearing halves.

5. **Radius ride.** This is a condition wherein the bearing rides the crankshaft fillet radius. The result is rapid wear at the bearing edge (Figure 50). Radius ride is most likely to occur after a crankshaft has been ground undersize and the shop has failed to properly "relieve" the radius.

6. **Fatigue failure.** Fatigue refers to stress or load on the bearing material, causing it to crack and break (Figure 51). The major cause of bearing fatigue is load concentration on one surface section rather than over the entire surface. This can be caused by bearing cap shift, lugging, "hot rodding," or wrong bearing material selection.

Some Other Bearing Wear Patterns

Refer to Figures 52, 53, and 54 for examples of some other bearing wear patterns that may be encountered.

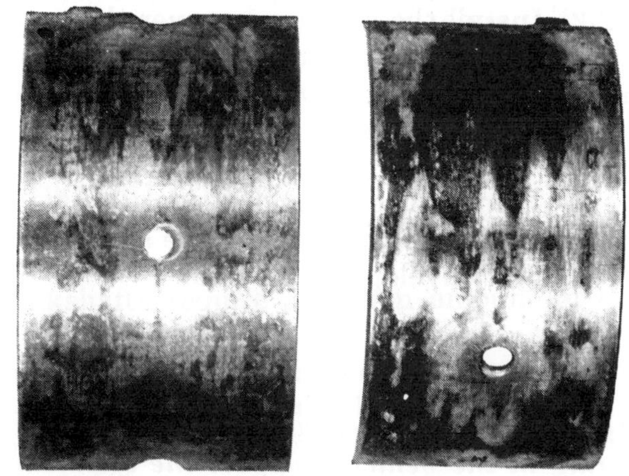

FIGURE 53
These bearings were loose in their seats. Note the heavy coating of black varnish (oxidized oil) on the bearing back surfaces.

1. Mount a dial indicator on the engine block. Position the indicator stem against the crankshaft flange.

2. Pry the crankshaft fully to the rear of the engine.

3. Set the dial indicator to zero.

4. Pry the crankshaft fully forward and read the end play on the dial indicator.

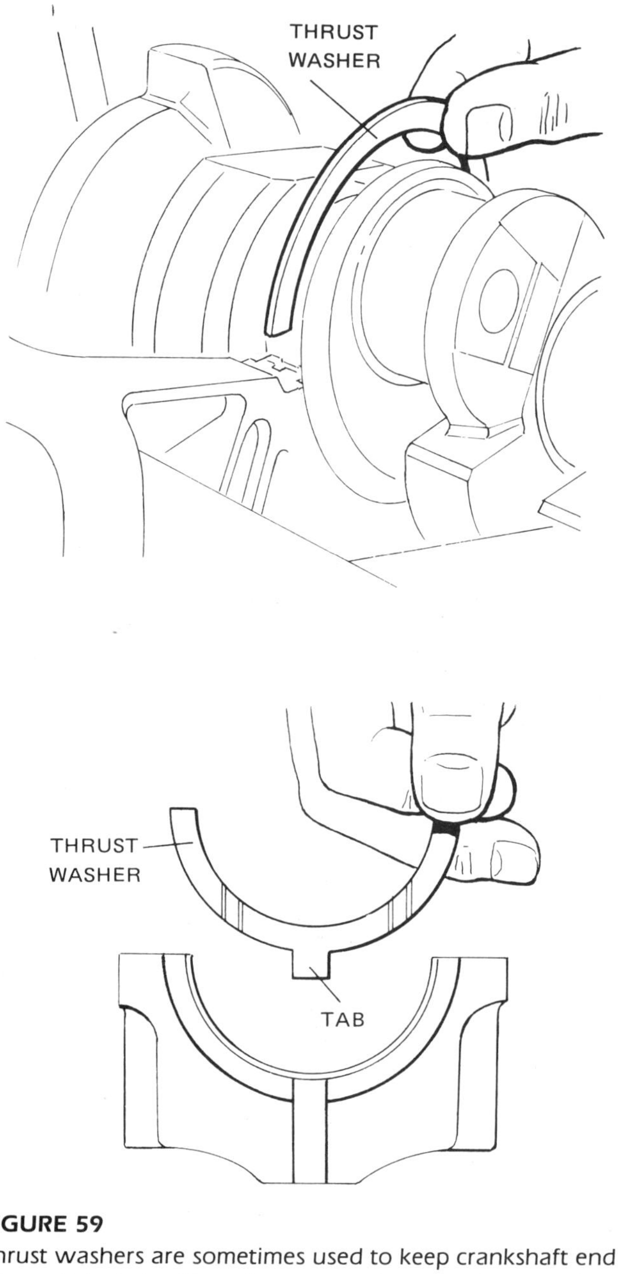

FIGURE 59
Thrust washers are sometimes used to keep crankshaft end play within limits. The tab prevents thrust washer rotation.

FIGURE 60
A feeler gauge is sometimes used to check crankshaft end play. It is preferable, however, to use a dial indicator.

If end play is too great, replace the thrust bearing (or washers). If end play is insufficient, check for raised metal or dirt on the thrust surfaces of the crankshaft and bearing. Also, check to be sure that the bearing is properly installed and that it is the correct one for the engine (Figure 61).

Seating the Thrust Bearing

A thrust bearing that is being installed must be seated (aligned) properly. To do this,

1. Set the thrust bearing and cap in place, leaving the bolts finger tight.

2. Pry the crankshaft forward against the thrust bearing surfaces (Figure 62).

3. Holding the crankshaft forward, tighten the main bearing cap bolts to the specified torque.

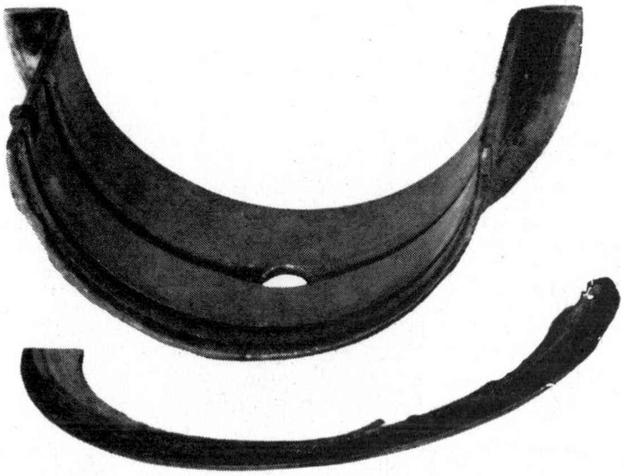

FIGURE 61
Using the wrong thrust bearing or "riding" the clutch pedal can cause this damage.

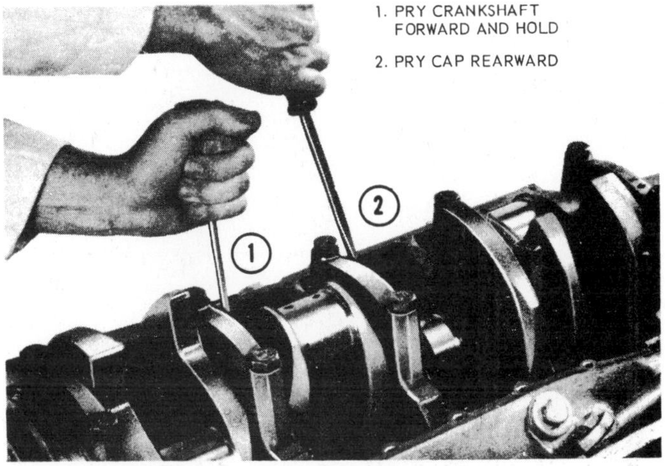

FIGURE 62
Seating the crankshaft thrust bearing.

PISTON AND RING SERVICE

Clean the pistons thoroughly using solvent and a soft bristle brush. Carefully inspect for fractures at the ring lands, skirts, and pin bosses (Figure 63).

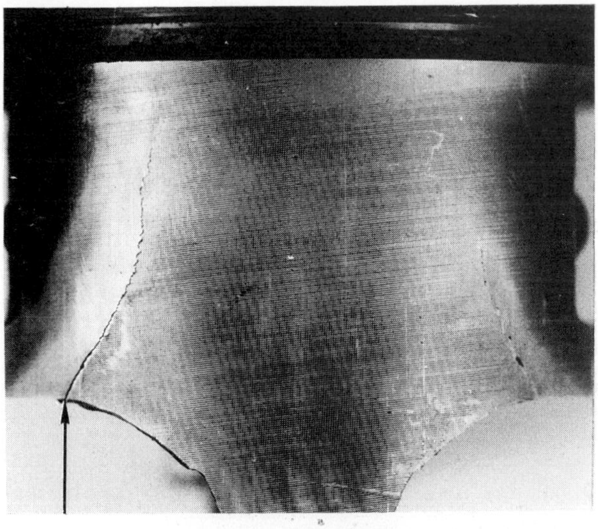

FIGURE 63
Cracked piston skirt.

NOTE: If wear on the piston skirt is uneven, the connecting rod may be bent and should be checked for alignment (Figure 64).

FIGURE 64
The diagonal wear pattern extending across the skirt of this piston indicates connecting rod misalignment. (Courtesy of Dana Corporation)

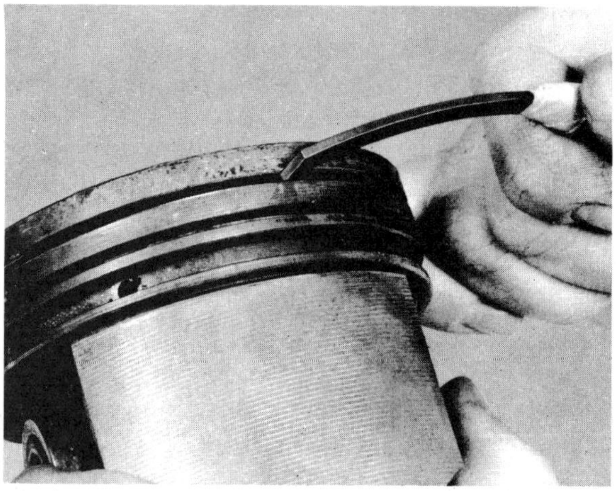

FIGURE 65
Piston ring grooves can be cleaned with a groove cleaner tool or the end of a broken ring filed to a sharp edge. Clean carefully; do not scratch or remove metal from groove sides. (Courtesy of Dana Corporation)

FIGURE 66
Cleaning oil drain holes in the piston. Do not enlarge the drain holes.

Cleaning Piston Ring Grooves

If the pistons pass inspection, clean the ring grooves with a special tool or the end of a broken ring (Figure 65). Probe and open all oil drain holes or slots in the piston (Figure 66).

Measuring Piston Diameter

Use an outside micrometer to measure piston diameter. *Piston diameter is generally checked at the centerline of the pin bore and at 90 degrees to the pin (Figure 67).*

Piston clearance is equal to cylinder diameter minus piston diameter. For example, if the cylinder bore measures 3.875″, and the piston measures 3.8725″, the clearance is .0025″. Compare this figure to the specified piston-to-cylinder bore clearance to determine if the piston clearance is proper.

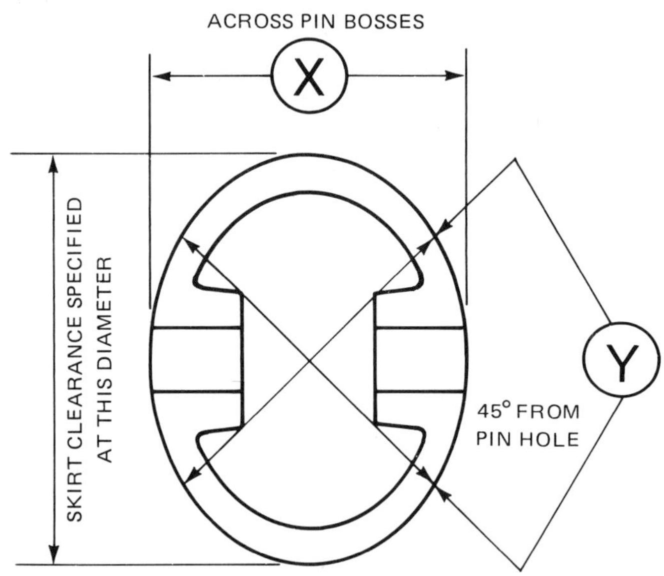

CAM	DIAMETER REDUCED AT X	DIAMETER REDUCED AT Y
A	.005 TO .007	.001 TO .002
B	.006 TO .008	.003 TO .004
C	.009 TO .011	.008 TO .010
D	.012 TO .014	.010 TO .012
E	.013 TO .015	.0065 TO .0085

FIGURE 67
Typical piston skirt diameter contours.

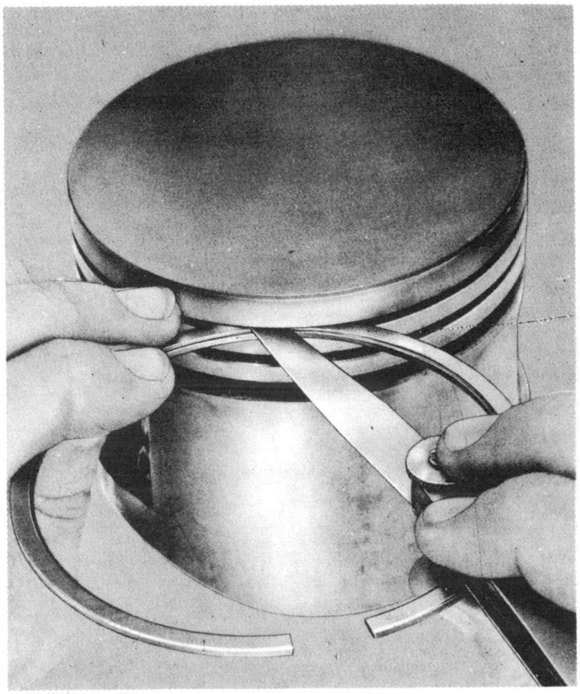

(a)

(b)

FIGURE 68
(a) Checking for excessive top groove wear with a .006" feeler gauge and a new ring.
(b) Top ring and piston groove side wear.

FIGURE 69
Using a long feeler gauge to check piston ring end gap.

Measuring Piston Ring Side Clearance

Always measure ring side clearance before "okaying" a piston for reuse (Figure 68[a] and [b]). Insert a new compression ring into the top ring groove. Use a feeler gauge as shown and check the clearance all around the groove. If the clearance is over .006", the piston should be replaced.

Measuring Piston Ring End Gap

If piston ring ends butt together, the ring may break or the cylinder wall may be scored. To be sure the rings have end-gap clearance, check them as follows:

CHART 1

RING GAP RECOMMENDATIONS

Ring Size	Gap Opening: Not Less Than	Gap Opening: Can Be As Much As
3-3-31/32″	.012″	.020″
4-5″	.015″	.023″
5-1/16-6″	.018″	.033″

1. Lubricate the ring face with engine oil.

2. Position the ring in the cylinder by using an inverted piston to push the ring to the lower limit of ring travel.

3. Measure the gap between the ends of the ring with a feeler gauge (Figure 69). Maximum end-gap clearances are shown in Chart 1. If the end-gap clearance is less than specified, try another ring set or file the ring ends (Figure 70).

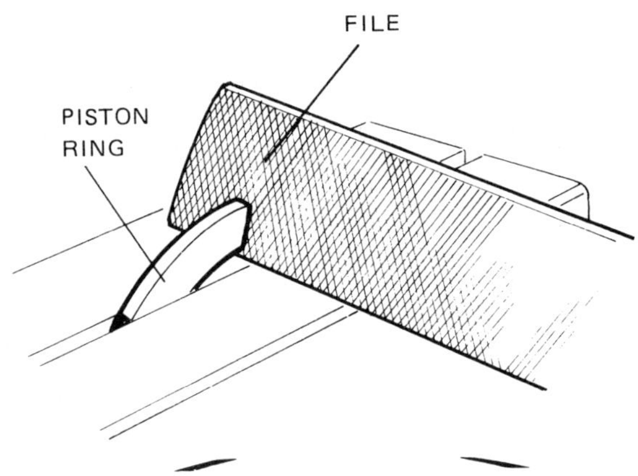

FIGURE 70
If ring end gap is too small, you can secure the ring in a soft-faced vise and remove metal by careful filing.

Installing Piston Rings

The oil ring should be installed on the piston before the compression rings are installed. Assemble as follows:

1. Wrap the expander-spacer around the ring groove so that the gap is at the rear of the piston in line with the pin (Figure 71). *Be sure the expander-spacer ends are butted together and not overlapped.*

2. Spiral the top rail into position (Figure 72[a]).

3. Spiral the bottom rail into position (Figure 72[b]).

4. Use a ring expander to assemble the compression rings (Figure 73). Observe the markings to be sure you install the rings right side up. Study Figure 74.

5. Stagger the ring ends in order to minimize loss of compression through the open gaps.

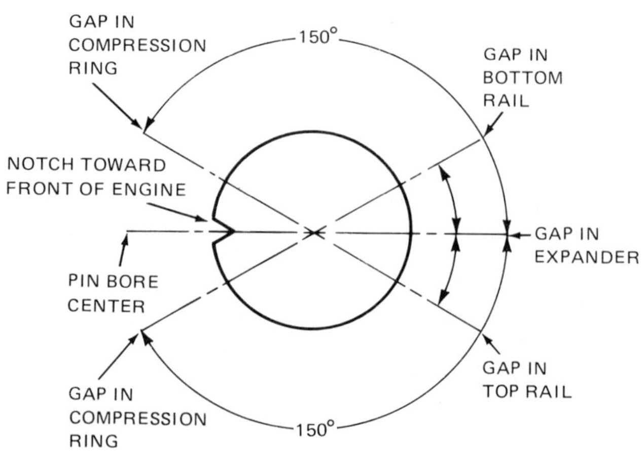

FIGURE 71
Recommended ring gap positions.

PISTON PIN AND CONNECTING ROD SERVICE

Piston pins on today's engines are either press-fitted into the connecting rod, or are fullfloating and held by retainers in the piston (Figure 75[a] and [b]).

Press-Fit Piston Pins

To properly remove a press-fit pin, special fixtures and a press are required (Figure 76[a] and [b]). Inspect the pin for scuffing and wear. Check the pin hole diameters in the rod and piston against specifications (Chart 2). If pin-to-piston clearance is excessive, or if there is not enough interference with the connecting rod, use a larger (oversize) pin. The pin hole surfaces in the piston and rod must be precision honed to fit the oversize pin.

When installing a press-fit pin, follow this procedure:

1. Apply a light coat of lubricant to the pin and mating surfaces of the piston and rod (Figure 77).

2. Position the connecting rod and piston in their proper assembled relationship. Refer to a shop manual for proper positioning (Figure 78).

3. Complete the installation with a press and the proper fixtures.

COMMENT: An alternate way to install press-fit pins is the pin furnace method (Figure 79).

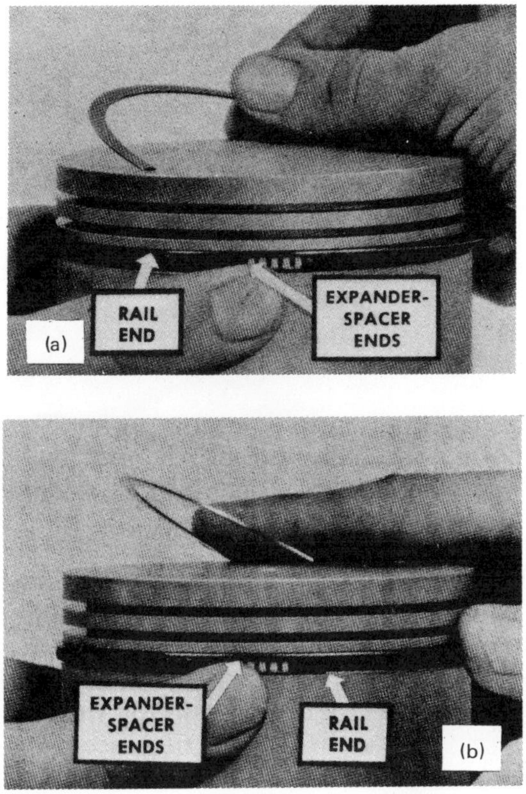

FIGURE 72
(a) When installing a set of piston rings, begin by placing the expander-spacer in the oil ring groove with the ends of the expander-spacer above the solid portion of the groove bottom. Keep the ends of the expander-spacer butted. Then install the steel rail on the top side of the expander-spacer. (b) Installing the remaining steel rail on the lower side of the expander-spacer.

Piston Pin Bushings

A free-floating pin generally has a bronze bushing pressed into the "eye" (small end) of the connecting rod. Repair is made by pressing a new bushing into place and then expanding it (Figure 80). The bushing is then trimmed with a cutter and honed to the proper size.

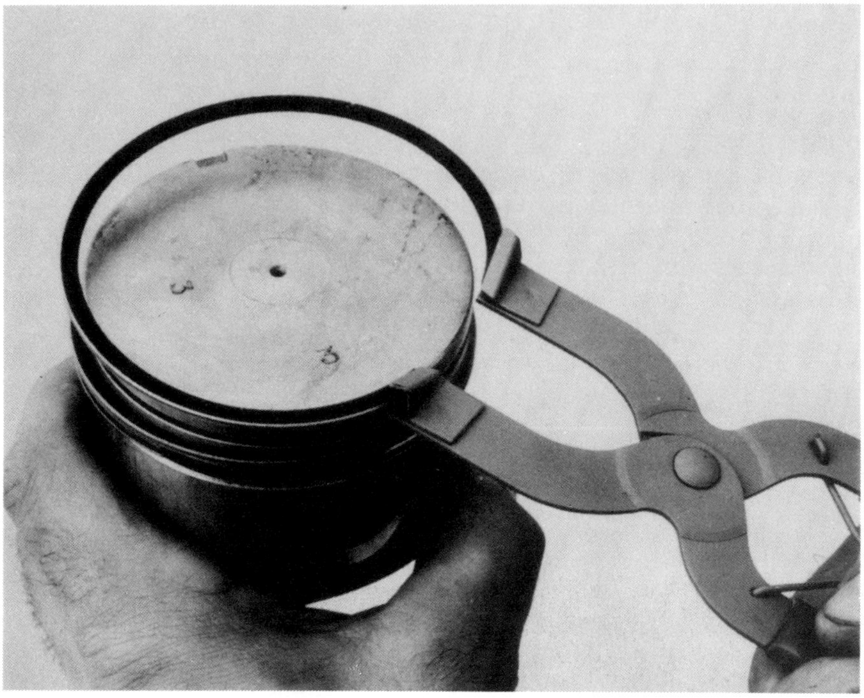

FIGURE 73
Using a ring expander. Never install by hand, or spiral or over-expand a compression ring. Doing so can prevent it from moving freely in and out of its groove. The result may be oil consumption and reduced compression.

CHART 2

Precision Pin Fits on Engines with $\frac{3}{4}$″ to $1\frac{1}{4}$″ Diameter Pins

Description	Aluminum Piston	Cast Iron Piston	Connecting Rods
Full floating	.0001 to .0003″ clearance	.0003 to .0005″ clearance	.0003 to .0005″ clearance (all pressure fed, .0005 to .0007″ clearance)
Oscillating in piston (no bushing)	.0003 to .0005″ clearance	.0006 to .0008″ clearance	clamped in rod
Oscillating in piston (press fit in rod)	.0003 to .0005″ clearance		−.0008 to −.0012″ press fit

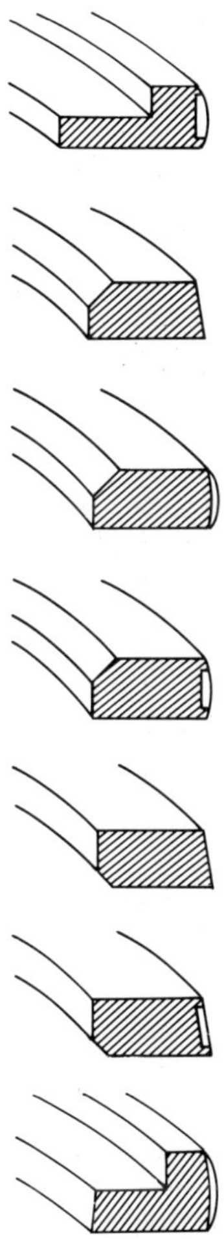

FIGURE 74
Various piston ring cross-section configurations. Always check for markings to be sure you are installing the rings right side up.

Connecting Rod Big-End Bore Measurement

Measure the big-end bore diameter for taper and out-of-roundness (Figure 81[a] and [b]). If either taper or out-of-roundness is greater than specified, recondition or replace the connecting rod.

Connecting Rod Alignment

Connecting rods are normally checked for bend and twist (Figure 82). Misaligned rods can cause abnormal stress on the rod bearings, rings, pins, and pistons (Figure 83).

Figure 84 shows a connecting rod being straightened (aligned) with a twisting fixture. Severely misaligned rods are usually replaced.

INSTALLING PISTON ASSEMBLIES

The following procedure is recommended for installing the piston assembly:

1. Position the crankshaft rod journal at BDC.

2. Slip rubber or plastic boots over the rod bolts. This prevents nicking of the crankshaft rod journal by the threads as the rod is installed (Figure 85).

3. Dip the piston head in oil so that all the rings and the pin are well lubricated.

4. Tighten a ring compressor around the rings (Figure 86). Be sure it is square and tight all around the piston.

5. Place the assembly atop the cylinder, with the rod and piston skirt in the bore. Be sure the piston is facing in the right direction. Tap the piston through the compressor and into the cylinder (Figure 87).

6. Plastigage® and lubricate the rod bearing halves.

7. After the rod is seated on the crankshaft, check to see that the upper bearing half is still in position. Install the rod cap with its bearing. Torque the rod nuts. For maximum

safety and performance, rod nuts should never be reused. Nuts are deformed each time they are reused. This causes increased assembly friction and decreased loading strength. Even though a torque wrench shows proper loading, the rod caps will actually be loose when the old nuts are used.

8. Rotate the engine and measure connecting rod side clearance (Figure 88).

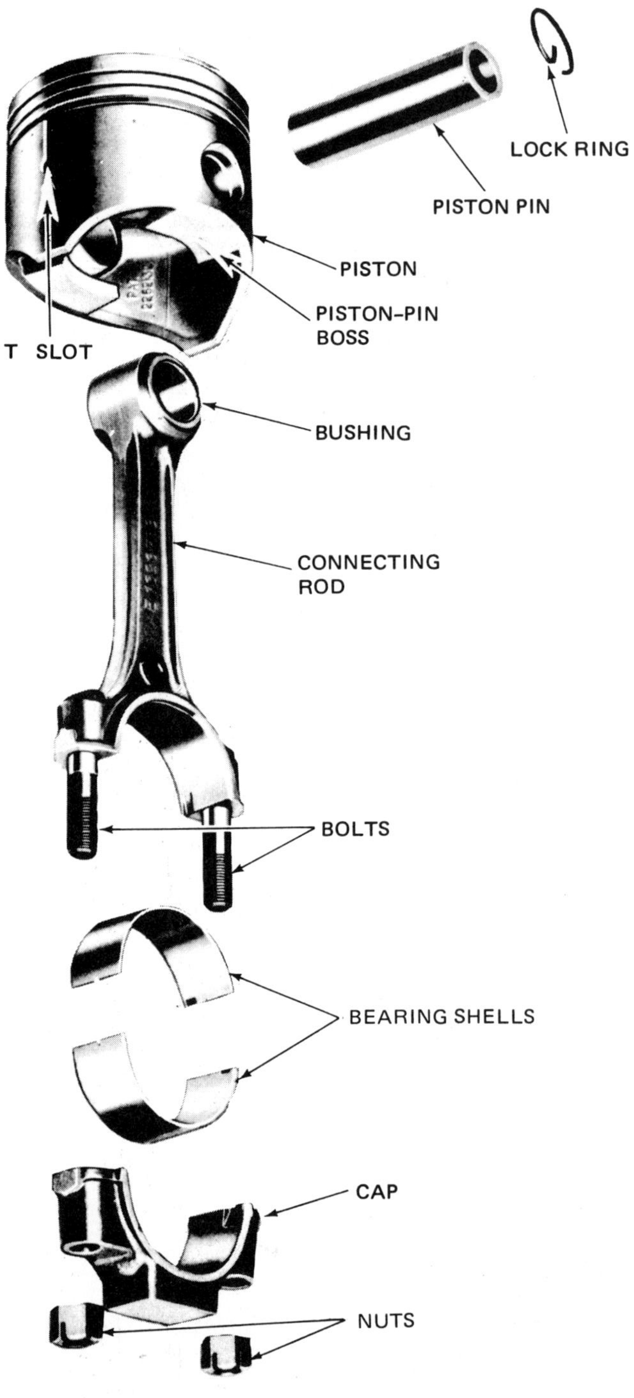

(a)

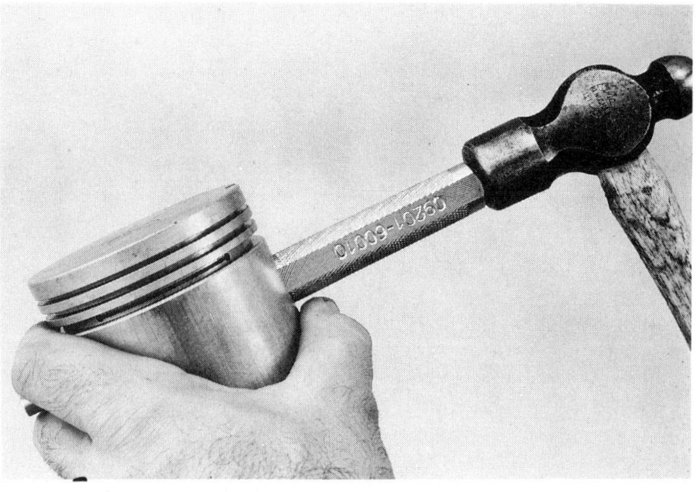

(b)

FIGURE 75
(a) Exploded view of a full-floating piston pin assembly. (Courtesy of Chrysler Corporation) (b) Tapping out a full-floating piston pin.

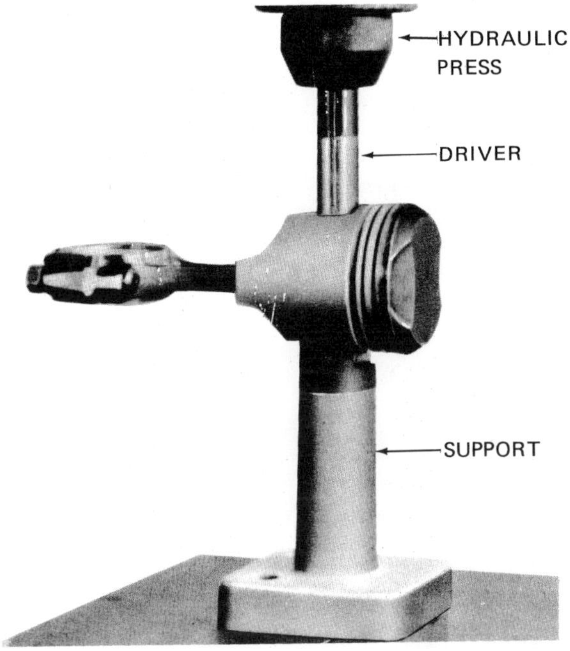

(a)

(b)

FIGURE 76
(a) Setup for removing a press-fit piston pin. (b) This damage resulted because improper press fixtures were used.

FIGURE 77
Special press-fit lubricant.

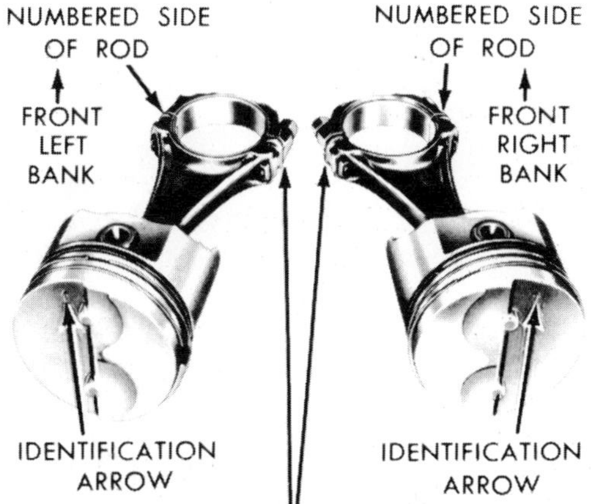

FIGURE 78
Factory identification marks are used to insure the correct assembling relationship among pistons, connecting rods, and caps. (Courtesy of Ford Parts and Service Division)

FIGURE 79
A commercially available gas-fired heater used to heat the rods for piston assembly. (Courtesy of ABS Products)

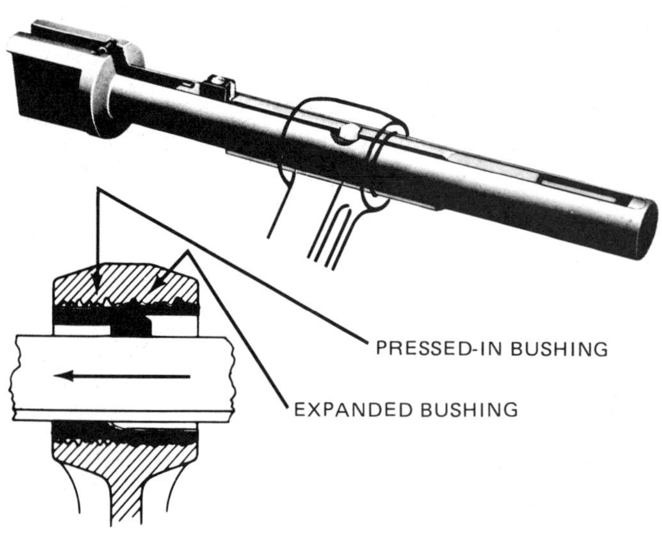

FIGURE 80
Expanding a pin bushing after installation. (Courtesy of Sunnen Products Company)

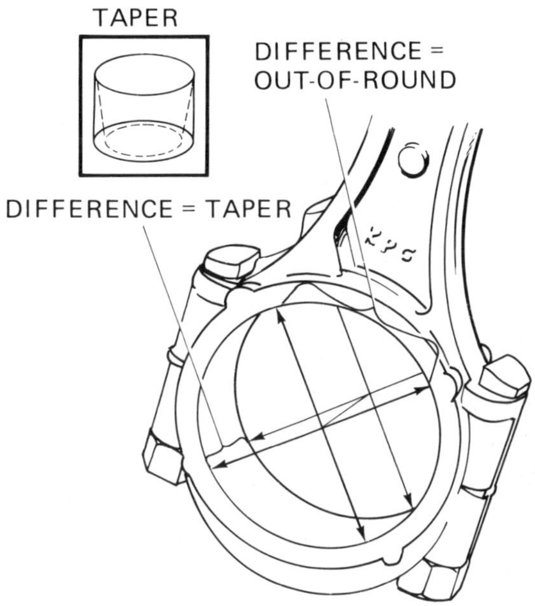

(a)

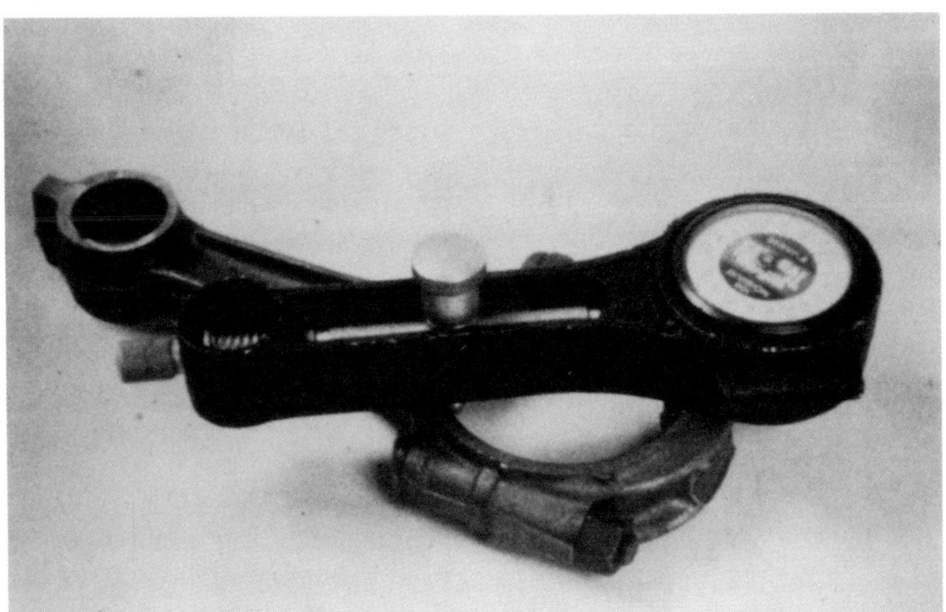

(b)

FIGURE 81
(a) Connecting rod big-end bore measurement points. The difference in the end diameters is the taper. In the cross diameters, the difference is the out-of-round. (Courtesy of Ford Parts and Service Division) (b) Measuring connecting rod big-end concentricity with a special dial gauge.

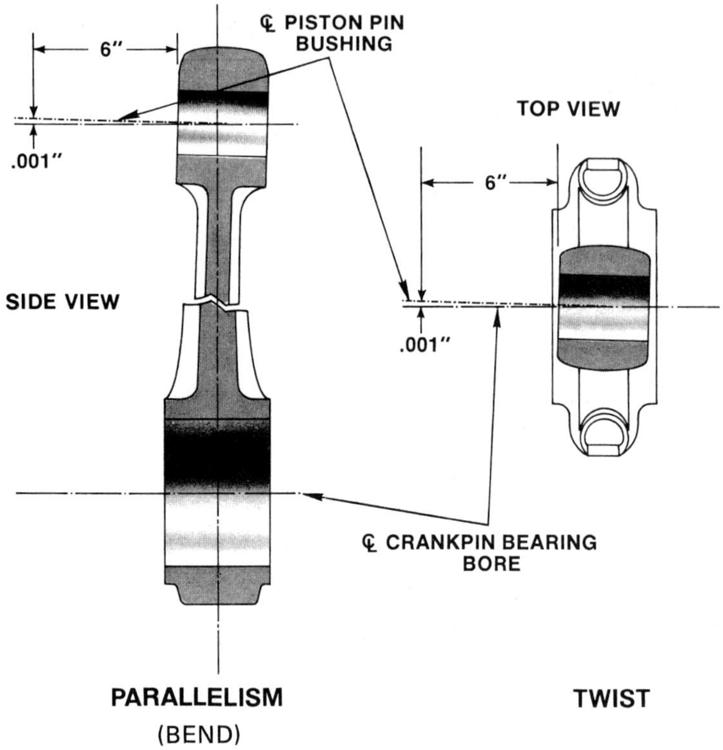

FIGURE 82
Connecting rod alignment specification limits. (Courtesy of Clevite Industries Inc.)

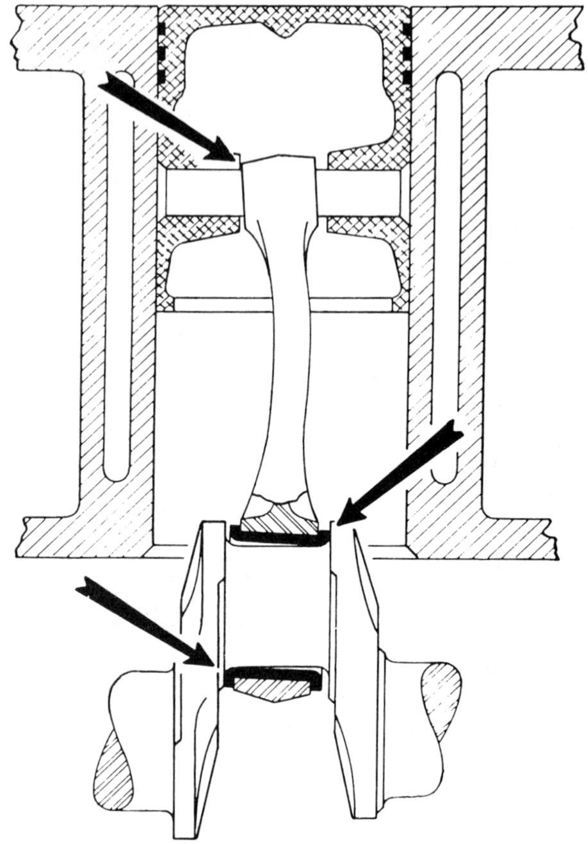

FIGURE 83
Pressure areas caused by a bent connecting rod. (Courtesy of Federal-Mogul Corporation)

SERVICING CRANKSHAFT VIBRATION DAMPER

The hub of the crankshaft vibration damper (harmonic balancer) on most engines extends into the front cover oil seal. A grooved hub allows the seal to leak (Figure 89).

Most crankshaft vibration dampers (harmonic balancers) can be removed by using a puller as shown in Figure 90. Once removed, check the condition of the rubber. If it has no spring when jabbed, is not completely intact, has pieces missing, or is severely cracked, purchase a new vibration damper. Remove the key from the crankshaft snout and check its fit in the vibration damper hub. If there is any keyway wear, replace the damper (Figure 91).

> **SHOP TIP**
> Vibration damper hub repair sleeves are available for a number of engines. They can eliminate an oil leakage problem caused by a grooved damper hub.

FLYWHEEL SERVICE

When the flywheel is exposed or removed in service, inspect it carefully. Check it for excessive face runout, dishing, hardspots, scoring, cracks, and ring gear wear (Figure 92).

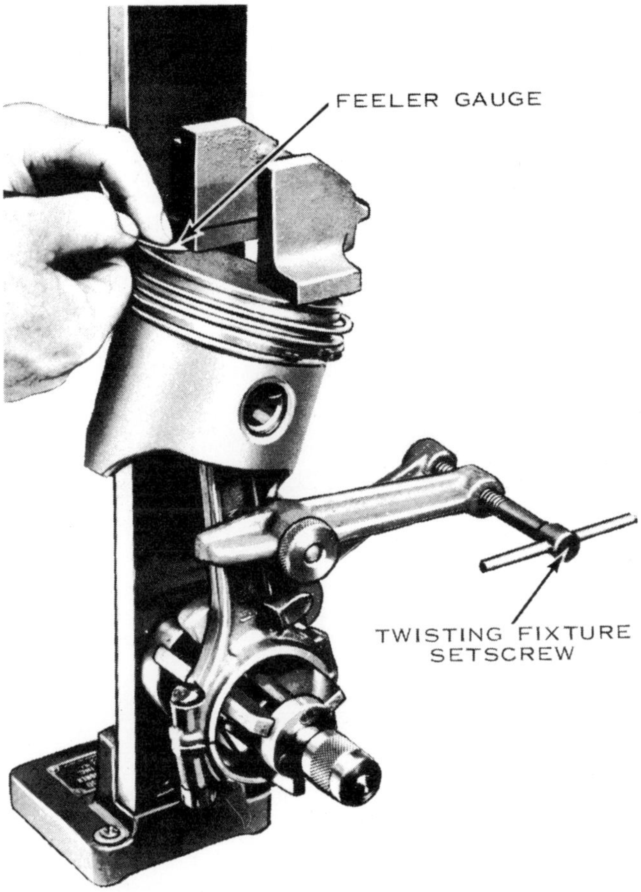

FIGURE 84
To check a connecting rod for twist, turn the piston as far as possible on its wrist pin, and then measure as shown. The twisting fixture can be used to straighten the rod.

FIGURE 85
Rod boots protect the crankshaft.

FIGURE 86
Use a good ring compressor tool to install the piston and ring assemblies in the block. Many ring jobs have been ruined by bad compressor tools. (Courtesy of Ford Parts and Service Division)

Measuring Flywheel Runout

Figure 93 shows a setup for measuring flywheel face runout. Runout in excess of factory specification can cause engine vibration, clutch chatter, and premature clutch assembly failure.

If runout is excessive, remove the flywheel and check for raised metal between it and the crankshaft mounting flange. Use a fine file to remove any nicks or burrs.

FIGURE 87
Tapping the piston assembly into its bore.

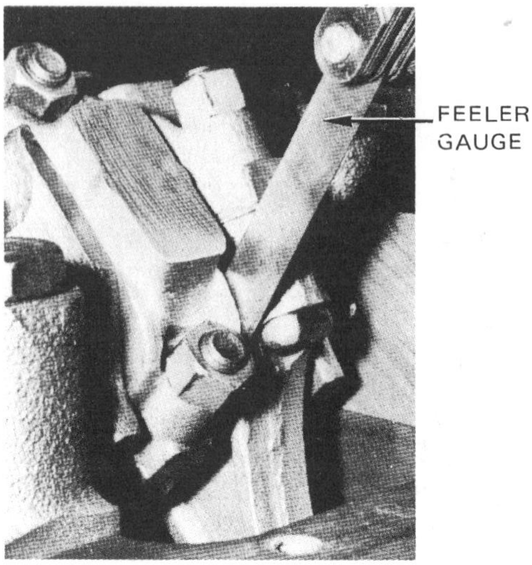

FIGURE 88
Checking connecting rod side clearance with a feeler gauge.

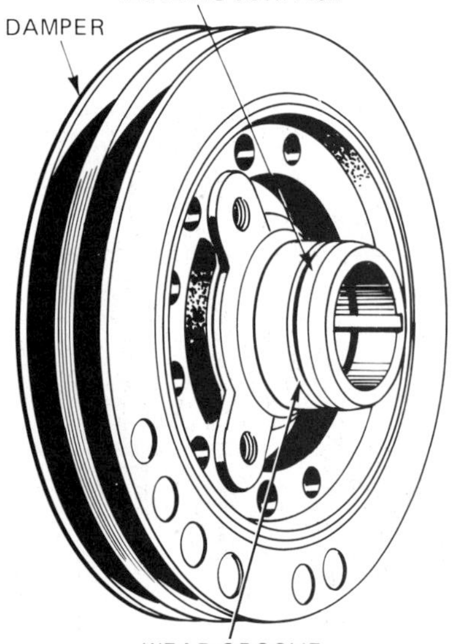

FIGURE 89
The front cover oil seal lip surface is shown on this vibration damper. If this surface becomes grooved, an oil leak can occur.

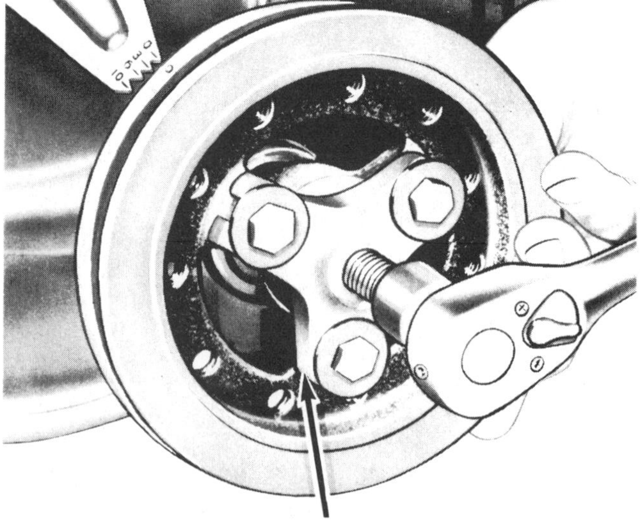

FIGURE 90
Pulling off a harmonic balancer. (Courtesy of Ford Parts and Service Division)

FIGURE 91
This vibration damper needs to be replaced.

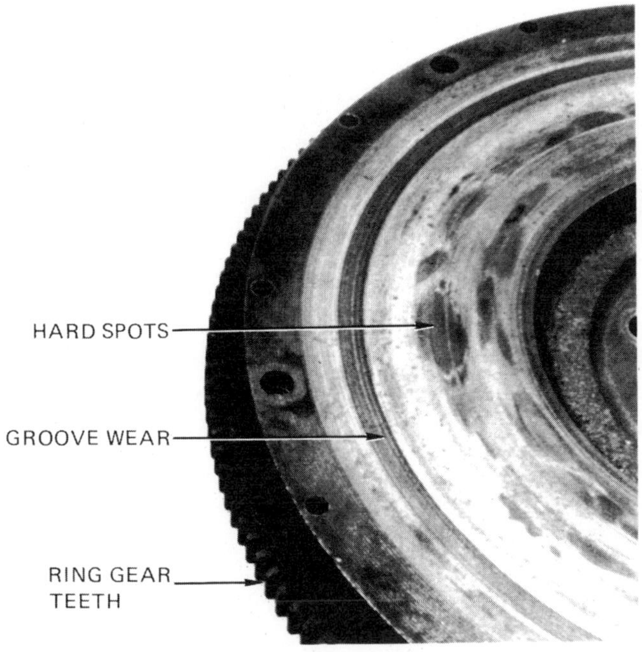

FIGURE 92
Badly grooved and hard spotted flywheel.

FIGURE 93
Checking flywheel face runout. (Courtesy of Ford Parts and Service Division)

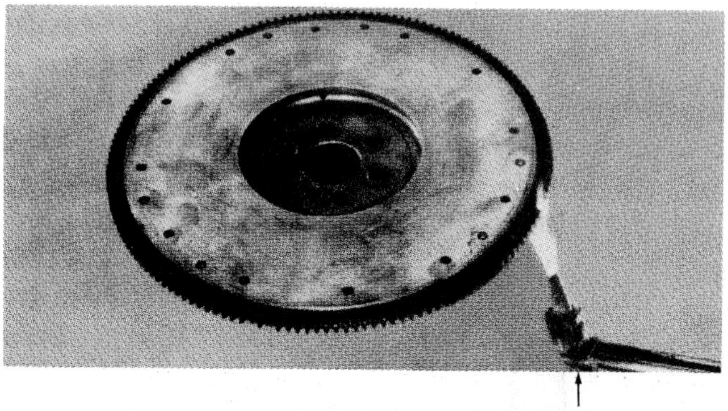

FIGURE 94
Removing ring gear. (Courtesy of Ford Parts and Service Division)

FIGURE 95
Pilot bearing (or bushing) bore in the end of a crankshaft.

If there are no nicks or burrs, check the runout of the mounting flange. If the flange is at fault, it can be "trued" in a machine shop. If the flywheel is at fault, have it resurfaced.

NOTE: Resurfacing operations should always "reference off" the crankshaft mounting flange of the flywheel.

Ring Gear Wear

Worn ring gear teeth means that weight has been lost. This can affect engine balance.

To remove a damaged ring gear, heat it with a torch (Figure 94). When the gear is thoroughly hot, knock it off with a hammer and punch.

CAUTION: Never heat a replacement ring gear hotter than 500°F, or the teeth will lose their hardness.

Pilot Bushing or Bearing

In engines used with manual transmissions, a pilot bushing or bearing is installed in a bore in the end of the crankshaft (Figure 95).

Replace the bushing or bearing if it is worn or loose in its bore (Figure 96). When installing a new bearing or bushing, first coat it with a small amount of antiseize lubricant. Then drive it in with a suitable driver.

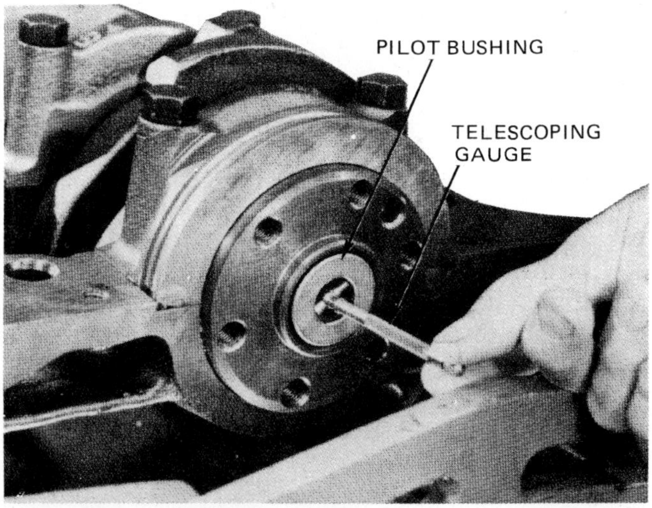

FIGURE 96
Measuring pilot bushing ID wear. (Courtesy of Ford Parts and Service Division)

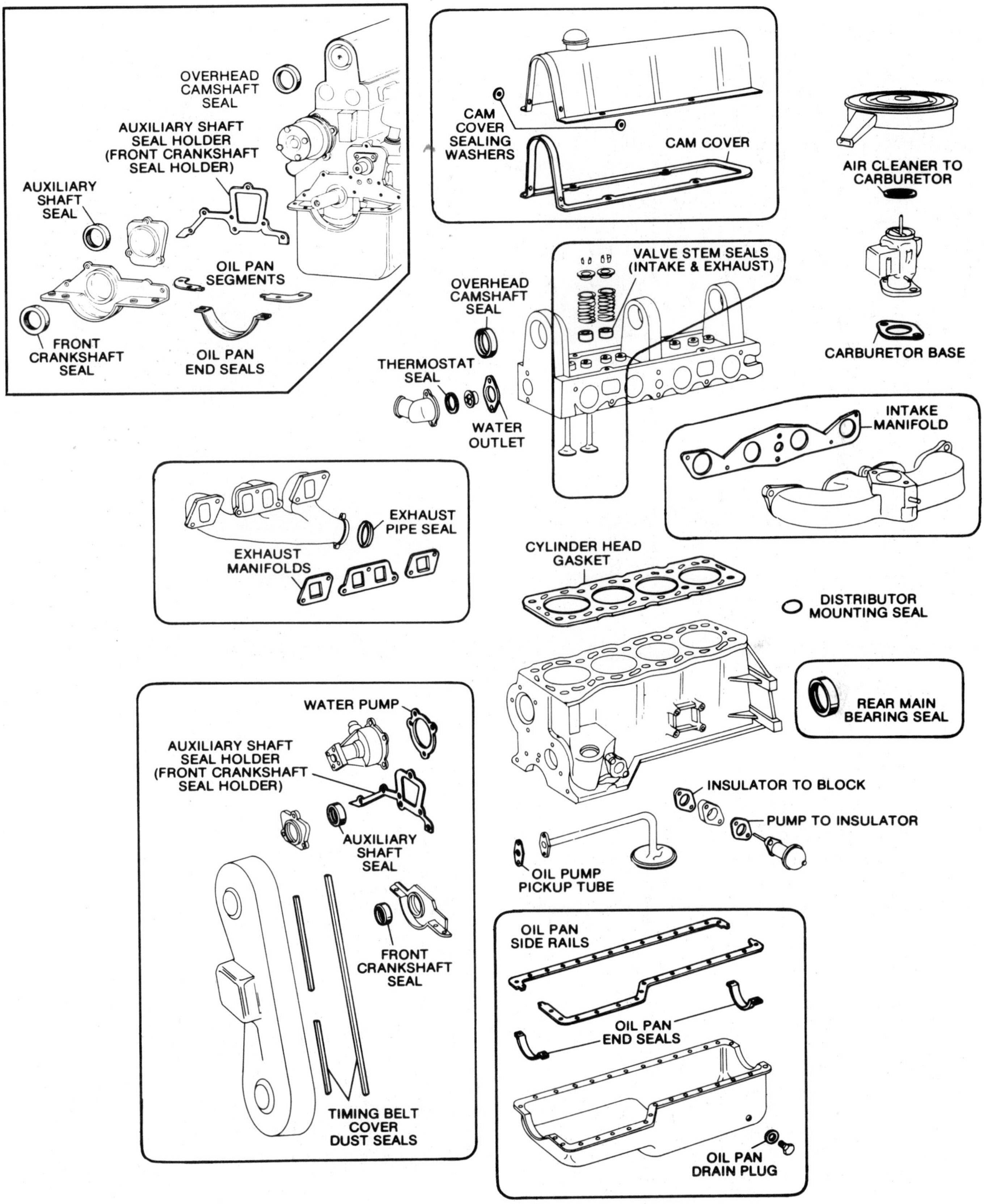

FIGURE 97

Pinto OHC engine. (Courtesy of Ford Parts and Service Division)

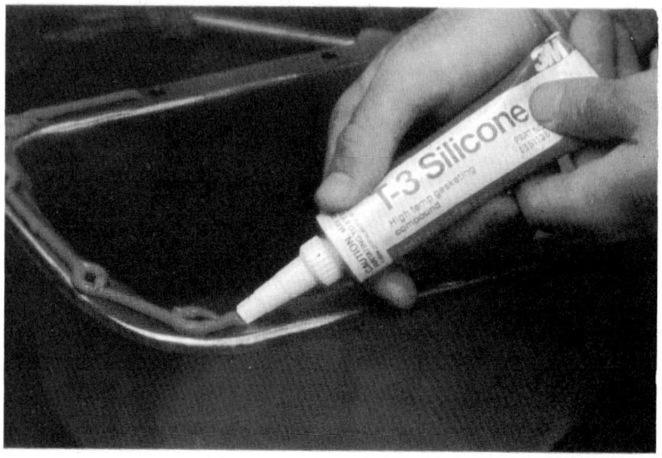

FIGURE 98
RTV gasket compound. (Courtesy of Automotive Trades Division/3M)

FIGURE 99
Anaerobic adhesive/sealant. (Courtesy of Automotive Trades Division/3M)

> **SHOP TIP**
> You can usually remove a pilot bushing by packing the ID with wheel bearing grease. Then select a bolt that just fits the bushing ID. Place the bolt into the grease, and hit the bolt head with a hammer. The bushing will be forced out hydraulically.

GASKETS AND SEALS

Figure 97 shows a variety of the engine gaskets and seals for a typical four-cylinder OHC engine.

RTV Silicone Rubber

RTV (room temperature vulcanizing) silicone rubber is often used for formed-in-place gaskets. RTV usually comes in a tube and is applied like toothpaste (Figure 98). The humidity in the air acts as a curing agent for this material.

Anaerobic Compound

This formed-in-place gasket material cures in a manner opposite to that of RTV. An anaerobic adhesive/sealant is pictured in Figure 99.

Anaerobic gasket material cures when confined under pressure between rigid metal parts. Loctite Corporation originally developed anaerobic material for securing threaded fasteners.

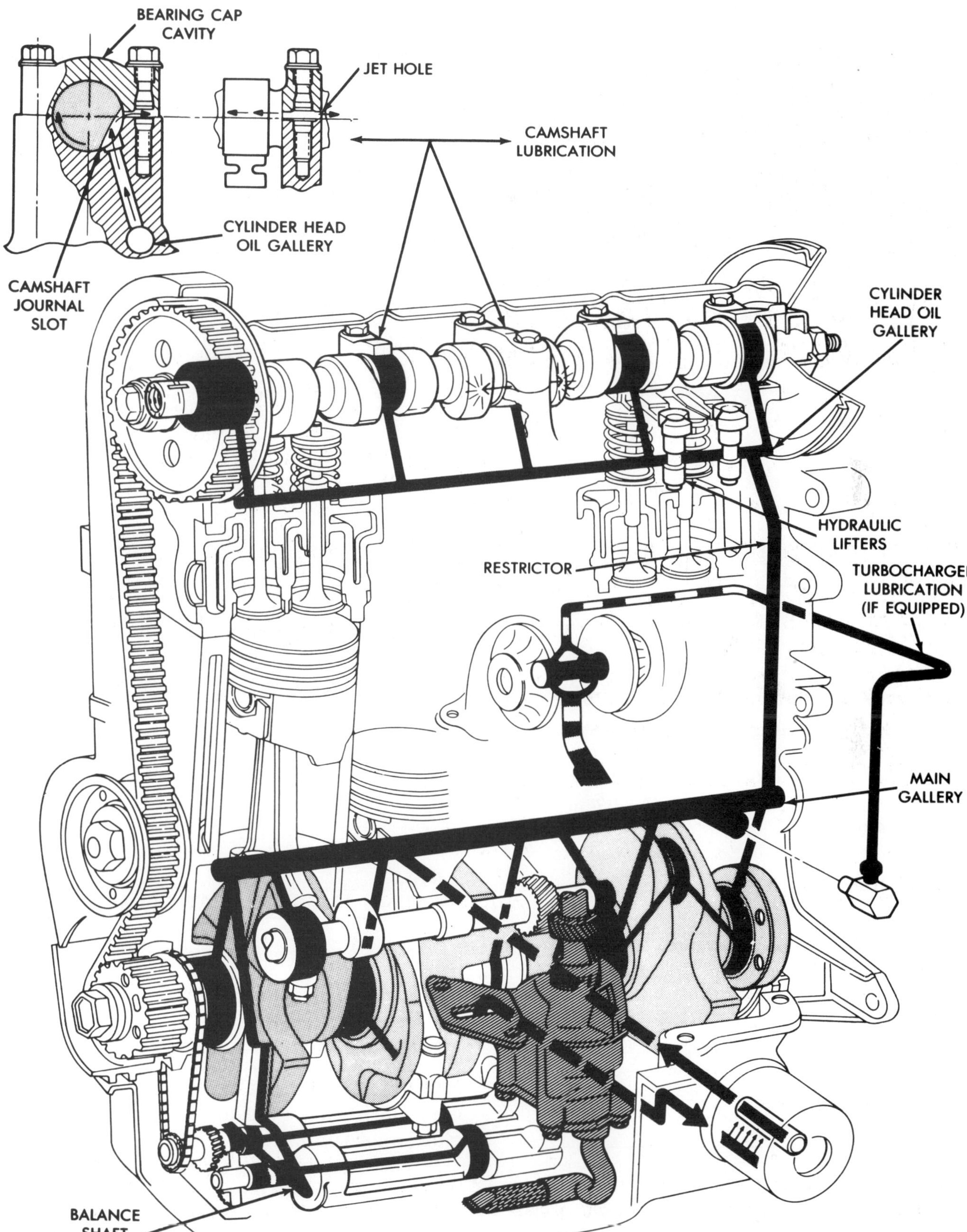

FIGURE 100

Engine lubrication system for the Chrysler 2.2/2.5 liter engine. Notice the two balance shafts and their locations. (Courtesy of Chrysler Corporation)

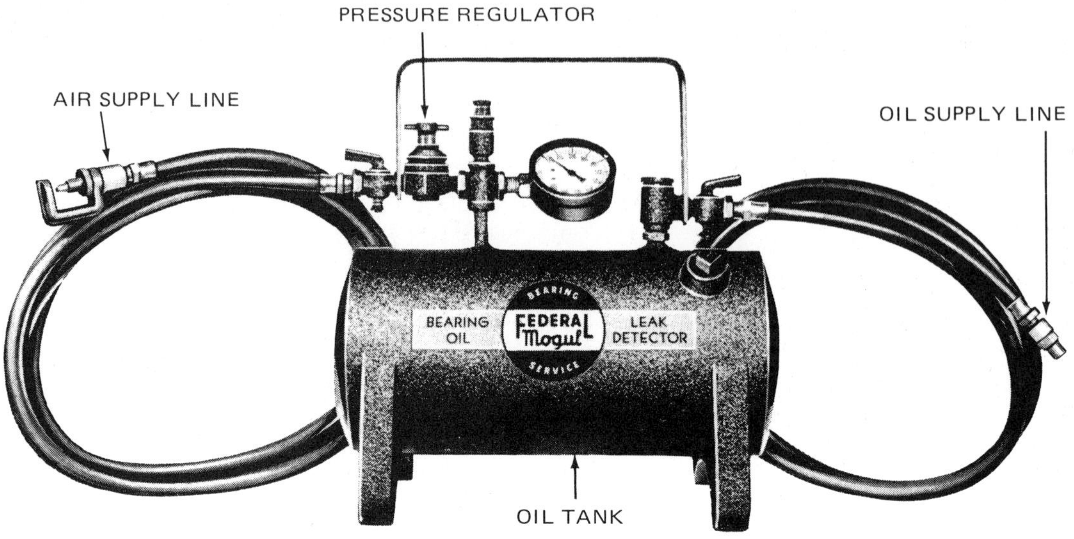

FIGURE 101
Engine prelubrication. (Courtesy of Federal-Mogul Corporation)

WARNING: Certain brands of RTVs "poison" an oxygen sensor. The fumes given off during curing cause a powdery white coating to form on the probe end, making it inoperative. Always read the product label to see if the RTV meets specifications for use in oxygen-sensor-equipped engines.

AUXILIARY SHAFTS

Some engines are equipped with balance shafts installed in a carrier attached to the crankcase. The shafts are interconnected to rotate in opposite directions (Figure 100). This serves to counterbalance the engine reciprocating masses.

Installation

During installation, crankshaft-to-balance shaft timing must be established. This is generally done by lining up alignment dots.

PRIMING THE LUBRICATION SYSTEM

An engine prelubrication tank is shown in Figure 101. The tank is filled with oil and attached to the air supply line. The oil supply line is screwed into an oil pressure port on the outside of the block. Oil can then be supplied to the engine lubrication system, even though the engine is not running.

The engine prelubrication tank allows an overhauled engine to be primed with oil prior to being started. It is extremely important to avoid "dry starts." The time lag between first starting the engine and oil being delivered to the bearings can be, in some cases, more than a minute.

TRUE CASE HISTORY

Vehicle	Triumph Spitfire
Engine	Four-cylinder, 1300 cubic centimeters
Mileage	Unknown
Problem	After repeated clutch assembly replacements, the web would rip out of the clutch without warning and for no obvious reason.
Cause	Excessive crankshaft end play created by worn out thrust bearings. The worn bearings allowed the flywheel back-and-forth shift to break the clutch hub from the disc.

TRUE CASE HISTORY

Vehicle	Pontiac Grand Am equipped with TBI
Engine	Four-cylinder, 2.5 liter
Mileage	35,371
Problem	1. Engine backfire, missing, and rough idle. 2. Shop replaced spark plugs, exhaust valves, oxygen sensor, coil pacs, ignition module, and fuel injector. None of this cured the problem.
Cause	Excessive crankshaft end play was interrupting the crank sensor signal.
Cure	Replaced the #5 crankshaft (thrust) bearing.

ASE STUDY QUESTIONS

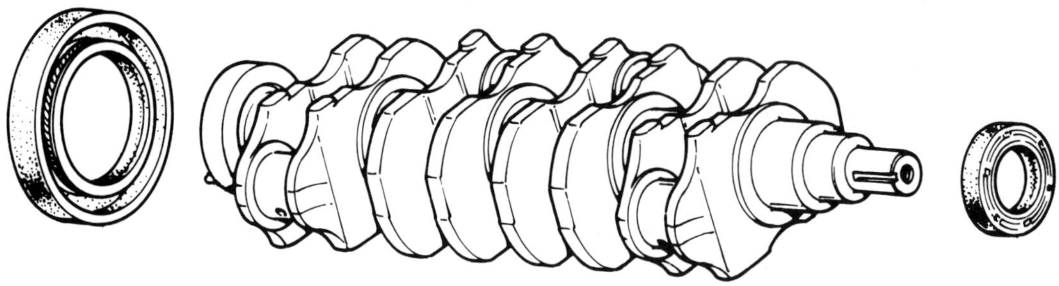

1. **Technician A** says that the oil seals shown above are positioned backwards.

   Technician B says that the oil seals shown above should be installed dry.

   Who is right?

   a. A only
   b. B only
   c. Both A and B
   d. Neither A nor B

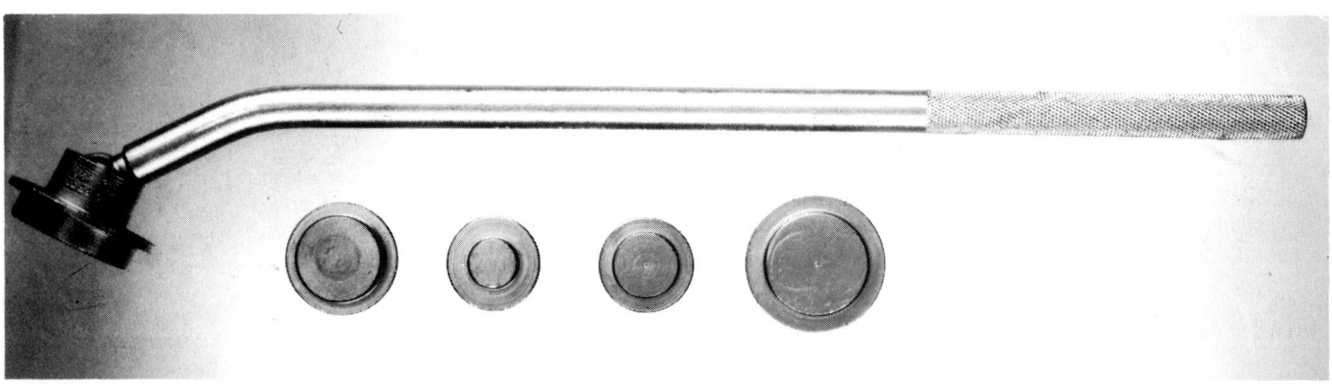

2. The tool pictured above is used for installing

   a. Cam bearings
   b. Core plugs
   c. Pilot bushings
   d. Oil seals

3. **Technician A** says that the thread damage shown above can be caused by over torquing.

   Technician B says that the thread damage shown above can be caused by cross-threading.

   Who is right?

   a. A only
   b. B only
   c. Both A and B
   d. Neither A nor B

4. The tool shown above is being used to

   a. Remove the cylinder ridge
   b. Bore the cylinder
   c. Pull the cylinder sleeve
   d. Install the cylinder liner

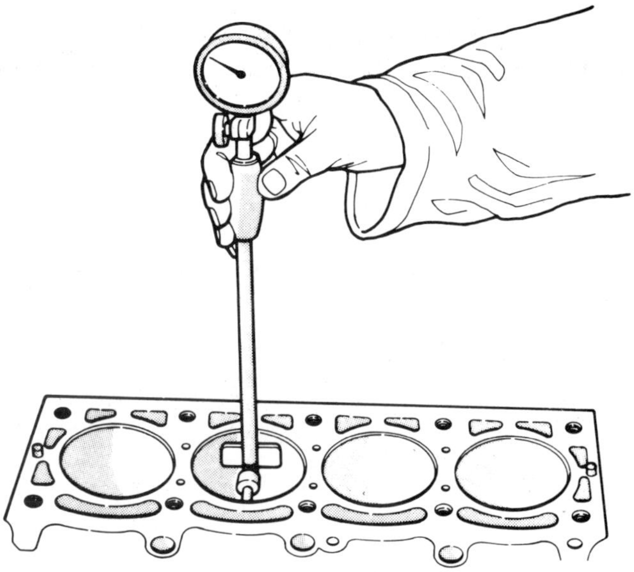

5. The gauge pictured above checks

   a. Bore out-of-roundness
   b. Bore taper
   c. Both a and b
   d. Neither a nor b

6. **Technician A** says that a crosshatch pattern should be present after using the tool shown above.

   Technician B says to wash the cylinders with solvent after using the tool shown above.

   Who is right?

   a. A only
   b. B only
   c. Both A and B
   d. Neither A nor B

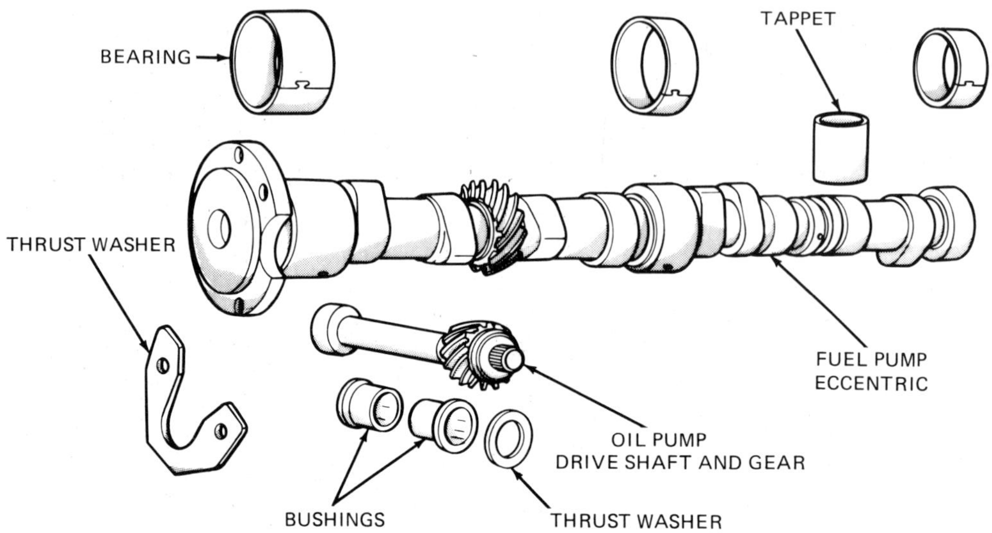

7. The bearings shown above need to be installed.

   Technician A says to install each bearing to proper depth.

   Technician B says to install each bearing using a driver.

   Who is right?

   a. A only
   b. B only
   c. Both A and B
   d. Neither A nor B

8. **Technician A** says that the set-up shown above is used to measure thrust wear.

   Technician B says that the set-up shown above is used to measure the fillet radius.

   Who is right?

   a. A only
   b. B only
   c. Both A and B
   d. Neither A nor B

9. The dial gauge shown above is positioned to check:

   a. Crankshaft snout roundness
   b. Crankshaft end play
   c. Both a and b
   d. Neither a nor b

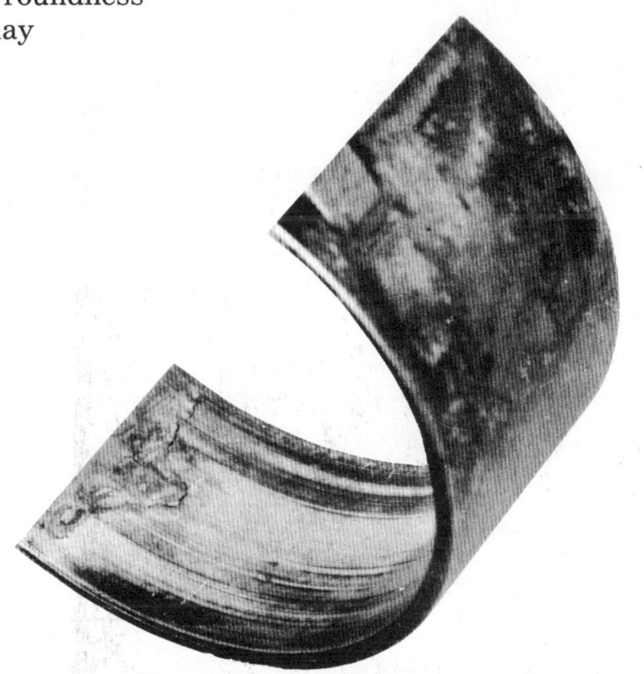

10. A bearing half shows localized damage near the joint face (see the above picture).

    Technician A says that this could be caused by applying excessive torque.

    Technician B says that this could be caused by insufficient crush.

    Who is right?

    a. A only
    b. B only
    c. Both A and B
    d. Neither A nor B

11. Which of the following could have caused the piston damage shown above?

    a. Over-advanced timing
    b. Adjusting the distributor by "ear"
    c. Cooling system malfunction
    d. All of the above

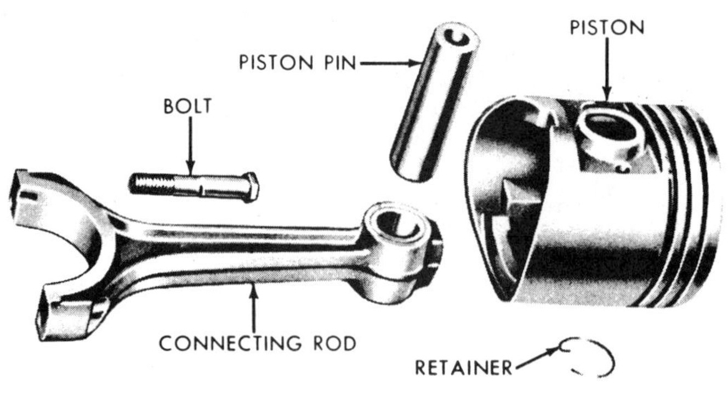

12. Refer to the above illustration.

    Technician A says that this piston pin is pressed into the connecting rod eye (small end).

    Technician B says that this piston pin oscillates in the piston.

    Who is right?

    a. A only
    b. B only
    c. Both A and B
    d. Neither A nor B

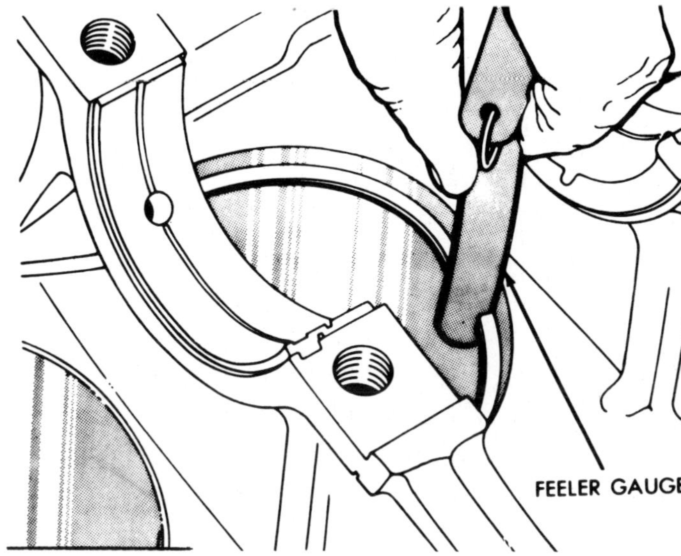

13. What measurement check is being performed in this picture above?

    a. Piston clearance
    b. Cylinder out-of-round
    c. Cylinder taper
    d. Piston ring gap

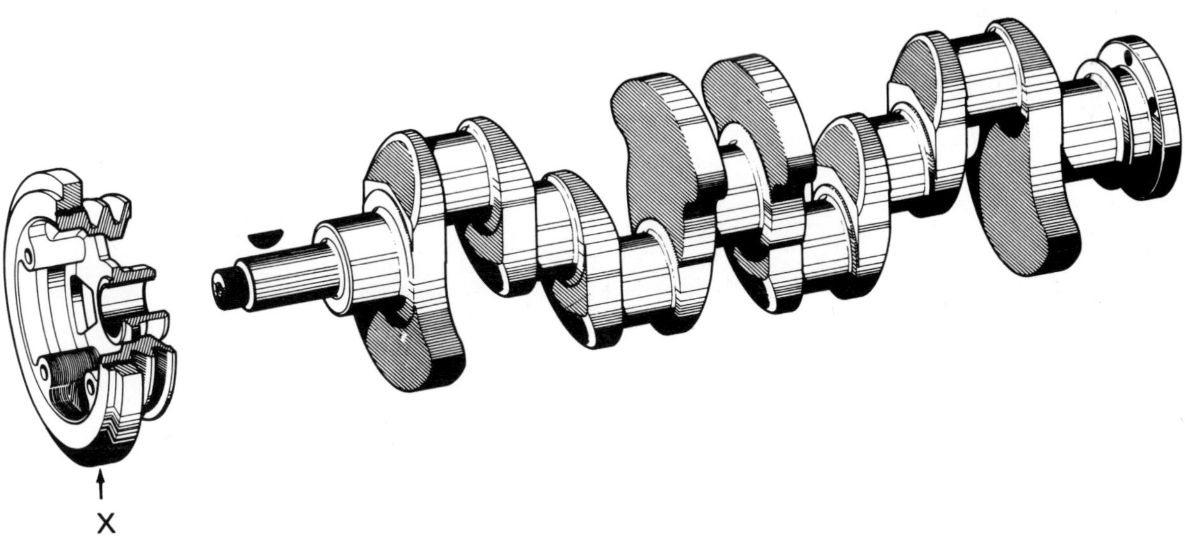

14. **Technician A** says that a broken crankshaft can result if part X above is defective.

    Technician B says that a sheared woodruff key can result if part X above fits loosely.

    Who is right?

    a. A only
    b. B only
    c. Both A and B
    d. Neither A nor B

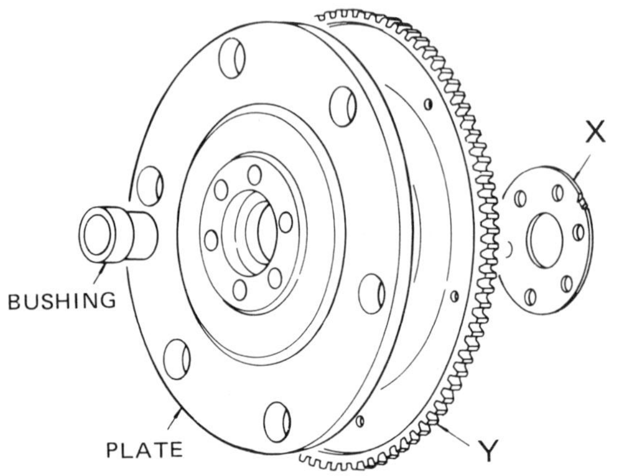

15. The flywheel shown above has excessive face runout after being installed.

    Technician A says that this could be caused by a nick or burr on part X.

    Technician B says that this could be caused by rough spots on part Y.

    Who is right?

    a. A only
    b. B only
    c. Both A and B
    d. Neither A nor B

16. A flywheel ring gear has worn teeth.

    Technician A says that this can affect engine balance.

    Technician B says that repair can be made by "truing" the crank flange.

    Who is right?

    a. A only
    b. B only
    c. Both A and B
    d. Neither A nor B

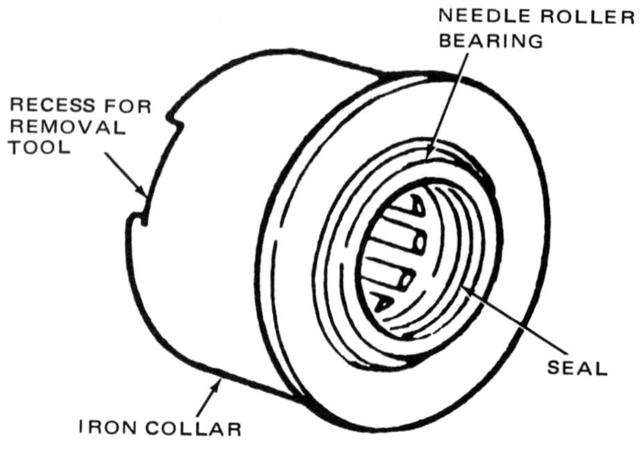

17. **Technician A** says that the bearing shown above should fit loosely in its bore.

    Technician B says that the bearing shown above should be lubricated when installing.

    Who is right?

    a. A only
    b. B only
    c. Both A and B
    d. Neither A nor B

18. **Technician A** says that RTV silicone sealer is used to secure threaded fasteners.

    Technician B says that anaerobic sealer fumes can "poison" an oxygen sensor.

    Who is right?

    a. A only
    b. B only
    c. Both A and B
    d. Neither A nor B

19. Refer to the engine shown at the right.

    Technician A says that shaft X and shaft Y must be timed.

    Technician B says that shaft X and shaft Y rotate in opposite directions.

    Who is right?

    a. A only **c.** Both A and B
    b. B only **d.** Neither A nor B

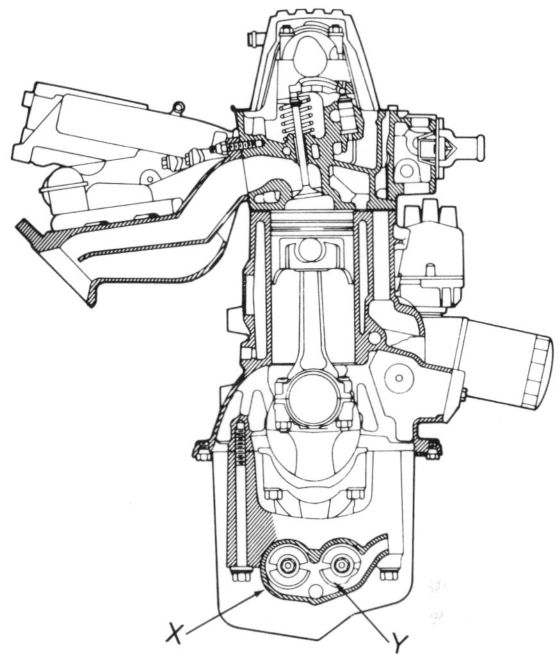

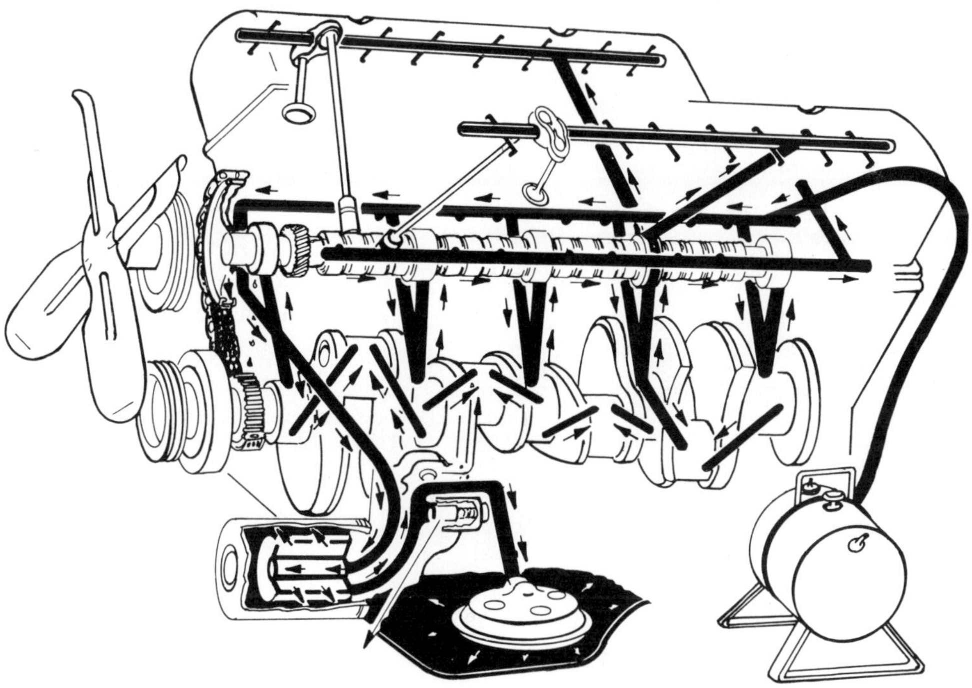

20. **Technician A** says that the device shown above is used to prevent a "dry start" after an engine overhaul.

    Technician B says that the device shown above is used to charge the lubrication system.

    Who is right?

    a. A only **c.** Both A and B
    b. B only **d.** Neither A nor B

SECTION REVIEW QUESTIONS

1. How can a cork oil pan gasket be stretched to fit?

2. What are load spreader washers?

3. How should a rope-type rear main seal be prepared prior to installation?

4. How can a feeler gauge be used to determine if an oil seal is fully seated in its bore?

5. A block deck warped more than ____________″ should be resurfaced.

6. How should main bearing bore alignment be checked?

7. What is a Heli-Coil®?

8. What can result if the ring ridge is not removed before taking a piston out of its bore?

9. Define cylinder taper. What is considered the maximum allowable taper?

10. Define cylinder out-of-roundness. What is considered the maximum allowable out-of-roundness?

11. Outline the procedure for deglazing a cylinder wall with a hone.

12. What is a torque plate?

13. After honing, what should be used to clean the cylinder bores?

14. Outline the basic procedure for installing a set of camshaft bearings.

15. What is the maximum allowable out-of-roundness for a crankshaft journal?

16. What are the usual crankshaft bearing journal undersizes?

17. What are sludge trap plugs?

18. Why is it important to check the fillet radii at the end of the rod and main bearing journals?

19. Describe the "instant keyway" process for repairing a crankshaft.

20. What is a "dry start"?

21. Describe a bearing fatigue failure.

22. Define bearing crush.

23. What is radius ride?

24. What is Plastigage®? How is it used?

25. What methods are used to keep (control) crankshaft end play within limits?

26. What is the procedure for seating (aligning) a crankshaft thrust bearing?

27. If the wear pattern on a piston skirt is uneven, what might be bent?

28. Describe the procedure generally used for measuring piston diameter.

29. Compression ring side clearance should not exceed what amount?

30. If piston ring end-gap clearance is less than specified, what should be done?

31. When installing piston rings, why should the end gaps be staggered?

32. Why should the ends of an oil ring expander-spacer be butted together and not overlapped?

33. Describe a full-floating piston pin.

34. What is a pin furnace?

35. Misaligned rods can cause what problems in an engine?

36. How can connecting rod side clearance be measured?

37. How can you prevent nicking of the crankshaft rod journals by the rod bolts when installing piston assemblies?

38. What things should you look for when inspecting a vibration damper?

39. Excessive flywheel runout can cause what problems?

40. What is the purpose of a pilot bushing?

41. What is RTV? Where is it used?

42. What is anaerobic compound? Where is it used?

43. What are balance shafts?

Section D

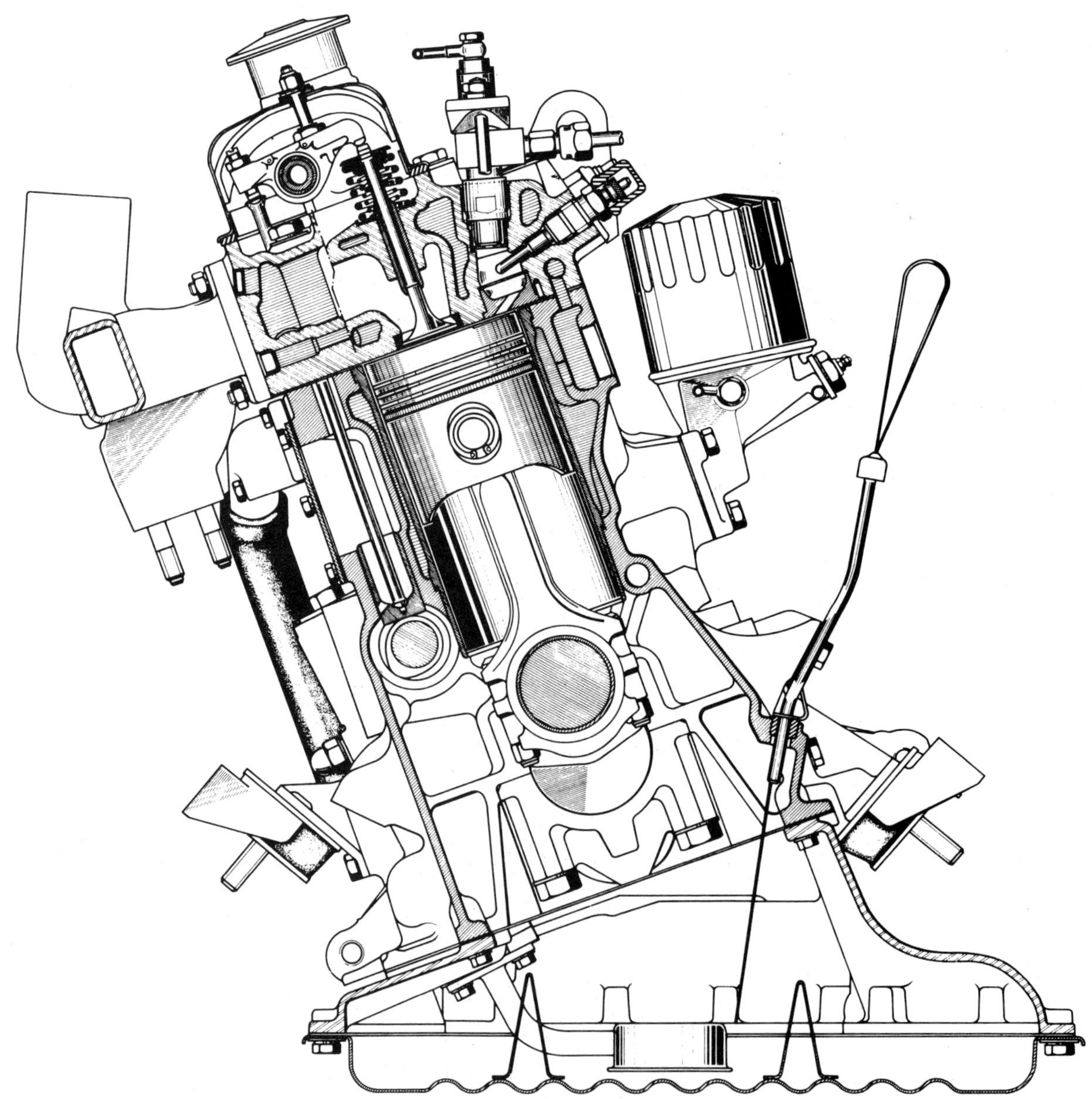

LUBRICATION AND COOLING SYSTEMS DIAGNOSIS AND REPAIR

After studying this section, and with hands-on practice on a "live" vehicle, you should be able to perform the following tasks:

- Perform oil pressure tests; determine needed repairs.
- Inspect, measure, repair, or replace oil pumps (includes gears, rotors, and housing), pressure relief devices, and pump drives.
- Perform cooling system tests; determine needed repairs.
- Inspect, replace, and adjust drive belts and pulleys in accordance with manufacturers' specifications.
- Inspect and replace engine cooling and heater system hoses.
- Inspect, test, and replace thermostat, bypass, and housing.
- Inspect coolant; drain, flush, and refill cooling system with recommended coolant.
- Inspect, test, and replace water pump.
- Inspect, test, and replace radiator, pressure cap, and expansion tank.
- Clean, inspect, test, and replace fan (both electrical and mechanical), fan clutch, fan shroud, and cooling-related temperature sensors.
- Inspect, test, repair, or replace auxiliary oil coolers.

LUBRICATION SYSTEM

The main purpose of the lubrication system is to provide a steady supply of oil, usually under pressure, to the moving parts of the engine (Figure 1[a] and [b]). Proper oil lubrication and cooling are essential to long engine life.

System Components

The principal components of the lubrication system, along with a brief description of the function of each, are as follows:

1. Oil pan. The oil pan is bolted to the bottom of the engine block and serves as a reservoir (sump) for the engine oil. It is shown in Figure 2.

2. Dipstick. The dipstick is used to measure the level of oil in the sump and is marked to show whether or not oil should be added.

> **SHOP TIP**
> Do not always assume that a satisfactory reading on the dipstick is a guarantee of proper crankcase oil level. Incorrect dipsticks are not as rare as you might think. If in doubt, check length and markings against specifications in a lubrication guide.

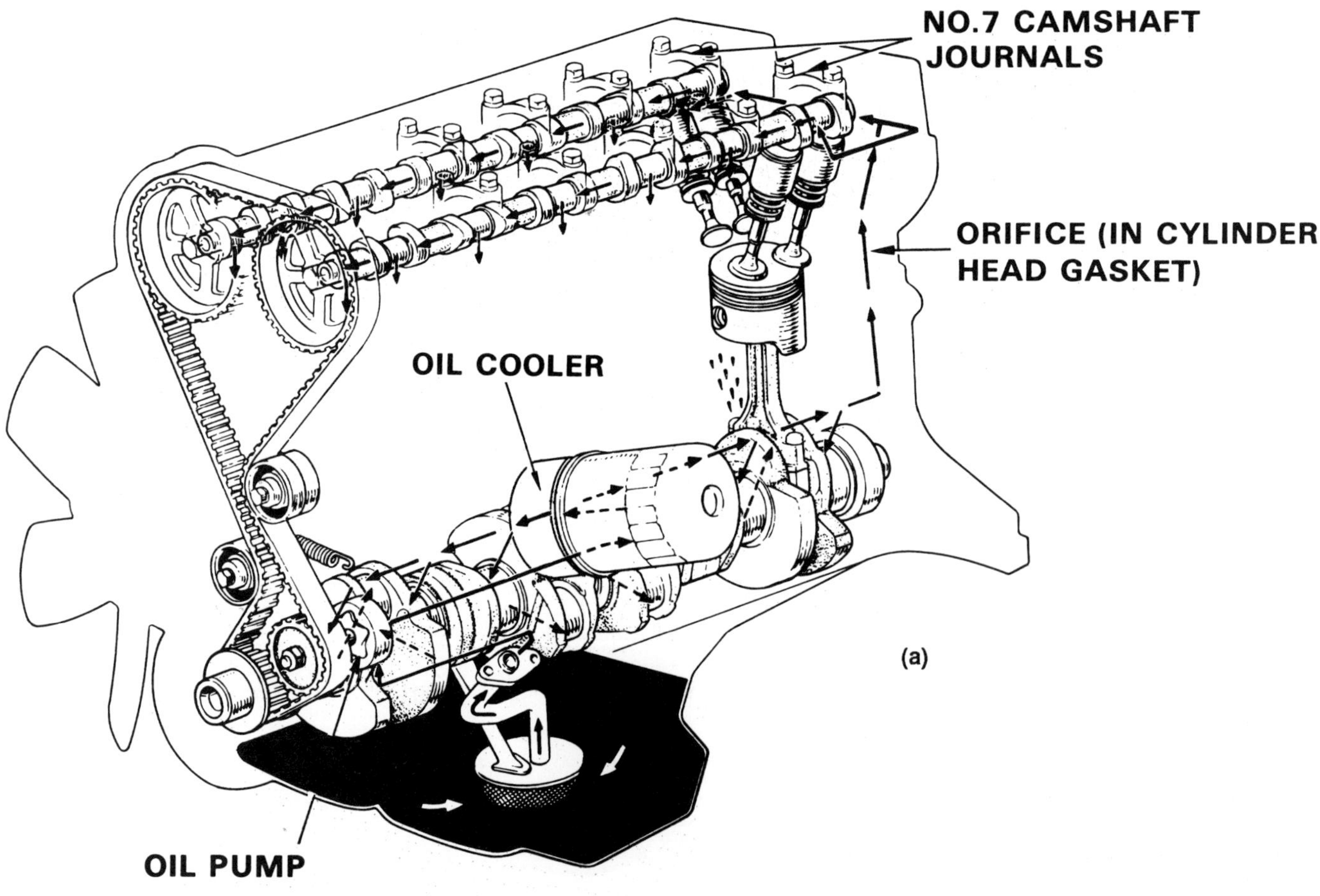

FIGURE 1
(a) Engine lubrication system. (Courtesy of Toyota Motor Sales, U.S.A., Inc.)

3. **Pickup screen.** The pickup screen is submerged in the oil in the oil pan. It is the inlet from the oil to the oil pump (Figure 3).

4. **Oil pump.** The oil pump provides the "push" for circulating oil under pressure through the engine (Figure 4).

5. **Oil filter.** The oil filter traps the smaller particles of metal, dirt, and insolubles carried by the oil so that they do not recirculate in the engine. This keeps the oil clean to reduce engine wear and promote longer life of the oil (Figure 5).

6. **Oil pressure indicator.** Usually, this indicator is a warning light that comes on if the oil pressure is low. Sometimes a pressure gauge is used instead of a warning light.

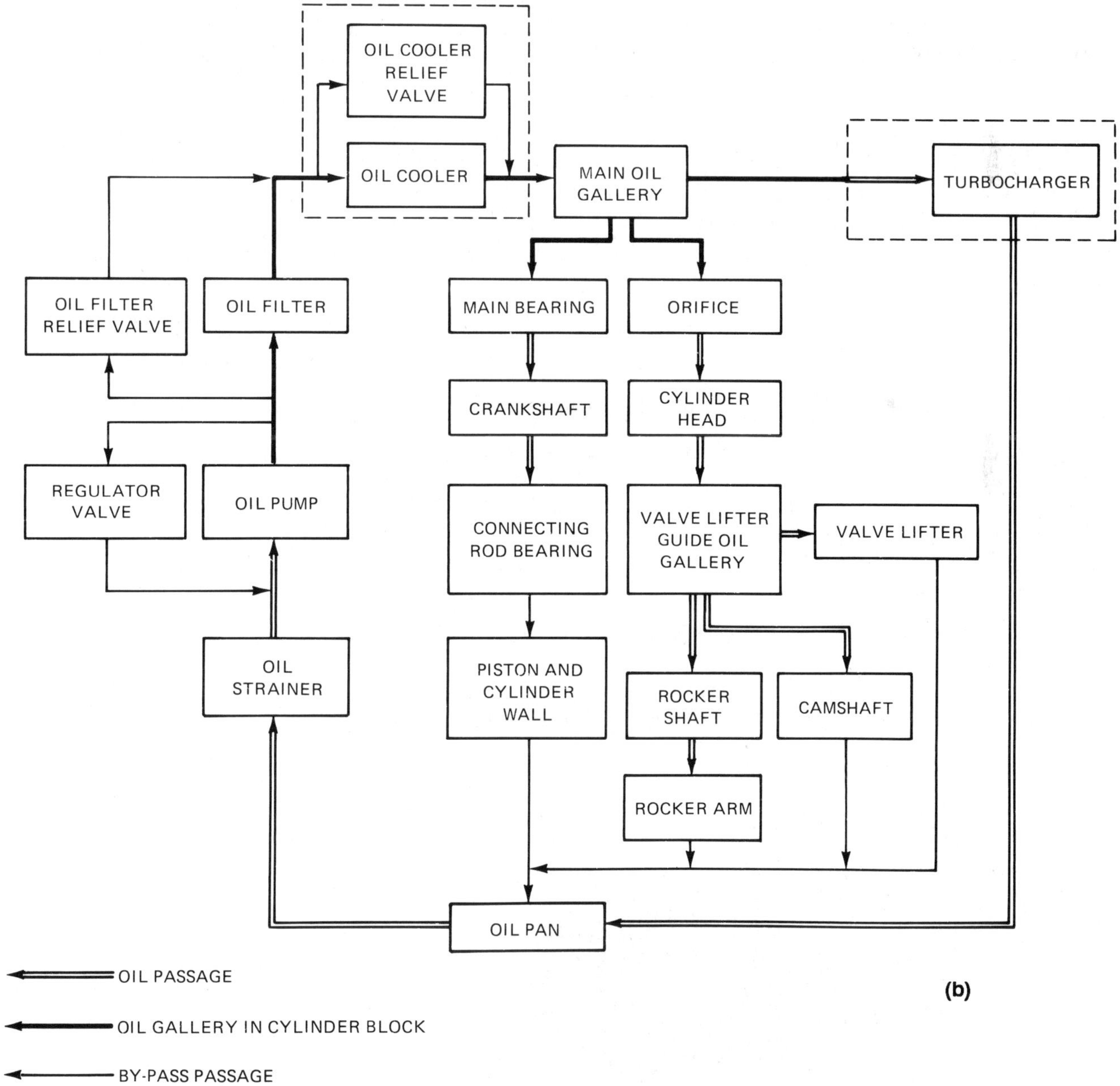

FIGURE 1 (Continued)
(b) Flow chart for a typical lubrication system.

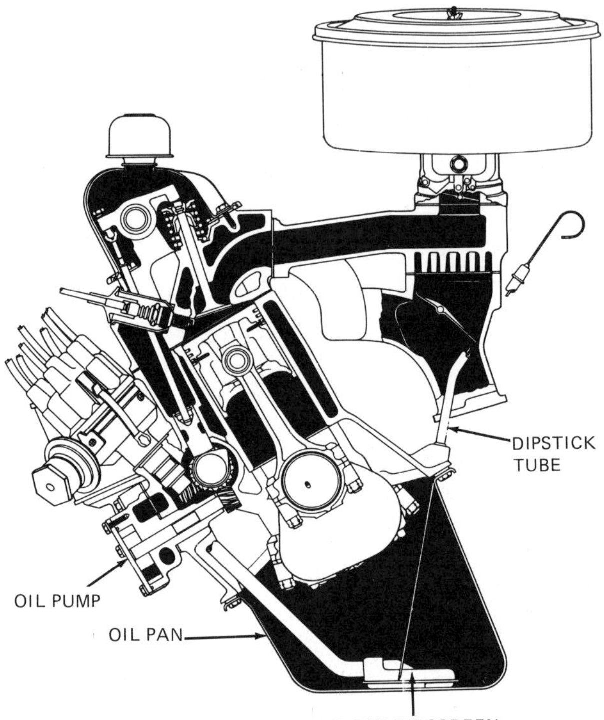

FIGURE 2
Oil stored in the oil pan is drawn up by the oil pump. (Courtesy of Chrysler Corporation)

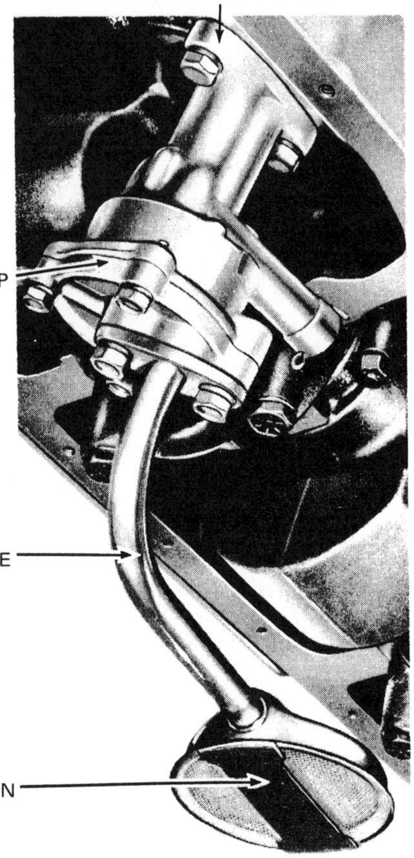

FIGURE 3
Details of an oil pump and pickup tube. (Courtesy of Ford Parts and Service Division)

OIL PRESSURE TESTS

If the oil pressure warning light stays on with the engine running, and oil level is satisfactory, perform an oil pressure test:

1. Remove the oil pressure sender unit and install an oil pressure gauge (Figure 6).

2. Start the engine and compare against specifications (Chart 1).

3. If the oil pressure is within specifications, replace the oil pressure sender and recheck.

4. If oil pressure is not within specifications, inspect the oil pump and drive, bearings, or for some unintended opening in the lubrication system (Figure 7).

Refer to Chart 2 for a listing of specific lubrication system problems.

FIGURE 4
Various oil pumps. (Courtesy of Sealed Power Corporation)

FIGURE 5
Oil leaving the oil pump flows to the oil filter, where metal wear particles, dirt, and carbon are trapped. The filter on the right is "loaded" and needs replacing.

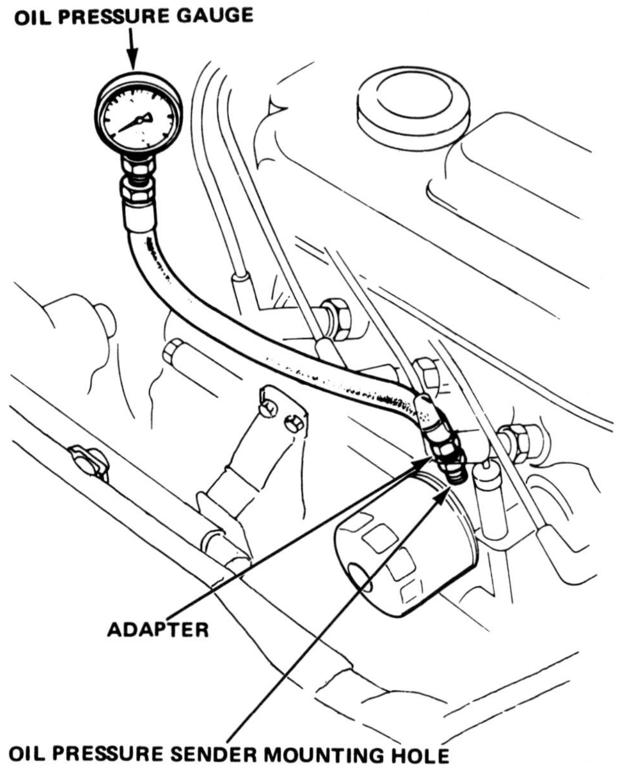

FIGURE 6
A master oil pressure gauge has been installed on this Honda Civic engine. Specifications are 48 to 60 psi at 3000 rpm, and 21 psi minimum at idle. (Courtesy of American Honda Motor Co., Inc.)

Testing the Oil Pressure Sending Unit

If the oil pressure warning lamp does not go on:

1. Disconnect the wire lead from the sending unit and ground it.

2. Turn the ignition switch to on. The oil pressure light should go on.

3. If the light does not go on, check the fuse, bulb, wires, and connections.

Checking for a Clogged Oil Filter

When a full-flow oil filter becomes clogged, its bypass (relief) valve opens (Figure 13[a] and [b]). This permits oil to pass around the filter and enter the engine. To check for clogging, feel the outside temperature of the filter housing with your hand. The housing should feel evenly hot with the engine at normal operating temperature. If the housing is cool, replace the oil filter.

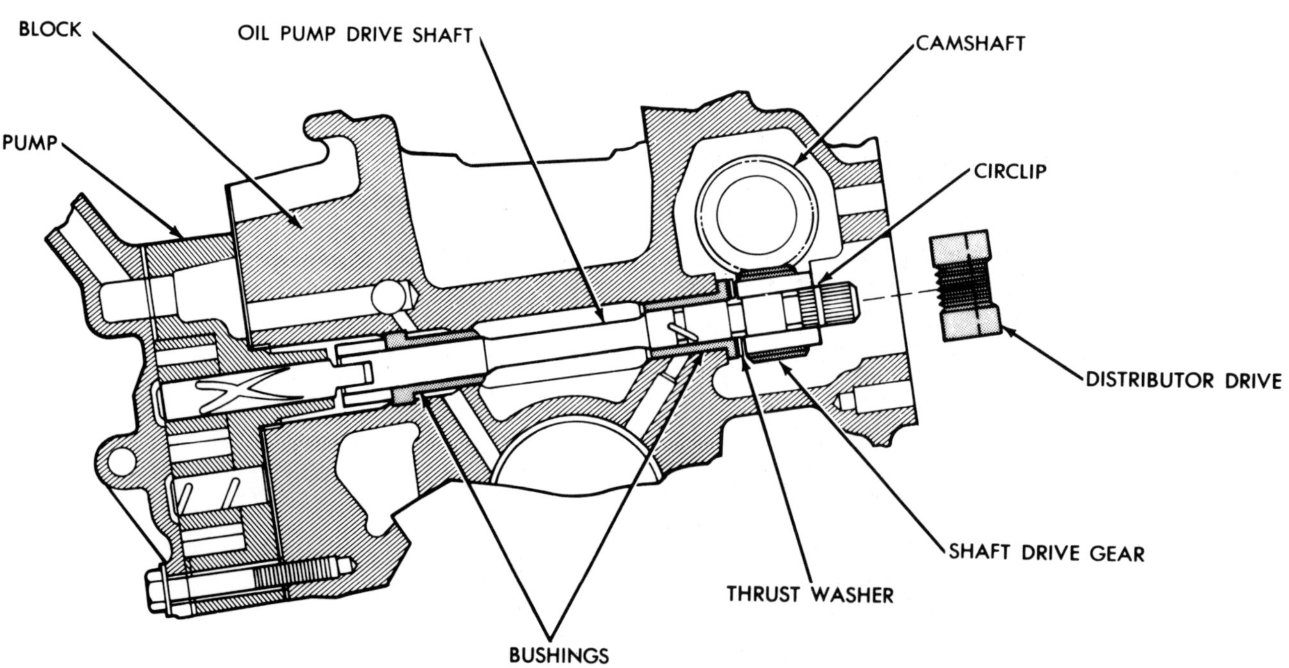

FIGURE 7
This oil pump driveshaft, driven by the camshaft, passes through the crankcase to drive the oil pump and distributor. (Courtesy of Chrysler Corporation)

CHART 1

Engine Type	Oil Pressure at idle (psi)	Oil Pressure at 3000 rpm (psi)
1.4L (85 CID)*	6	45
1.6L (98 CID) Inc. Turbo*	6	45
1.6L (98 CID)	7	40
1.7L (105 CID)	6	45
2.0L (122 CID)*	6	45
2.2L (135 CID) Inc. Turbo	4	25
2.3L (140 CID) Diesel*	6	45
2.5L (153 CID) Inc. Turbo	4	50
2.6L (159 CID) Inc. Turbo*	6	45
3.7L (225 CID)	4	30
5.2L (318 CID)	6	30
5.7L (360 CID)	6	30

*Engines produced by Mitsubishi Motors Corporation for Chrysler

CHART 4-2
An example of factory oil pressure specifications. Chrysler recommends at least 10 psi of oil pressure per 1000 rpm. Therefore, the minimum oil pressure for an engine operating at 700 rpm is 7 psi.

FIGURE 8
A stuck open relief ball will cause little or no oil pressure.

CHART 2

TROUBLESHOOTING CHART FOR LUBRICATION SYSTEM PROBLEMS

Oil Pressure Drop

1. Low oil level
2. Inaccurate oil pressure gauge or sending unit
3. Clogged oil filter
4. Worn parts in oil pump
5. Thin or diluted oil
6. Excessive bearing clearance
7. Stuck relief valve or ball in oil pump (Figure 8)
8. Weak relief valve spring
9. Loose, bent, or cracked oil pump suction tube
10. Internal oil leak (missing or loose front oil gallery plug)
11. External oil leak

External Oil Leak

1. Broken or improperly seated fuel pump gasket
2. Broken or improperly seated valve cover gasket
3. Loose oil filter
4. Loose or stripped oil pan drain plug
5. Loose rear oil gallery plug
6. Loose or unseated rear camshaft plug
7. Broken or improperly seated oil pan gaskets (Figure 9)
8. Damaged or improperly seated timing chain cover gasket and/or seal
9. Loose or damaged oil pressure sending unit
10. Damaged or improperly seated rear main seal (Figure 10)

Erratic Oil Pressure

1. Air leak on suction side of oil pump
2. Overfilled crankcase (allowing crankshaft to churn the oil and form air bubbles)

Excessive Oil Pressure

1. Stuck closed relief valve in oil pump
2. Excessively stiff relief valve spring
3. Improper grade of oil

No Oil Pressure

1. Stuck open oil pump relief valve
2. Damaged pump drive (Figure 11)
3. Obstructed pickup screen or tube (Figure 12)

Burst Oil Filter

1. Stuck closed oil pump relief valve

Oil Pump Lockup

1. Foreign particles drawn into pump

Twisted or Broken Pump Drive

1. Foreign particles drawn into pump

Noisy Valves

1. High or low oil level in crankcase
2. Thin or diluted oil
3. Low oil pressure

Connecting Rod or Main Bearing Noise

1. Insufficient oil supply
2. Thin or diluted oil
3. Excessive bearing clearance

Oil Pressure Light Does Not Light When Ignition Switch is On and Engine is NOT Running

1. Burned out bulb
2. Open light circuit wiring
3. Defective oil pressure sending unit

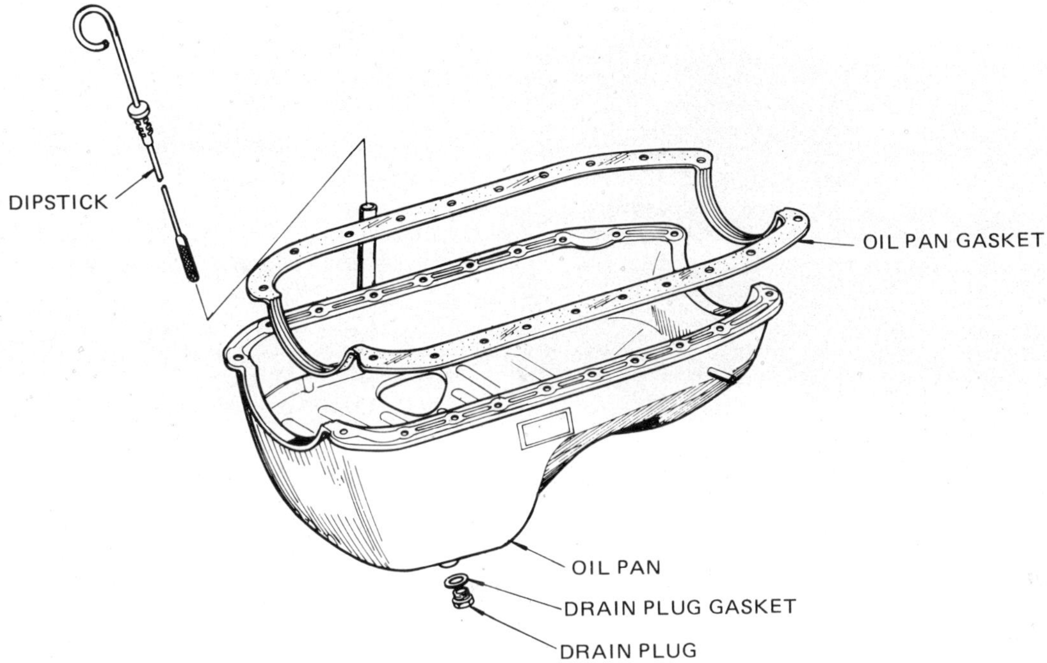

FIGURE 9
External oil leaks after an engine overhaul are often caused by improperly installed oil pan gaskets.

FIGURE 10
Full circle rear main oil seals must fit squarely and tightly into the retainer. Otherwise, an external oil leak can result.

FIGURE 11
Twisted intermediate shaft caused by solid debris entering the oil pump. (Courtesy of TRW Automotive Aftermarket Division. Reprinted with permission)

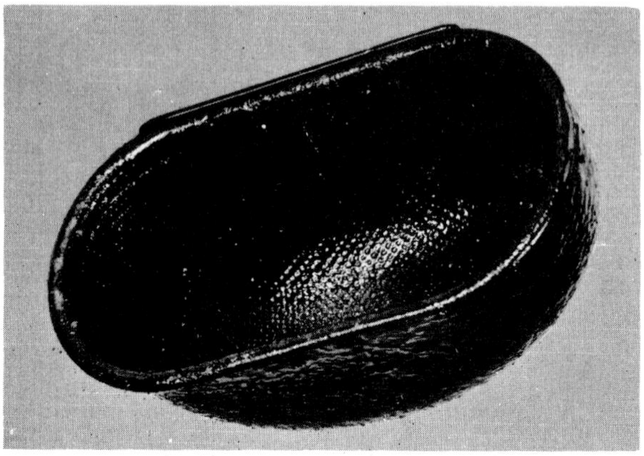

FIGURE 12
Clogged oil pump screen. (Courtesy of Dana Corporation)

OIL PUMP DESIGN

Several types of oil pumps are used in automotive engines:

1. Gear (Figure 14).
2. Rotor (Figure 15).
3. Hypo-cycloid (Figure 16).

Pump Drives

The engine drives the oil pump in one of several ways. A common method is for the distributor to operate the pump by means of a drive shaft (Figure 17).

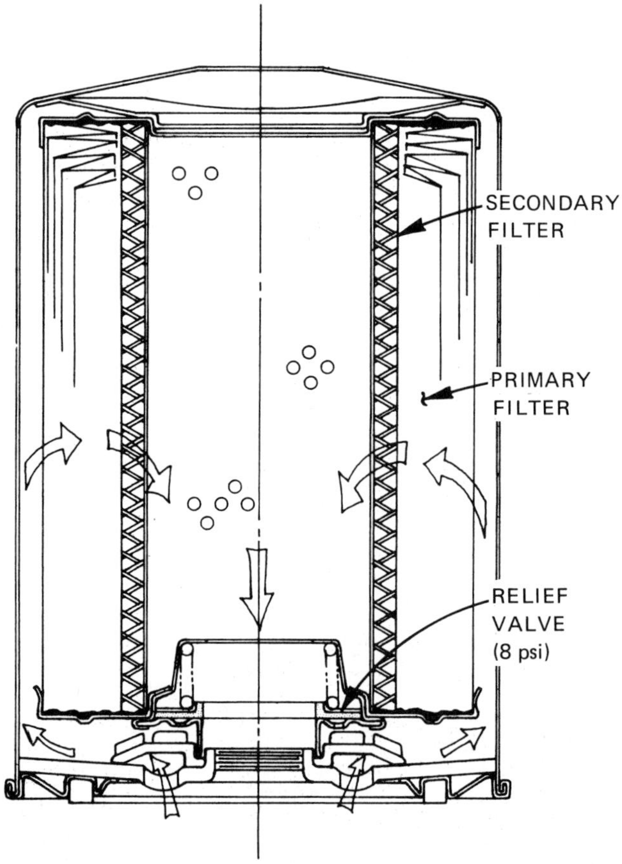

(a)

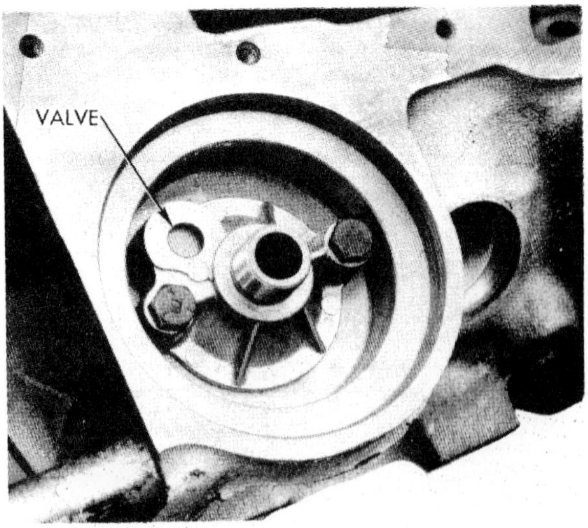

(b)

FIGURE 13
(a) Typical oil filter circuit. (b) One type of in-block oil filter bypass valve. (Courtesy of Chevrolet Motor Division, General Motors Corporation)

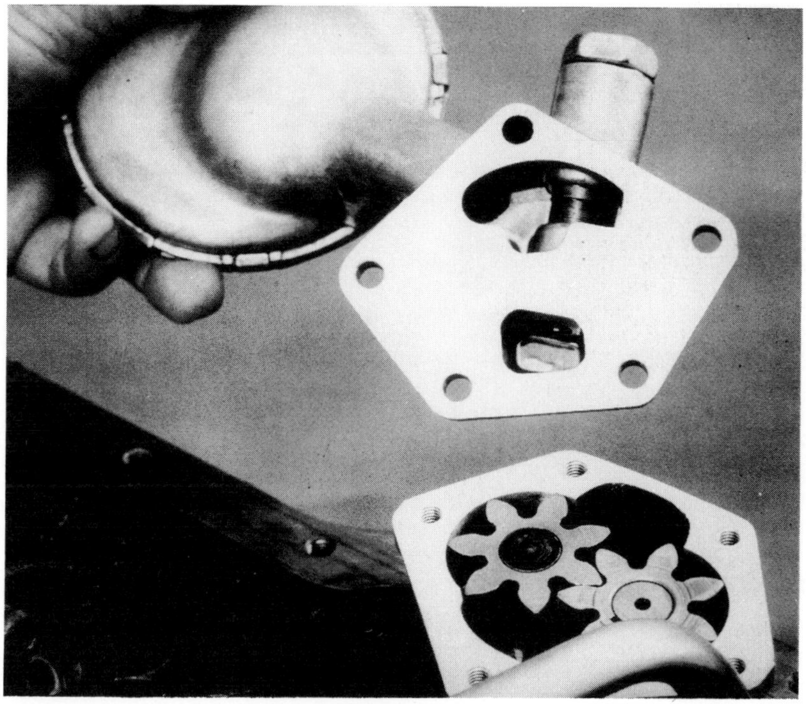

FIGURE 14
Gear type pump.

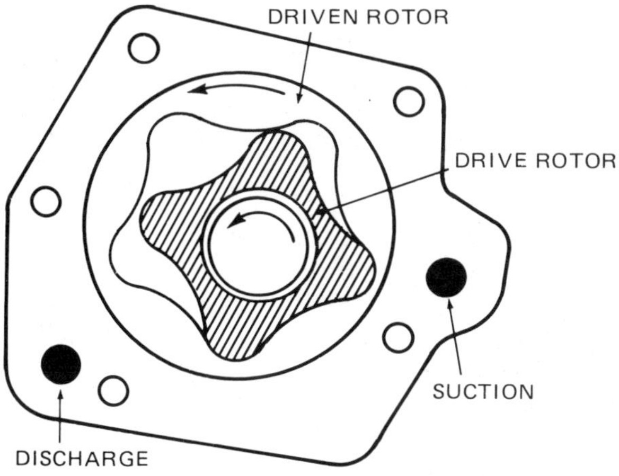

FIGURE 15
Rotor type pump.

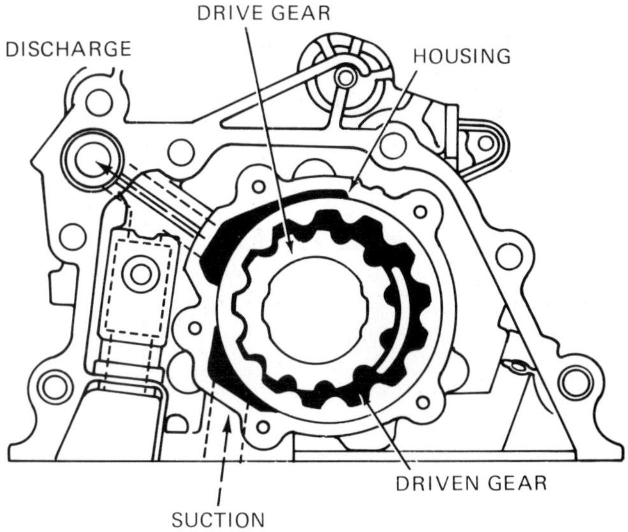

FIGURE 16
Hypo-cycloid type pump driven directly by the crankshaft.

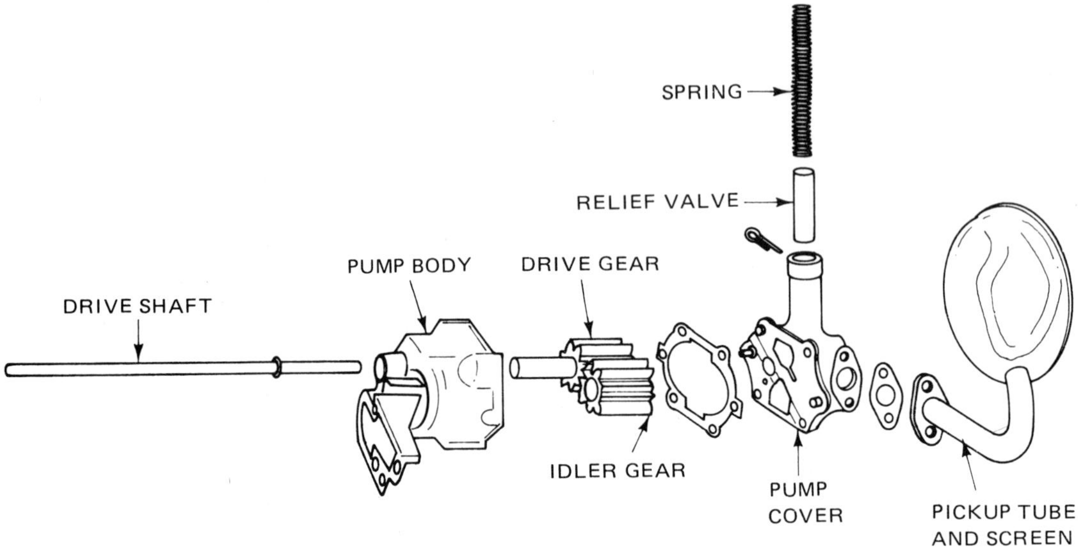

FIGURE 17
Gear type pump turned by a hexagon-shaped drive shaft. (Courtesy of Ford Parts and Service Division)

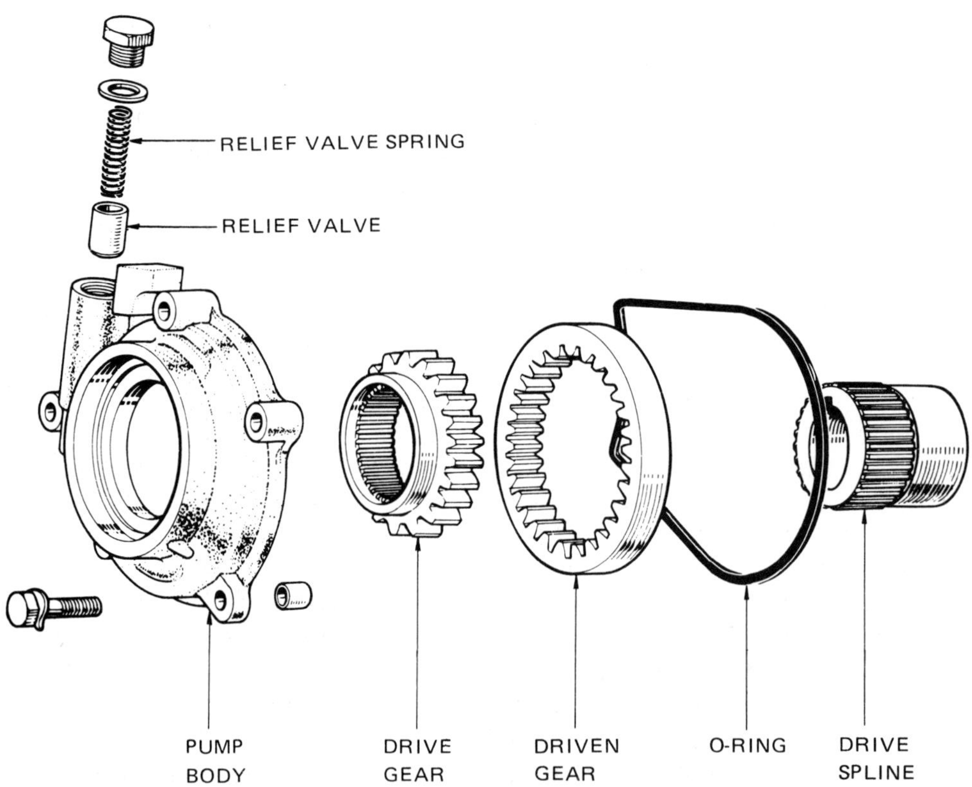

FIGURE 18
A crankshaft-driven oil pump with its component parts. (Courtesy of Toyota Motor Sales, U.S.A., Inc.)

NOTE: In some installations, the drive shaft has a gear that meshes with a spiral gear on the camshaft.

Another method of driving the oil pump is through a direct coupling with the crankshaft (Figure 18).

Pressure Relief Valve

Automobile engine oil pumps are usually of the positive displacement type. This means that maximum pressure must be limited. This is accomplished by using a pressure relief valve (Figure 19). If a pressure relief valve was not used, system pressure would become dangerously high.

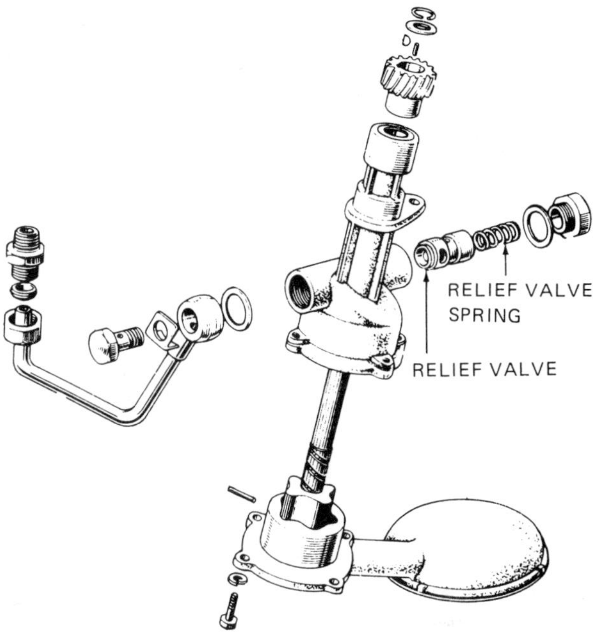

FIGURE 19
An oil pump showing the pressure relief valve assembly. (Courtesy of Toyota Motor Sales, U.S.A., Inc.)

OIL PUMP SERVICE

The oil pump should be removed for inspection whenever the engine is torn down, or when oil contamination is noticed. It should be checked for a sticking relief valve and/or wear by measuring clearances anytime the engine has low oil pressure.

Check for scoring and wear of the rotors or gears, pump housing, and cover plate. See Figure 20(a), (b), (c), and (d). Inspect the relief valve piston or ball for scoring and free movement. The relief valve assembly and cover can usually be replaced individually. Rotors or gears must be replaced in sets only.

> **SHOP TIP**
> If oil level is correct but aeration is suspected, remove the oil sending unit. Install a fitting, a petcock, and a length of transparent hose. Run the engine at a fast idle for five minutes with the petcock closed. Then open the petcock with the engine running at about 1000 rpm and direct the oil into the oil filler opening. Watch for air bubbles in the transparent hose. The presence of bubbles indicates the pump is picking up air.

OIL PUMP INSTALLATION TIPS

1. Always examine the pump drive for visible damage or rounding of hexagonal ends (Figure 21).

2. Check the pump drive end play. Excessive end play can cause the gears or rotors to rub against the inside of the pump cover, ruining the oil pump.

3. Closely check bolt-on or threaded-in pickup tubes. A mismounted flange, loose threads, or cracked pickup tube can cause aeration.

4. Avoid hammering directly on the oil pump when installing a press-in-type pickup screen. This can distort the pump body and cause internal seizure.

5. In some engines the oil pump is an integral part of the timing chain cover. These pumps are not self-priming. Fill the pump gear cavity with petroleum jelly prior to installation.

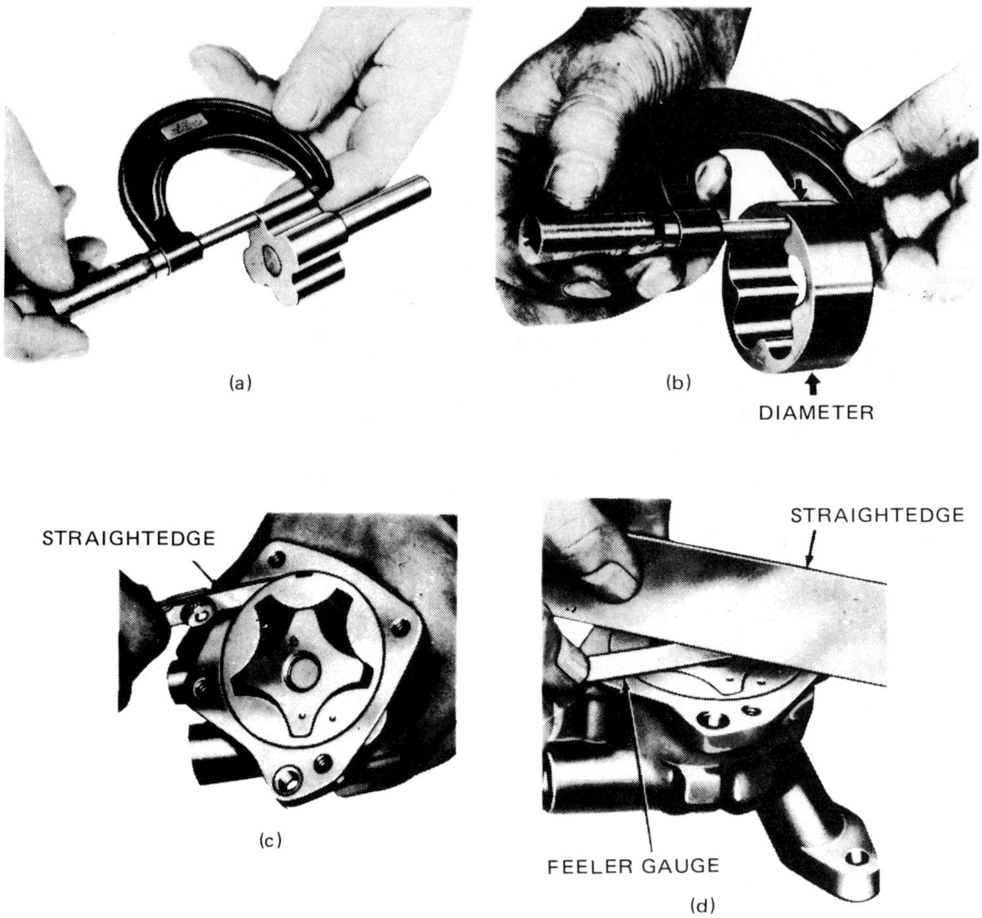

FIGURE 20
(a) Inner rotor thickness should be checked for minimum dimension. (b) Outer rotor thickness should be checked for minimum dimension. (c) Outer rotor to pump body wall must be held to minimum clearance. Not more than .012" is the normal limit. (d) Clearance over rotors is critical. One manufacturer limits space to .004".

AUXILIARY OIL COOLERS

Some engines use an auxiliary oil cooler to reduce latent engine heat (Figure 22). Heat is removed from the oil as air flows through the cooler core.

> **SHOP TIP**
> If there is any metal failure inside an engine equipped with an oil cooler, it is imperative that the cooler be replaced when making repairs. Due to cooler core construction, it is practically impossible to flush out trapped metal particles that can later break free and be released into the oil.

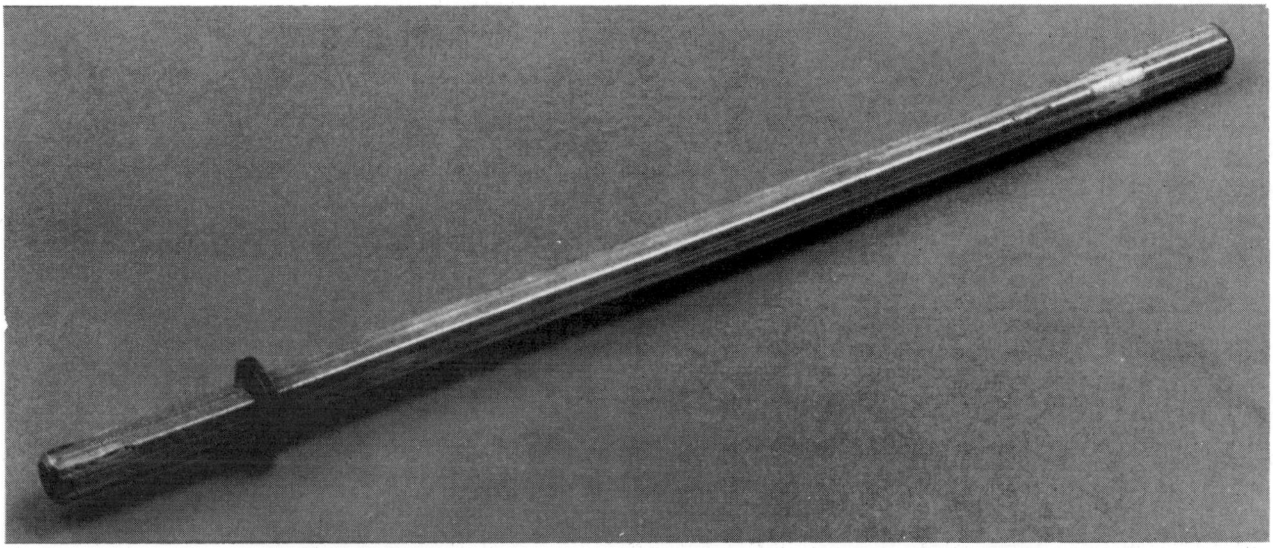

FIGURE 21
Note rounding of the hexagonal ends on this pump drive.

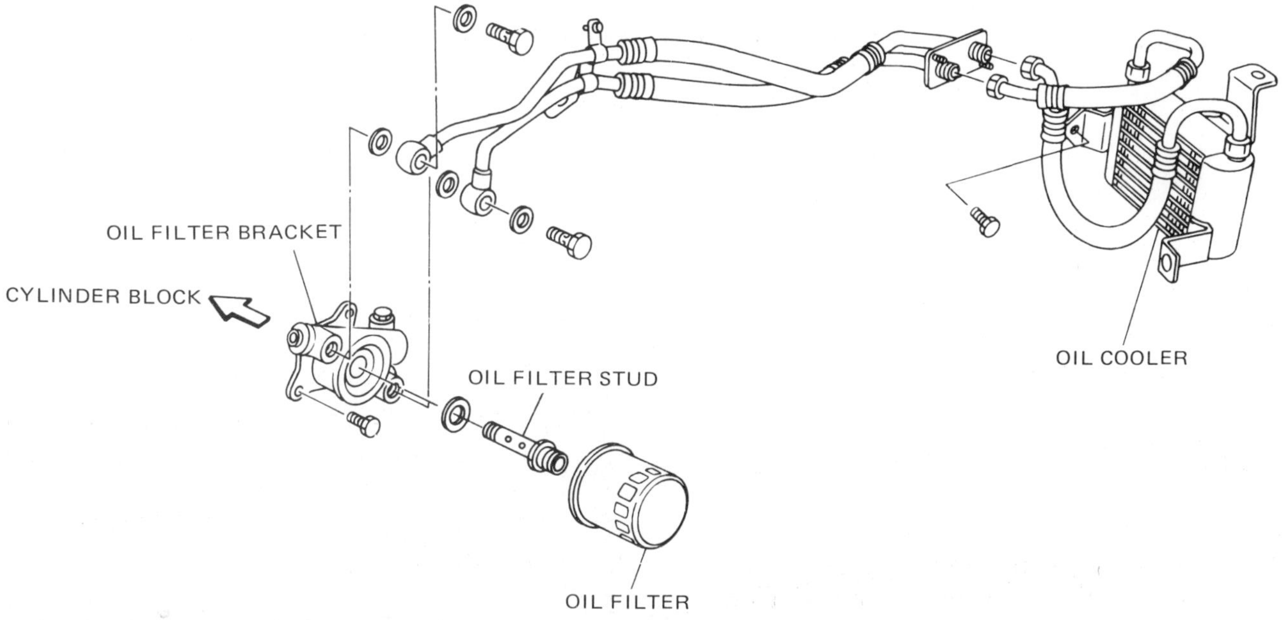

FIGURE 22
Oil cooler and attaching parts. (Courtesy of Nissan Motor Company, Ltd.)

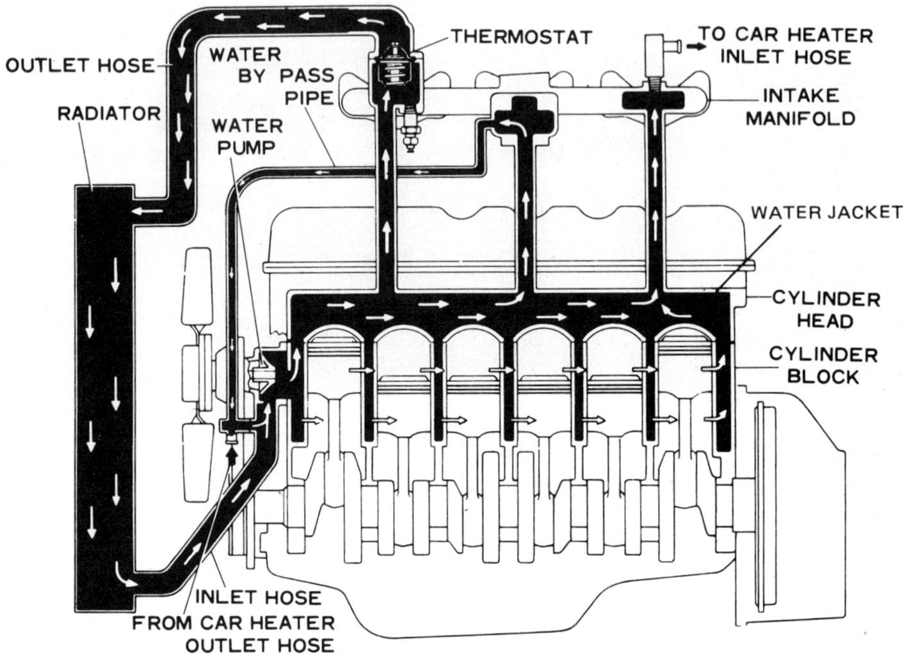

FIGURE 23
Engine cooling system. (Courtesy of Toyota Motor Sales, U.S.A., Inc.)

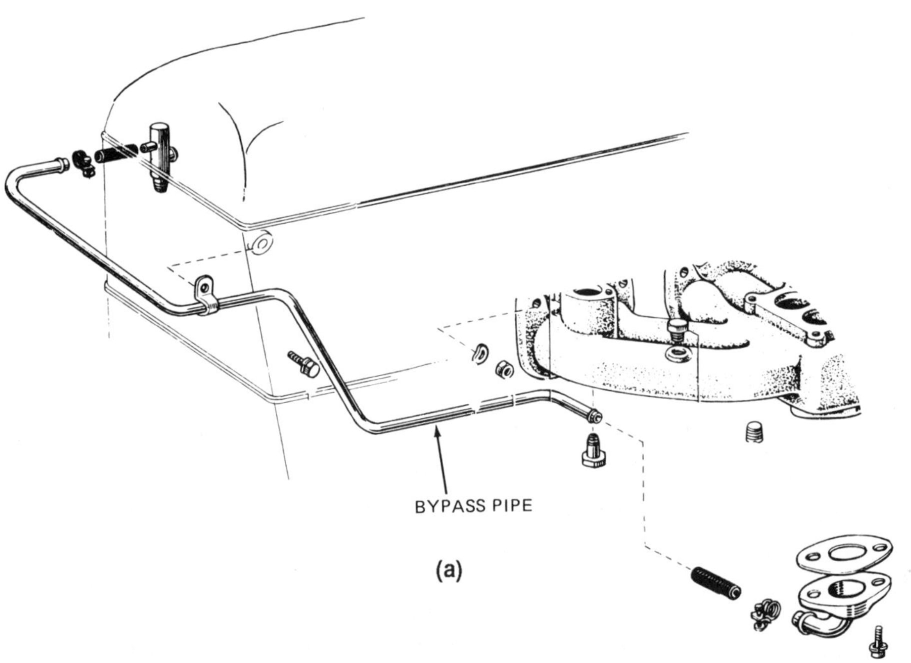

(a)

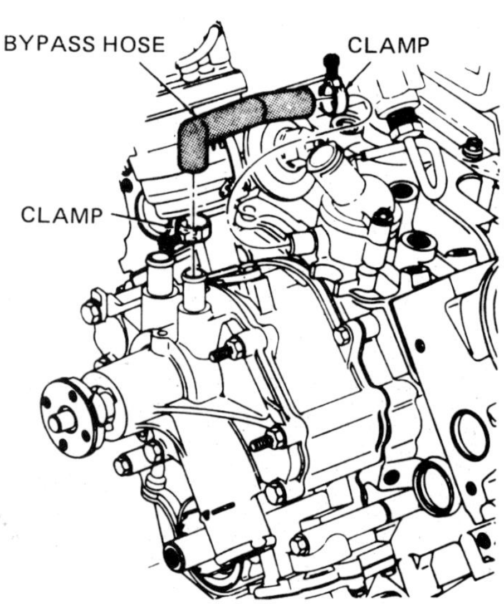

(b)

FIGURE 24
(a) Water bypass pipe. (Courtesy of Toyota Motor Sales, U.S.A., Inc.) (b) Water bypass hose. (Courtesy of Ford Parts and Service Division)

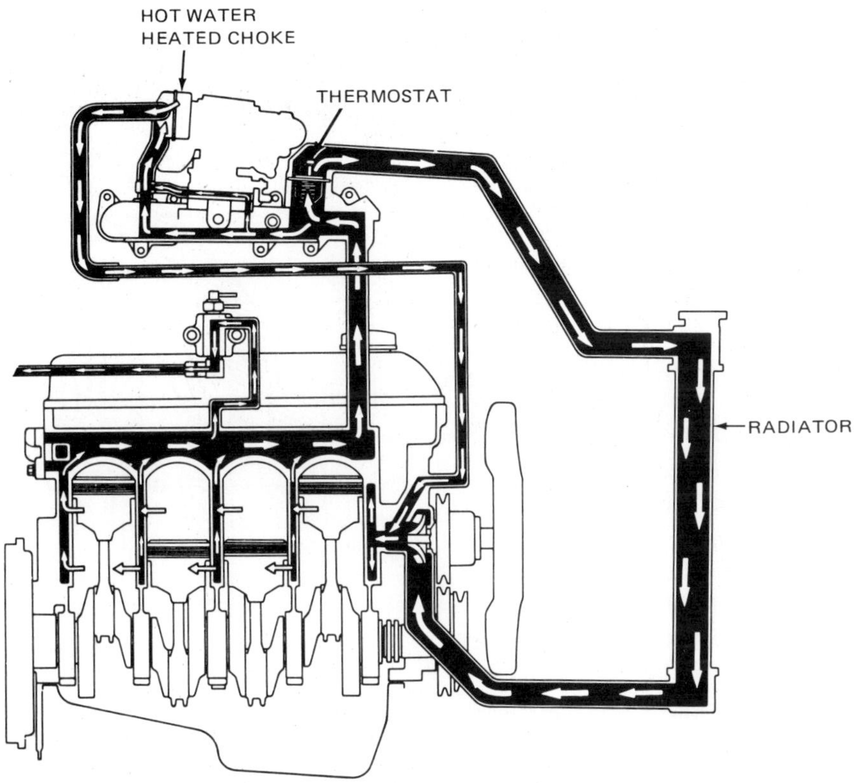

FIGURE 25
Cooling system flow in one popular import engine.

COOLING SYSTEM OPERATION

An automobile engine produces heat that must be removed quickly from the engine to avoid severe damage. The most popular method of removing this heat is to surround the heat-producing areas in the engine with "jackets" containing an ethylene glycol-based coolant (Figure 23). When this coolant reaches the radiator, it releases its heat into the atmosphere.

Transferring excess heat through the radiator, however, is only part of the function of a cooling system. An important additional function is bringing the engine to operating temperature quickly and evenly. A quick warm-up reduces blowby into the crankcase, reduces emissions, and decreases fuel consumption.

Bypass Operation

As soon as a cold engine is started, coolant circulation begins absorbing heat from the areas around valves and cylinders. Since the thermostat remains closed during warm-up, the coolant recirculates by passing through a bypass line (or hose) to the water pump and back into the engine block (Figure 24[a] and [b]). Thus, the engine block, heads, and intake manifold temperatures rise together, minimizing stress from uneven expansion.

Thermostat Operation

As operating temperature is approached, action begins in the thermostat. Depending on individual vehicles, a coolant temperature of 180° to 190° F generally starts to open the thermostat. This allows coolant flow into the radiator, through the radiator core, and back to the water pump (Figure 25).

The operating range of thermostats is about 25 degrees. One that starts to open at 180° F should be fully open at about 205° F (Figure 26).

If a thermostat is removed, coolant can move too fast in the cooling system. This can destroy the water pump and the radiator core by cavitation. Cavitation occurs when a liquid contains bubbles. These bubbles can become a destructive force.

When the water pump doesn't have a thermostat providing a restriction to push against, bubbles form. When these bubbles pop, they can cause damage. They attack the metal water pump impeller with the force of a 200-mph water jet. Little pieces of metal flake off, eventually destroying the pump and/or its housing.

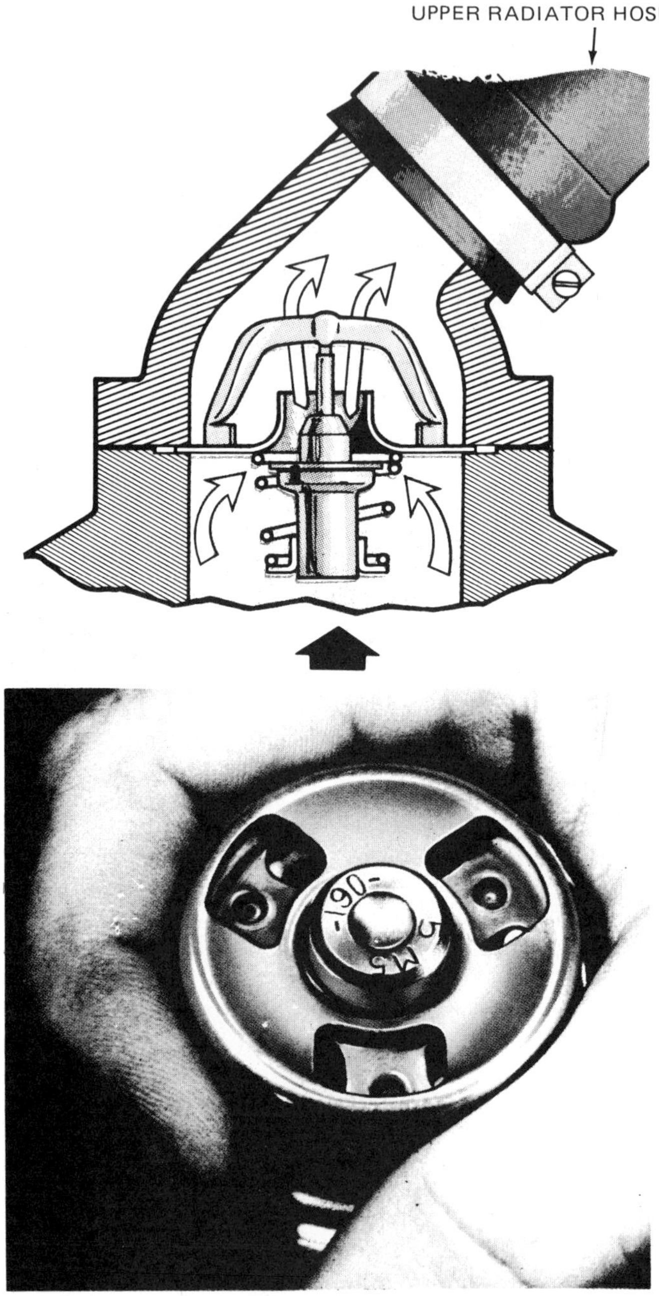

FIGURE 26
Thermostat operation.

Transmission Oil Cooler

One of the additional loads placed on many radiators is the cooling of automatic transmission fluid (ATF). Multiplying torque produces a great deal of heat in the converter.

NOTE: It is not unusual for ATF to reach a temperature of 250° to 300° F.

The transmission cooler, often located in the lower radiator tank, transfers heat to prevent ATF deterioration, and thus protects the transmission from damage. See Figures 27 and 28.

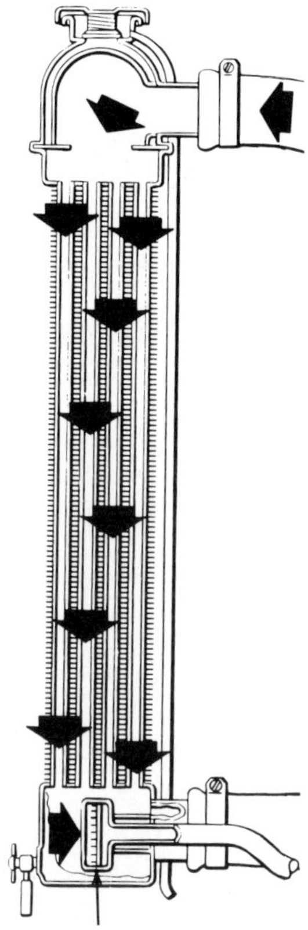

FIGURE 27
Transmission intercooler. (Courtesy of Ford Parts and Service Division)

Radiator Cap

The cooling system is designed to operate under positive pressure. Pressuring the cooling system as the coolant nears its boiling point permits the system to operate at a higher temperature without boiling off the coolant. Also, a greater temperature difference between the coolant and the outside air speeds up the transfer of heat from the coolant to the air.

The radiator cap allows the system to be pressurized. For every pound of pressure exerted on the coolant, the coolant boiling point is raised by approximately 3 degrees F.

Let's see what the combination of a proper mixture of coolant and 14 pounds per square inch (psi) of pressure will do to the boiling point of the coolant. Water boils without pressure at 212° F. Water with the proper mixture of antifreeze (without pressure) boils at 223° F. If the system is pressurized to 14 psi, that temperature is increased by 42 degrees F ($3 \times 14 = 42$). As a result, the coolant will not boil until it reaches 265° F. By using coolant and pressure, the boiling point is increased by 53° F over that of plain water.

If the pressure in the system gets higher than the rating of the cap, the pressure relief valve will be forced off its seat to relieve the pressure (Figure 29[a]).

As the coolant begins to cool off, it contracts. When the system cools and contracts, coolant is drawn into the radiator from a coolant recovery tank in order to eliminate air. On some systems, a radiator cap vacuum valve opens to admit air into the radiator tank (Figure 29[b]). If the vacuum produced by the contraction of the coolant is not relieved, the radiator hoses will collapse.

Fluid Fan Drive Clutch

On most automobiles, a constantly running fan is an unnecessary burden over 40 miles per hour. Because of this, many engines incorporate a silicone fluid drive clutch to control fan speed in relation to cooling demands (Figure 30).

Flex Fans

In some cases, a flex fan is used instead of a fluid drive clutch (Figure 31). Flexible blades vary the amount of air being drawn through the radiator, automatically increasing the pitch and moving more air at low engine speeds.

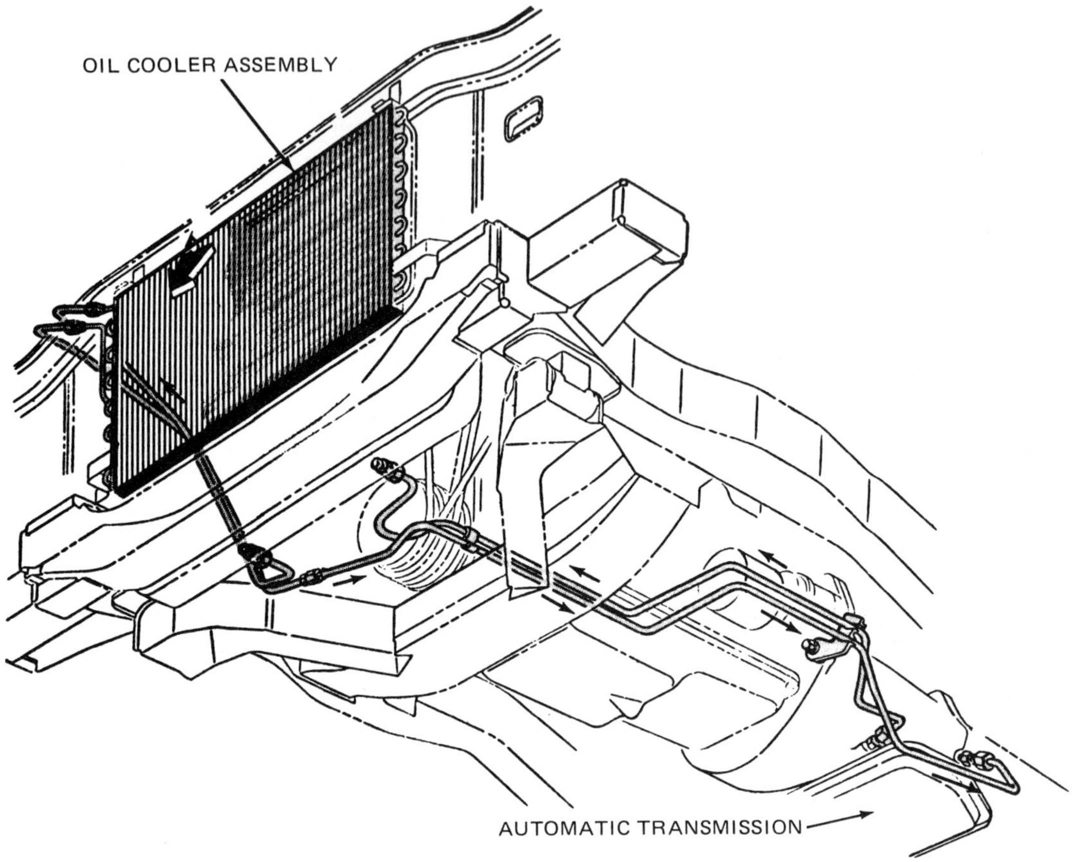

FIGURE 28
Auxiliary transmission oil cooler. (Courtesy of Chrysler Corporation)

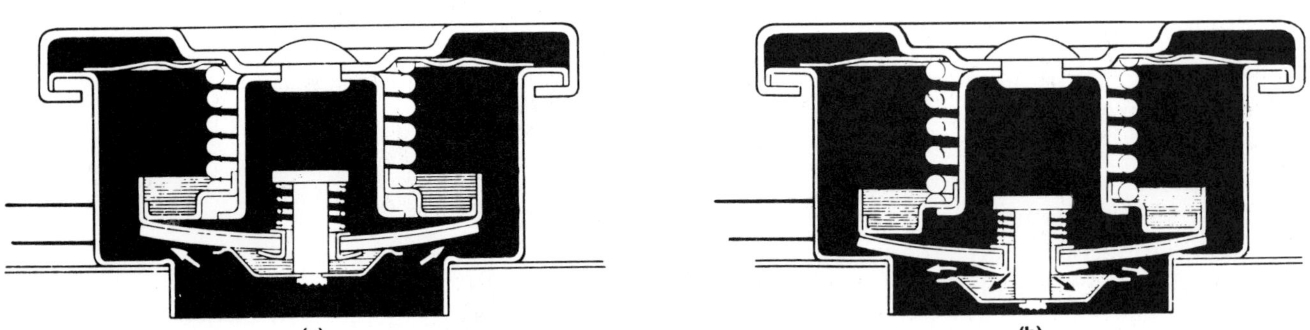

FIGURE 29
(a) Radiator cap pressure relief valve operation. (b) Radiator cap vacuum valve operation.

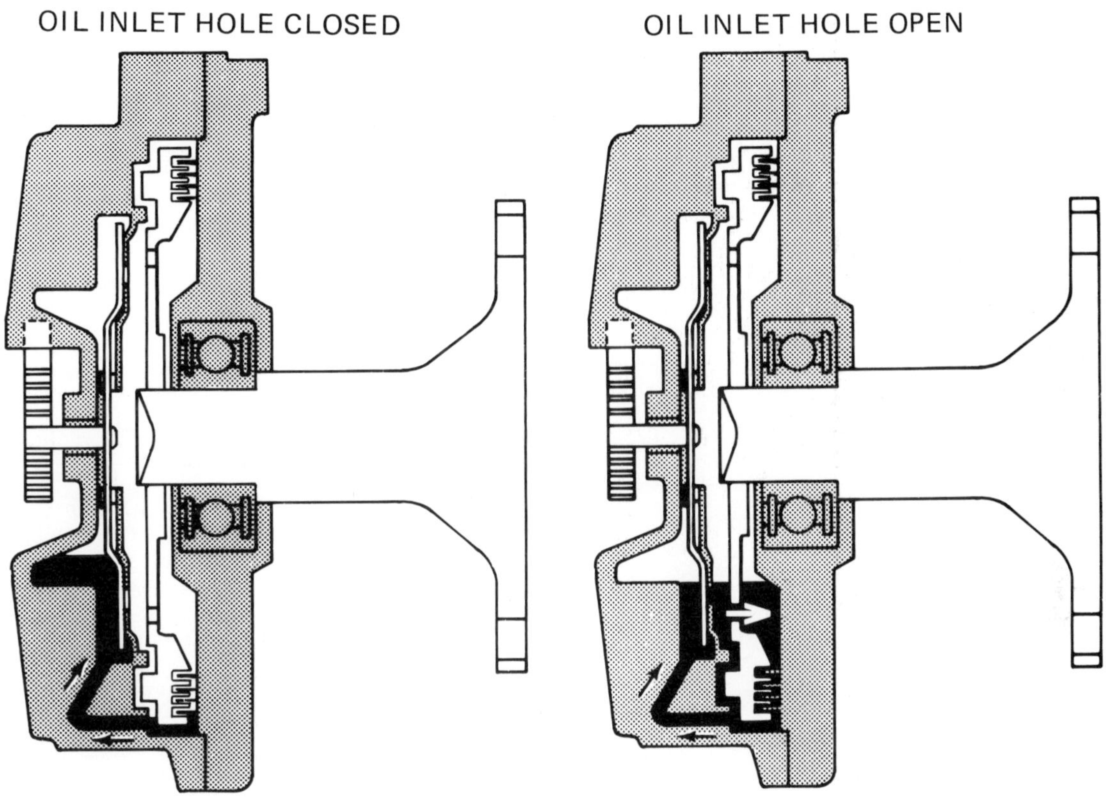

FIGURE 30
Fluid fan drive clutch.

Electric Fan Drives

Underhood temperatures are often quite high after engine shutdown. This heat remaining in the engine compartment can cause a temporary rich air-fuel mixture, resulting in hard restarting and stumble. Most electric fan drives incorporate temperature-actuated coolant switches that allow the fan to continue running after engine shutdown to cool the coolant and lower underhood temperatures.

An additional benefit is that the fan operates only as required by engine temperature conditions (Figure 32). This eliminates the engine power loss that occurs when a fan is driven continuously, as is a conventional mechanical fan.

FIGURE 31
Flex fan. (Courtesy of Chrysler Corporation)

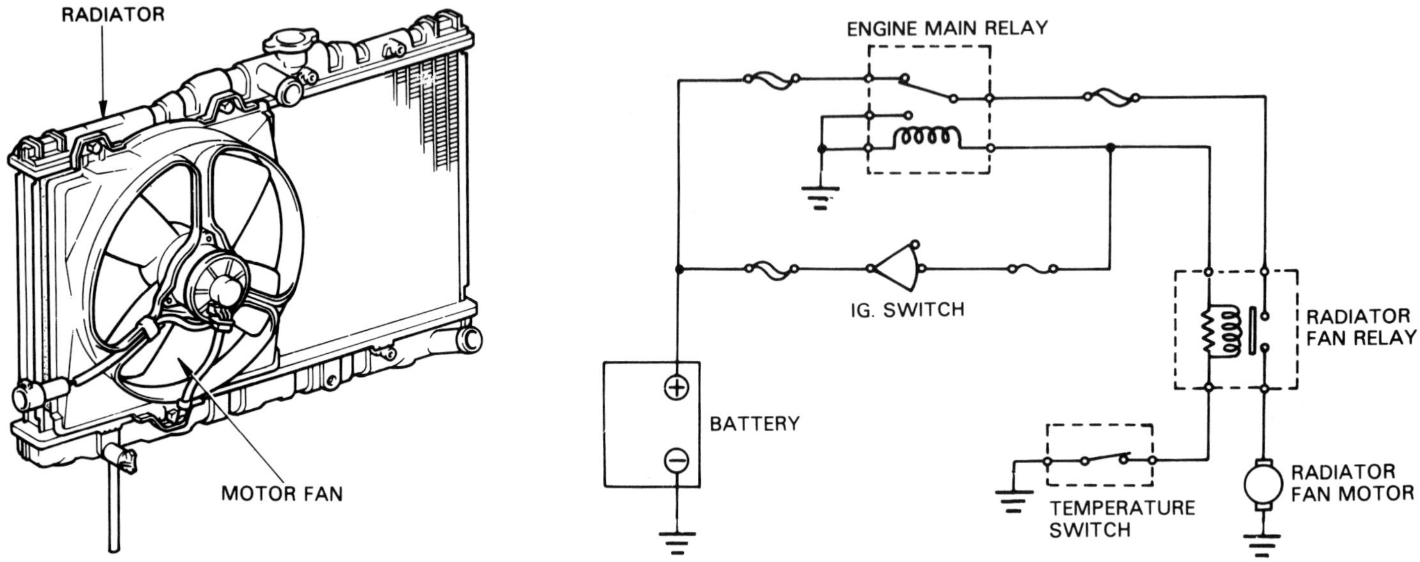

FIGURE 32
Today, electrically-driven fans are used primarily because of the transveral mounting of engines.

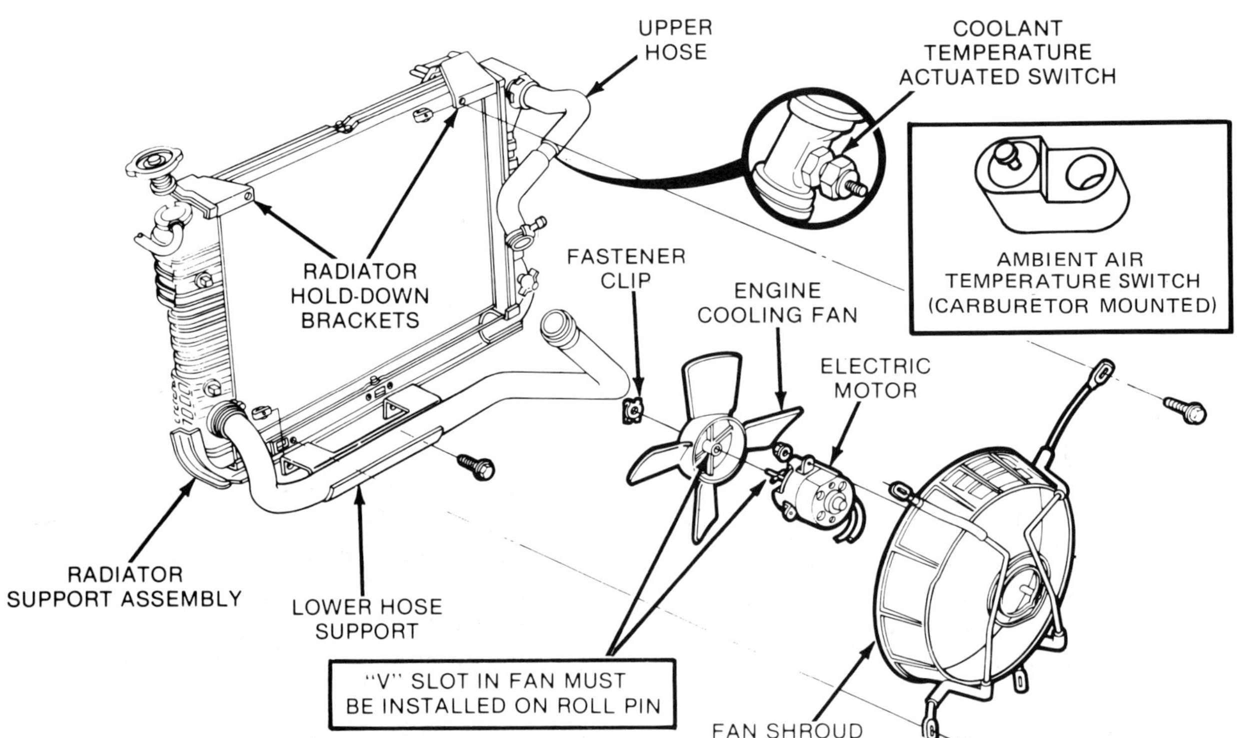

FIGURE 33
The fan shroud acts as a short wind tunnel.

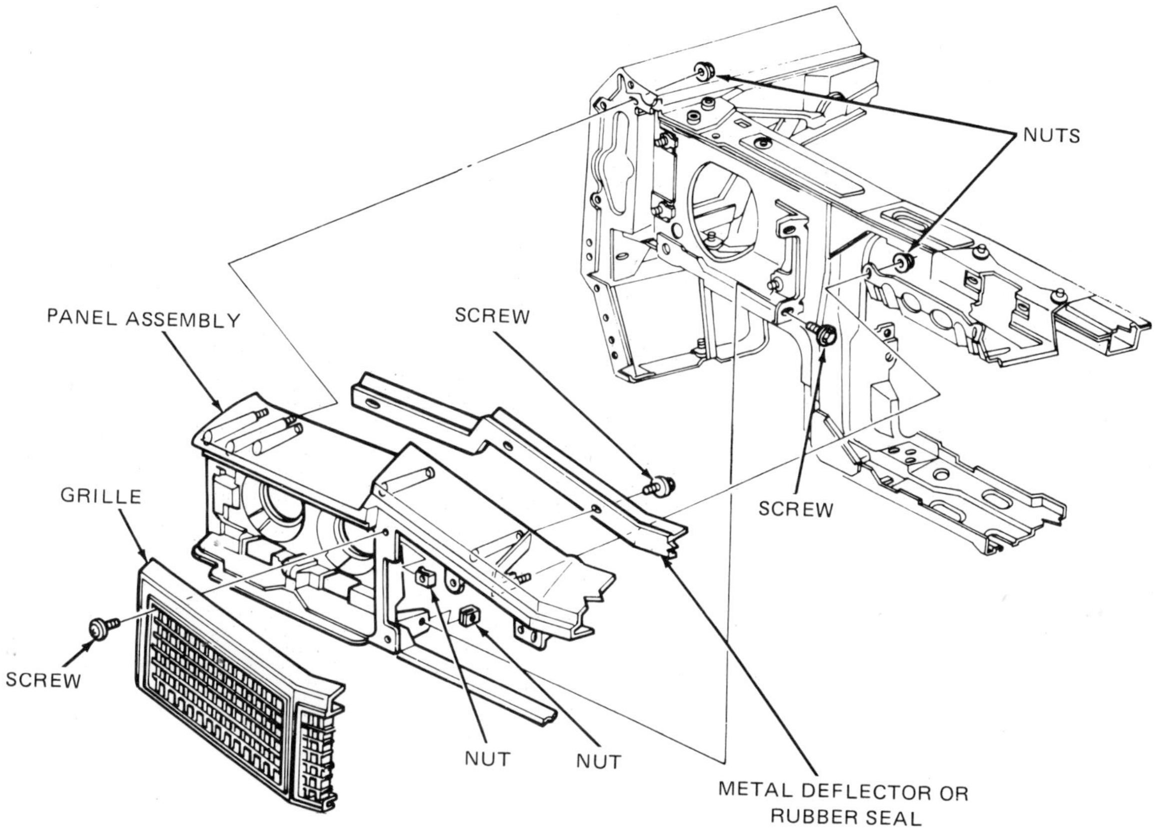

FIGURE 34
Radiator top yoke seal.

Shrouding

In many cases, placing a fan close behind the radiator provides ample air flow to dissipate heat from the radiator. On many cars (usually with air conditioning), however, the fan needs additional help. This help is provided by placing the fan in a shroud (Figure 33).

The shroud prevents the fan from pulling air from anywhere except through the radiator. Without shrouding, a fan can pull air from alongside, above, or below the radiator, without contributing to radiator cooling.

SHOP TIP

Many people do not realize the importance of the rubber seal between the radiator top yoke and the hood (Figure 34). This seal increases air flow through the radiator by preventing "ram" air from going over the top of the radiator. An overheating problem on an air-conditioned car is sometimes caused when this seal is missing or damaged.

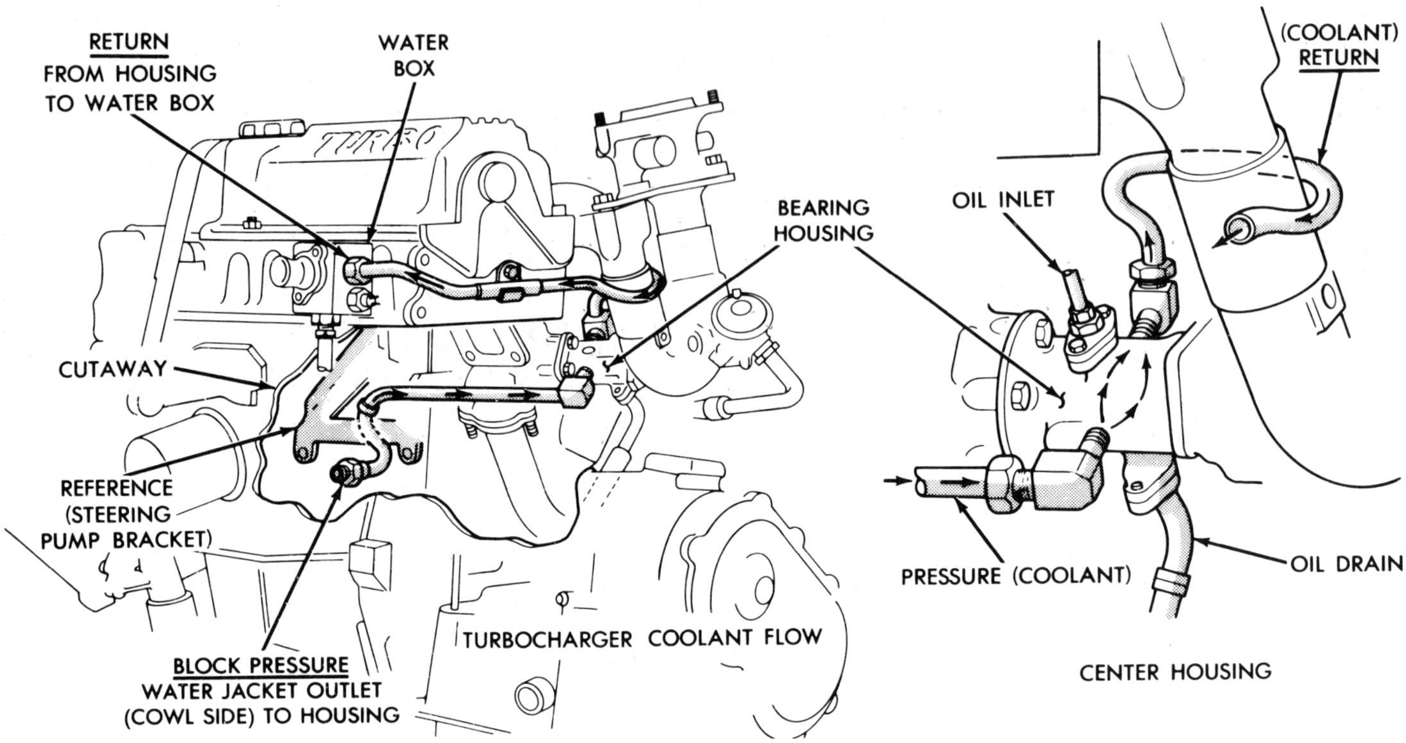

FIGURE 35
Turbocharger bearing housing coolant routing for the Chrysler 2.2 liter engine. (Courtesy of Chrysler Corporation)

FIGURE 36
Ethylene glycol-based antifreeze.

38. Bent or damaged air conditioning condenser fins
39. Incorrect fan spacing (Figure 43)
40. Fan installed backwards
41. Head gasket installed backwards
42. "Frozen" automatic transmission stator
43. Leaking automatic transmission cooler
44. Blocked coolant passages in cylinder head
45. Mixing premixed 50/50 antifreeze with water

Temperature Too Low—Slow Engine Warmup

1. Stuck open thermostat
2. Faulty heater controls
3. Inaccurate temperature gauge
4. Faulty temperature sending unit

Coolant Recovery System Inoperative

1. Low coolant level
2. Leak in system
3. Defective pressure cap
4. Clogged or leaking overflow tube
5. Restricted recovery bottle vent

Noise

1. Fan contacting shroud
2. Worn water pump (Figure 44)
3. Worn fan clutch
4. Loose or glazed belt(s)
5. Broken, cracked, or loose water pump pulley
6. Reversed heater hose connections
7. Internal coolant restriction
8. Head gasket installed backwards
9. Incorrect belt alignment

No Coolant Flow Through Heater Core

1. Collapsed or restricted heater hose
2. Restricted heater core (Figure 45)
3. Restricted return inlet in water pump
4. Restricted outlet in thermostat housing
5. Faulty or improperly adjusted heater control valve

Corrosion

1. Use of water containing high concentration of lime and minerals (Figure 46[a] and [b])
2. Use of antifreeze for extended length of time
3. Small leaks at water pump or hoses allowing air to be drawn in
4. Electrolysis
5. Combustion gas entering the cooling system

"Hot" Indicator Light Not Lit When Cranking Engine

1. Burned out bulb
2. Open light circuit
3. Defective ignition switch

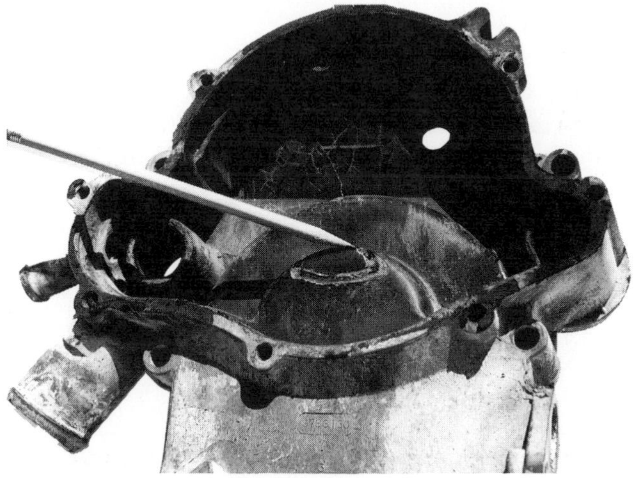

FIGURE 39
This aluminum front cover water pump cavity is "pinholed" from electrolysis. Pinholing allowed coolant to leak into the timing chain area and into the crankcase.

FIGURE 40
Some cylinder heads have threaded-in water plugs. They can leak or rust through and fill the crankcase with coolant.

FIGURE 41
A rusted-through thermal vacuum valve may allow coolant to be drawn into the intake manifold. This problem can make you think the engine has a blown head gasket.

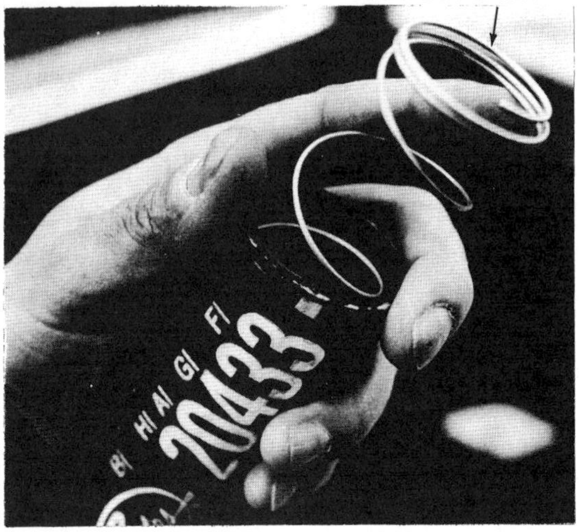

FIGURE 42
Without this support spring, the water pump suction at higher speeds can cause the lower hose to collapse and block coolant flow. (Courtesy of The Gates Rubber Company)

COOLING SYSTEM SERVICE

Under continued vehicle operation, ethylene glycol loses its effectiveness. For this reason, many car manufacturers recommend changing the coolant every 12 months.

Every several months, test the coolant with a hydrometer (Figure 47). Check the freezing point in winter months and the boiling point in summer months. If the coolant is low, add a 50% concentration of antifreeze and water. Do not add only water, because water dilutes the concentrate.

FIGURE 43
A missing fan spacer can cause engine overheating.

FIGURE 44
Badly corroded and worn water pump impeller shaft and bushings. (Courtesy of Dana Corporation)

FIGURE 45
Clogged heater core tubes.

Testing for External Leaks in the Cooling System

To test the system for external leaks, use a cooling system pressure tester to pressurize the system to the rating on the radiator cap (Figure 48). *Do this when the system is cold to avoid being burned.*

CAUTION: Do not exceed the rated pressure of the original cap, since damage to the radiator or heater core may result.

Be sure the seal on the tool and the cap seat in the radiator filler neck are in good condition. If the system will not hold pressure, check for external leaks at:

1. Hoses and connections.
2. Water pump.
3. Thermostat housing.
4. Core plugs (Figure 49).
5. Drain cocks.
6. Radiator seams.
7. Heater core.

If the leak is not external, it must be in either the engine or the transmission cooler.

Testing for Internal Leaks in the Cooling System

Internal coolant leakage is a much more serious problem than is external leakage. This type of leakage can damage the engine through rust, gum formation, oil dilution, hydrostatic lock, overheating, performance loss, and cylinder head and/or block erosion.

To verify and locate a combustion gas leak, a cooling system pressure tester is often used (refer to Section A of this book).

CORROSIVE WATER AREAS

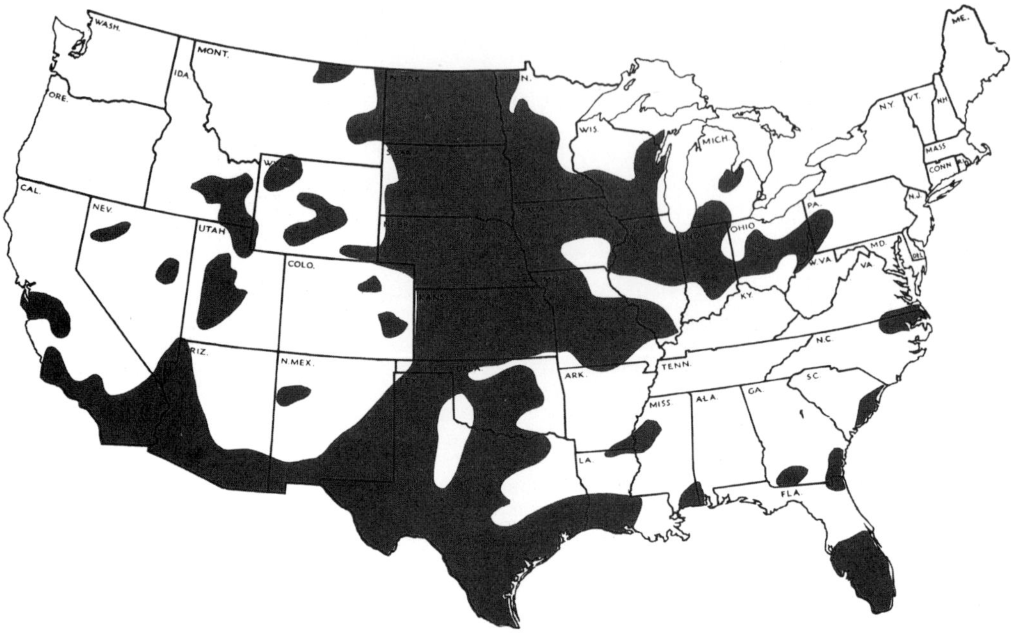

(a)

HARD WATER AREAS

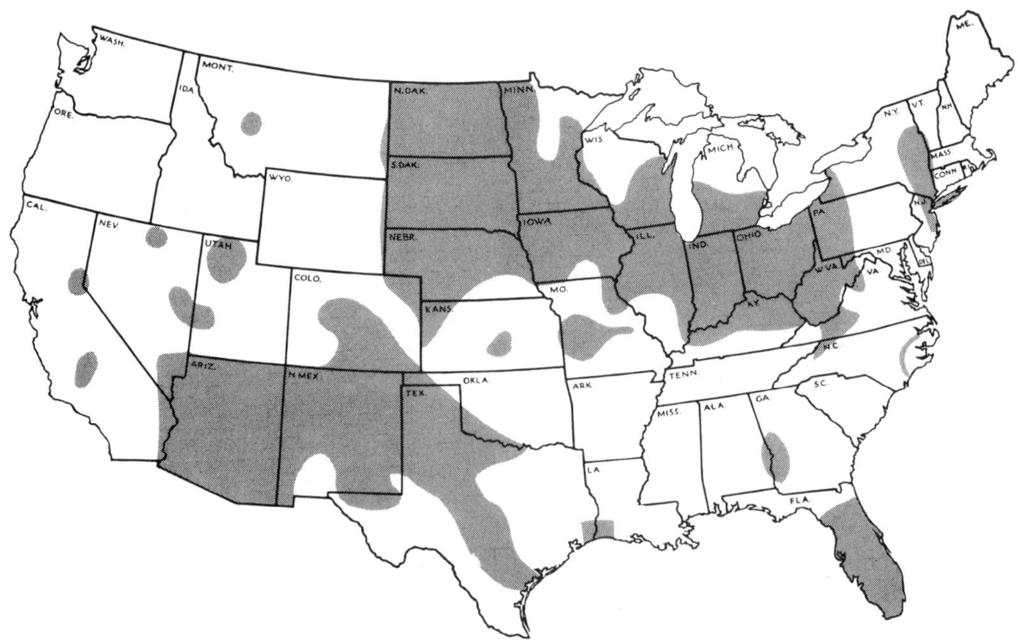

(b)

FIGURE 46

(a) If you operate in these areas, extra care should be taken to protect the cooling system against rust and corrosion. (Courtesy of Detroit Diesel) (b) If you operate in these areas, water hardness is undesirable and can cause scale buildup (magnesium and calcium). (Courtesy of Detroit Diesel)

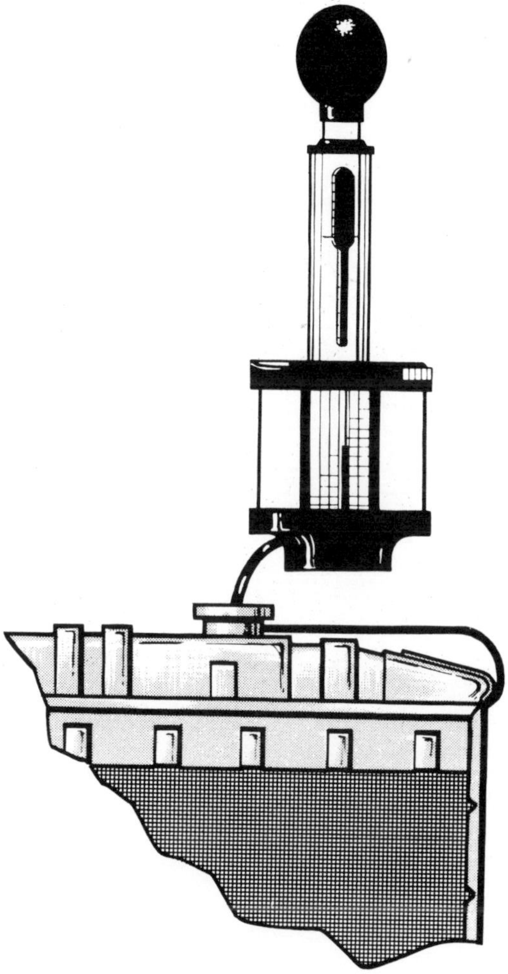

FIGURE 47
Coolant condition and antifreeze check.

FIGURE 48
Cooling system pressure tester.

NOTE: If the pressure tester gauge needle jumps when the engine is accelerated, the bad cylinder may be located by disconnecting the spark plug wires one at a time. The gauge fluctuations will become noticeably smaller when the wire to the leaking cylinder is disconnected.

WARNING: Pressure often builds up fast. Any excessive amount of pressure built up by continuous engine operation must be released to a safe pressure point. Never permit system pressure to exceed 18 psi.

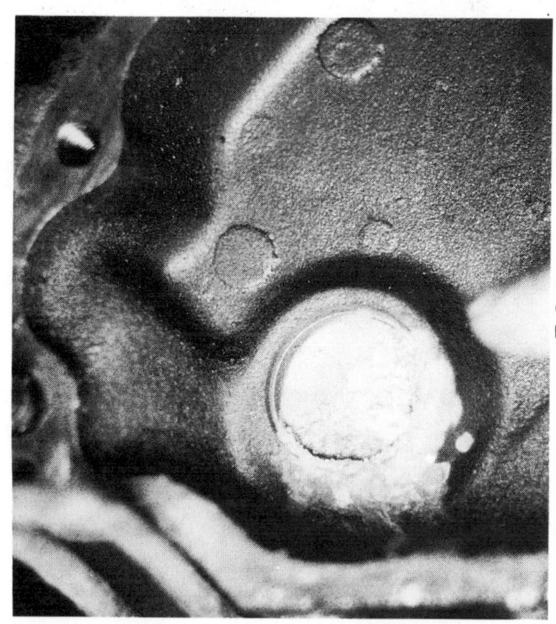

FIGURE 49
This rusted-through core plug is located at the back of the block. Replacement requires removal of the converter or flywheel. Such problems can be avoided by using brass plugs.

FIGURE 50
A crack in the head or block can be checked by introducing compressed air into each cylinder with the piston at TDC firing position and checking for bubbles at the thermostat housing.

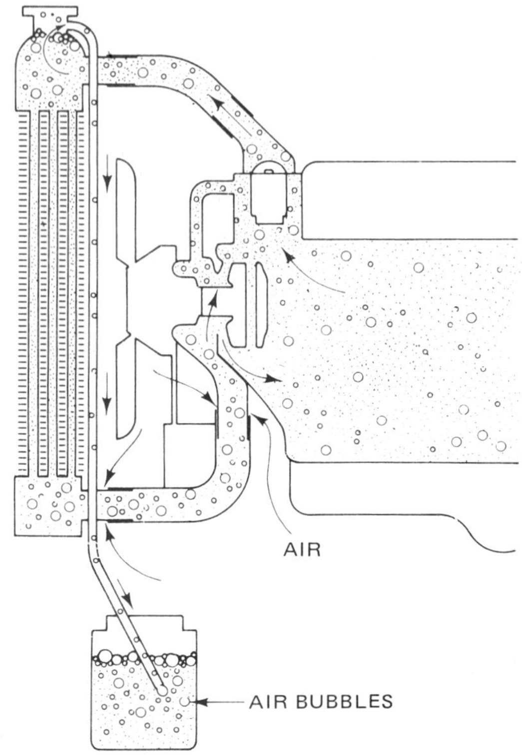

FIGURE 51
Testing for the presence of combustion gas bubbles coming from the overflow tube. (Courtesy of John Deere and Company Technical Services)

FIGURE 52
This product can seal cracks in a block or head. *Follow instructions exactly as printed on the can.*

SHOP TIP
A crack in the head or block can often be checked by introducing compressed air into each cylinder with the piston at TDC on compression. Then the coolant at the thermostat housing or at the radiator filler neck opening should be checked for bubbles (Figure 50).

Another way to test for combustion gas leakage into the coolant is to use a combustion leak tester (refer to Section A of this book).

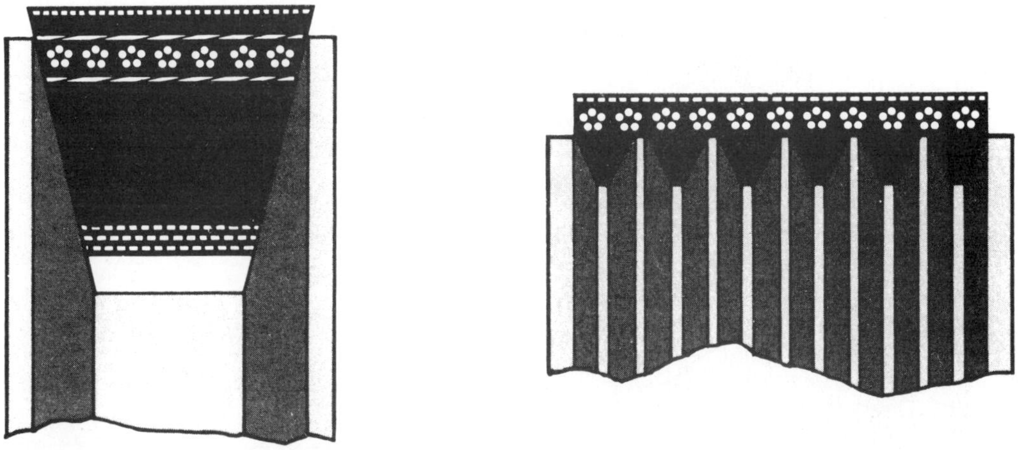

FIGURE 53
Comparison of a conventional V-belt (left) and a micro V-belt (right) in their respective pulleys. (Courtesy of The Gates Rubber Company)

Air from the atmosphere can be sucked into the cooling system through leakage at the inlet side of the water pump. This accelerates rust formation and interferes with coolant circulation. If such a condition is suspected, place the end of the overflow tube into a container of water (Figure 51). Operate the engine between 1500 and 2000 rpm for about 30 seconds. Air bubbles indicate an air leak (water pump seal, bypass hose, or lower radiator hose) or combustion gas leakage.

SHOP TIP
Sometimes a small crack in a block or head can be repaired by using a metallic block sealer (Figure 52).

Testing the Oil Cooler for Leaks

1. Disconnect both oil cooler lines at the radiator.

2. Connect a pressure gauge to one cooler connection and a shut-off valve to the other. Close the valve.

3. Connect a source of air pressure to the valve.

4. Coat all fittings with oil.

5. Open the shut-off valve and apply (up to 100 psi) air pressure. Oil bubbles will reveal fitting leaks. Repair all leaks.

6. Close the shut-off valve. The pressure gauge reading will drop if the cooler is leaking.

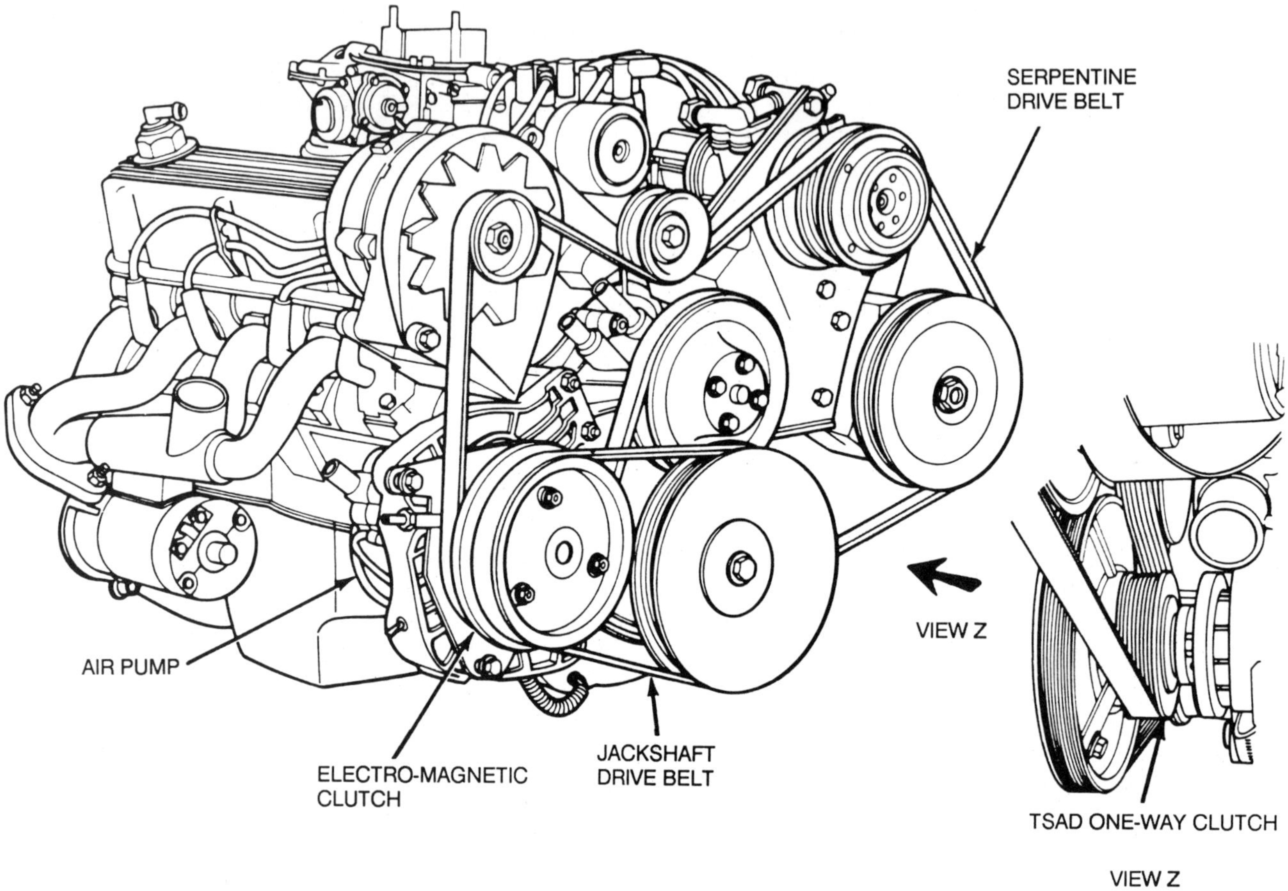

FIGURE 54
Serpentine drive belt system. (Courtesy of Ford Parts and Service Division)

DRIVE BELTS

In most automobiles, belts drive:

1. The alternator.

2. The water pump.

3. The fan.

4. A number of accessories (power steering pump, air conditioning compressor, and air pump).

Today, V-belts and V-ribbed drive belts are used (Figure 53[a] and [b]).

V-belts drive through contact at the sides, not on the inside diameter. The friction between the sides of the belt and the sides of the pulley groove transmits power from one pulley to the other.

Both sides of a V-ribbed belt can be used to transmit power. Such a belt set-up is called a *serpentine drive* (Figure 54).

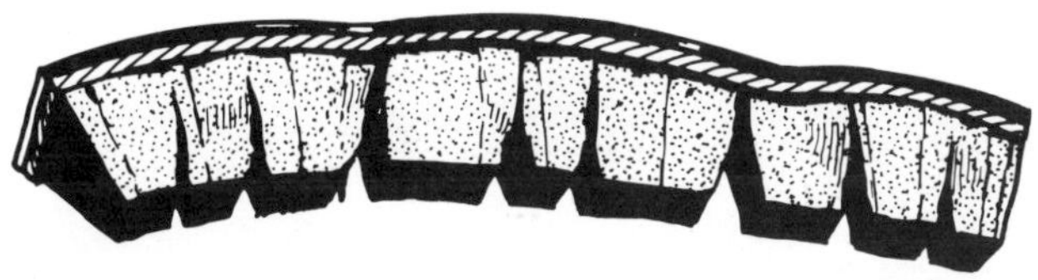

FIGURE 55
Badly cracked belt. (Courtesy of The Gates Rubber Company)

FIGURE 56
Glazed belt sidewall. (Courtesy of The Gates Rubber Company)

Belt Inspection

Belts should be inspected periodically for wear and proper tension. To inspect a belt, first be sure the engine is off. If the car has a thermostatically controlled electric fan, disconnect it from the battery. Then twist the belt so the sidewalls and bottom are visible.

NOTE: When checking, remember that belts shown to be defective may have been damaged by misalignment, a bad pulley, or some faulty mechanical component.

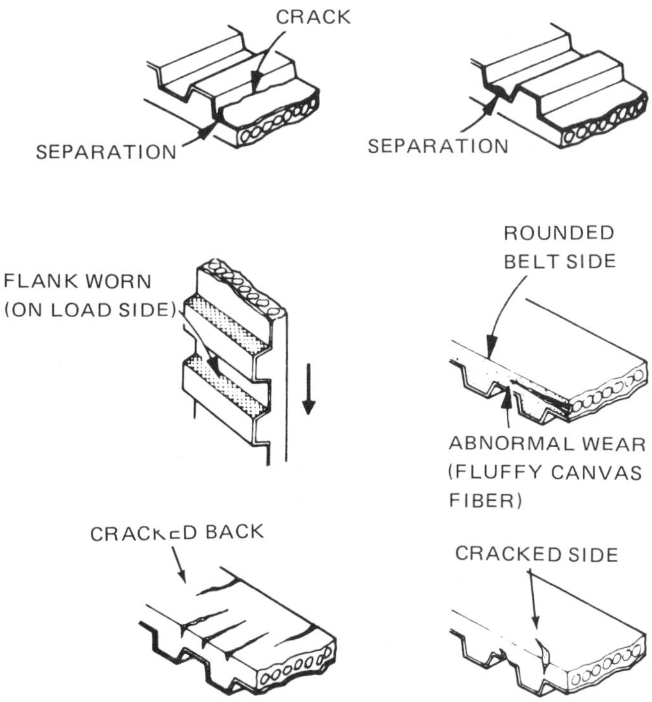

FIGURE 57
V-ribbed belt inspection areas.

If a belt exhibits any of the following signs, it should be replaced:

1. **Cracks** (Figure 55). Deep base cracks that appear at regular intervals are usually caused by the belt turning around a too-small pulley. Smaller, irregular cracks usually indicate that the belt has been in service a long time.

2. **Glazed or Shiny Sidewalls** (Figure 56). The friction caused by a belt operating too loosely causes the belt sidewalls to become slick and shiny. Grease and oil on a pulley can also cause this condition.

3. **Gouged Edge.** A gouged edge on a belt can be caused by a damaged pulley or interference with something while operating, or by pulley/belt misalignment.

4. **Jagged or Streaked Sidewalls.** This is usually the result of foreign material in the pulley (sand, gravel, or paint), or a rough pulley wall.

5. **Rib Damage** (Figure 57). V-ribbed pulleys must be clean at all times. Foreign objects in the pulley can damage the rib side of the belt.

SHOP TIP

Always be sure that V-ribbed belts are aligned correctly, since it is difficult to see the fit on the pulleys (Figure 58).

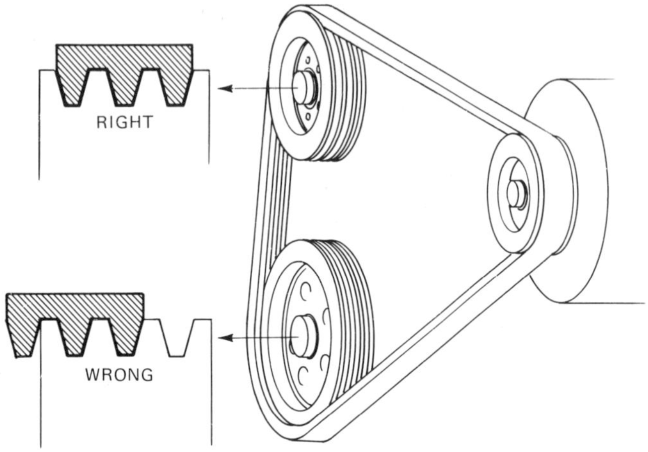

FIGURE 58
It is easy to make an alignment error when installing V-ribbed belts.

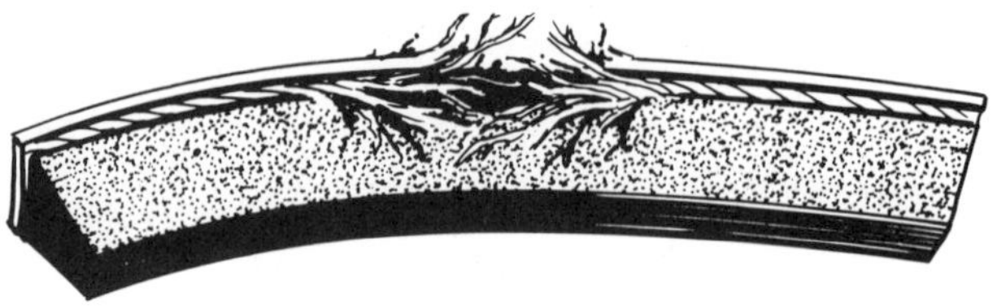

FIGURE 59
Damage caused by prying on a belt during installation. (Courtesy of The Gates Rubber Company)

6. Ply Break (Figure 59). Sometimes a foreign object in the pulley can cut into the belt and break the tensile cords. Tensile break can also occur if a belt is forced or pried on during installation.

Drive Belt Troubleshooting

Refer to Chart 4 for some additional drive belt problems.

> **SHOP TIP**
> Belts frequently develop a "squish" sound at engine idle. Diagnose by squirting water on the suspected belt(s) to try and eliminate the noise temporarily. Cleaning the belt and pulley will sometimes correct the problem. If not, replace the belt.

CHART 4

TROUBLESHOOTING CHART FOR DRIVE BELT PROBLEMS

Belt Squeal When Accelerating Engine
1. Loose belt
2. Glazed belt

Belt Squeak at Idle
1. Loose belt
2. Dirt and/or paint imbedded in belt
3. Misaligned pulleys
4. Non-uniform or worn belt
5. Non-uniform or worn pulley

Insufficient Accessory Output Due to Belt Slippage
1. Loose belt
2. Excessively glazed or worn belt

Belt Rolls Over in Groove or Jumps Off
1. Broken cord in belt
2. Loose belt
3. Misaligned pulleys
4. Eccentric pulley
5. Foreign object present between belt and pulley
6. Dual belts not matched

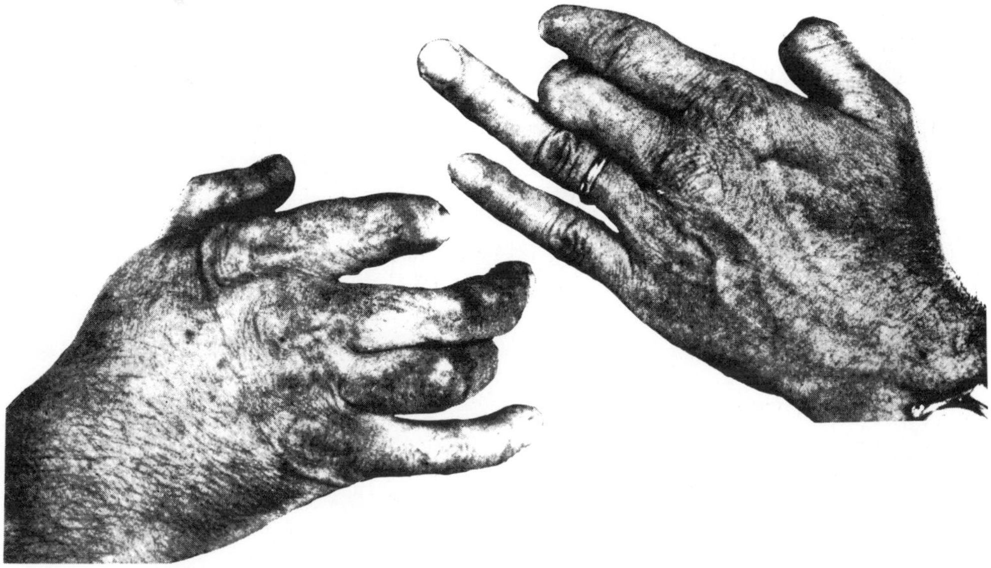

FIGURE 60
Accidental engine crank that occurred while a belt was being installed has caused this to happen more than once. (Courtesy of The National Safety Council)

Belt Installation

On most car models, belt installation should follow these steps:

1. For your own safety, disconnect the car's battery (Figure 60).

2. Check pulley alignment before removing the old belt (Figure 61).

3. Examine the entire drive belt routing before you loosen or take anything apart. A diagram of the routing of serpentine belts is often found under the hood.

4. Loosen the necessary bolts on the accessories and/or pulleys that act as belt tension adjusters (Figure 62).

5. Reduce belt tension by pushing accessories or pulleys toward the engine block until the belt is loose enough to remove with your hands.

6. Inspect the belt pulleys. Worn pulleys should be replaced (Figure 63). Remove any rust, oil, grease, or burrs from pulley grooves.

7. Install the new belt. Be sure the new belt is the correct size.

NOTE: Dual belts must be replaced in matched sets.

> **SHOP TIP**
> If a belt seems too short, try installing it in different sequences onto the pulleys. With some belts, there may be only one installation sequence that works.

8. Tension the belt by pulling accessories or pulleys away from the engine block. Tighten the bolts.

9. Run the engine 15 to 20 minutes to allow the belt to "seat"; then readjust the tension.

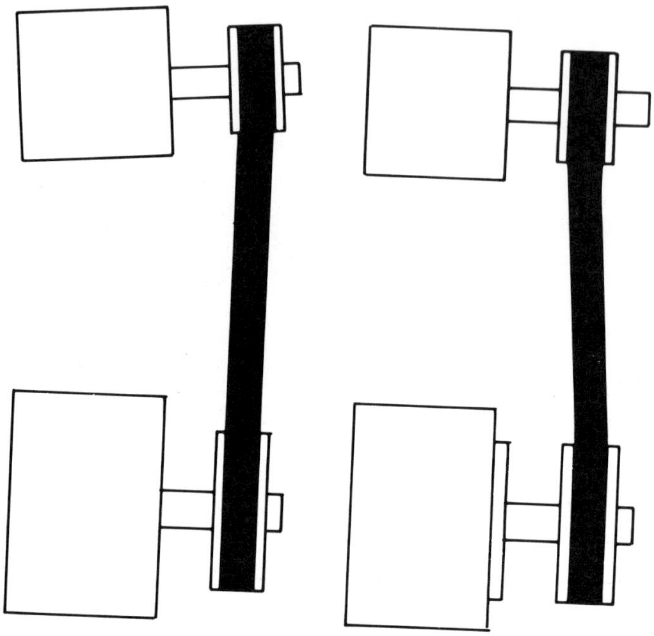

FIGURE 61
Belt misalignment can be due to non-parallel pulley shafts or pulleys located improperly on shafts.

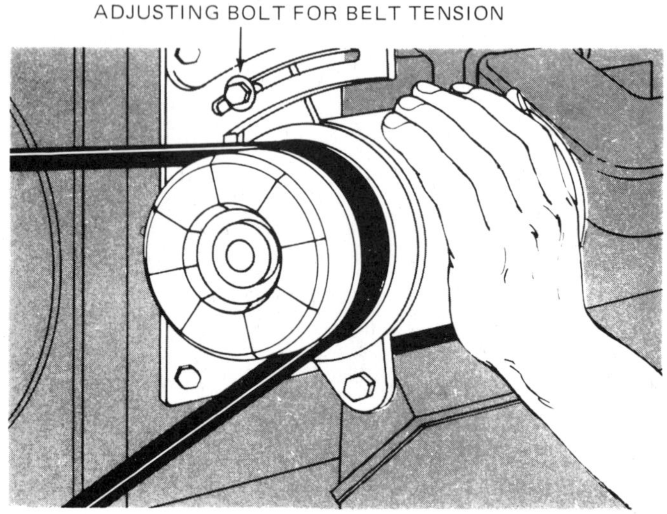

FIGURE 62
Typical belt tension adjustment mechanism. (Courtesy of The Gates Rubber Company)

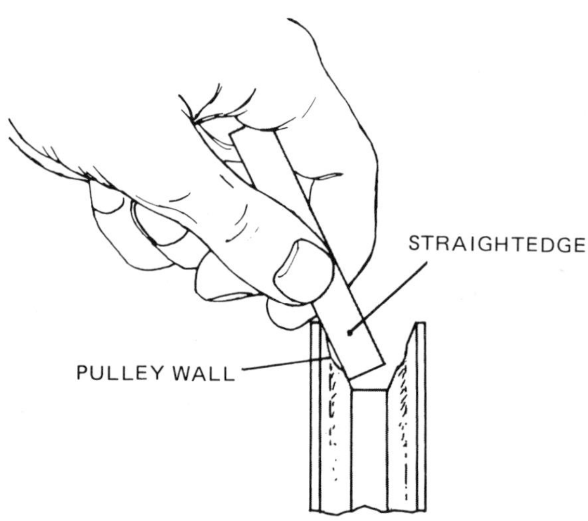

FIGURE 63
Worn pulleys are often hard to identify. They can, however, usually be detected with a straightedge on the inside pulley wall or by feeling with a fingernail.

Satisfactory performance of belt-driven accessories depends on proper belt tension. Three tensioning methods are used:

1. **Belt tension gauge method** (Figure 64).

2. **Torque equivalent method.** Adjustable accessory brackets are sometimes provided with a 1/2-inch square hole or torque pry point for installing a torque wrench and exerting a specific torque on the belt (Figure 65).

3. **Belt deflection method.** Place a straightedge across two adjacent pulleys and adjust belt tension with a push-pull force of 10 pounds applied at midpoint until the specified deflection is obtained.

For example, Chart 5 shows the belt tension specifications for the 2.2/2.5L engines used in Chrysler Corporation front wheel drive cars.

CAUTION: *Always be sure a replacement drive belt is sized correctly both by length and width. A slipping belt can have disastrous results!*

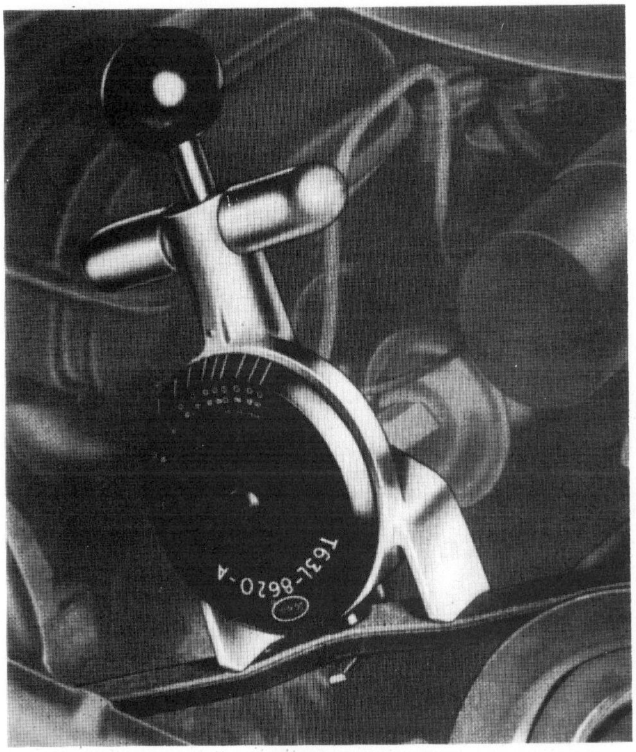

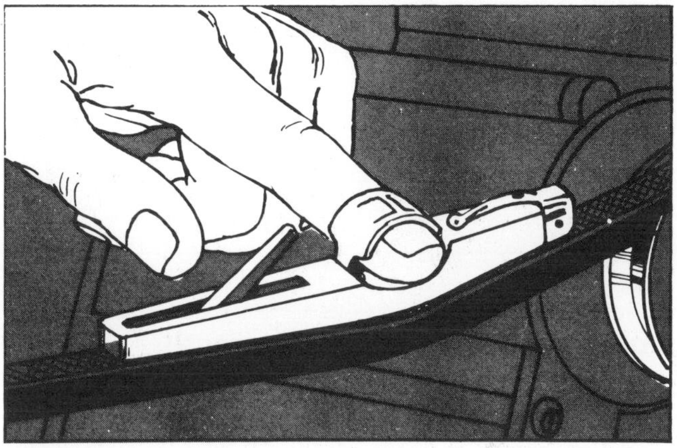

FIGURE 64
Contrary to what many people think, you *cannot* get an accurate belt tension check by pressing your thumb against the belt between its pulleys. Most modern belts run under far greater tension than this old "rule of thumb" method detects. Here are two styles of tension gauges that can be used to give you a correct reading.

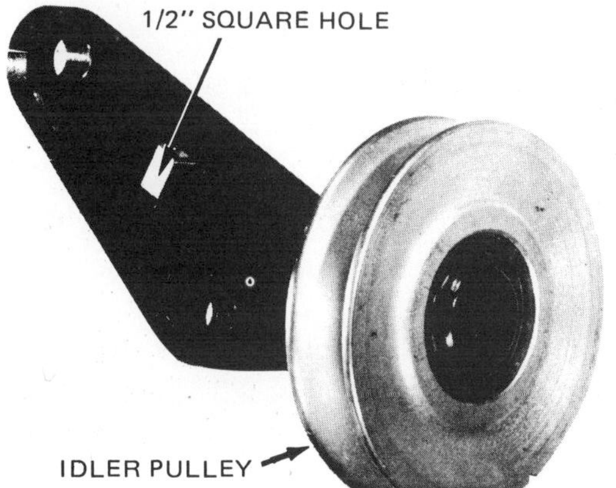

FIGURE 65
This idler puller bracket has a torque wrench pry point hole for belt tensioning.

HOSES

Hoses are a key part of the entire cooling system. They are exposed to both cold to hot temperature extremes and are subjected to constant flexing.

As a result, hoses can become clogged, cracked, and damaged by hardening (Figure 66[a] and [b]). At the other extreme, hoses can become soft and swollen, which can lead to rupturing. Periodically make a visual inspection of all hoses and check their condition by squeezing. A good hose feels firm, not soft and mushy.

NOTE: Many authorities consider it a good maintenance practice to replace all hoses every two years on today's vehicles.

CHART 5

Accessory Drive Belt		Gauge	Deflection	Torque
Air Conditioning	New	105 lb.	8mm (5/16 in.)	54 N·m (40 ft. lbs.)
Compressor	Used	80 lb.	9mm (7/16 in.)	41 N·m (30 ft. lbs.)
Air Pump	New	—	5mm (3/16 in.)	61 N·m (45 ft. lbs.)
	Used	—	6mm (1/4 in.)	47 N·m (35 ft. lbs.)
Alternator/Water Pump	New	115 lb.	3mm (1/8 in.)	149 N·m (110 ft. lbs.)
Poly "V"	Used	80 lb.	6mm (1/4 in.)	108 N·m (80 ft. lbs.)
Power Steering Pump	New	105 lb.	6mm (1/4 in.)	102 N·m (75 ft. lbs.)
	Used	80 lb.	11mm (7/16 in.)	75 N·m (55 ft. lbs.)

Connect Heater Hoses Correctly

Normal coolant flow through the heater flow control valve is in a direction that tends to open the valve (Figure 67). If flow through the valve is reversed, the coolant flow will act against the valve when it is open. This can cut off the heat supply to the heater core at intermediate or higher engine speeds.

SHOP TIP

Reversed heater hoses can cause certain heater valves to bounce back and forth between spring force and flow force. The noise sounds very much like a rod bearing knock.

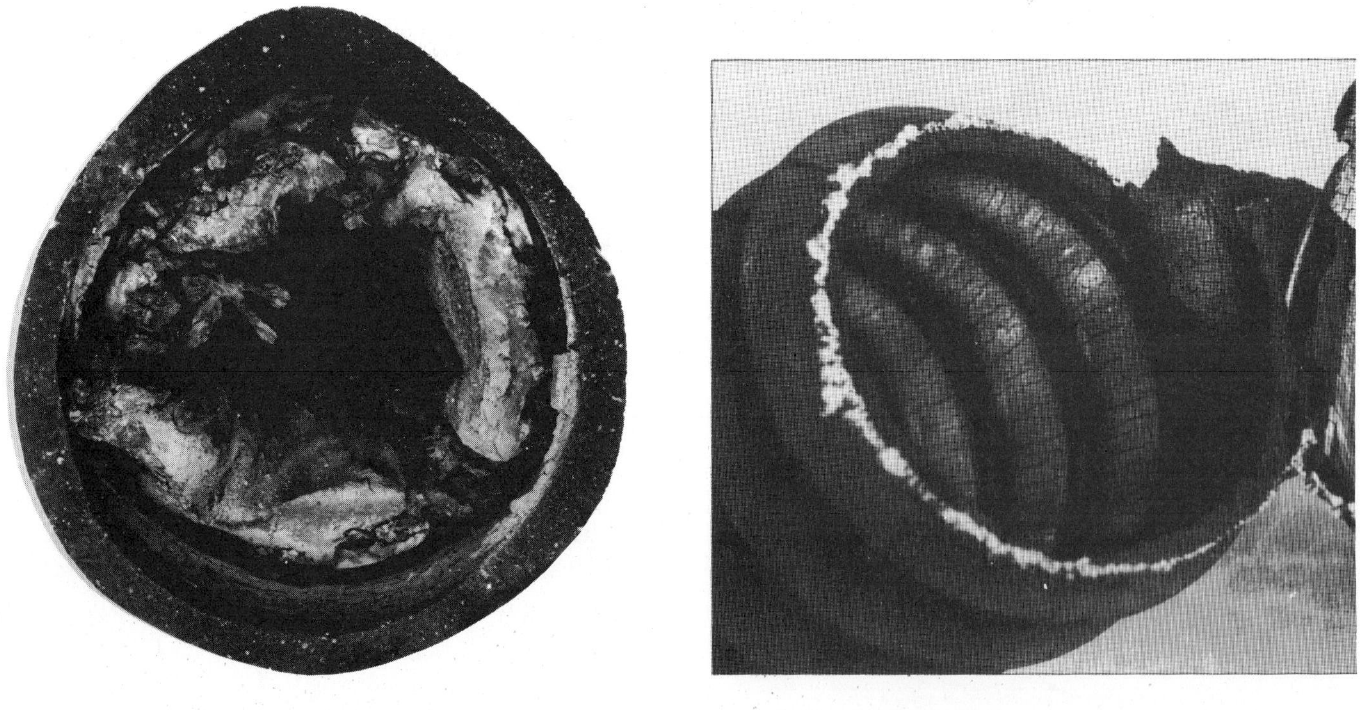

(a) (b)

FIGURE 66

(a) Badly clogged radiator hose. (b) A hose isn't necessarily good just because it looks good on the outside. (Courtesy of The Gates Rubber Company)

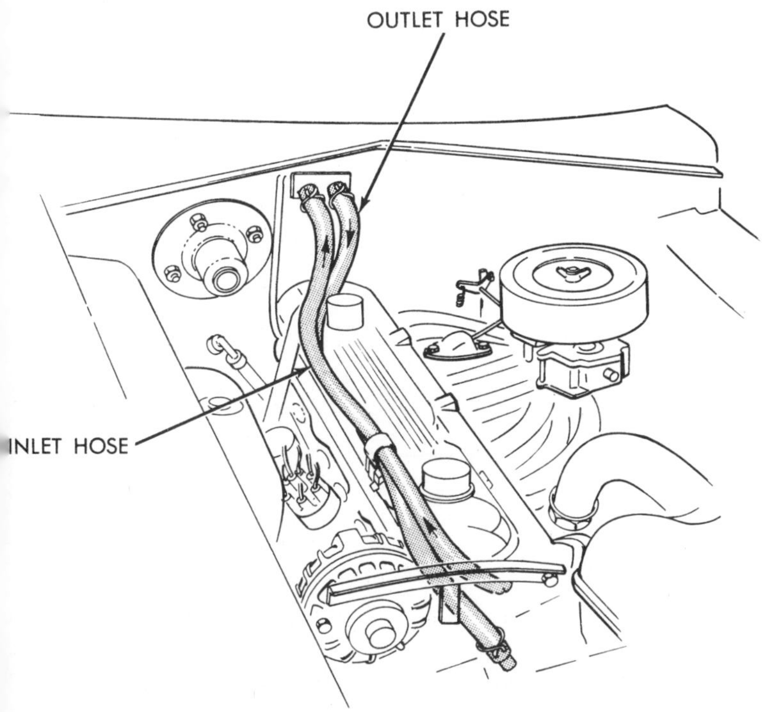

FIGURE 67
Typical heater hose routing. (Courtesy of Chrysler Corporation)

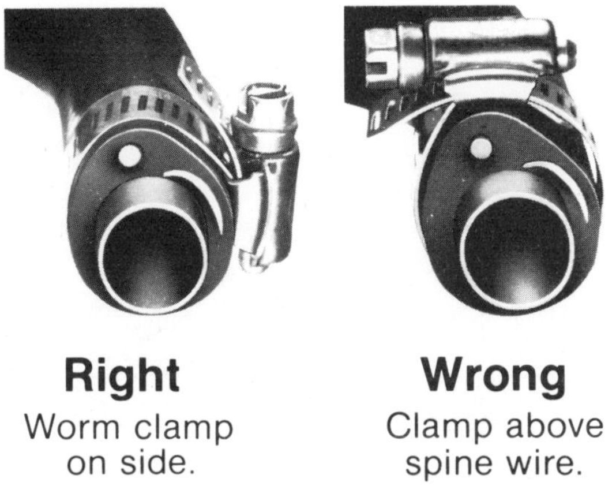

FIGURE 68
Heater hose clamp installation. (Courtesy of The Gates Rubber Company)

SHOP TIP

When installing a heater hose that has a spine wire, care must be taken not to position the screw mechanism of the clamp directly above the spine wire (Figure 68).

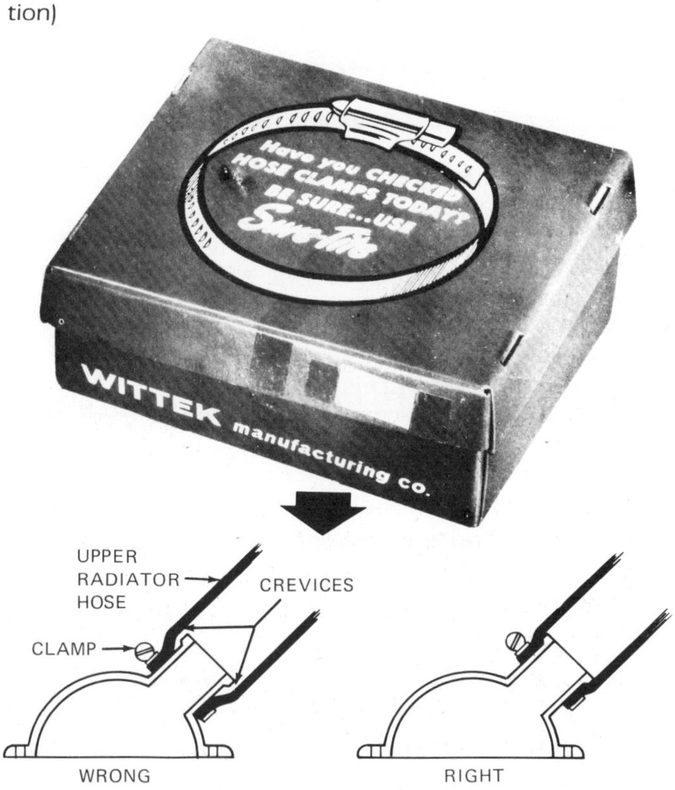

FIGURE 69
An incorrectly positioned hose clamp on the thermostat housing will often corrode and ruin the housing neck. See Figure 71.

Installing New Radiator Hoses

1. Locate the radiator drain petcock (if so equipped) and drain the coolant. When replacing an upper radiator hose, drain out only 3 or 4 quarts of coolant. When replacing a lower hose, drain the entire contents of the radiator.

2. Remove the clamps on the old hose. If the clamps are badly corroded or difficult to remove, cut them off with tin snips and replace with new clamps.

3. Pull the old hose off at the connections. If the hose is stuck, cut it off with a sharp knife.

CAUTION: Inserting a screwdriver under a stuck hose to break it loose from the radiator can damage metal nipples and break plastic nipples.

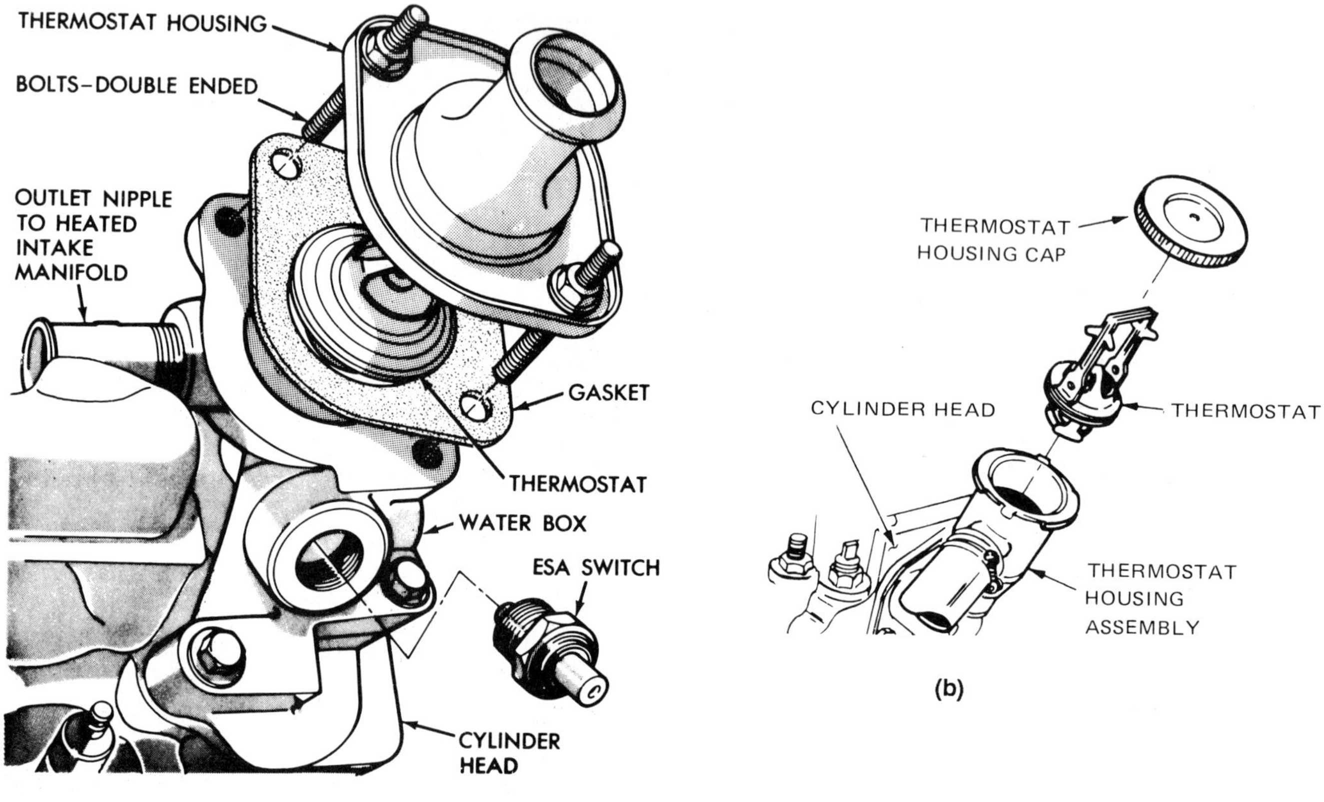

FIGURE 70

(a) Thermostat housing and water box assembly for the Chrysler 1.6 liter engine. (Courtesy of Chrysler Corporation) (b) The thermostat in this Pontiac engine is retained by the radiator cap (thermostat housing cap). (Courtesy of Pontiac Motor Division, General Motors Corporation)

4. Clean off the engine and radiator inlet and outlet nipple tubes with coarse sandpaper. Be sure the radiator tubes are not bent, dented, or cracked.

5. Position new clamps 1/4″ to 1/2″ from the ends of the new hose, making sure they are behind the ridge on the radiator outlet nipple (Figure 69).

6. Tighten the clamps. *CAUTION: Overtightening can cut into the hose.*

7. Close the radiator petcock and refill the system with coolant. Start the engine, bring it to operating temperature, and check for leaks.

REPLACING A THERMOSTAT

1. Drain several quarts of coolant from the system and remove the thermostat housing bolts. Tap the housing lightly with a hammer to loosen the gasket seal. Pull the housing off and remove the thermostat.

NOTE: Be sure to note how the thermostat is installed (Figure 70[a] and [b]).

2. Clean off the old gasket and scrape the gasket surfaces clean.

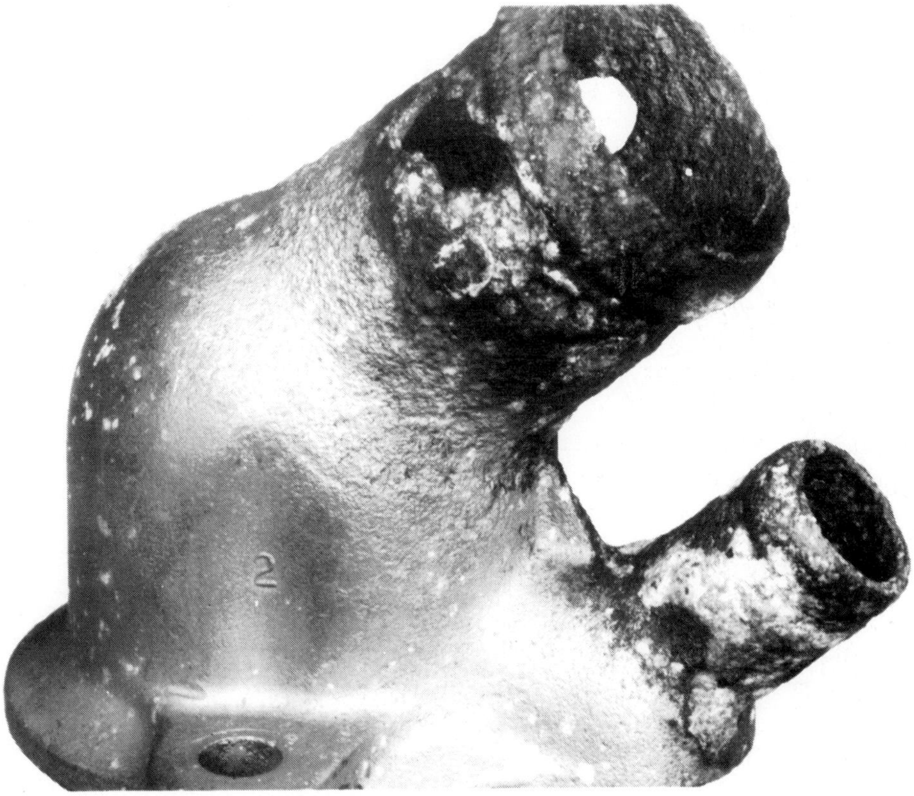

FIGURE 71
Badly corroded thermostat housing.

3. Check the thermostat housing for corrosion and the mounting flange for flatness (Figure 71).

4. Place the new thermostat with the spring and actuator pointing away from the radiator (or down into the engine). It is possible to install a thermostat upside down. The result is an overheated engine.

COMMENT: *It is not a good practice to remove the thermostat and not replace it. This can cause engine sludge formation, cavitation damage, and a reduction in fuel economy.*

5. Install a new gasket (if used) with sealer.

6. Replace the thermostat housing bolts and tighten them evenly. Refill the system with coolant and start the engine. Bring the engine to operating temperature and check for leaks.

CAUTION: *Be sure the thermostat remains in place prior to tightening the housing bolts. Many housings have been cracked by thermostats that have slipped out of place.*

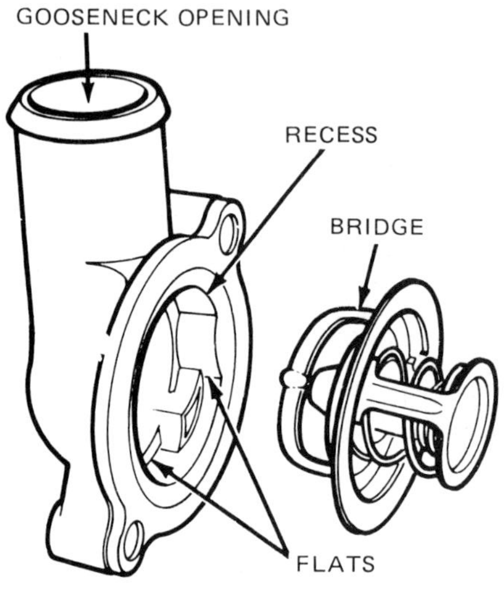

FIGURE 72
Vertically mounted thermostat.

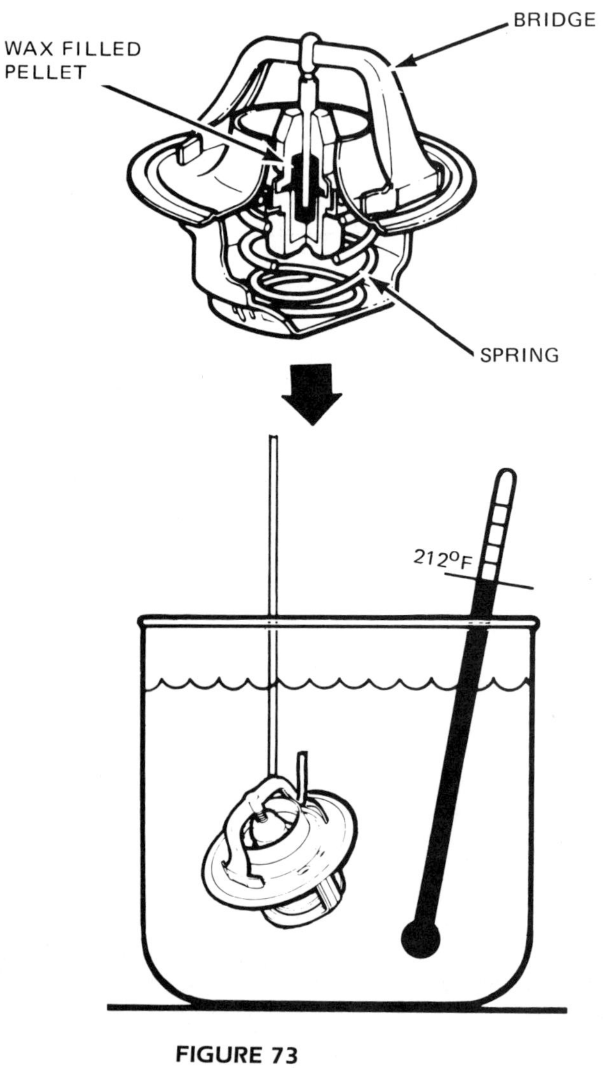

FIGURE 73
Testing a thermostat.

> **SHOP TIP**
> Vertically mounted thermostats can easily slip out of position during installation (Figure 72). This can be prevented by looping a rubber band through the thermostat bridge and pulling on it through the gooseneck opening. Retrieve the rubber band by cutting it after installing the thermostat.

TESTING A THERMOSTAT

An accurate thermometer and a means of gradually warming a pan of water are needed to test a thermostat. The procedure is as follows:

1. Push the thermostat open with your thumb just enough to push a piece of nylon thread or shim stock between the valve and its seat.

2. Suspend the thermostat in a pan of water by the nylon thread or shim stock (Figure 73). Be sure the thermostat does not touch the pan surfaces.

3. Gradually heat the thermostat. As soon as the thermostat starts to open, it will release the nylon thread and drop to the bottom of the pan. At that moment, read the temperature on the thermometer. The temperature should be within a few degrees of the thermostat rating (usually stamped on the thermostat).

CLEANING THE COOLING SYSTEM

Every cooling system should be completely drained and thoroughly flushed once a year. If the system is extremely dirty, reverse flushing is suggested (Figure 74).

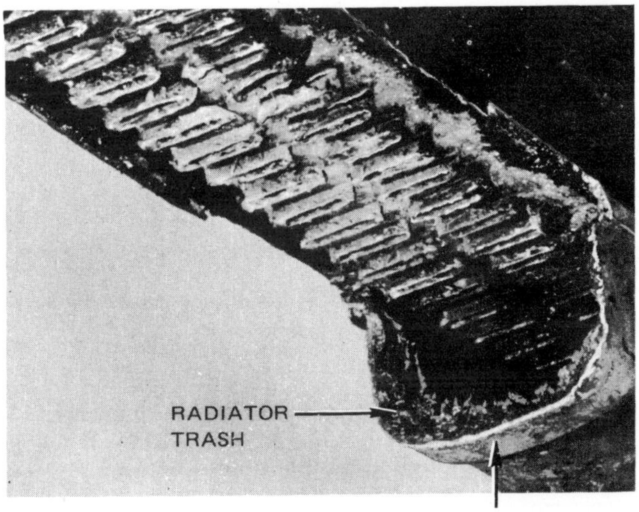

FIGURE 74
An accumulation of trash (bits of rubber, lime, and scale) is apparent in the bottom of this cutaway radiator tank.

FIGURE 75
Test for radiator core clogging. (Courtesy of Dana Corporation)

FIGURE 76
Cooling system cleaner.

Checking for a Clogged Radiator Core

Here is a simple test for restriction of coolant flow through the radiator:

1. Disconnect both upper and lower radiator hoses and plug both hose outlets on the radiator.

2. Fill the radiator to the neck and remove the plug from the lower hose nipple.

3. Water flow should be free and fast (Figure 75). The radiator should empty quickly. Cleaning and flushing are required if water flows at less than lower hose outlet capacity.

Flushing the Radiator

Rust, scale, and oil or grease deposits in the radiator cannot be removed simply by flushing with water. Many chemical cleaners are available that emulsify oil and grease, loosen scale and remove rust. Deposits are then held in suspension so that they may be flushed out.

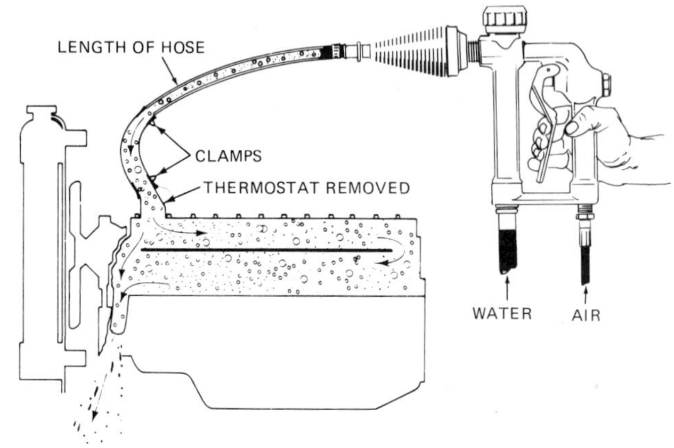

FIGURE 77
Reverse flushing.

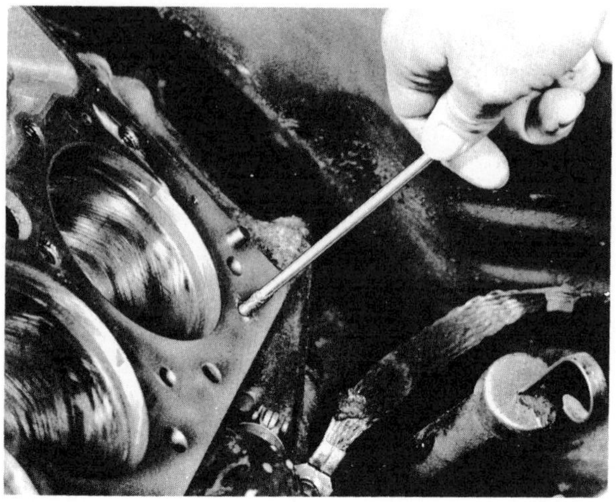

FIGURE 78
Probing for engine block deposits. (Courtesy of Dana Corporation)

Some chemical cleaners can be kept in the cooling system for some time before they are flushed out. Acid-type cleaners must be neutralized following the cleaning step and must be completely flushed out. Always follow the manufacturer's recommendations for the particular cleaner being used (Figure 76).

Reverse Flushing

Reverse flushing of the cooling system is the forcing of water through the cooling system, using air pressure in a direction opposite to that of the normal flow of water. The basic procedure for reverse flushing an engine is as follows:

1. Drain the radiator and remove the hoses at the radiator.

2. Remove the thermostat and reinstall the housing.

3. Install a flushing gun to the hose coming from the thermostat (Figure 77).

4. Connect the water hose of the gun to a pressure water source, and the air hose of the gun to a pressure air source.

5. Turn on the water.

6. When the cylinder block is filled, turn on the air (up to 20 psi) in short blasts.

7. Allow the cylinder block to fill between air blasts.

8. Continue this procedure until the water from the engine runs clean.

COMMENT: *When radiator and block flushing is necessary, the heater and its hoses should also be cleaned and flushed. This prevents rust and deposits from being carried back into an engine and radiator that have been cleaned.*

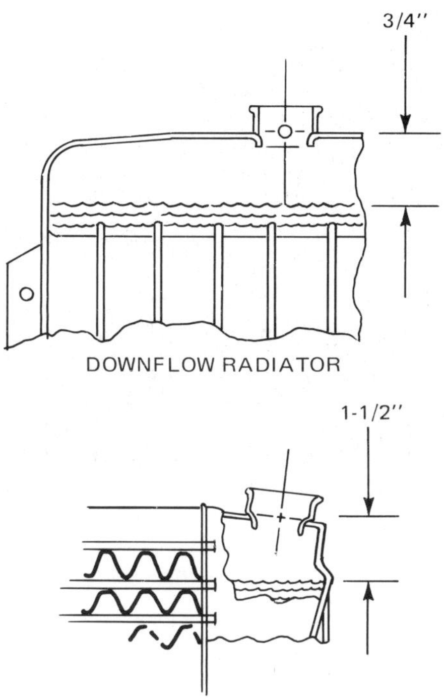

FIGURE 79
Proper coolant level for a downflow and a crossflow radiator.

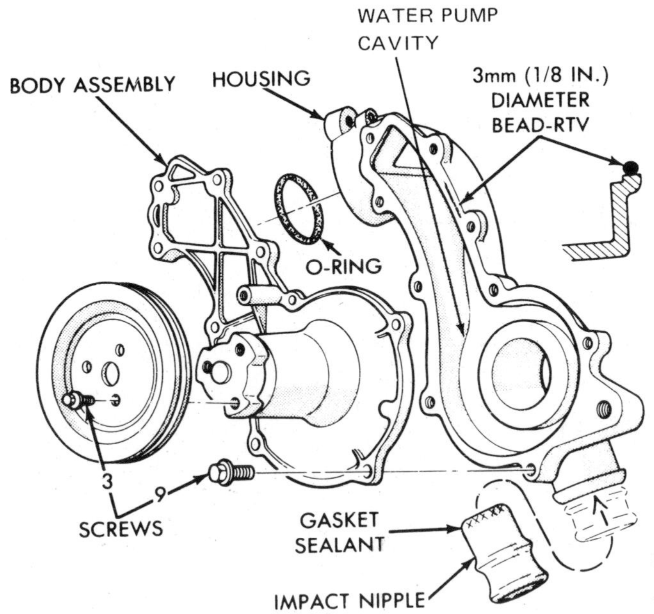

FIGURE 80
This water pump is mounted using RTV sealer instead of a paper gasket. Some others require use of anaerobic compound. (Courtesy of Chrysler Corporation)

Deposits in the Block

Deposits of rust, scale, and grease can be checked with a piece of welding rod. Insert the rod into the block through a water passage and probe for the deposits (Figure 78). If deposits are felt, remove the core (expansion) plugs and flush the block.

To clean out badly clogged engine water jackets, flush through the core plug holes with a small flexible tube attached to the pressure gun nozzle.

Although not all deposit formation in the cooling system can be prevented, serious deposit buildup can be avoided. Hard water contains minerals that leave deposits. Therefore only soft water (without salt) should be mixed with coolant.

FILLING THE COOLING SYSTEM

Once the cooling system is drained for any reason, refilling it with the correct mixture of coolant is an important service procedure. Doing it correctly is vital in order to prevent the entrapment of air and inadequate coolant level. Here is a recommended procedure:

1. Check the tightness of all hose clamps.

2. Position the heater temperature selector lever (or knob) in the maximum heat position.

3. Fill the cooling system until the coolant is below the radiator cap seat.

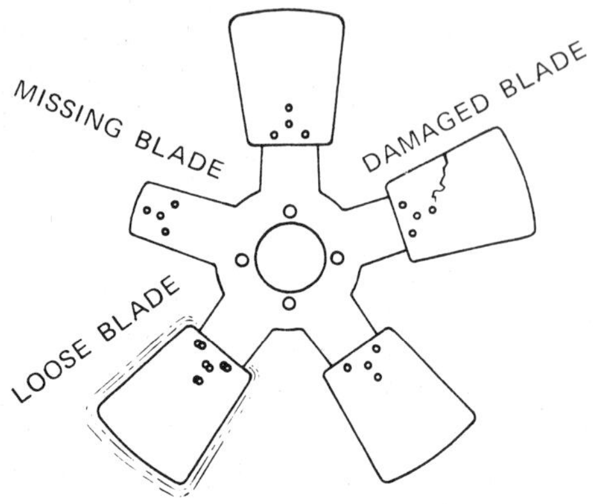

FIGURE 81
ALWAYS REPLACE A DAMAGED FAN! Check for bent blades by laying the fan assembly on a flat surface with the leading edge facing down. Replace if there is more than .090″ total clearance between the blade tip touching the flat surface and opposing blade(s).

FIGURE 82
Example of a damaged water pump shaft caused by an out-of-balance (bent) fan.

4. With the radiator cap off, start and run the engine until the thermostat opens. You can determine when this occurs by "feeling" the upper radiator hose for hot coolant flow.

CAUTION: *Wear a face shield if you are going to look down into the filler neck. If a trapped air pocket/bubble were to be released by the opening thermostat at this time, you could be showered with hot coolant.*

5. Stop the engine and add coolant to the radiator to adjust the coolant level as shown in Figure 79.

NOTE: *On some newer cars (especially imported models), a particular cooling system bleeding procedure must be followed to insure that all trapped air is removed. Otherwise, the engine may overheat, or it may never go into closed loop operation.*

NOTE: *Due to the construction of crossflow radiators, the coolant level cannot be accurately checked when the engine is running.*

WATER PUMPS

The water pump can be checked for leakage and for a loose or noisy bearing. This is done by loosening the drive belt and grasping the fan with your hand. There should be no radial movement of the water pump shaft and pulley. Feel for any roughness by rotating the fan and pump shaft. A water pump that exhibits leaks, shaft play, or bearing roughness should be replaced.

SHOP TIP

A quick test sometimes used to determine whether or not a water pump is working properly is to see if the heater warms up properly. A defective pump will not be able to circulate heated coolant through the long heater hose.

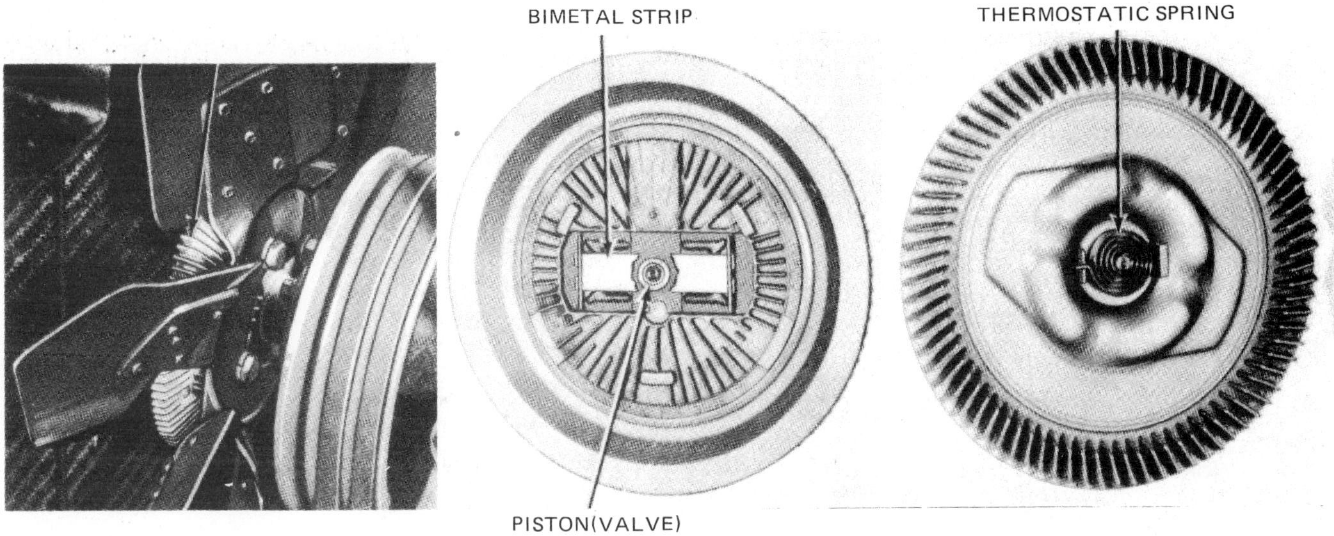

FIGURE 83
A thermostatic strip or coil attached to the fan drive controls the amount of fluid in the clutch. (Courtesy of Ford Parts and Service Division)

Water Pump Replacement

Follow this general procedure to replace a water pump:

1. Drain the cooling system.

2. Disconnect the coolant hoses attached to the water pump.

3. Loosen the alternator/water pump drive belt.

4. Remove the water pump pulley.

5. Remove the water pump mounting bolts.

6. Remove the water pump.

NOTE: It may be necessary to pry on the side of the water pump casting to loosen it from the engine.

7. Clean off any old gasket material or sealant from the water pump mounting surface.

8. Inspect and clean the water pump impeller cavity (if it is the exposed type).

9. Position a new water pump gasket in place, or apply special sealant to the water pump mounting surfaces (Figure 80).

10. Place the pump in position and replace the mounting bolts.

11. Turn the pump shaft by hand to be sure it rotates freely.

12. Check the fan clutch.

13. Carefully inspect the fan (Figure 81).

COMMENT: Water pump shaft damage can be brought about by a bent fan blade (Figure 82).

14. Adjust the drive belt.

15. Refill the system and check for coolant leaks.

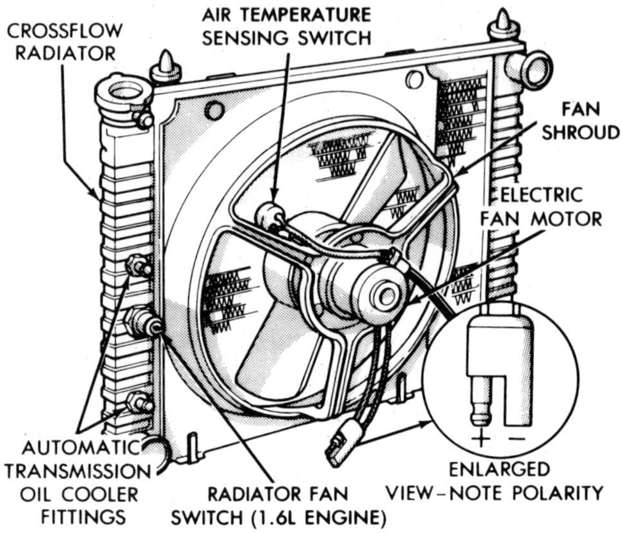

FIGURE 84
Note the electrical switches pictured here. When troubleshooting, they can be bypassed by jumping connectors. (Courtesy of Chrysler Corporation)

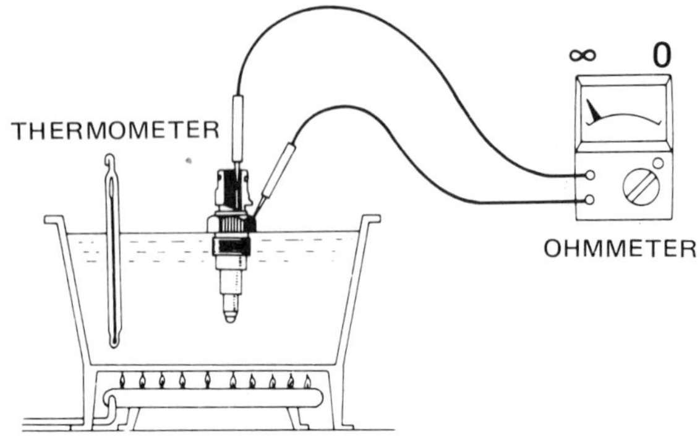

FIGURE 85
Checking a temperature switch.

5. **Shaft vibration.** The fan clutch is mounted on the water pump between the water pump shaft and the fan. Any vibration can mean a worn bearing in the water pump or fan clutch.

When Should a Fan Clutch Be Replaced?

Factory air conditioned vehicles often come equipped with a thermostatically controlled fan clutch bolted to a rigid fan (Figure 83). If any of the following circumstances exists, the fan clutch must be replaced:

1. **Engine overheating.** The fan spins freely in both directions and does not engage when the engine is overheating.

2. **Excessive fan clutch play.** Manually move the fan back and forth. More than 1/4″ play at the blade tips indicates the clutch bearing is worn.

3. **Greasy buildup.** Check for greasy dirt from fluid leakage around the outer seal and at the center thermostat spring.

4. **Bearing failure.** If the fan cannot be rotated by hand, or if hand rotation makes a harsh rumbling sound bearing failure exists.

SHOP TIP

Proper operation of a fan clutch may be checked with a timing light. First, chalk consecutive numbers from one to however many blades the fan may have. Connect the timing light to any cylinder and strobe the numbered fan blades with the engine idling at 1500 rpm and at 2500 rpm. If the numbers run forward, the fan is turning faster than the engine and is not slipping.

TESTING AN ELECTRIC FAN MOTOR

To test an electric fan motor, disconnect the fan motor wire connector and connect it to a good 12-volt battery. Observe correct polarity as shown in Figure 84. If the fan runs normally, the motor is working properly. If not, replace the motor.

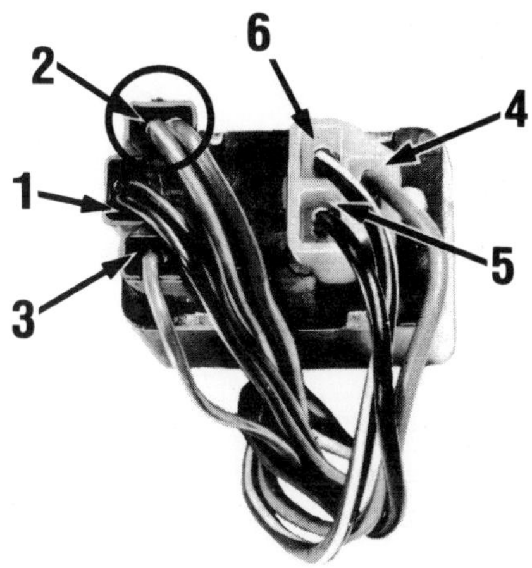

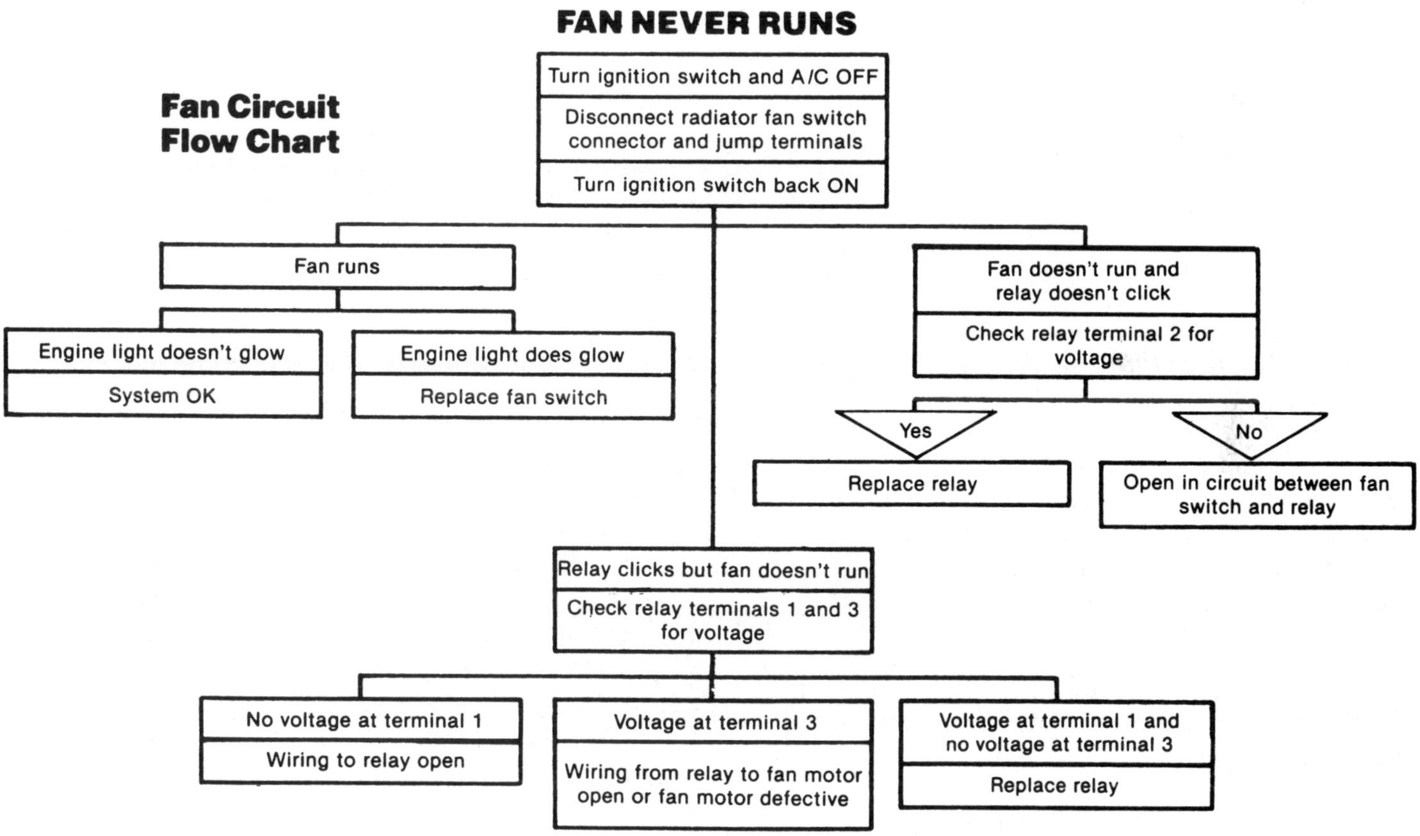

FIGURE 86

Typical fan circuit flow chart. If the fan never runs and the relay does not click, check terminal 2 for voltage. (Courtesy of Chrysler Corporation)

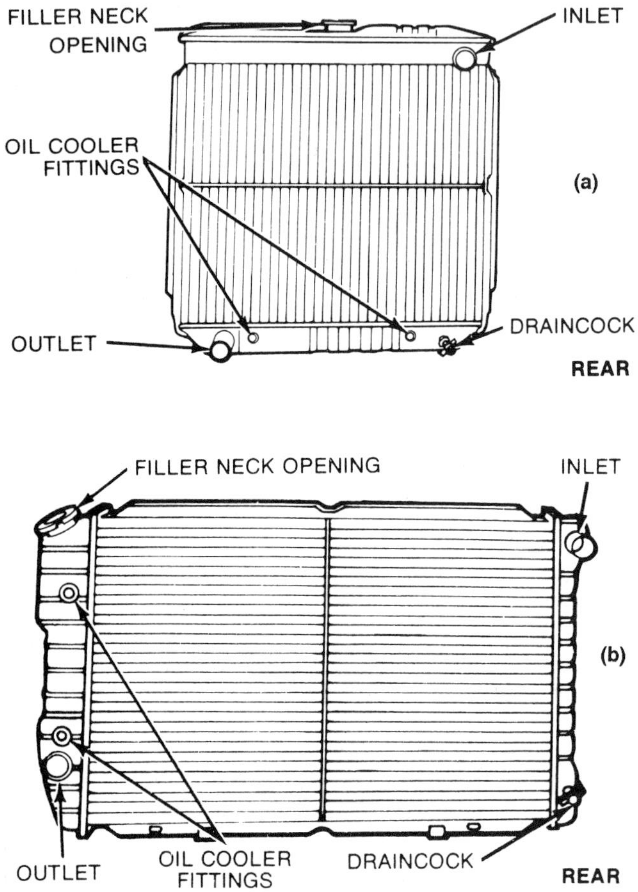

FIGURE 87
(a) Construction of a downflow type radiator. (Courtesy of Ford Parts and Service Division) (b) Construction of a crossflow type radiator. (Courtesy of Ford Parts and Service Division)

TESTING AN ELECTRIC FAN MOTOR TEMPERATURE SWITCH AND RELAY

Using an ohmmeter, the temperature switch can be checked for continuity at the specified temperature (Figure 85). If continuity is not as specified, replace the switch.

When checking a condition wherein the fan never runs and the relay does not click, check the specified relay terminals for voltage. Figure 86 shows a fan circuit flow chart.

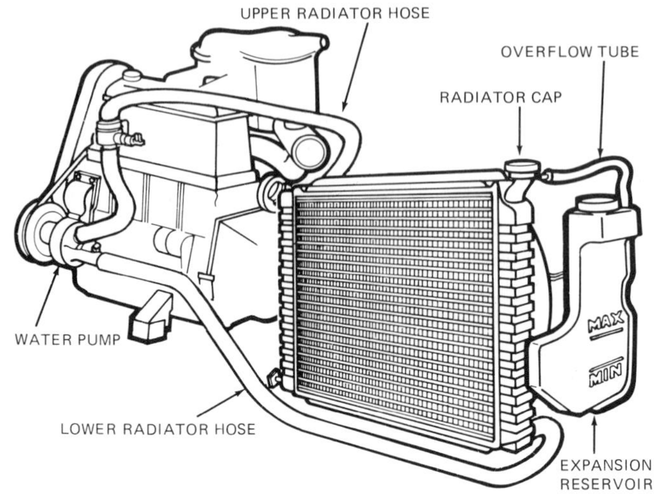

FIGURE 88
Transverse mounted engine showing the radiator and expansion reservoir. (Courtesy of Chrysler Corporation)

RADIATORS

Two types of radiators are in current use:

1. **The downflow type.** In this type of radiator, the tubes are arranged for vertical flow of the coolant. Tanks are placed at the top and bottom (Figure 87[a]).

2. **The crossflow type.** In this type of radiator, the tubes are arranged horizontally for side-to-side flow of the coolant (Figure 87[b]).

Coolant Expansion System

When coolant in the radiator expands to the overflow point, it passes through the pressure relief valve in the radiator cap and into an expansion reservoir (Figure 88). When the coolant in the radiator cools down, it contracts. This condition causes the coolant to be pulled from the expansion reservoir through the vacuum (vent) relief valve in the radiator cap, back into the radiator. Briefly, then, the main advantages of this sytem are:

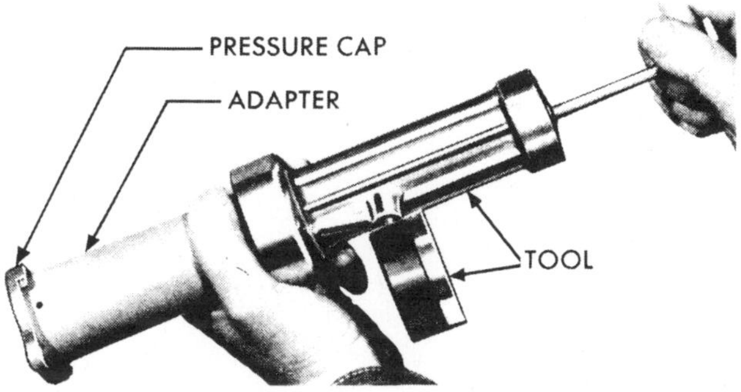

FIGURE 89
Testing a radiator pressure cap.

FIGURE 90
Radiator cap vacuum (vent) valve. On this cap it should hang freely as shown.

1. The liquid coolant that is expelled during after-boil is captured by the expansion reservoir and thus does not spill out on the pavement.

2. Coolant is recovered, since coolant in the expansion reservoir transfers back to the radiator.

Testing Coolant Recovery System

Have coolant level between "MIN" and "MAX" marks on the coolant expansion (recovery) tank with the radiator pressure cap installed. After an engine warm-up and cool-down cycle, open the radiator drain (if so equipped). Coolant should be drawn from the reserve tank into the radiator.

If this test does not draw coolant from the recovery tank, check for leaks at the radiator filler neck nipple, recovery tank hose, and radiator filler neck to radiator cap seals. Also, check the recovery tank nipple and tank overflow tube for links or obstructions.

Testing Radiator Pressure Cap

The radiator pressure cap should be pressure tested to assure that the cap will hold system design pressure and open at the specified pressure. The testing procedure is as follows:

1. Dip the sealing gasket on the pressure cap into water. Fasten the cap to a cooling system pressure tester tool (Figure 89).

2. Working the tool plunger, bring the pressure to the cap rating. If the cap fails to hold the pressure, replace the cap.

3. Remove the cap from the pressure tester tool and check the vacuum valve at the bottom of the cap (if so equipped). On some caps, it should hang freely (Figure 90). On other caps, it is spring loaded and should snap closed after being pulled open.

COMMENT: Always be sure the filler neck opening is free of nicks, rust, scale, or deposits before installing a new cap.

SHOP TIP

When troubleshooting an overheating complaint, warm up the engine and check for cold spots on the radiator. Cold spots indicate the radiator core is plugged.

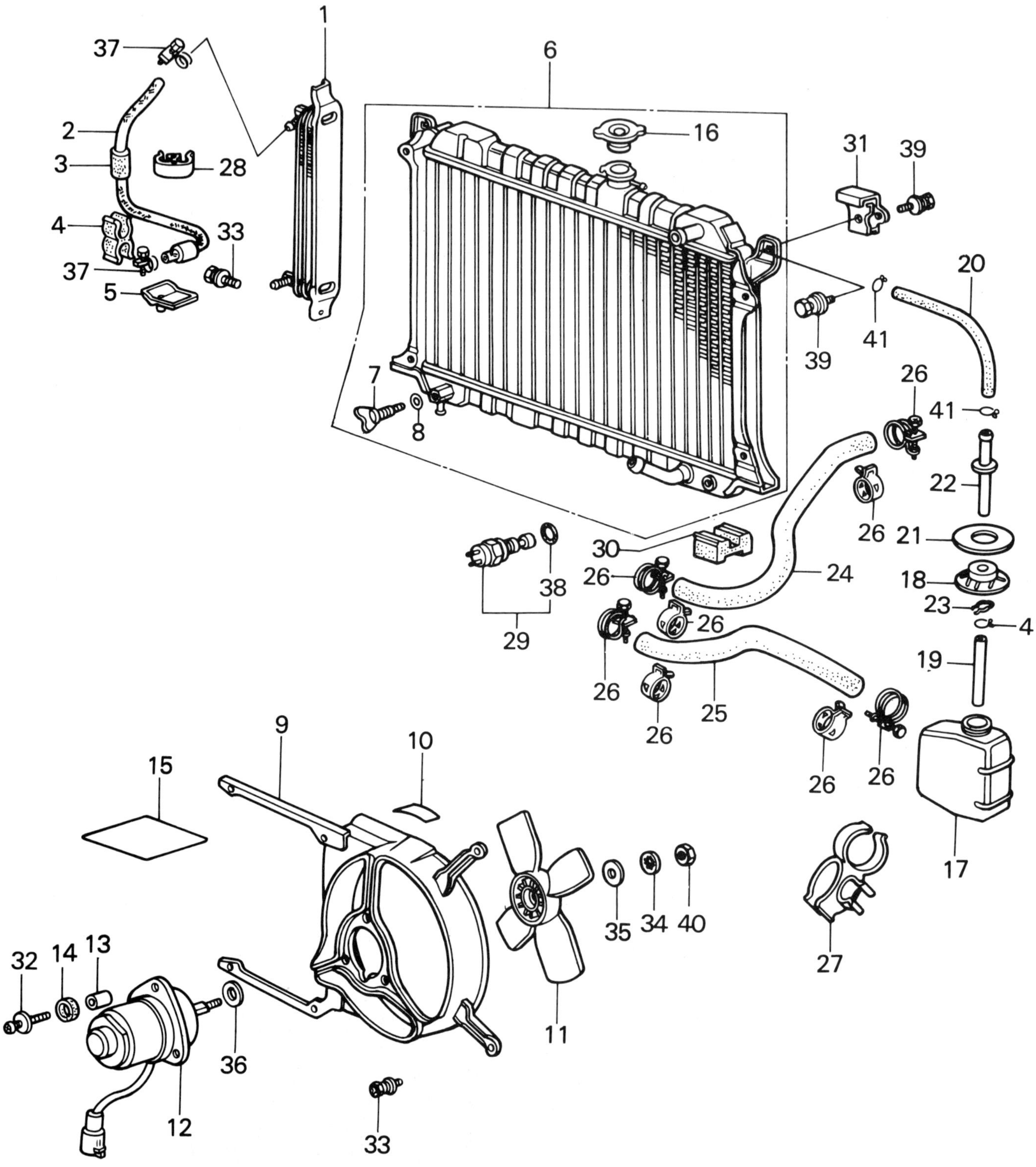

FIGURE 91
Locate the fan (11) and the fan support assembly (9). How many other parts can you identify?

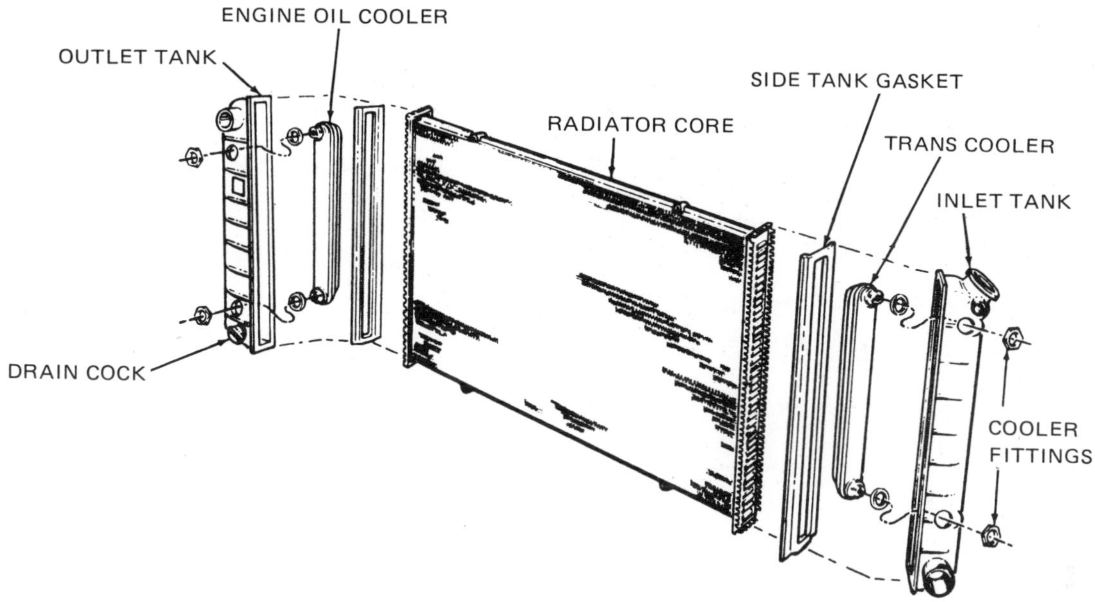

FIGURE 92
Disassembled aluminum core radiator with plastic tanks. These tanks are subject to steam erosion and often require replacement. (Courtesy of Buick Motor Division, General Motors Corporation)

Radiator Removal

1. Drain the coolant.

2. Remove hose clamps and hoses from the radiator.

3. Remove the coolant recovery system tank-to-radiator.

4. Remove the fan and fan support assembly (Figure 91).

5. Remove the radiator mounting brackets and/or bolts.

6. Remove automatic transmission lines (if so equipped).

7. Lift the radiator from the engine compartment.

CAUTION: Do not damage the radiator cooling fins or water tubes during removal.

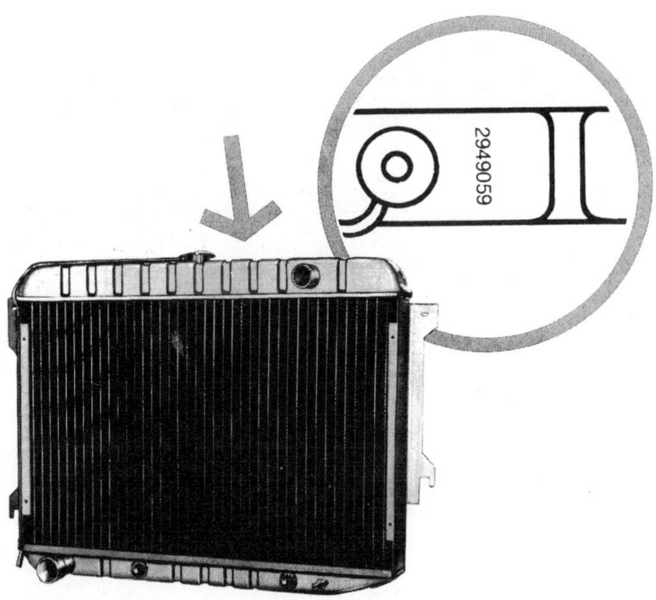

FIGURE 93
Radiator identification. (Courtesy of Modine Manufacturing Company. Copyright 1988)

SHOP TIP

When pulling the radiator out of a car, link the two automatic transmission lines that run into the radiator with a plastic hose and two clamps. This prevents leakage and allows the car to be moved.

Radiator Disassembly

Figure 92 shows a disassembled aluminum core radiator. The plastic tanks and gaskets are held onto the core by bent-over locking tabs.

Radiator Installation

1. Seat the radiator into the mounting brackets.

2. Tighten the mounting brackets and/or bolts.

3. Attach the automatic transmission lines (if so equipped) and the lower radiator hose.

4. Install the fan and fan support assembly.

5. Install the upper radiator hose (including the coolant recovery hose).

6. Refill the cooling system.

SHOP TIP

To facilitate ordering of radiators, original equipment manufacturer (OEM) numbers are located on the radiator. See Figure 93.

TRUE CASE HISTORY

Vehicle	Chevrolet
Engine	Eight-cylinder, 396 CID
Mileage	900 miles on a just-installed exchange engine.
Problem	Noisy lifters during deceleration and during hot idle. Once the engine is back up to speed, the lifters quiet down.
Cause	Big-block Chevrolets tend to aerate the engine oil. Under full oil pressure, the resulting air is compressed in the front portion of the lifter galleys. When decelerating and during hot idle there is lower oil pressure, which allows this air to expand back into the galleys. This cuts off oil to the lifters for the front two cylinders, causing noisy lifters.
Cure	Drill a 1/16″ bleed hole in the center of each front galley plug. These holes allow air to escape without losing more than about 5 pounds of pressure at hot idle.

TRUE CASE HISTORY

Vehicle	Chevrolet M-Van
Engine	Six-cylinder, 4.3 liter
Mileage	15,346
Problem	Squealing noise from belt tensioner area at idle. The problem was more severe when the engine was cold.
Cause	Misaligned power steering pump and a loose a/c compressor brace bolt.
Cure	1. Shimmed power steering pump to achieve proper alignment. 2. Replaced serpentine belt and tightened compressor brace.

TRUE CASE HISTORY

Vehicle	Pontiac 6000
Engine	Six-cylinder, 2.8 liter
Mileage	16,794
Problem	1. Oil pressure light intermittently came on while making a turn. 2. The dealer installed 4 sending units and exchanged the instrument cluster. None of this cured the problem.
Cause	A loose fit between the oil pickup tube/screen and oil pump. When the oil pan was removed, the tube/screen fell out of the pump.

ASE STUDY QUESTIONS

1. An engine has erratic oil pressure and hydraulic valve lifter noise. This occurs only at speeds above 40 mph.

   Technician A says that this could be caused by an air leak.

   Technician B says that this could be caused by an overfilled crankcase.

   Who is right?

   a. A only **c.** Both A and B
   b. B only **d.** Neither A nor B

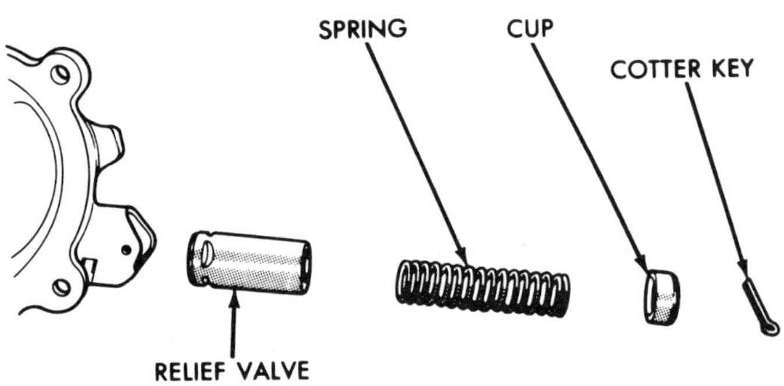

2. The spring shown above is accidentally left out during an engine repair.

   Technician A says that this will cause a blown oil filter.

   Technician B says that this will result in little or no oil pressure.

   Who is right?

   a. A only **c.** Both A and B
   b. B only **d.** Neither A nor B

3. A cooling system is being checked with the tester shown above. Gauge pressure rises when the engine is started.

   Technician A says that this could be caused by a combustion chamber crack.

   Technician B says that this could be caused by clogged tubes in the radiator core.

   Who is right?

   a. A only **c.** Both A and B
   b. B only **d.** Neither A nor B

4. The water pump/alternator belt on the engine shown to the right needs to be replaced.

   Technician A says to remove the air pump belt.

   Technician B says to adjust the tension at pivot bolt "E."

   Who is right?

   a. A only **c.** Both A and B
   b. B only **d.** Neither A nor B

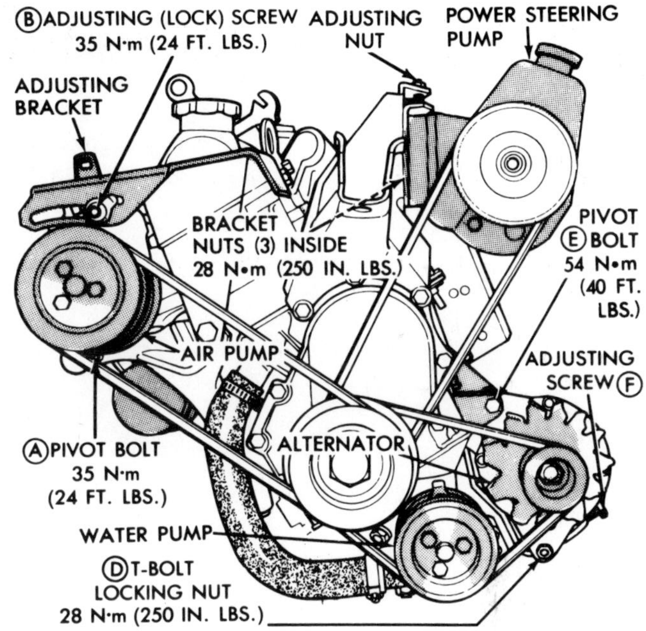

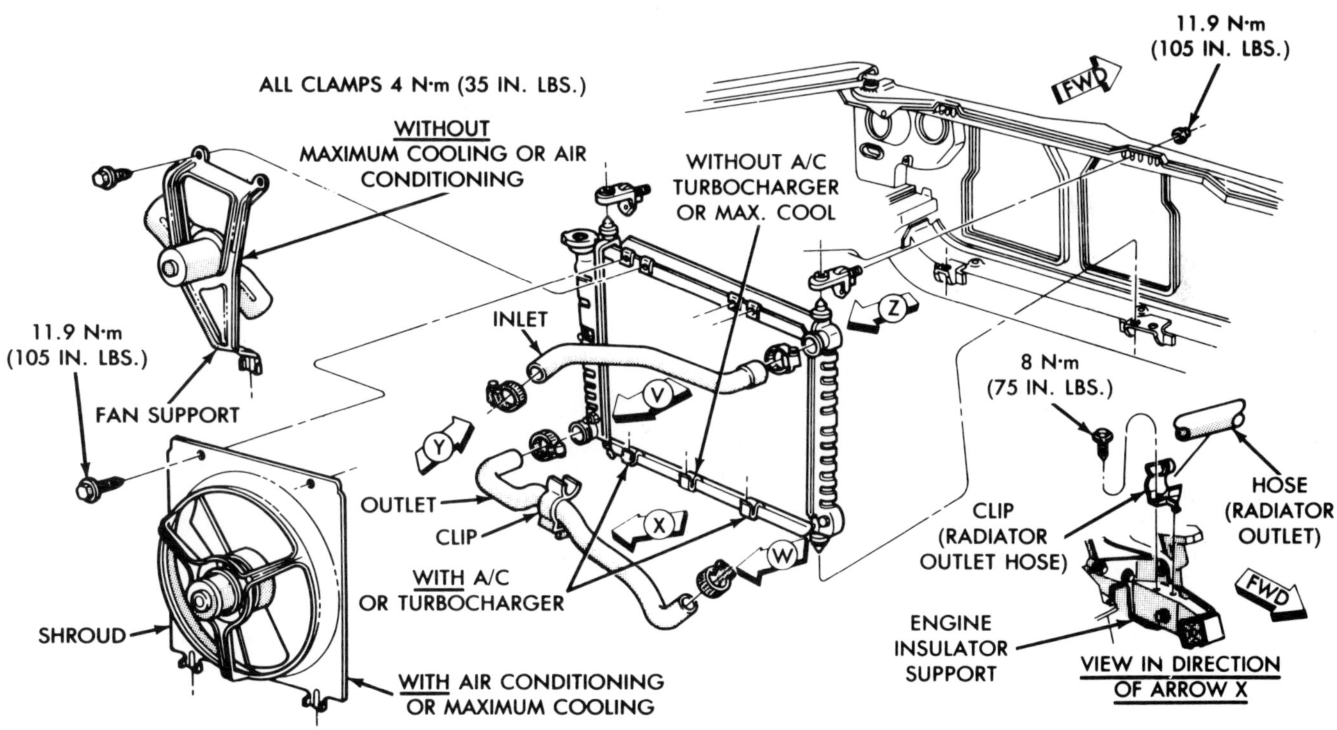

5. Refer to the above front wheel drive vehicle cooling system layout.

   Technician A says that the lower radiator hose should be clipped to the front engine insulator support.

   Technician B says that the use of a torque wrench is recommended when tightening the radiator hose clamps.

   Who is right?

   a. A only **c.** Both A and B
   b. B only **d.** Neither A nor B

6. Which of these would be the MOST LIKELY cause of the damage shown above?

   a. Wrong thermostat
   b. Out-of-position thermostat
   c. Excessive gooseneck torque
   d. Upside-down thermostat

7. Due to the configuration of certain engines, air can be trapped in the cylinder head during a coolant drain and refill.

   Technician A says that this can cause a cracked head.

   Technician B says to unscrew the coolant temperature sending unit sufficiently to bleed the air.

   Who is right?

   a. A only **c.** Both A and B
   b. B only **d.** Neither A nor B

8. A water pump is noisy.

   Technician A says that this could be caused by a corroded bearing.

   Technician B says that this could be caused by a defective seal.

   Who is right?

   a. A only **c.** Both A and B
   b. B only **d.** Neither A nor B

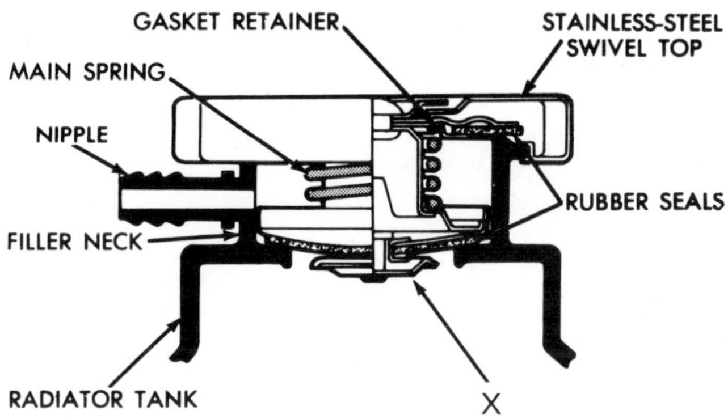

9. Valve X shown above is stuck closed.

Technician A says that this can cause the radiator hoses to collapse on cool-down.

Technician B says that this can prevent coolant return to the radiator on cool-down.

Who is right?

a. A only
b. B only
c. Both A and B
d. Neither A nor B

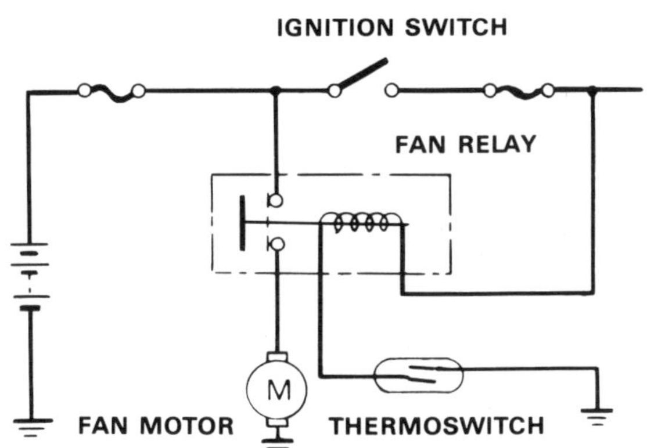

10. The electric fan motor shown above runs continuously.

Technician A says that this could be caused by melted-together fan relay contacts.

Technician B says that this could be caused by a defective thermo switch.

Who is right?

a. A only
b. B only
c. Both A and B
d. Neither A nor B

11. An engine is being torn down because of a badly spun rod bearing.

Technician A says to flush the oil passages in the block and head.

Technician B says to replace the oil cooler core.

Who is right?

a. A only
b. B only
c. Both A and B
d. Neither A nor B

SECTION REVIEW QUESTIONS

1. List the principal components of a typical lubrication system.
2. How should an oil pressure sending unit be tested?
3. How can you quickly check for a clogged oil filter?
4. What types of oil pumps are used in automotive engines?
5. What is the purpose of an oil pump pressure relief valve?
6. What on-car test can you make to verify oil aeration?
7. In some engines the oil pump is an integral part of the timing chain cover. What needs to be done to these pumps prior to installation?
8. What is an oil cooler?
9. What is the purpose of a cooling system bypass hose?
10. What can happen if a thermostat is removed from the cooling system?
11. Describe an automatic transmission oil cooler.
12. Why are cooling systems pressurized?
13. What is a fluid fan drive clutch?
14. What is a flex fan?
15. An electric fan continues to run after engine shutdown. Is this normal? Explain your answer.
16. What is a fan shroud?
17. What coolant mix is generally used in an engine?
18. How can you test a cooling system for external leaks?
19. Describe two methods that can be used to test a cooling system for internal leaks.
20. How can an automatic transmission oil cooler that is located in the radiator be tested for leaks?
21. What is a serpentine drive belt?
22. When examining a drive belt, what should you look for?
23. How should new drive belts be "seated"?
24. List three belt tensioning methods.
25. Reversed heater hoses can cause what problem(s)?
26. When installing a heater hose that has a spine wire, what precaution must be taken?

27. When installing a vertically mounted thermostat, what can be done to prevent it from slipping out of position?

28. How can a thermostat be tested?

29. How can you test for restricted coolant flow through the radiator?

30. Describe the procedure for reverse flushing an engine.

31. On some cars a particular cooling system bleeding procedure must be followed. Why?

32. What should you look for when examining the water pump on an engine?

33. When should a fan clutch be replaced?

34. How can a timing light be used to check a fan clutch?

35. How can an electric fan motor be tested?

36. What two types of radiators are in current use?

37. Describe the operation of a coolant expansion system.

38. How can a coolant expansion system be tested?

39. How is a radiator pressure cap tested?

40. How are some aluminum core radiators disassembled?

Section E

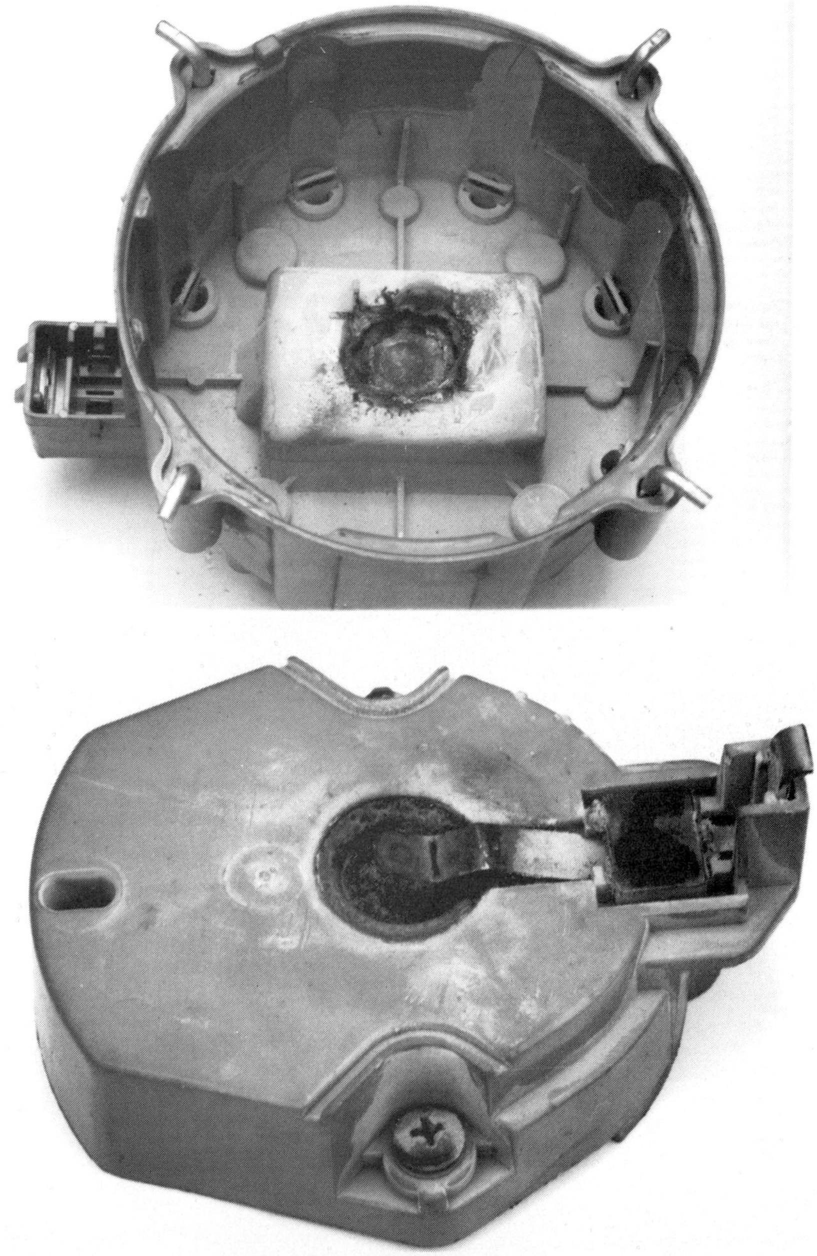

IGNITION SYSTEM DIAGNOSIS AND REPAIR

After studying this section, and with hands-on practice on a "live" vehicle, you should be able to perform the following tasks:

- Diagnose no starting, hard starting, and engine misfire on vehicles with electronic or point-type ignition systems; determine needed repairs.
- Inspect, test, repair, or replace ignition primary circuit wiring and components.
- Inspect, test, and service distributor (including drives, shaft, bushings, cam, breaker plate, and vacuum and mechanical advance/retard units).
- Inspect, test, adjust, service, repair or replace ignition points and condenser.
- Inspect, test, service, repair or replace ignition system secondary circuit wiring and components.
- Inspect, test, and replace ignition coil.
- Check and adjust ignition system timing and timing advance.
- Inspect, test, and replace electronic ignition system pick-up sensor or trigger devices.
- Inspect, test, and replace electronic ignition system control unit (module).

POINT-TYPE IGNITION SYSTEM

A typical point ignition system consists of the distributor assembly, ignition coil, resistor assembly, spark plugs, and the wiring necessary to connect the components. See Figures 1 and 2.

The purpose of the ignition system is to make available, at the proper time, enough voltage to cause a spark to jump the spark plug gap for ignition of the air–fuel mixture in the cylinder. Two basic circuits are contained in the ignition system:

1. **Primary** (low voltage) circuit. As shown in Figure 2, this includes the battery, ignition switch, resistor, coil primary winding, contact points, and condenser.

2. **Secondary** (high voltage) circuit. This part of the ignition system is responsible for delivering the high-voltage spark to the spark plugs. As shown in Figure 2, this includes the coil secondary winding, rotor, distributor cap, high voltage wires, and spark plugs.

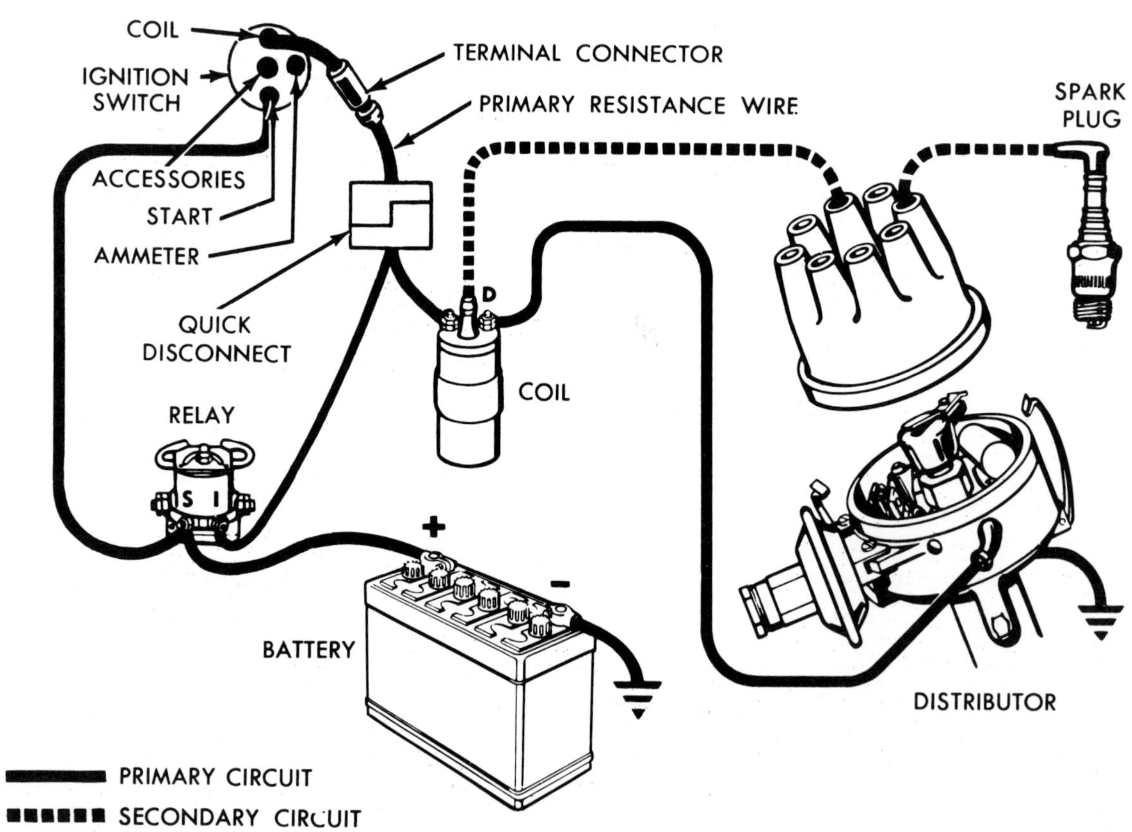

FIGURE 1
Point-type ignition system component layout. Note the primary resistance wire. (Courtesy of Ford Parts and Service Division)

ELECTRONIC-TYPE IGNITION SYSTEM

An electronic ignition system is shown in Figure 3. System components include:

1. Battery.
2. Ignition switch.
3. Resistor or resistance wire (only on some cars).
4. Coil.
5. Distributor.
6. Spark plugs.
7. Control unit (electronic module).

There are many similarities between electronic ignition and point-type ignition components (Figure 4). The functions of the battery, ignition switch, coil, and secondary circuit are the same in both systems.

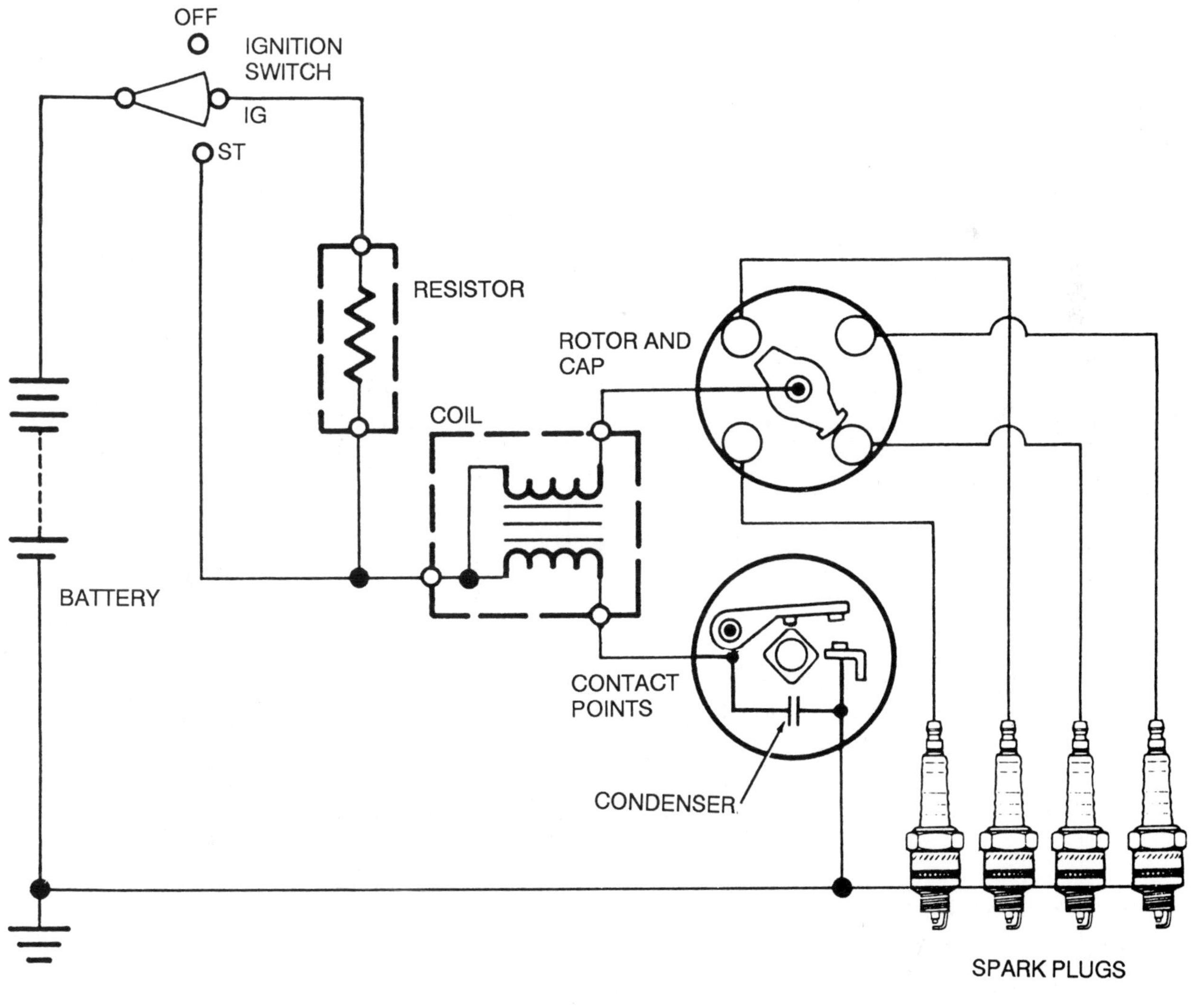

FIGURE 2
Schematic of a typical point-type ignition system.

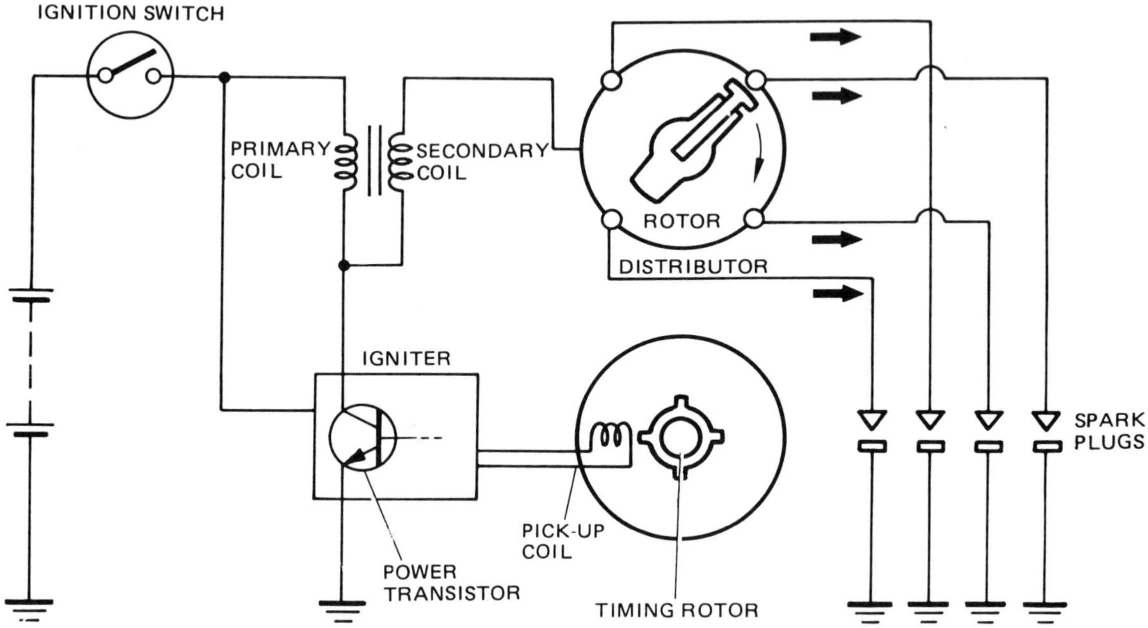

FIGURE 3
An import electronic ignition system. The igniter is the ignition module. (Courtesy of Toyota Motor Sales, U.S.A., Inc.)

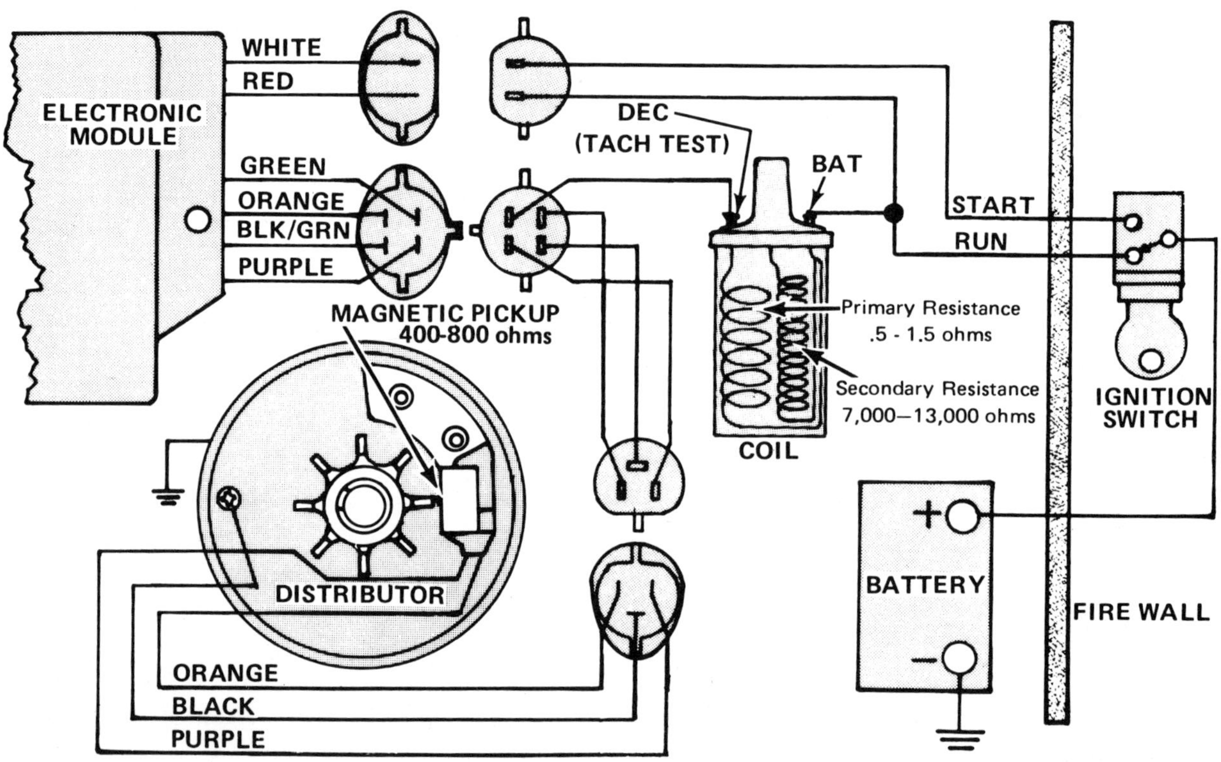

FIGURE 4
A domestic electronic ignition system component layout. (Courtesy of Ford Parts and Service Division)

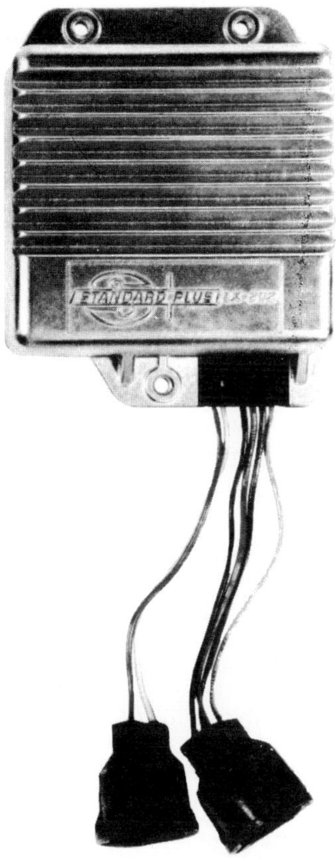

FIGURE 5
Electronic ignition control unit (module) for a Ford ignition system.

YEAR FIRST USED	DELCO PART NUMBERS	DESCRIPTION	SYSTEM USED
1974	D1906	1875990 (W, C, G, B)	ALL 4, 6 AND 8 CYLINDER ENGINES
1977	D1906	1894322 (W, C, G, B)	ALL 4 CYLINDER ENGINES
1978	D1918	1894308 (R, C, L, H, B)	ESS-(ELECTRONIC SPARK SELECTION) ESC-(ELECTRONIC SPARK CONTROL) Buick Turbo
1980	D1941	1976239 (R, C, L, H, B)	EMR ELECTRONIC MODULE RETARD 10° if cold start
1980	D1952 D1952 D1956	1977907 1978617 1976908 (C, B, R, E, N, P, B)	EST-(ELECTRONIC SPARK TIMING)
1981	D1957 D1961 D1958	1977858 1979107 1979571 (P, N E R, + C B)	EST (ELECTRONIC SPARK TIMING)
1982	D1959	1978778 (W, G, B, C)	MECHANICAL ADVANCE "Luv"
1983	D1967	1985342 (P, N E R, + C B)	EST (ELECTRONIC SPARK TIMING)

FIGURE 6
Various ignition modules for the General Motors high energy ignition (HEI) system. (Courtesy of Champion Spark Plug Company. All rights reserved.)

Coil

The coil in an electronic ignition system is essentially the same as that used in a point-type system, except it may have more secondary windings. As a result, a higher secondary voltage can be produced (40,000 volts or more).

Control Unit (Electronic Module)

The electronic control module is the "heart" of the electronic ignition system. The module is usually separate from the distributor, and its internal components are non-serviceable (Figures 5 and 6).

The dwell function is performed by the module, and is not adjustable. In some systems, the dwell is fixed. In other systems, it is not. The module controls the dwell by determining how long the primary current flows between interruptions.

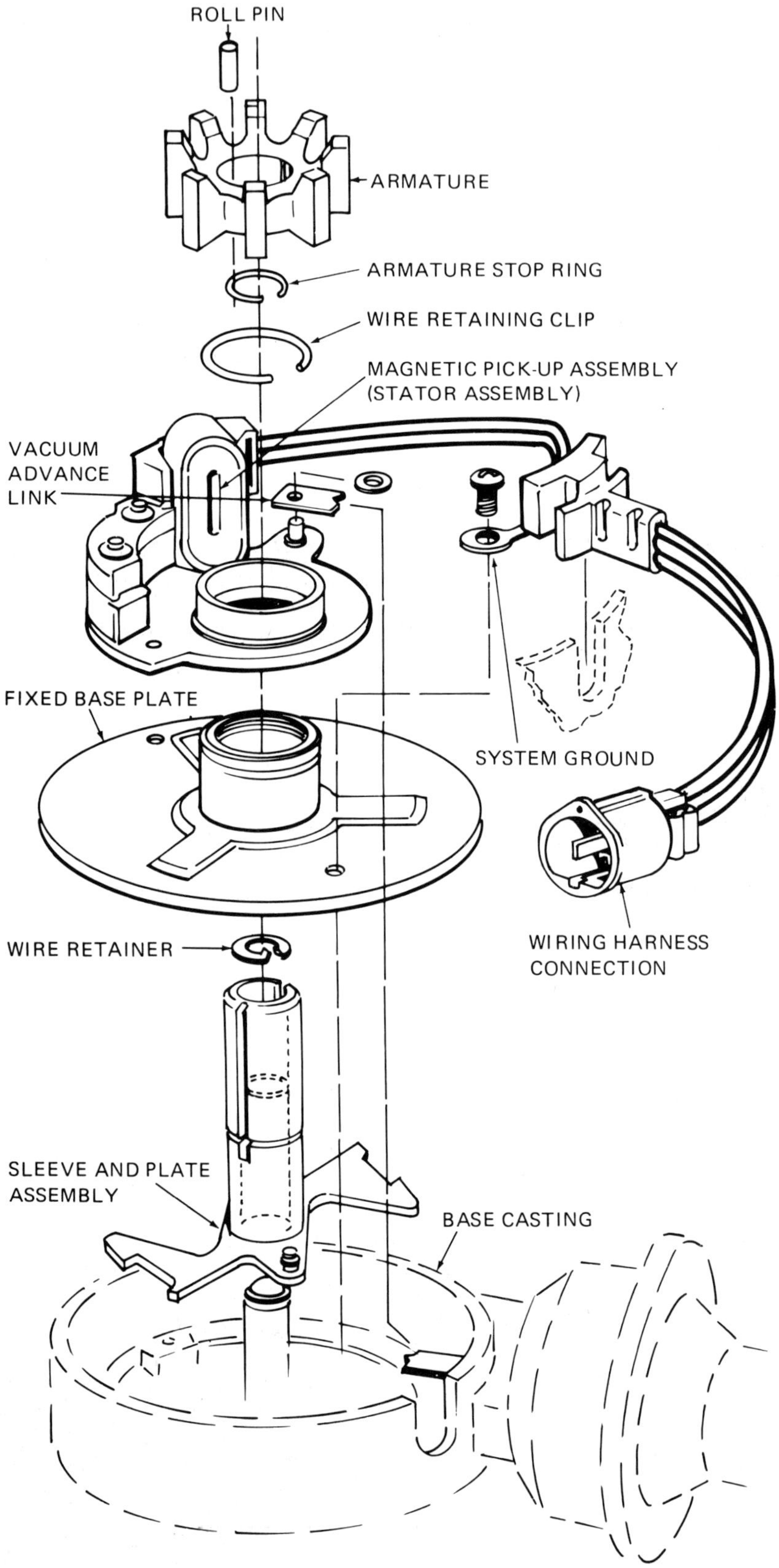

FIGURE 7
Disassembled view of an electronic ignition distributor. (Courtesy of Ford Parts and Service Division)

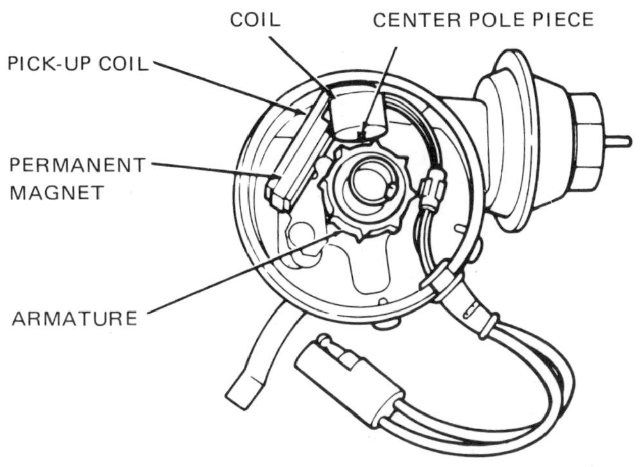

FIGURE 8
Chrysler electronic ignition distributor showing the primary current control components. (Courtesy of Chrysler Corporation)

Distributor

The electronic ignition distributor is different than a point-type distributor. A typical electronic ignition distributor is shown in Figure 7. It is made up of a:

1. Reluctor (often called an armature) that rotates with the distributor shaft.

2. Magnetic pickup coil assembly, which is fixed to a base plate.

Primary Current Control

The magnetic pickup coil assembly in many ignition systems consists of a permanent magnet and a coil that is wound around a center pole piece (Figure 8).

The permanent magnet provides a magnetic field between itself and the pole piece. This produces a magnetic signal in the pickup coil. The signal increases when one of the teeth of the rotating armature nears the pole piece, decreasing the air gap (Figure 9).

When the armature tooth passes the pole piece, the air gap increases again, and the magnetic signal grows weaker. Also, the induced voltage suddenly changes polarity (from + to −, or vice versa). At the instant this happens, the voltage is zero.

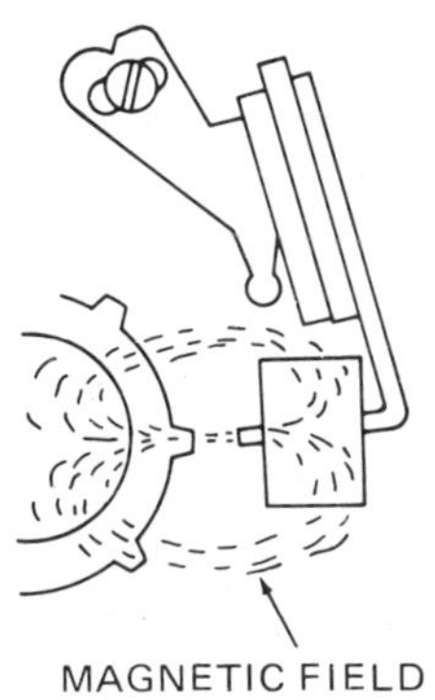

FIGURE 9
Ignition signal generation.

FIGURE 10
Cracked ignition coil.

CHART 1 TROUBLESHOOTING CHART FOR POINT-TYPE IGNITION SYSTEMS

Burned or Pitted Ignition Points

1. Dirt or oil on contact faces
2. Voltage regulator setting too high
3. Contact faces misaligned or gap too small
4. Incorrect or faulty condenser
5. Worn distributor shaft or bushings
6. Resistor not in circuit
7. Touching contact faces with the hands during installation
8. Wrong or defective coil (Figure 10)
9. Defective ignition switch
10. Worn breaker plate (Figure 11)

No Spark to Spark Plugs

1. Defective coil
2. Defective condenser
3. Defective rotor
4. Defective coil-to-distributor high-tension lead
5. Oily, dirty, or burned points
6. Points not opening
7. Loose or dirty primary wiring circuit connections
8. Defective primary resistor
9. Defective spark plug wires
10. Moisture in distributor cap
11. Breaker plate not grounded (Figure 12)

Weak or Intermittent Spark at Spark Plugs

1. Oily, dirty, or burned points
2. Defective condenser
3. Incorrectly set dwell
4. Discharged battery
5. Loose or dirty primary wiring circuit connections
6. Defective coil
7. Burned rotor and/or cap contacts
8. Defective spark plug wires
9. Weak point spring tension
10. Worn distributor bushings or bent shaft
11. Worn distributor cam

Engine Misfire

1. Fouled spark plugs (Figure 13)
2. Incorrect spark plug gap
3. Weak spark
4. Improper spark plug heat range
5. Cross-firing
6. Incorrect ignition firing
7. Worn breaker plate

Backfiring in Intake Manifold

1. Intake valve not seating properly
2. Carbon tracking in distributor cap (Figure 14)
3. Plug wires installed incorrectly
4. Lean mixture

Backfiring in Exhaust System

1. Turning key off and on while engine is running
2. Current flow interruption in primary circuit
3. Faulty points or condenser
4. Incorrect valve timing

Engine "Kicks" (Attempts to Run Backwards) During Cranking

1. Ignition timing too far advanced
2. Carbon tracking in distributor cap
3. Plug wires installed incorrectly (Figure 15)

When the module reads this "zero voltage," it shuts off primary current to the ignition coil. This collapses the magnetic field across the ignition coil, firing a spark plug. Then as the polarity reverses back, a power-switching transistor in the module starts the current flowing to the coil again.

COMMENT: From a purely technical standpoint, the power transistor is forward and reverse biased to turn on or off. The polarity applied biases the transistor, rather than "zero voltage."

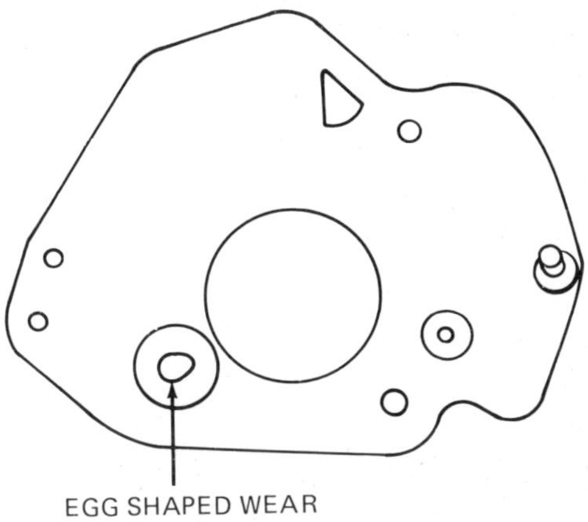

FIGURE 11
This breaker plate was removed from the distributor on a Ford. Five sets of contact points had been installed during 23,000 miles of driving. Can you see why?

IGNITION SYSTEM TROUBLESHOOTING

When troubleshooting the ignition system, remember that the condition of the battery, battery cables, starter solenoid, starter, and alternator can affect operation of the ignition system.

Charts 1 and 2 list some additional ignition system problems.

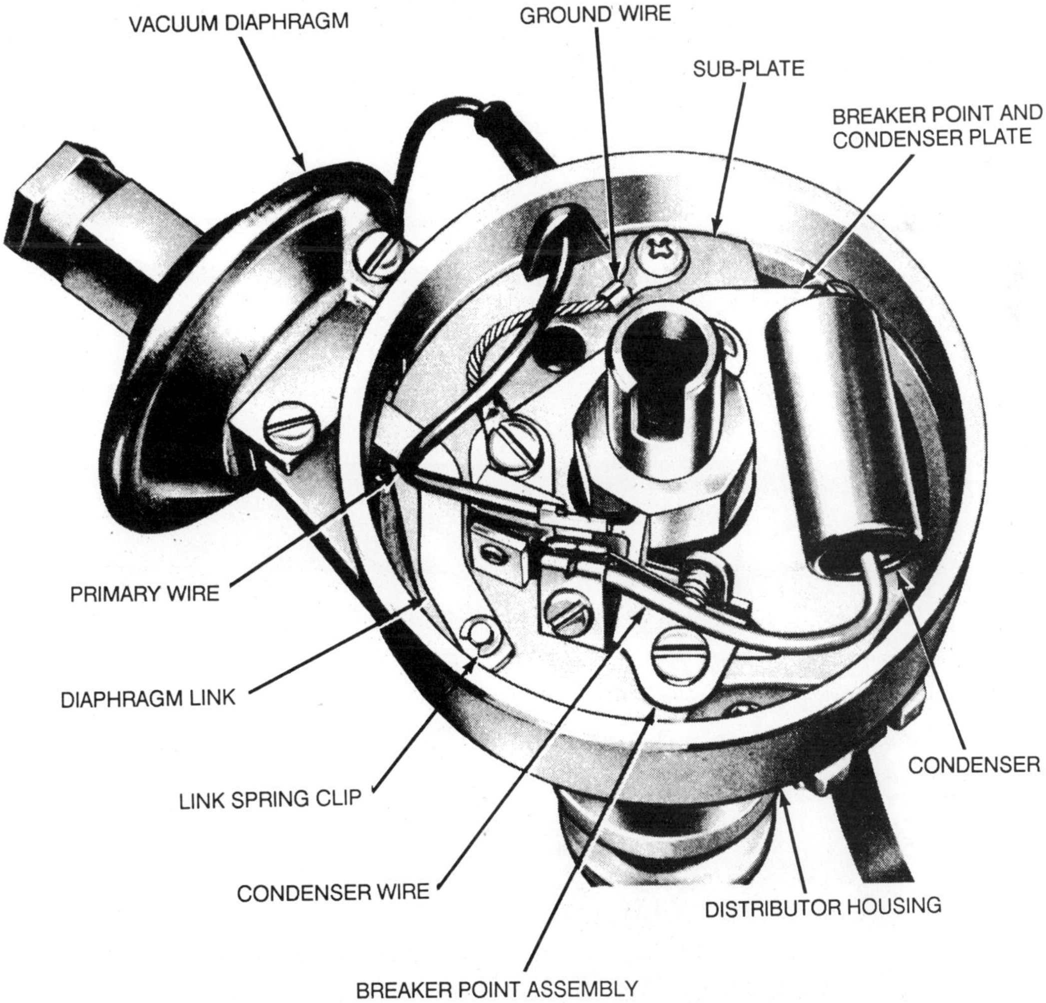

FIGURE 12
Some distributor breaker plates have an attached ground wire. (Courtesy of Ford Parts and Service Division)

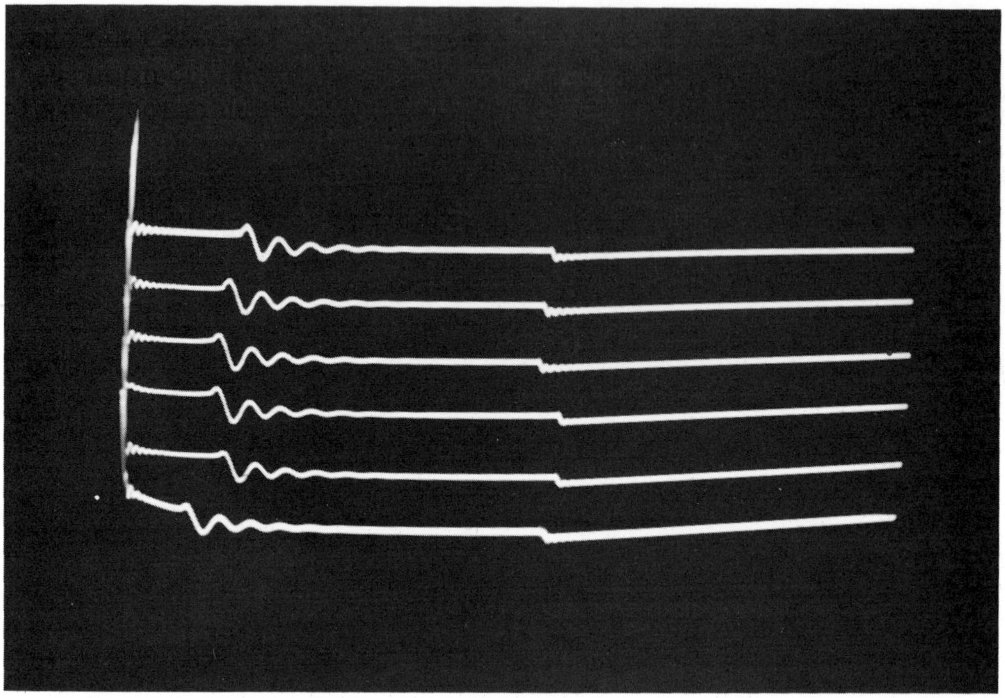

FIGURE 13
This engine had a misfire. The oscilloscope pattern shows a fouled spark plug.

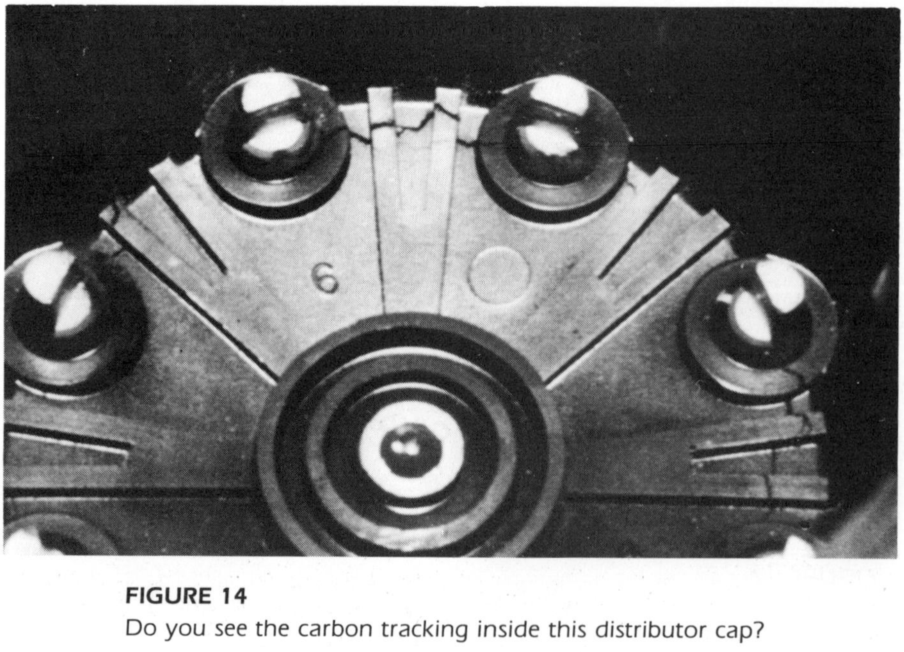

FIGURE 14
Do you see the carbon tracking inside this distributor cap?

CHART 2 TROUBLESHOOTING CHART FOR ELECTRONIC-TYPE IGNITION SYSTEMS

Engine Will Not Start (Fuel and Carburetion Known to be Good)

1. Defective dual ballast resistor (Figure 16)
2. "Punctured" rotor
3. Faulty ignition coil
4. Faulty pickup coil
5. Faulty control unit (module)
6. Faulty wiring (Figure 17)
7. Moisture in distributor cap

Engine Surges (Not Lean Carburetor)

1. Loose connections in wiring harness (Figure 18)
2. Loose pickup coil leads
3. Faulty ignition coil
4. Radio frequency interference (RFI)
5. Corroded power feed wires at starter solenoid
6. Wrong PROM installed

Engine Misses (Carburetion Known to be Good)

1. Defective spark plugs (Figure 19)
2. Defective spark plug wire(s)
3. Loose or corroded wiring connections (Figure 20)
4. Faulty pickup coil leads
5. Faulty control unit (module)
6. Defective armature in distributor (Figure 21)

Engine Hesitates and Stalls

1. Crank sensor and pulley misaligned
2. Corroded pickup coil connections
3. Loose B+ connection at ignition module

Engine Misses at Idle

1. Wrong PROM installed

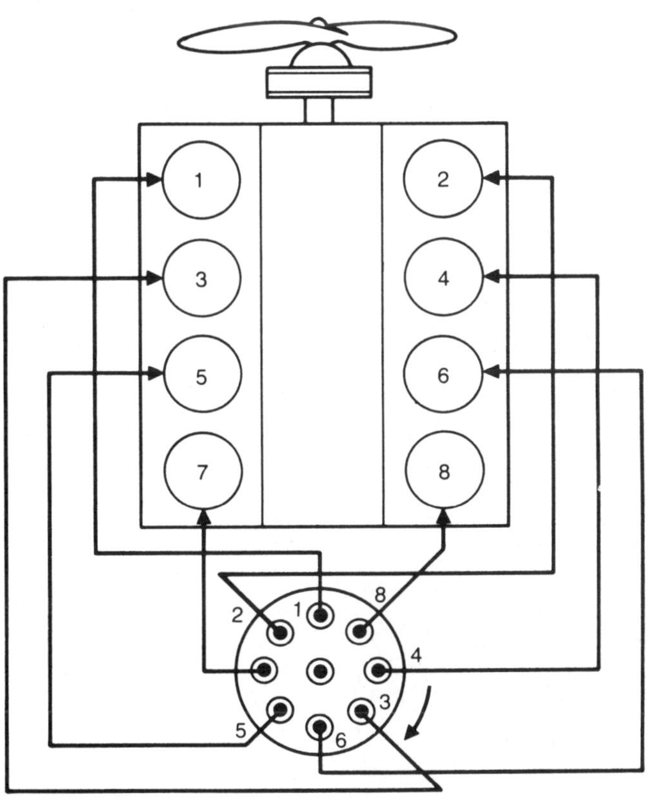

FIGURE 15
Spark plug wire routing for a 1, 8, 4, 3, 6, 5, 7, 2 firing order.

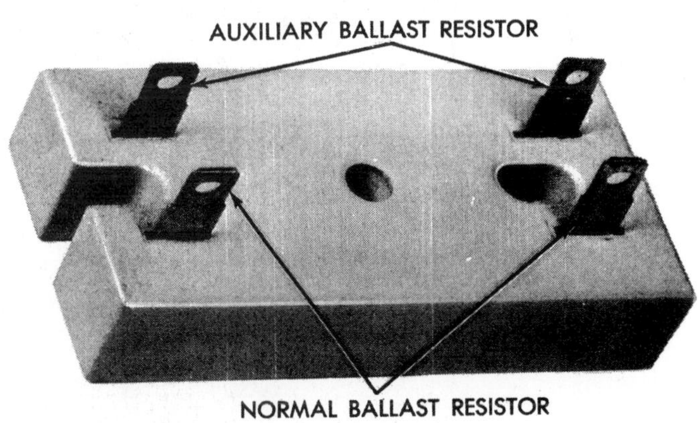

FIGURE 16
Dual ballast resistor. The auxiliary section of the resistor shown should be 4.75 to 5.75 ohms. The normal section should be .5 to .6 ohms.

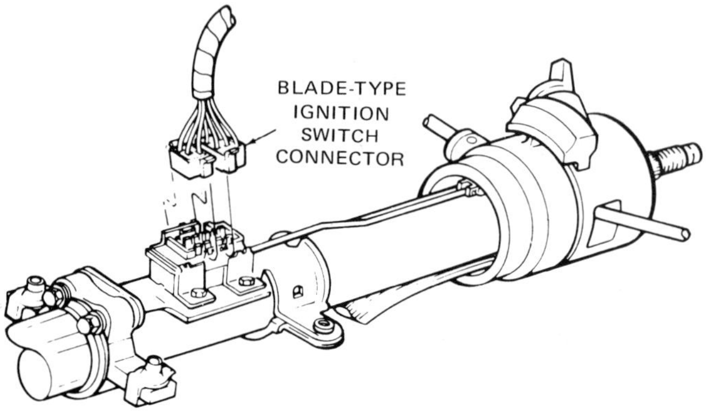

FIGURE 17
Corroded or poor ignition switch connections can cause an engine no-start condition. (Courtesy of Ford Parts and Service Division)

FIGURE 18
A loose connection at the battery feed (B+) wire to the distributor can cause engine surge.

Inspecting the Cap

Once you have removed the cap, check it for corroded electrodes (inserts) and a worn rotor contact button. Look for carbon tracks or other flaws. Inspect the towers for corrosion, burning, or cracks (Figure 22).

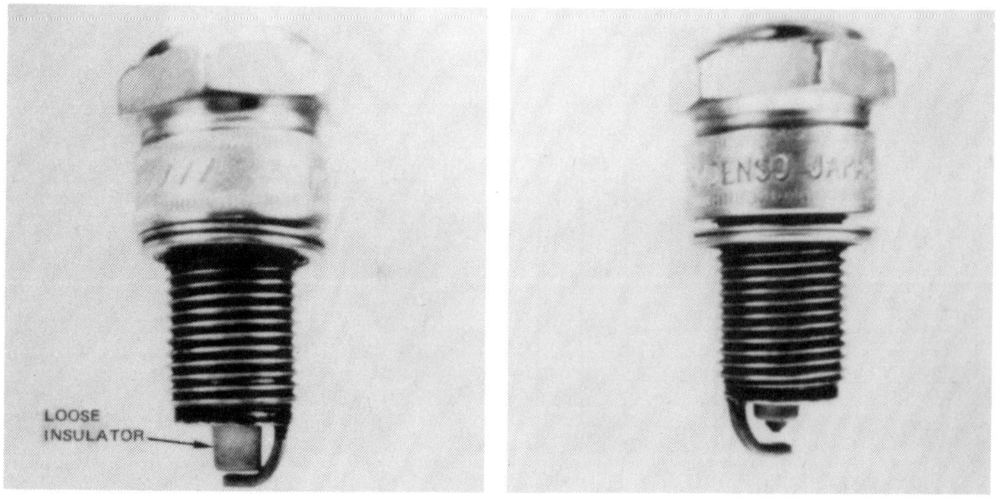

FIGURE 19
When examining spark plugs, always shake them to check for loose (broken) insulators. Using air tools to install plugs can separate the insulator as shown on the left.

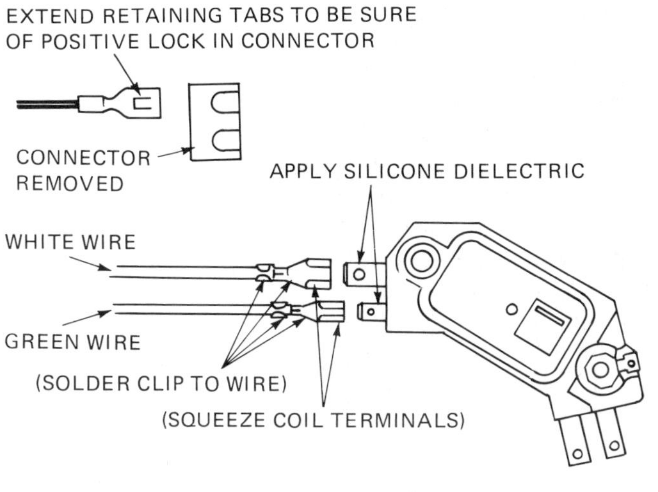

FIGURE 20
Electronic ignition system module connections must be clean and tight.

FIGURE 21
Broken trigger wheel (armature) from an AMC distributor. A "dead" cylinder was the result.

***COMMENT:** Mild corrosion in the towers may be cleaned out with a special wire brush (Figure 23). Caps with heavy corrosion must be replaced.*

Be sure to note the position of all spark plug cables before replacing the cap (Figure 24).

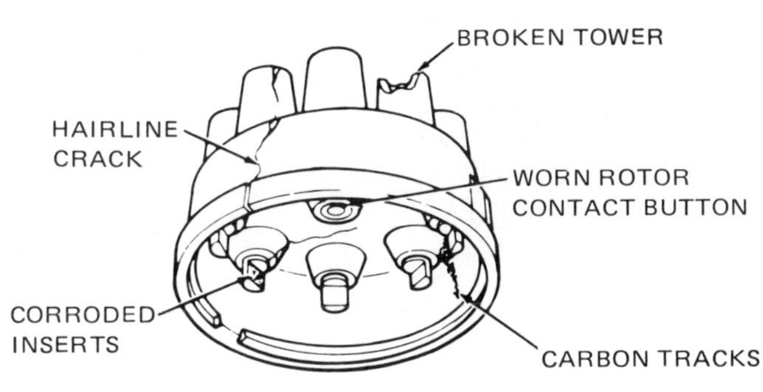

FIGURE 22
Distributor cap inspection.

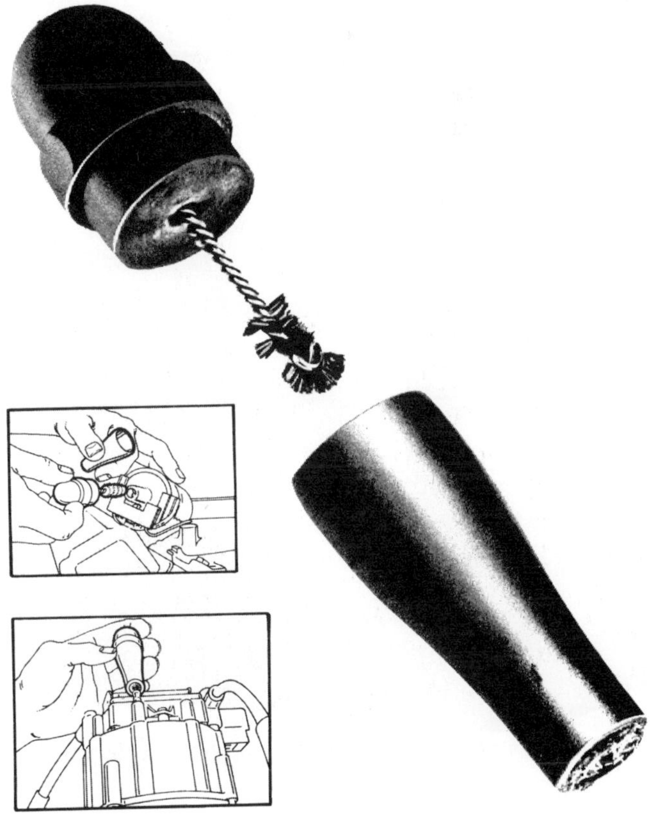

FIGURE 23
Distributor cap towers can be cleaned with this special brush. (Courtesy of Thexton Manufacturing Company, Inc.)

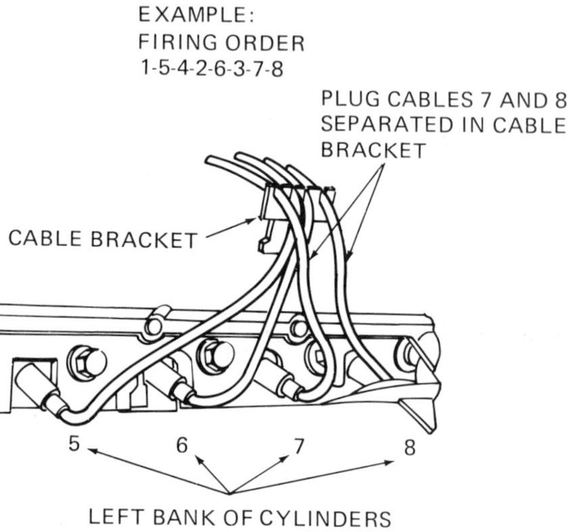

FIGURE 24
The position of plug cables in cable brackets is important in order to prevent induction cross-firing.

Inspecting the Rotor

Be sure the rotor is the correct one for the vehicle. Check the rotor for excessive burning of the tip. Inspect the contact spring (and resistance rod, if it has one) for corrosion, burning, or damage. If you see any of these problems, replace the rotor (Figure 25[a] and [b]).

Before installing the rotor, lubricate the felt wick in the center of the distributor shaft, if it is so equipped (Figure 26).

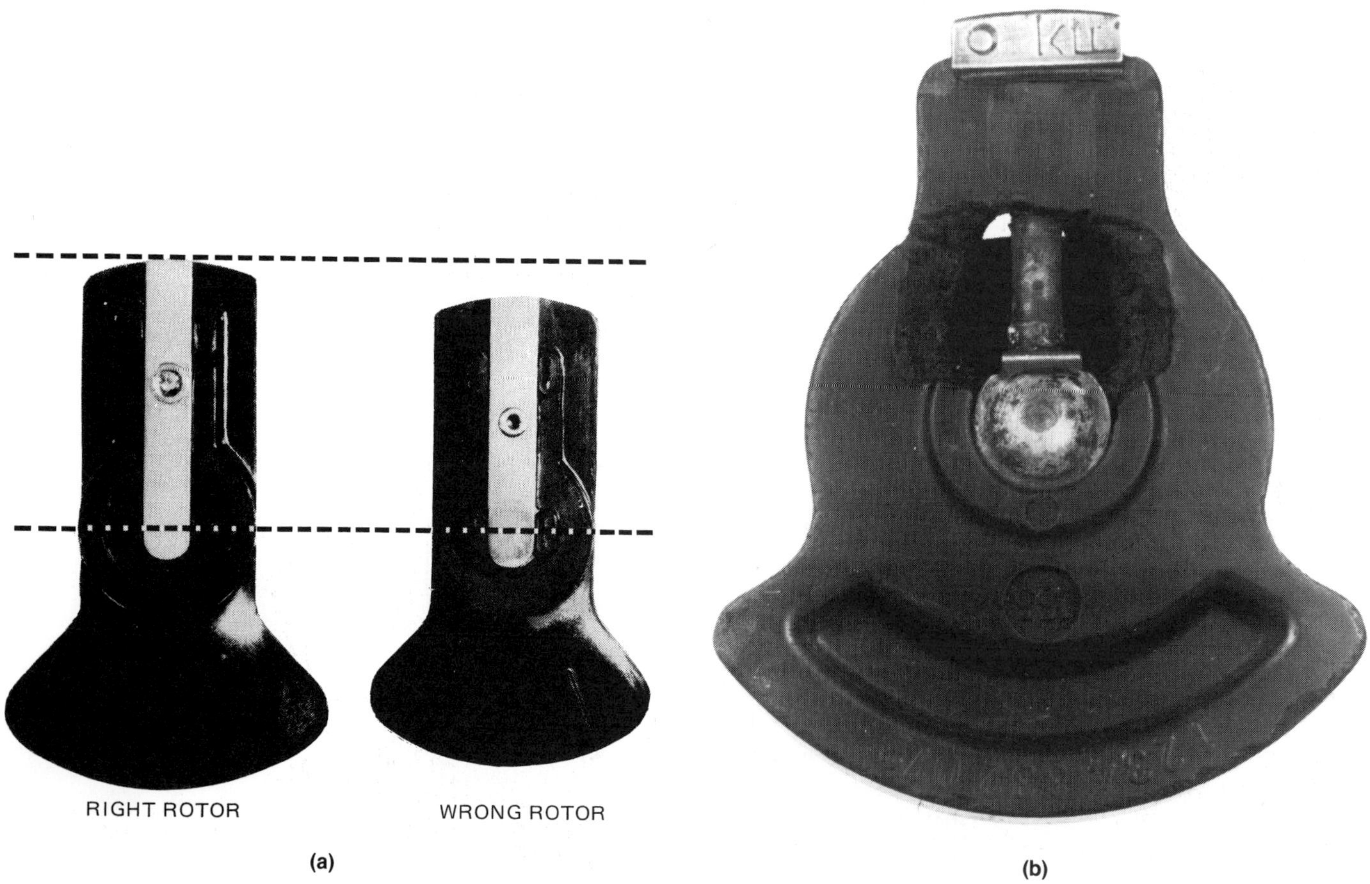

FIGURE 25
(a) Make sure the distributor has the right rotor. Wrong rotor application is not as rare as it may sound. (b) Burned through (punctured) rotor.

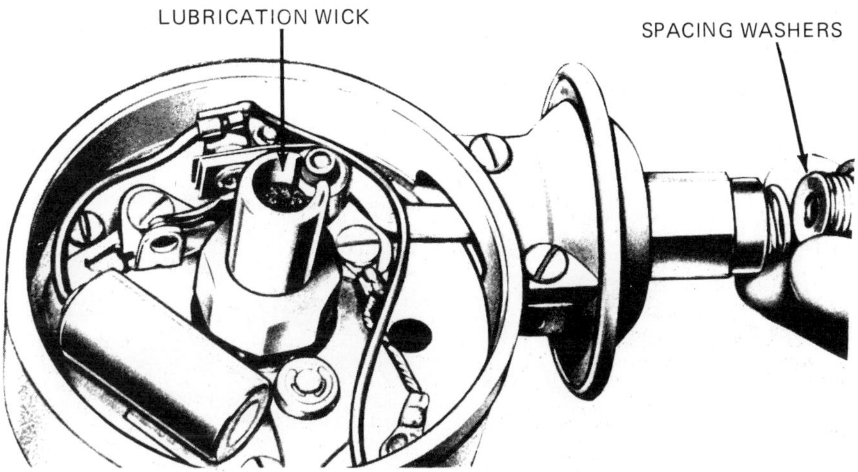

FIGURE 26
Exploded view of a distributor. Locate the felt lubrication wick. The wick should be lubricated periodically with several drops of oil.

SHOP TIP

Always check mechanical advance weight movement when removing the rotor. The rotor should spring back when turned by hand in the direction of rotation and released (Figure 27).

SHOP TIP

If a hard start condition on an older car is being caused by condensation in the distributor cap, the following remedy often works. Drill a 1/8-inch breather hole through the side of the distributor cap, just above the shoulder (Figure 28).

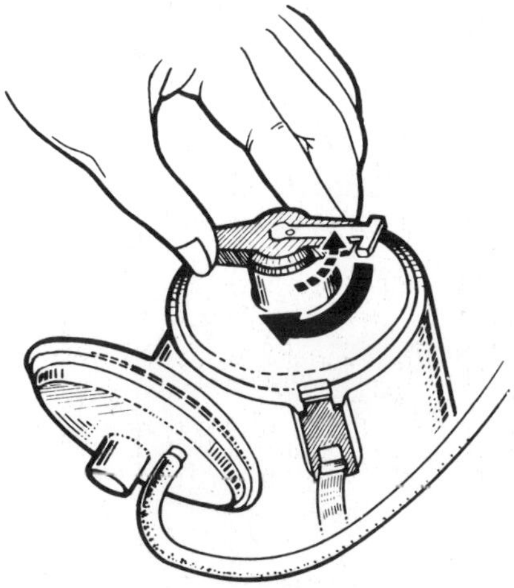

FIGURE 27
Quick check for "frozen" or sticking mechanical advance weights.

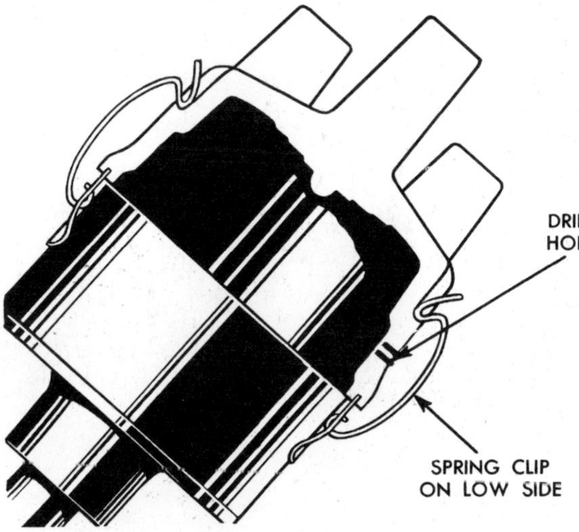

FIGURE 28
Drill hole in distributor cap as shown to solve condensation problem. Some caps are built with a vent hole.

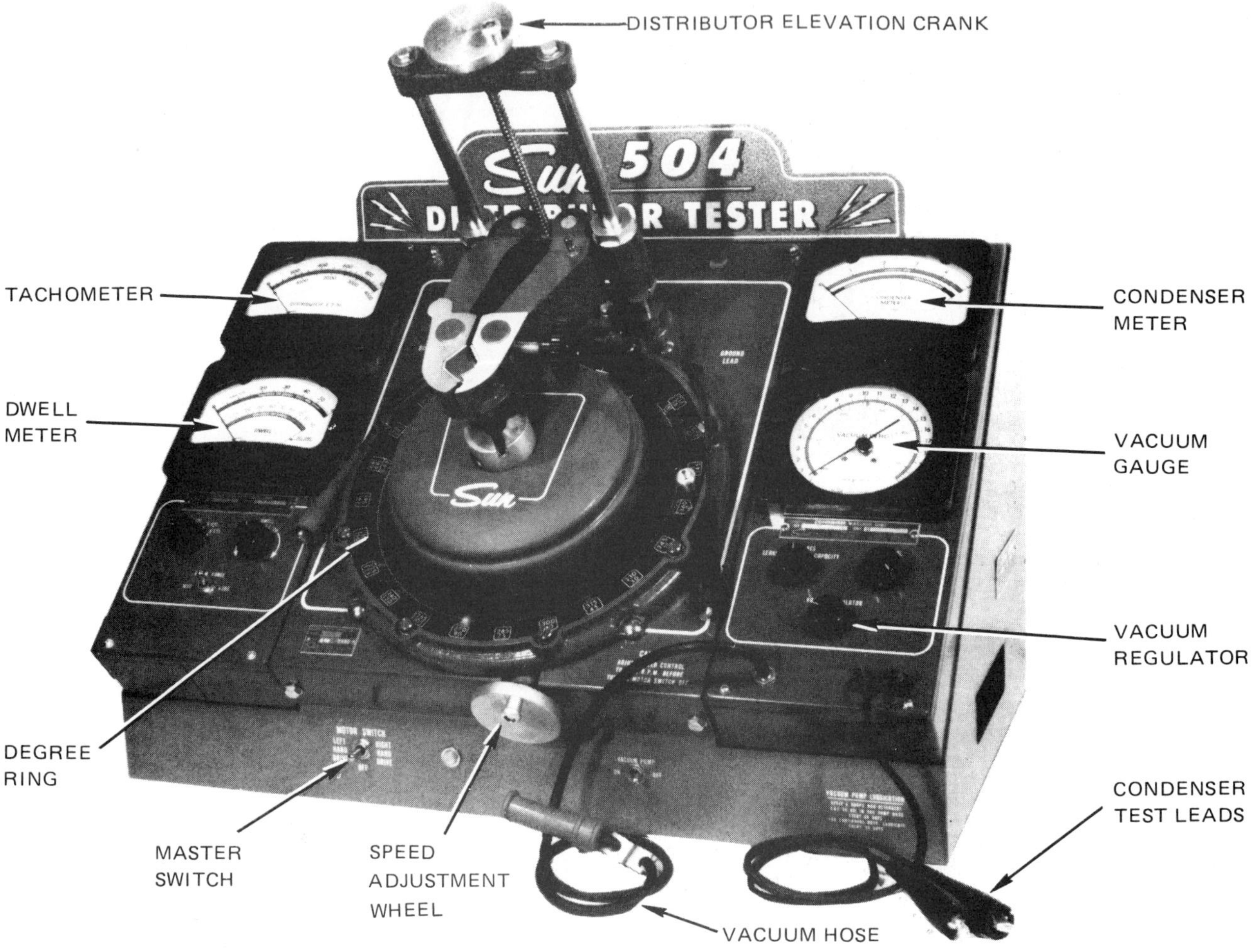

FIGURE 29
Distributor tester (sometimes called a synchograph).

FIGURE 30
Ball bearing breaker plates.

FIGURE 31
Worn vacuum advance diaphragm link hole.

Inspecting the Distributor

If diagnosis indicates that poor engine performance is caused by a defective or maladjusted distributor, the distributor can be removed from the engine and tested in a distributor tester (Figure 29).

Before placing the distributor in the tester, remove any accumulation of grease or dirt from the distributor housing and breaker plate assembly. Inspect as follows:

1. Check distributor shaft side clearance and end play.

2. Turn the distributor by hand and feel for bushing wear and smoothness of operation.

3. Inspect breaker plate for wear and smoothness of operation (Figure 30).

4. Be sure distributor cam is smooth and lightly lubricated.

5. Inspect condition of insulators and flexible leads.

6. Check ignition points (if so equipped) for alignment, pitting, and burning; check the rubbing block for wear.

7. Inspect vacuum advance linkage for alignment and wear (Figure 31).

Shaft and Bearing Wear Test

An excessively worn distributor shaft bushing can cause erratic and incorrect spark timing as engine rpm is changed. If excessive wear is indicated, both the distributor shaft and bushing must be replaced (Figure 32). Test as follows:

1. Install distributor in tester.

2. Adjust speed control until tachometer reads 500 rpm.

3. If possible, pull on top of distributor shaft with spring tension gauge until scale reads 16 ounces (Figure 33).

4. Observe dwell meter while tension is being applied to distributor shaft. *If dwell angle changes more than 3 degrees, excessive shaft and bushing wear is indicated.*

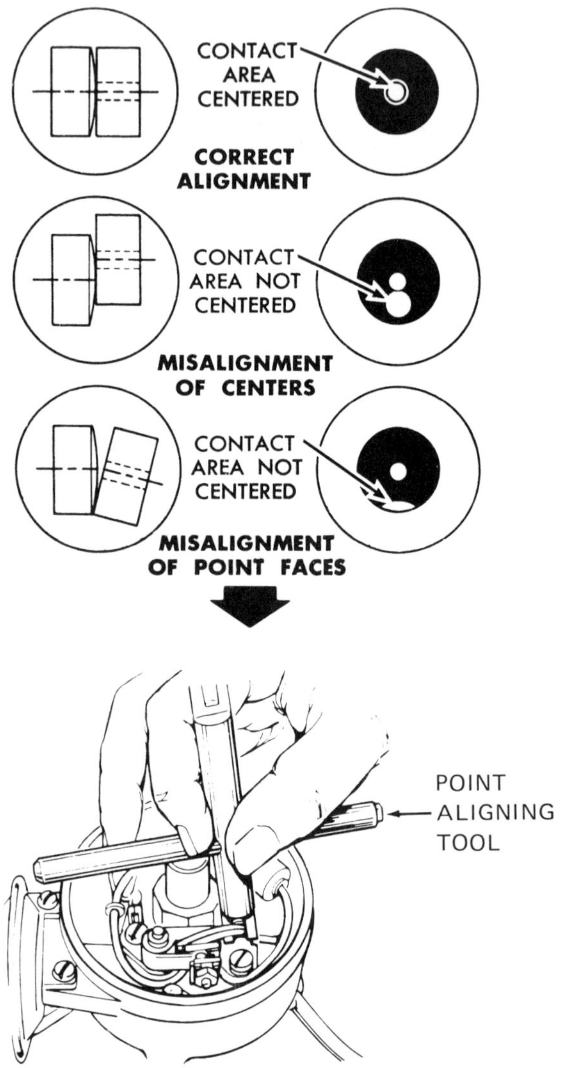

FIGURE 35
After replacing ignition points, always check face alignment and bend (align) the stationary point if required.

ELECTRONIC IGNITION MAINTENANCE

The only scheduled maintenance for an electronic ignition system is spark plug replacement (the interval varies). However, the coil, spark plug wires, distributor cap, rotor, and ignition timing should be checked if there are any performance problems.

Setting the Air Gap (Not Always Required)

Remove the distributor cap and crank the engine over until one of the reluctor teeth is perfectly aligned with the pickup coil (Figure 42). Then follow this procedure:

1. Loosen the pickup coil hold-down screw.
2. Insert a *brass* feeler gauge of the correct thickness between the reluctor tooth and the coil.
3. Adjust the air gap.
4. Tighten the hold-down screw.
5. Remove the feeler gauge.

NOTE: Check each reluctor tooth to be sure it is not hitting the pickup pole piece. If there is contact, the distributor shaft may be bent.

SPARK PLUG OPERATION

"Heat range" refers to a spark plug's ability to dissipate (or get rid of) combustion heat (Figures 43 and 44).

For proper engine performance, spark plugs must operate within a certain temperature range. "Hot" plugs can cause preignition. If the plugs are not hot enough, they will not burn off deposits on the insulator and electrodes. If the plugs run too "cold," carbon deposits will form on the insulator, causing fouling and misfiring.

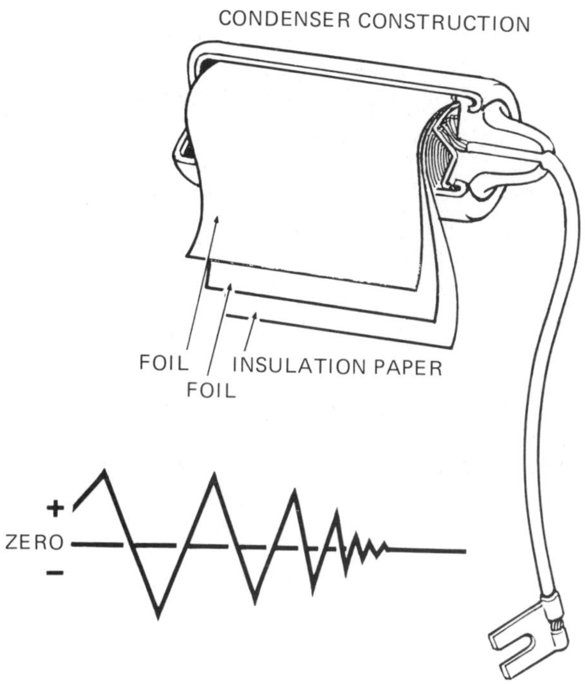

FIGURE 41
Condenser construction and voltage oscillations. (Courtes
of The Echlin Manufacturing Company)

NOTE: It is very important not to let any foreign particles fall into the cylinder when removing plugs, since this can cause damage to the engine.

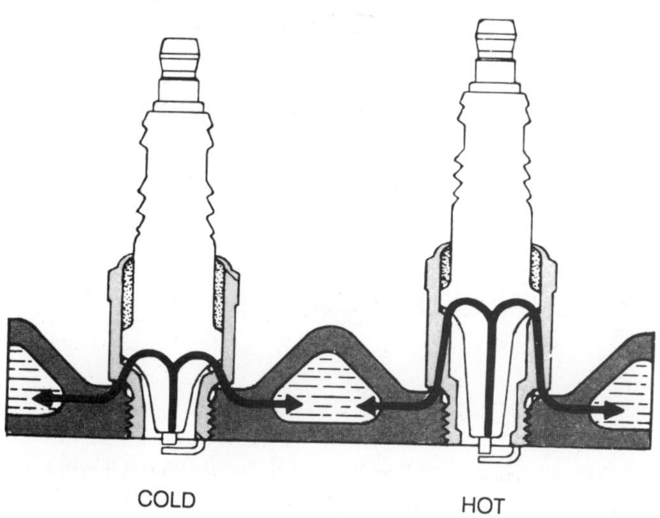

FIGURE 43
Example of spark plug heat range. (Courtesy of Champion Spark Plug Company. All rights reserved)

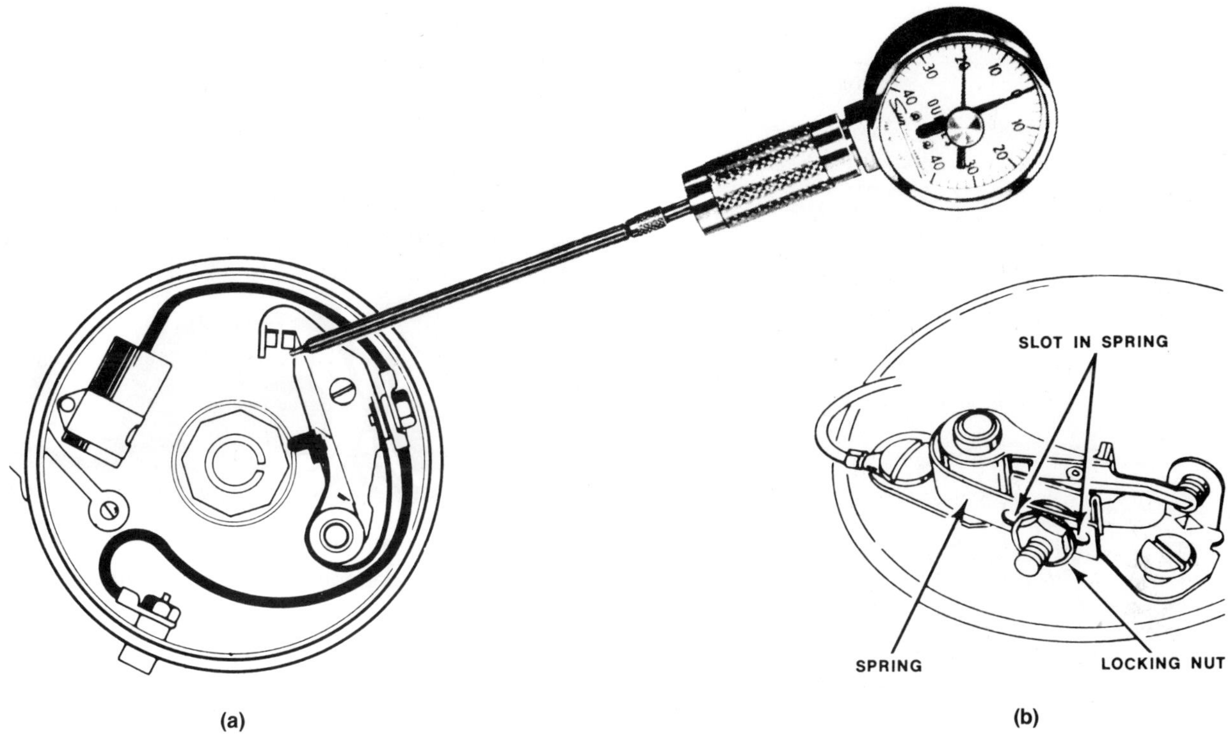

FIGURE 36
(a) Ignition point spring tension gauge. (b) Point spring tension can be changed by bending the spring, or by moving the spring along its slot.

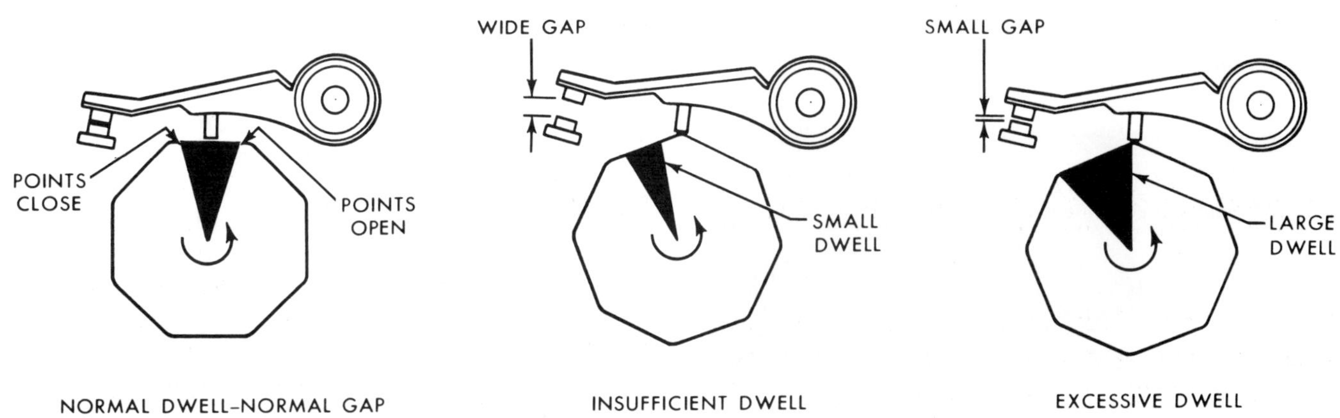

FIGURE 37
Point dwell.

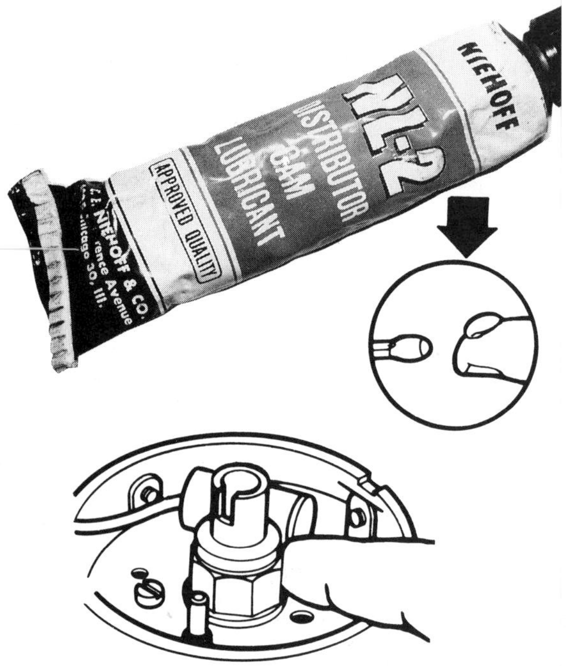

FIGURE 38
Lubricating the distributor cam.

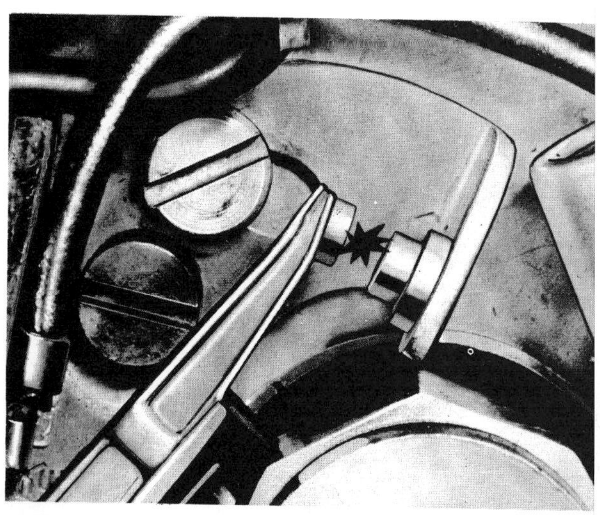

FIGURE 40
The condenser provides a "path of lesser resistance," w
reduces arcing at the points and speeds up the collaps
the magnetic field in the coil.

FIGURE 44
Antifoulers make the plug heat range much hotter and often burn up the plugs as shown.

Oil fouling covers the plug firing end with *wet, black deposits* caused by an excessive amount of oil reaching the cylinders. Check for worn rings, worn valve guides, damaged valve guide seals, or a ruptured fuel pump diaphragm. Temporary relief may be obtained by switching to "hotter" plugs. However, the only long-term cure is to remedy the cause.

Plugs fouled with *particles of aluminum* indicate that aluminum parts (pistons and/or heads) are disintegrating (Figure 46). This can be due to detonation, preignition, or overadvanced ignition timing.

FIGURE 45
(a) A spark plug wire should be removed from a spark plug boot by twisting and pulling the boot. (Courtesy of AC-Delco Division, General Motors Corporation) (b) Special pliers and pullers are available for removing boots from spark plugs. (Courtesy of AC-Delco Division, General Motors Corporation)

FIGURE 46
These aluminum plug deposits resulted when a piston head melted. (Courtesy of Champion Spark Plug Company. All rights reserved)

Preignition can cause extensive plug damage (Figure 47). When plugs indicate the presence of preignition, check the heat range of the plugs and the combustion chamber for carbon deposits and for cooling system problems.

Certain fuels tend to form scavenger deposits. The deposits are *crusty,* and *white or yellow* in color (Figure 48). For those fuels, this condition is normal.

Splashed fouling often occurs when new plugs are installed in an engine that has heavy deposits in the combustion chamber (Figure 49). The new plugs restore the firing to normal. This raises the combustion chamber operating temperature, and the accumulated deposits flake off and fuse to the plug insulator. Sometimes this is called spark plug "whiskering."

Overheated plugs produce a *white or gray blistered insulator* (Figure 50). This can result when plugs are too "hot" for the engine, or when an engine overheating problem is present.

Under certain atmospheric conditions, around a conductor carrying high voltage, a pale bluish glow (corona) can sometimes be observed. Corona may leave a nonconductive yellowish stain around the plug insulator, near the top of the shell (Figure 51). *Do not misinterpret this stain as being caused by plug leakage.*

Gapping Spark Plugs

New plugs should be properly gapped before they are installed. Use a round feeler gauge to measure the gap, and use a bending tool to adjust the side electrode (Figure 52). You should feel a slight drag on the feeler gauge when the gap is correct.

The gap on plugs used with electronic ignition systems may be wider than that on plugs used in point-type ignition systems. Gaps may be as great as .080 inch.

NOTE: Wide-gap spark plugs are manufactured for certain vehicles. Setting conventional plugs to the wide specifications may cause misfiring or hard starting.

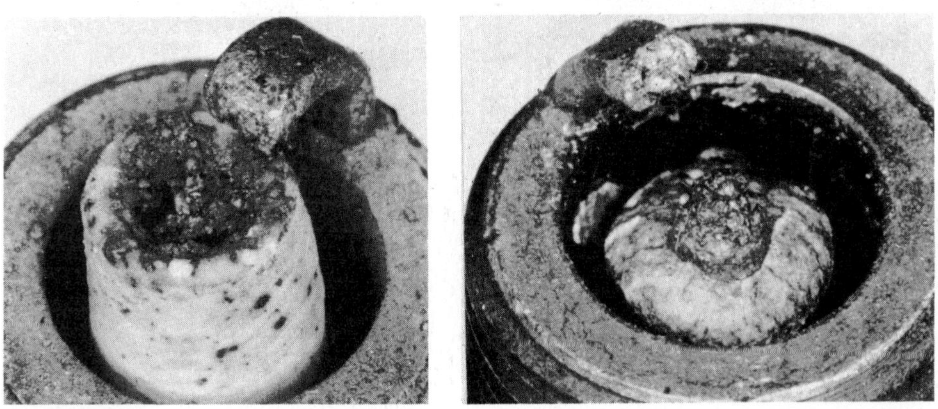

FIGURE 47
Preignition destroys plugs rapidly. (Courtesy of Champion Spark Plug Company. All rights reserved)

FIGURE 48
Scavenger deposits. (Courtesy of Champion Spark Plug Company. All rights reserved)

FIGURE 49
Splash fouled plug. (Courtesy of Champion Spark Plug Company. All rights reserved)

Installing Spark Plugs

Before putting the old plugs back in or installing new plugs, wipe the threads with a clean cloth and put a drop of oil on the threads. Then screw the plugs in by hand until they are finger tight.

Use a plug socket and ratchet wrench to final tighten the plugs. Tighten gasket-type plugs about 1/4 turn. Tighten tapered seat plugs about 1/6 of a turn. Refer to Chart 3.

FIGURE 50
Overheated spark plug. (Courtesy of Champion Spark Plug Company. All rights reserved)

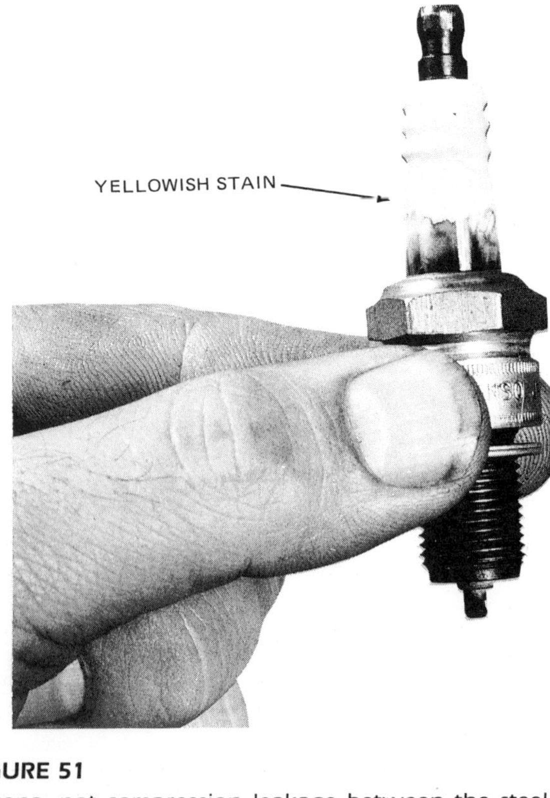

FIGURE 51
Corona, not compression leakage between the steel shell and insulator, caused these yellowish stains.

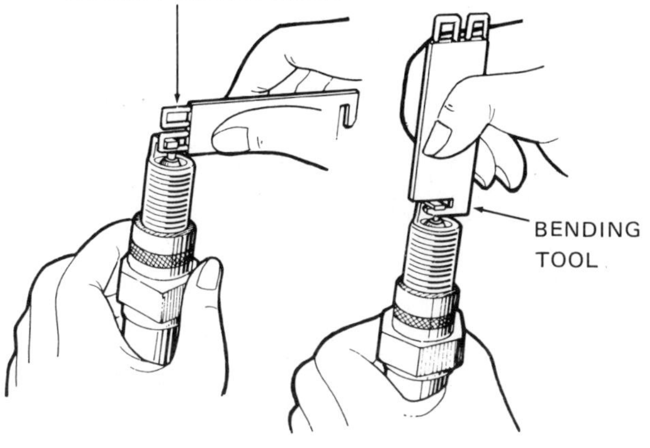

FIGURE 52
Plug gap checking and adjustment.

Checking Spark Plug Wires

Look for cracked, brittle, or damaged spark plug wires and boots (Figure 53). Bad wires cause misfiring, poor engine performance, reduced fuel economy, and hard starting.

If you suspect internal damage, perform an ohmmeter test (Figure 54). Replace all plug wires that exceed the resistance specifications.

Replacing Spark Plug Wires

When replacing spark plug wires, be sure to note the firing order of the plugs (Figure 55).

The high voltage in a spark plug wire can induce voltage in the plug wire next to it. Therefore, do not route wires from consecutively firing cylinders next to one another, or crossfire may result. On computer-controlled vehicles, correct spark plug wire routing is very important in order to prevent radio frequency interference from creating malfunctions in other electrical circuits.

SHOP TIP

When installing a new secondary wire, gently push the wire and nipple over the cap or coil tower. Then pinch the large diameter of the nipple to release trapped air (Figure 56). In addition, coat the ends of the secondary wires with dielectric silicone prior to installation. This helps ensure proper seating and prevents corrosion.

CHART 3 SAE-ISO RECOMMENDED INSTALLATION TORQUE

Plug Thread	Cast Iron Heads		Aluminum Heads	
	Pound Feet	Newton Meters	Pound Feet	Newton Meters
10mm Gasket Seat	7–11 lb. ft.	10–15NM	7–11 lb. ft.	10–15NM
12mm Gasket Seat	11–19 lb. ft.	15–25NM	11–19 lb. ft.	15–25NM
14mm Gasket Seat	26–29 lb. ft.	35–40NM	15–22 lb. ft.	20–30NM
14mm Tapered Seat	7–15 lb. ft.	9–20NM	7–15 lb. ft.	9–20NM
18mm Gasket Seat	32–38 lb. ft.	43–52NM	28–34 lb. ft.	38–46NM
18mm Tapered Seat	15–20 lb. ft.	20–27NM	15–20 lb. ft.	20–27NM
⅞″–18	35–43 lb. ft.	47–58NM	31–39 lb. ft.	42–53NM
If no torque wrench is available when installing tapered seat 14 and 18mm—tighten 1/16″ turn (snug) after finger tight.				

Use above values only when threads on spark plugs are clean, dry and smooth, the plug has been finger tightened and a new folded gasket used (except in the case of tapered seats).

FIGURE 53
Hard-type plug boots can crack or become carbon tracked.

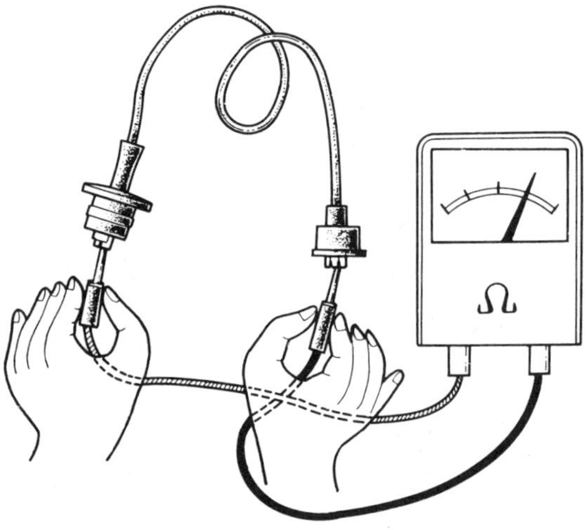

FIGURE 54
Checking plug wire resistance.

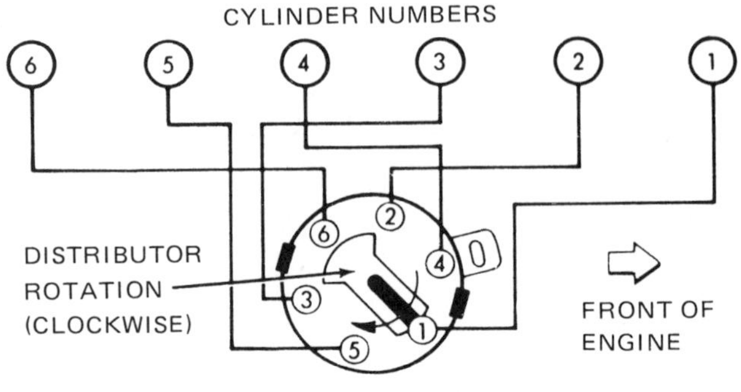

FIGURE 55
Typical six-cylinder firing order.

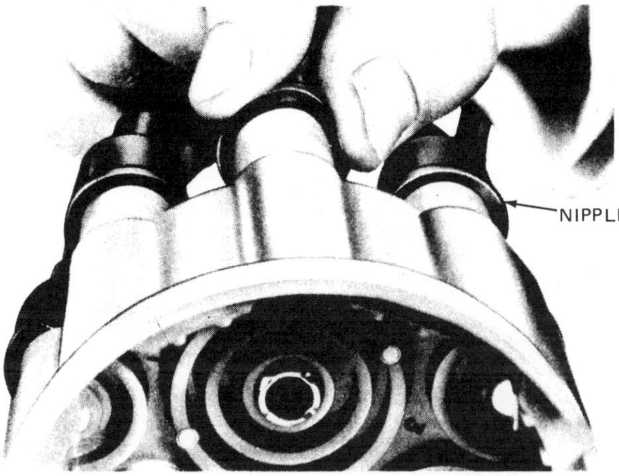

FIGURE 56
Always release trapped air when installing secondary cable ends in the distributor cap. (Courtesy of Chrysler Corporation)

IGNITION TIMING ADJUSTMENT

Engine power and exhaust emissions are directly related to ignition timing. Once initial (base) timing has been set, the rest of the timing curve is automatically controlled. This is accomplished by centrifugal and vacuum advance mechanisms, or by a computer (Figure 57[a] and [b]).

The timing light is connected to the ignition system as shown in Figure 58. The light flashes whenever the distributor sends a spark to the number one cylinder.

Initial timing is performed at a *specific engine rpm* when the centrifugal advance mechanism is not functioning. On some cars, the vacuum advance must also be *disconnected* so that it does not affect the initial timing adjustment.

NOTE: On computer-controlled vehicles, it may be necessary to disconnect, jump, or ground a specific electrical connector when setting initial timing.

With the timing light aimed at the timing marks, press the trigger and observe the position of the marks (Figure 59). The high-speed flashing of the light will "freeze" the motion so the marks can be read.

(a)

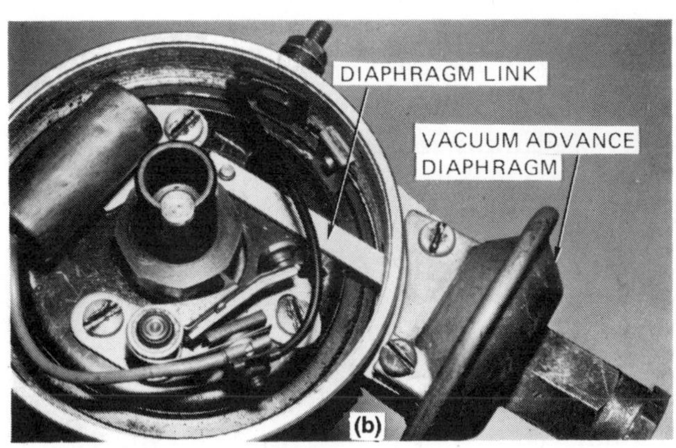

(b)

FIGURE 57
(a) Centrifugal advance weights and spring assembly. (Courtesy of Ford Parts and Service Division) (b) Vacuum advance assembly.

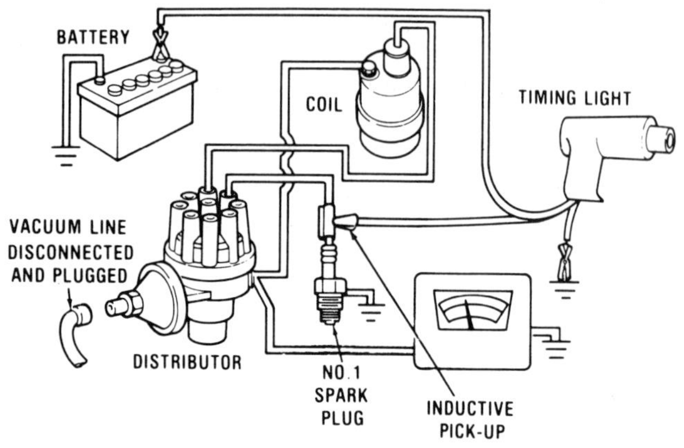

FIGURE 58
Timing light hook-up.

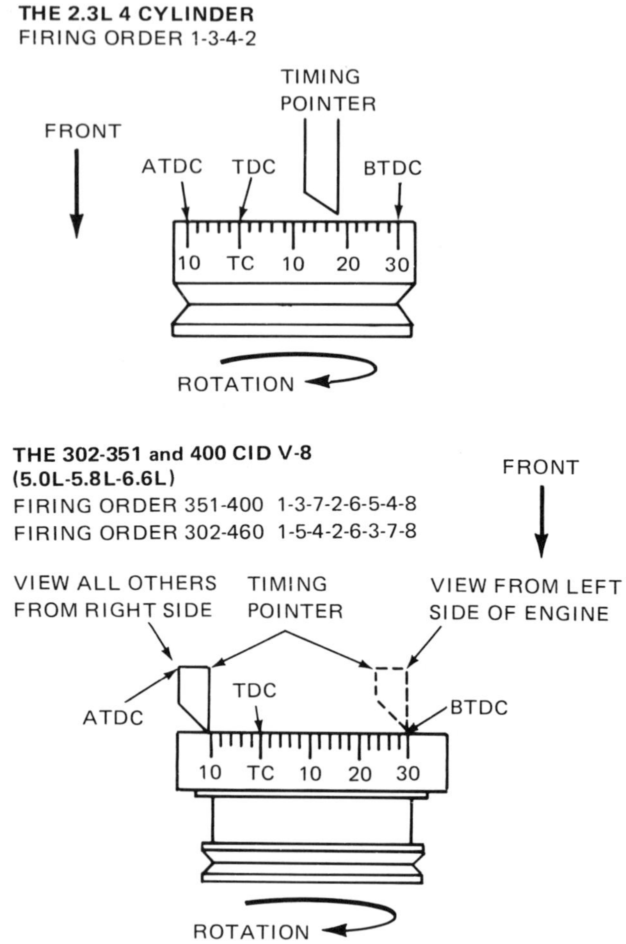

FIGURE 59
Timing mark examples. (Courtesy of Ford Parts and Service Division)

CAUTION: Be sure to view timing marks from a position that allows your eyes to be squarely in line with them. If marks are viewed from another angle, an error in timing may result.

To adjust timing (on the majority of cars), loosen the distributor hold-down clamp and turn the distributor until the timing marks are lined up. After adjusting, tighten the distributor hold-down clamp and recheck the timing.

Centrifugal Advance Weight Check

1. With the engine operating at idle rpm and the distributor vacuum line(s) disconnected, note timing mark position with a timing light.

2. Slowly increase and decrease engine speed from idle to 2500 rpm while observing the position of the timing mark.

3. The timing mark should move away from and return to its original position smoothly and without hesitation. If it does not, the centrifugal advance weights are "frozen" or are sticking (Figure 60).

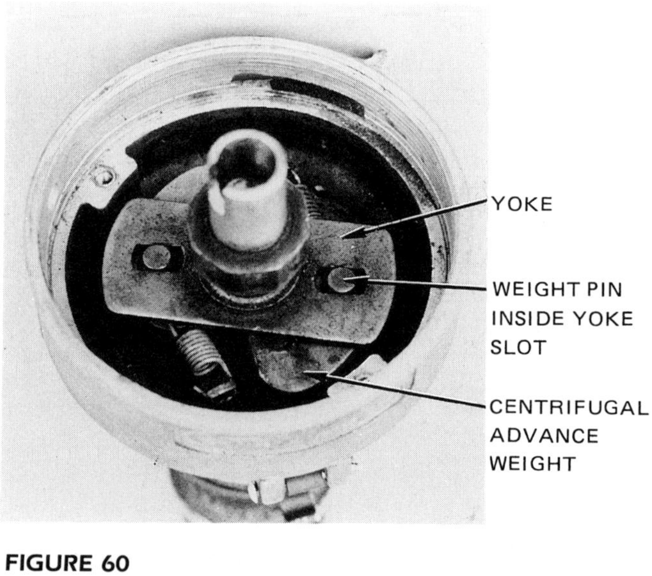

FIGURE 60
These advance weights were sticking because of yoke slot wear.

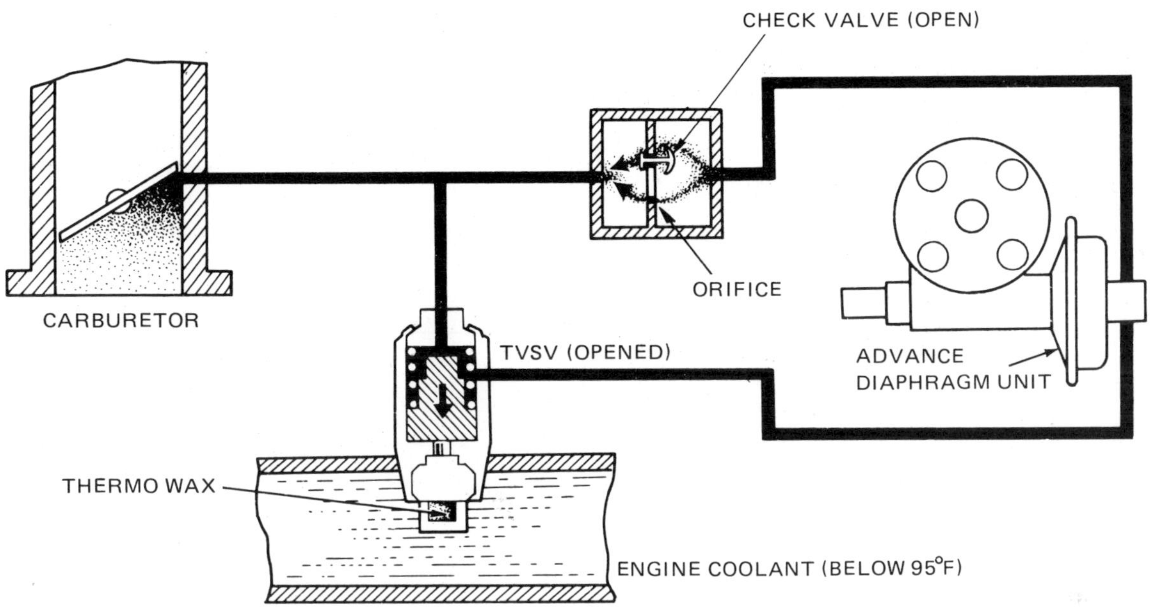

FIGURE 61
A spark delay system. When the engine is cold, the TVSV is open. This allows ported vacuum to act directly on the advance diaphragm so that there will be no vacuum delay. Which hose would you disconnect and apply vacuum to , in order to check the advance diaphragm?

Vacuum Advance Check

1. Disconnect the vacuum hose(s) to the distributor advance diaphragm unit (Figure 61) and connect a hand-operated vacuum pump to the hose fitting.

2. With the engine operating at 1000 rpm, apply vacuum to the advance diaphragm.

3. The timing mark should advance when viewed with a timing light.

4. If there is no advance, stop the engine and apply 18″ Hg vacuum to the advance diaphragm. There should be no leakage. If vacuum does not hold, replace the diaphragm.

NOTE: Many late-model vehicles do not have a centrifugal or vacuum advance mechanism. Spark timing is controlled by a computer.

Magnetic Timing Probe

Some engines require the use of a magnetic timing probe. The probe is triggered by a magnetic signal from the crankshaft vibration damper or by a special impulse ring on the crankshaft.

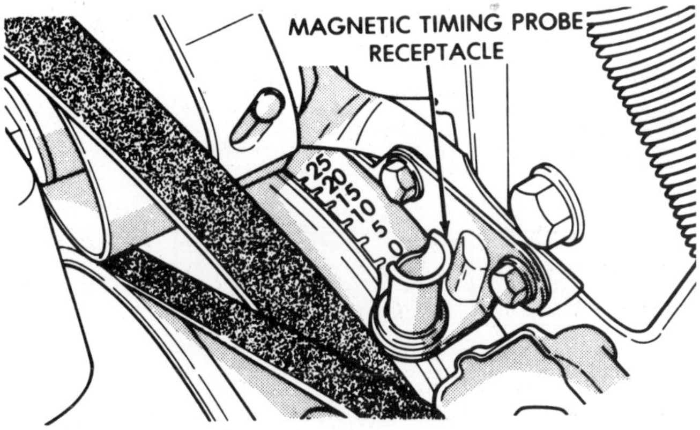

FIGURE 62
Receptacle for a timing light that uses a magnetic probe. (Courtesy of Chrysler Corporation)

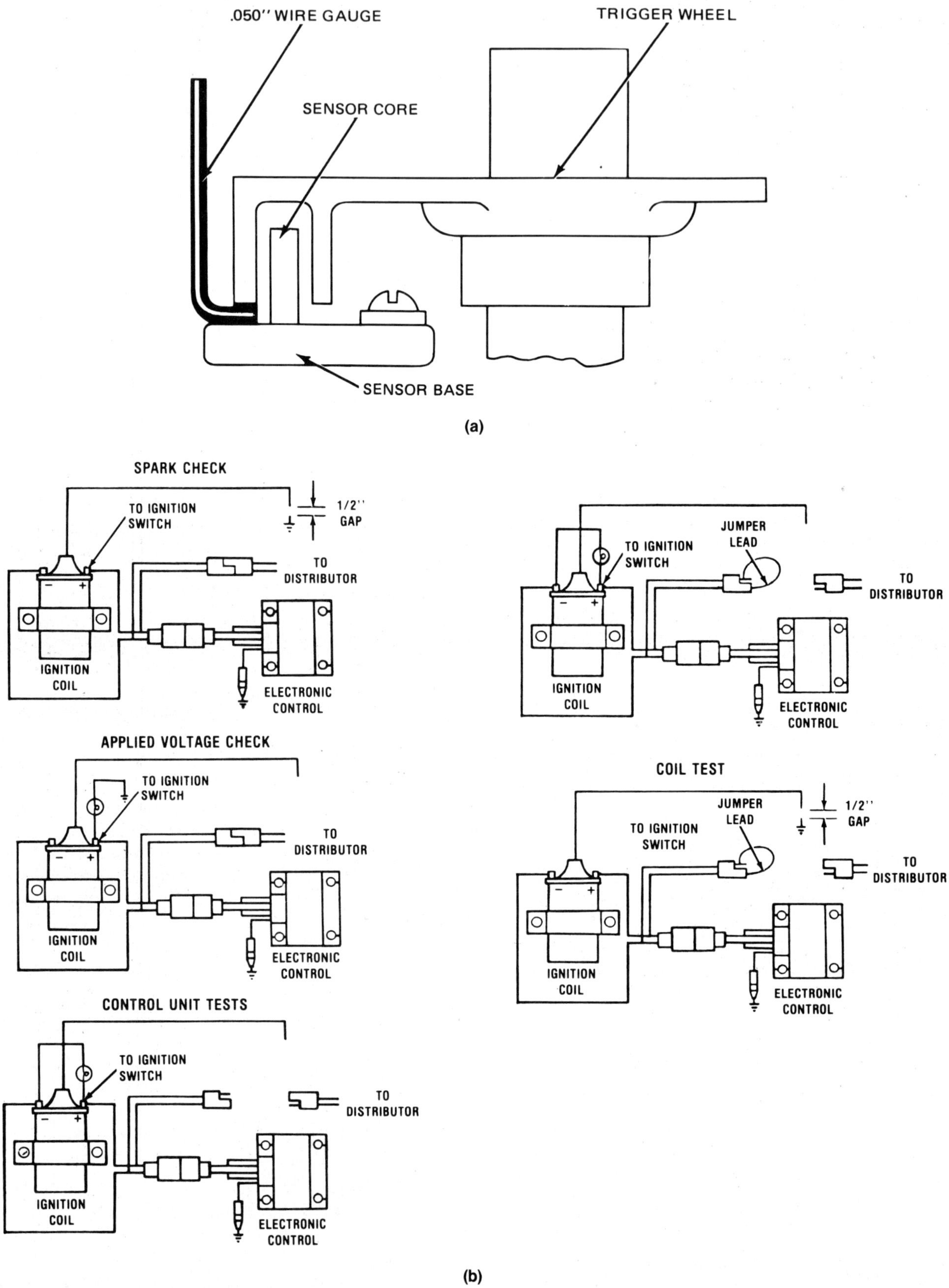

FIGURE 63

(a) Checking clearance between the trigger wheel and sensor base assembly. (Courtesy of American Motors Corporation) (b) BID ignition system checks. (Courtesy of American Motors Corporation)

Most oscilloscopes are equipped with this magnetic probe. The probe is inserted into a holder next to the vibration damper (Figure 62). The oscilloscope then displays the ignition timing on a meter (in degrees).

ELECTRONIC IGNITION DIAGNOSIS

The following procedures are designed to help you locate the cause of electronic ignition problems. The procedures are typical. If more specific tests are required for the vehicle you are working on, refer to a factory manual.

Begin by making these preliminary checks:

1. Fuel to the carburetor.

2. Battery cranking voltage (9.6 volts minimum).

3. Primary and secondary wiring (damage and loose connections).

4. Spark at spark plug while cranking engine.

AMERICAN MOTORS BREAKERLESS INDUCTIVE DISCHARGE SYSTEM (BID)

Refer to Figure 63 for the BID ignition system checks.

The BID system was introduced on American Motors Corporation (AMC) engines in 1975. The system includes the following parts:

1. Electronic control unit.

2. Sensor coil and trigger wheel.

3. Advance mechanism.

4. Distributor cap.

5. Rotor.

6. Coil.

7. Spark plug wires.

Spark Check

1. Remove wire from distributor cap and hold it 1/4 inch from ground.

2. Crank the engine. If spark occurs, ignition primary is satisfactory. The problem is in the cap, rotor, or spark plug wires.

Applied Voltage Check

1. If no spark occurs, connect a test lamp from the positive (+) terminal on the coil to ground.

2. Turn the ignition switch to both *on* and *start.* The test lamp should light in both positions. If not, the fault is in the circuit between the battery and coil.

Control Unit Test

1. If the test lamp lights in both ignition switch positions of the applied voltage check, connect lamp across positive (+) and negative (−) terminals of coil.

2. Unplug the sensor leads near the distributor and turn the ignition switch *on.* If the lamp does not light, check the control unit ground. If it is OK, the control unit is defective and must be replaced.

3. If the lamp lights in step 2, jump across the sensor lead terminals coming from the control unit. If the lamp does not go off, the fault is in the control unit.

Coil Test

1. If the lamp goes off in step 3, remove the lamp and hold the coil wire 1/2″ from ground.

2. Short across the terminals of the sensor lead; if a spark does not occur each time the terminals are shorted, the coil is faulty.

3. With the ignition *off*, use a ohmmeter to measure coil primary resistance between the positive (+) and negative (−) terminals. The meter should read 1.25 to 1.40 ohms.

4. Measure secondary resistance between the center tower and the positive (+) coil terminal. The meter should read 9000 to 15,000 ohms. If either test is not within specifications, replace the coil.

Sensor Test

1. Using an ohmmeter, measure sensor resistance at the distributor leads. The meter should read 1.6 to 2.4 ohms.

2. Tap the sensor lightly. Resistance should not vary.

3. Connect one ohmmeter lead to either sensor lead; touch other lead to center core of sensor. The ohmmeter should show an open circuit.

Troubleshooting and Service Hints

1. Clearance between the trigger wheel and the sensor base assembly must be .050″. Refer back to (Figure 63[a]).

2. If system dwell is not set according to specifications, check the trigger wheel air gap and/or check for a worn distributor shaft and bushings.

3. Sometimes the trigger wheel assembly binds on the inner distributor shaft and causes insufficient centrifugal advance.

CHRYSLER ELECTRONIC IGNITION SYSTEM (EXCEPT HALL EFFECT)

Chrysler electronic ignition was first used in production in 1972. The basic system consists of the following components:

1. Magnetic pulse distributor.
2. Electronic control unit.
3. Ballast resistor (single or dual).
4. Coil.
5. Distributor cap.

The system may be tested using a voltmeter and an ohmmeter with a 20,000 ohm/volt rating. Figure 64[a] shows the component layout for one type of Chrysler electronic ignition system.

Checking the Wiring Harness

1. Turn ignition switch *off*.

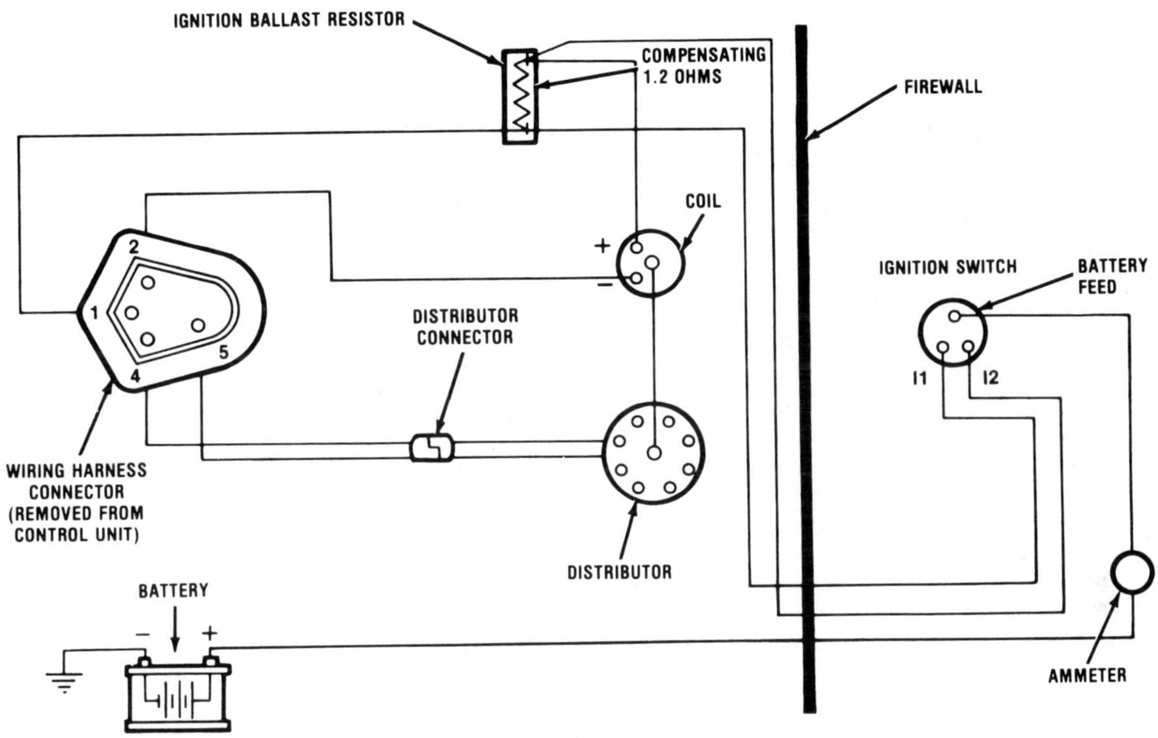

(a)

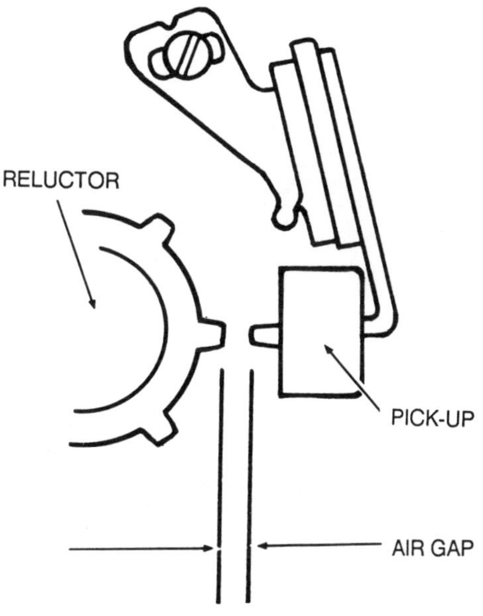

(b)

FIGURE 64
(a) A widely used Chrysler electronic ignition system. (Courtesy of Chrysler Corporation) (b) Clearance (air gap) between the pickup coil and reluctor must be set with a non-magnetic feeler gauge.

2. Remove the wiring harness connector from the control unit.

3. Turn the ignition switch *on.*

4. Connect the voltmeter negative lead to ground.

5. Connect the voltmeter positive lead to wiring harness connector cavity #1. Available voltage at the cavity should be within 1 volt of battery voltage.

6. Connect the voltmeter positive lead to wiring harness connector cavity #2. Available voltage at the cavity should be within 1 volt of battery voltage.

7. Turn ignition switch *off.*

Checking the Pickup Coil

1. Connect an ohmmeter to wiring harness connector cavities #4 and #5. The resistance reading should be between *150 and 900 ohms.*

If the readings are higher or lower than specified, disconnect the dual lead connector coming from the distributor. Using an ohmmeter, check the resistance at the dual lead connector. If the reading is not between *150 and 900 ohms,* replace the pickup coil assembly in the distributor.

2. Connect one ohmmeter lead to a good ground and the other lead to either connector of the distributor. The ohmmeter should show an open circuit. If the ohmmeter does show a reading, the pickup coil in the distributor must be replaced.

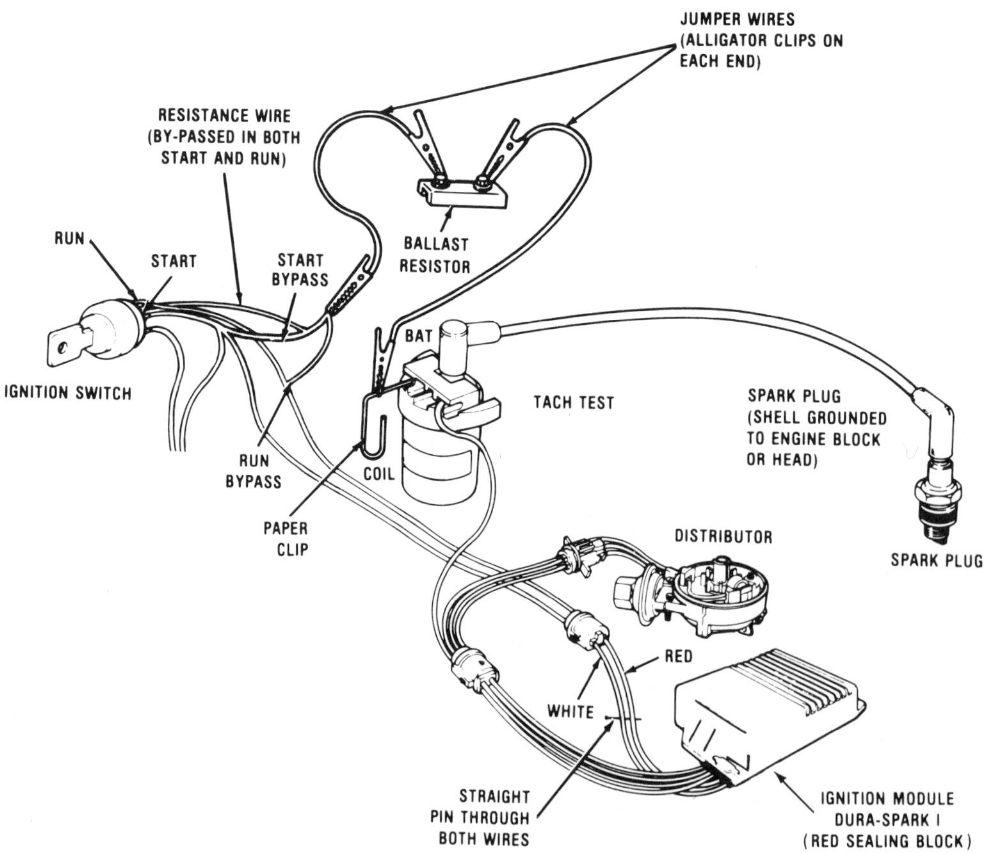

FIGURE 65
An early generation Ford Dura-Spark electronic ignition system. (Courtesy of Ford Parts and Service Division)

Checking the Control Unit Ground Circuit

Connect one ohmmeter lead to ground, and the other lead to the control unit connector pin #5. The ohmmeter should show continuity between ground and the connector pin. If continuity does not exist, tighten the bolts holding the control unit to the bulkhead (firewall). Then recheck. If continuity still does not exist, replace the control unit.

Troubleshooting and Service Hints

1. When replacing a reluctor, make sure that it is installed with the arrow facing the direction of rotation.

2. Clearance between the pickup(s) and reluctor teeth is adjustable (Figure 64[b]). Use a non-magnetic (brass) feeler gauge and set the clearance according to specifications. Make sure that there is no more than a .002″ variation on any reluctor tooth. If variation is greater than .002″, the distributor shaft may be worn or bent.

FORD ELECTRONIC IGNITION SYSTEM

Ford's solid-state ignition (SSI) system was introduced in 1973 on some Mark IV models. Dura-Spark was introduced in 1977 and thick-film integrated (TFI) ignition in 1982. Even though these Ford systems are called by different names, they have similar components and operate in a similar manner. These components are:

1. Electronic control module.
2. Coil.
3. Distributor cap.
4. Ignition resistor wire (not used on some systems).
5. Magnetic pickup coil (stator).
6. Armature.
7. Advance mechanism.
8. Spark plug wires.

The first step in testing a Ford solid-state electronic ignition system is to determine the type of ignition system the vehicle is equipped with. The easiest way to identify it is by the color of the sealing block on the ignition module (Figure 65). The following tests are for a Dura-Spark system.

Tap Test

1. Remove the distributor cap and rotor. Crank the engine to align one tooth of the armature with the pole piece in the pickup coil.

NOTE: If the system is a Dura-Spark I, a 1.3-ohm ballast resistor must be placed in series with the primary circuit at the BAT terminal of the ignition coil.

2. Remove the coil wire from the distributor. Connect it to a "dummy" spark plug or hold it 1/4 inch from the cylinder block or head.

3. Turn the ignition switch to *run* and tap the distributor body with a screwdriver handle. There should be a spark at the spark plug or coil wire terminal.

Coil Condition

Check the ignition coil for internal shorts or opens and for specified primary and secondary resistance values (Figure 66[a]).

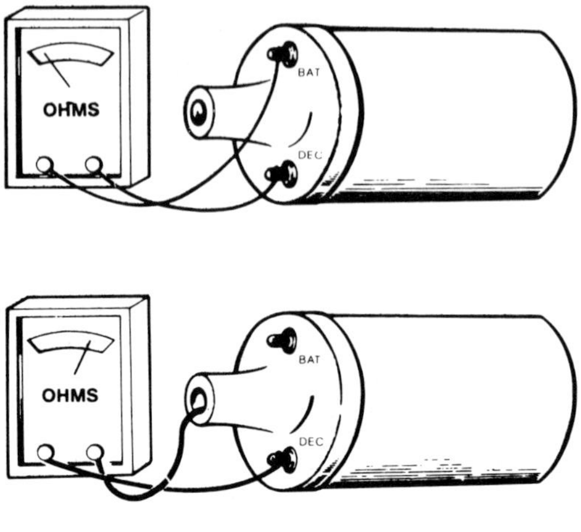

(a)

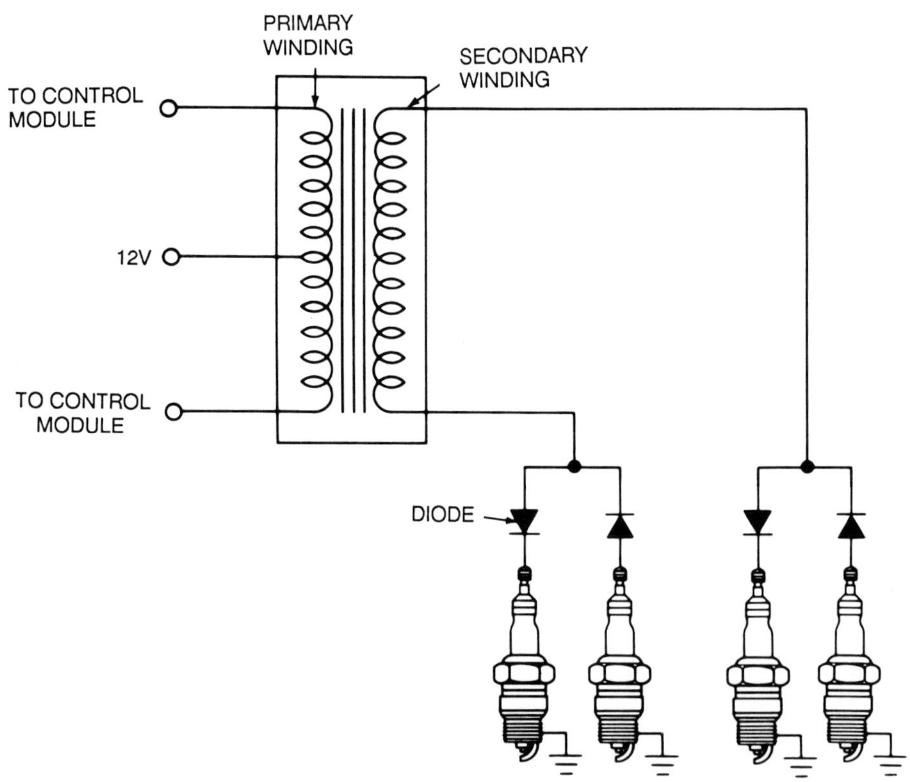

(b)

FIGURE 66

(a) Ford electronic ignition system coil tests. (Courtesy of The Echlin Manufacturing Company) (b) Schematic of the Ford distributorless ignition system for a four-cylinder engine with a firing order of 1-3-4-2. Two plugs are fired at the same time; one in the cylinder that is on its compression stroke and the other, a "waste spark," in the companion cylinder on the exhaust stroke. (Courtesy of Ford Parts and Service Division).

Pickup Coil Test

Connect an ohmmeter to the parallel blades at the distributor connector. Resistance should be 400 to 1000 ohms. A reading less than 400 indicates a short. An infinity reading indicates an open. With the ignition switch off, check the resistance between the purple wire at the distributor and ground, and between the orange wire and ground. In each case, the resistance should be over 70,000 ohms.

Troubleshooting and Service Hints

1. The small magnetic pole piece in the center of the pickup coil sometimes cracks. If it is cracked, there are two north and south poles instead of one. This causes erratic firing of the spark plugs.

2. When a spark plug wire or the high-tension coil wire is removed, apply silicone grease to the inside of the boots before reconnecting.

Ford Electronic Engine Control Systems

All electronic engine control (EEC) systems contain an electronic control assembly (ECA), a variety of engine sensors, an ignition module, and a unique distributor. The distributor's only function is to distribute secondary voltage to the spark plugs. Timing is controlled by the ECA.

Ford Distributorless Ignition

Ford has recently developed a distributorless ignition system. This system uses a double-ended ignition coil that fires two plugs at once (Figure 66[b]).

GENERAL MOTORS HIGH ENERGY IGNITION (HEI) SYSTEM

The high energy ignition system was introduced as an option on General Motors V-8 engines in 1974. It became standard on all engines in 1975. AMC began using GM's HEI system in 1980. The system consists of these components:

1. A special large distributor housing.

2. Advance mechanism (not used in some HEI systems).

3. Electronic module.

4. Pickup coil (includes pole piece).

5. Timer core (trigger wheel).

6. Distributor cap.

7. Rotor.

8. Coil (internally or externally mounted).

9. Radio noise suppression capacitor (inside the module on some models).

10. Spark plug wires.

The following troubleshooting information applies to a conventional HEI system.

Initial Voltage Checks

1. Make a secondary output check using an oscilloscope. A conventional HEI system should produce at least 25,000 volts (25KV) during cranking or engine idling. If it does not, follow the procedures in step 2.

2. If the engine misses or will not start, make voltage checks at the diagnostic connector (if so equipped). Refer to Figure 67.

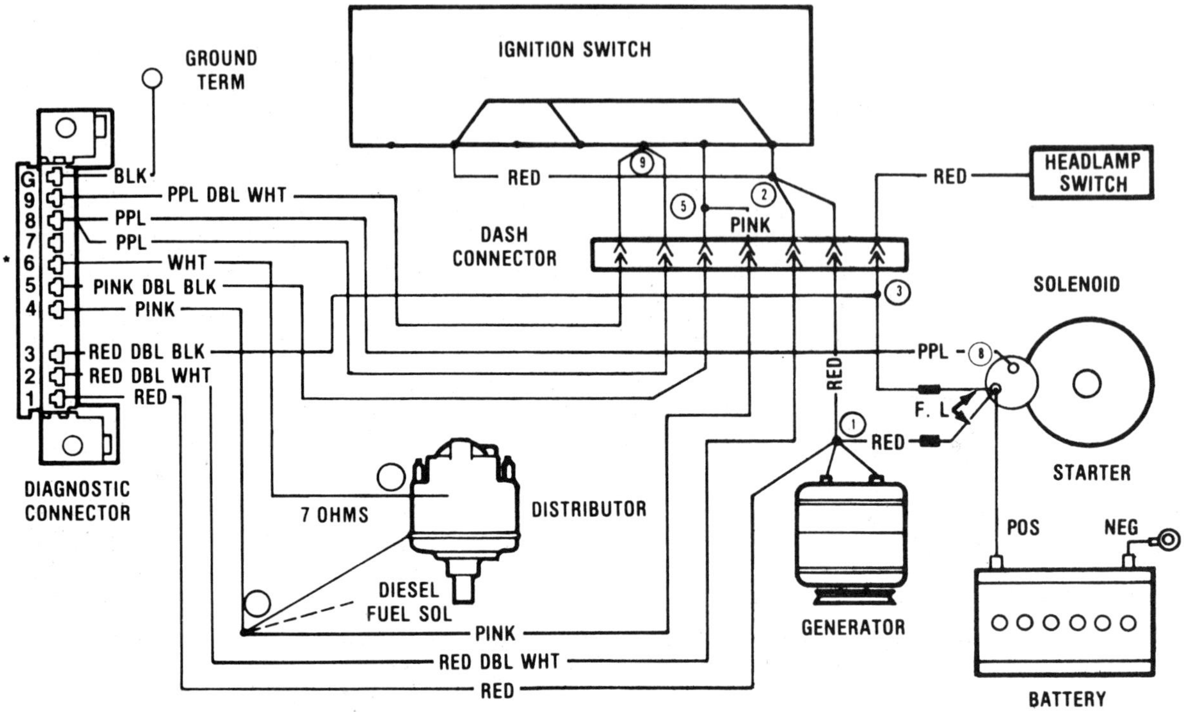

*TERMINAL 6 IS NOT USED WITH DIESEL

VOLTMETER CONNECTIONS	VOLTS CRANKING	VOLTS IDLING
1. 6 and G	7 or more	9.6 or more — Check HEI distributor.
	Under 7	Go to test 2.
2. 4 and G	7 or more	9.6 or more — 7 volts or more at HEI battery terminal, check the HEI distributor.
	Under 7	Go to test 3.
3. 5 and G	7 or more	9.6 or more — Check for open circuit or loose connection in wire from battery terminal on distributor (fuel solenoid on diesel) through engine dash connector to ignition switch.
	Under 7	Replace ignition switch.

FIGURE 67
General Motors diagnostic connector checks on a conventional HEI system. Many HEI cars came equipped with this orange colored diagnostic connector box in order to quickly access test points.

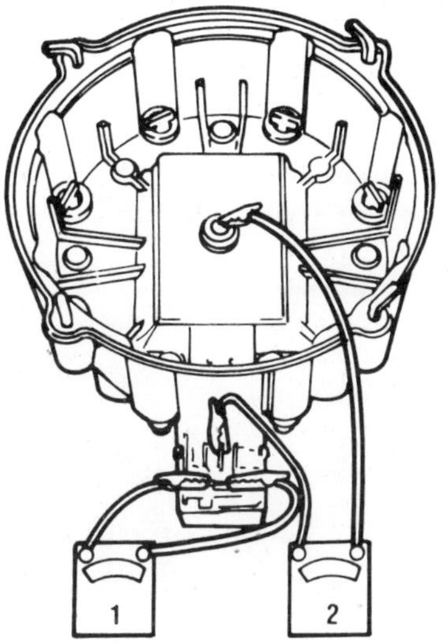

FIGURE 68
HEI system V-8 and V-6 coil test points. (Courtesy of BWD Automotive Corporation)

Ignition Coil Test

1. Connect an ohmmeter as shown in the number 1 meter hookup in Figure 68. Resistance should be less than 1 ohm for this HEI coil winding. If it is greater than 1 ohm, replace the coil.

2. Connect an ohmmeter as shown in the number 2 meter hookup in Figure 68. Resistance should be 6000 to 30,000 ohms for this HEI coil winding. Replace the coil if not in this range.

Pickup Coil Test

1. Connect an ohmmeter as shown in the number 1 meter hookup in Figure 69. The ohmmeter should show an open circuit. Replace the pickup coil if the ohmmeter shows a reading.

2. Connect an ohmmeter as shown in the number 2 meter hookup in Figure 69. Pull on the terminals and move wires around to check for bad connections. *Resistance should be 500 to 1500 ohms.*

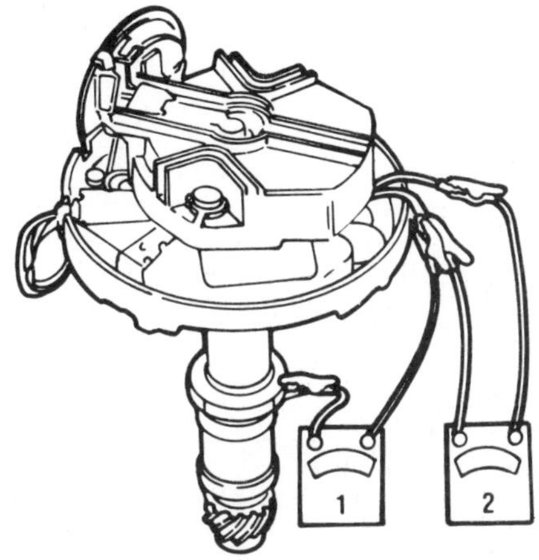

FIGURE 69
HEI pick-up coil test points. (Courtesy of BWD Automotive Corporation)

Module Test

The module can be checked with GM tester J-24642 or equivalent. Figure 70[a] gives details for making a quick HEI module test.

Troubleshooting and Service Hints

1. Incorrect dwell (or break current) can cause poor fuel economy, rough running, and backfire. A fixed dwell could be caused by a defective module or coil.

2. Poor ground connection(s) at the coil-to-distributor cap screws can shorten module life (Figure 70[b]).

GENERAL MOTORS COMPUTER-CONTROLLED-COIL IGNITION SYSTEM

The computer-controlled-coil ignition (C^3I) system was introduced on Buick's 3.8 liter turbocharged V-6 engine in 1984. The heart of the system is an electronic coil module that replaces the standard distributor and coil (Figure 70[c]).

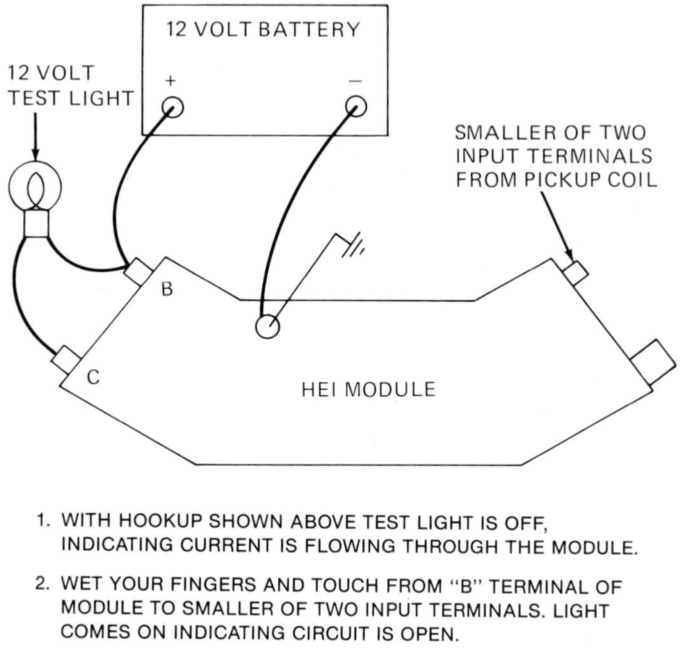

(a)

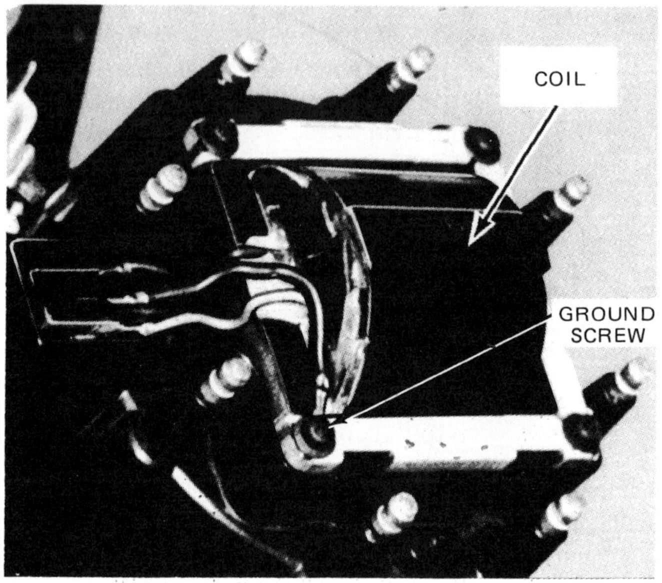

(b)

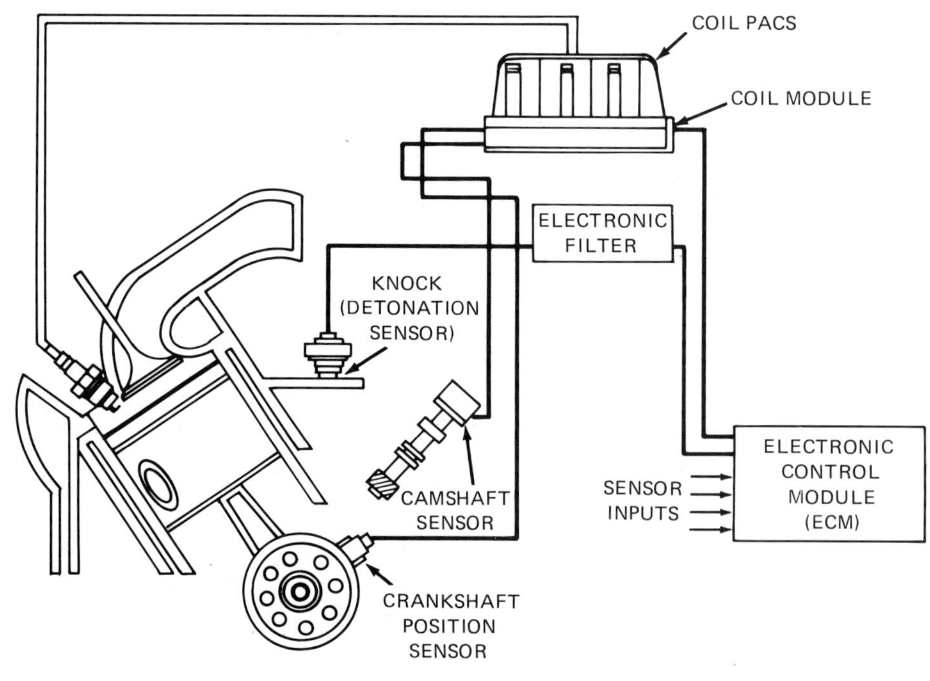

(c)

FIGURE 70

(a) HEI module test. This test checks the make and break circuit only, not the dwell function. (b) HEI coil mounted to cap. (c) General Motors C³I system. The heart of the system is the electronic coil module (often referred to as the "black brick"). The cam sensor is adjusted by using a voltmeter. If not correct, off-idle hesitation and fuel economy problems can result. (Courtesy of Buick Motor Division, General Motors Corporation)

In 1987, the C^{3}I system evolved into the direct ignition system (DIS). This system is a distributorless ignition system, consisting of two or three separate ignition coils, an ignition module, and a crankshaft sensor.

HALL EFFECT SWITCH

A Hall Effect switch is a special electronic device that consists of a fixed-position magnet, steel pick-up, and transducer (Figure 71). When a distributor rotor vane passes in the narrow space between the magnet and the electronic circuit, it alternately cuts off and passes the magnetic field from the magnet to the circuit. If magnetism is allowed to strike the circuit, it will result in a voltage that turns a transistor on (Figure 72). When the magnetic field is cut off by a vane, the transistor turns off and current flow in the coil primary circuit is interrupted. This produces high voltage at the spark plug.

Since the Hall Effect switch is either on or off, it produces a pulsating DC wave with sharp corners that looks similar to a square wave. *This exact on/off condition supplies more precise timing information to the ECM.* As a result, Hall Effect distributors are being found in more and more new vehicles.

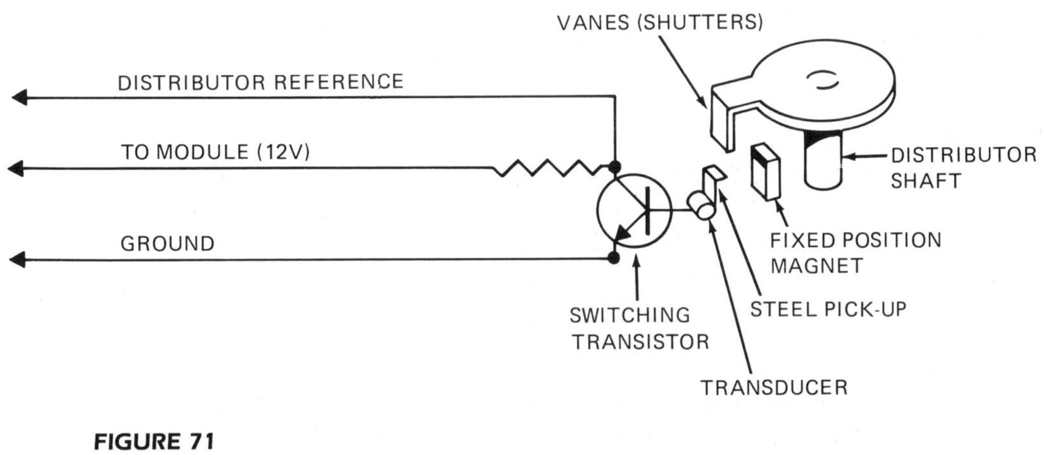

FIGURE 71
The Hall Effect switch is an electronic device that puts out a voltage signal which is controlled by the presence or absence of a magnetic field on an electronic circuit. The distributor reference signal tells the electronic control module (ECM) which cylinder is next to fire.

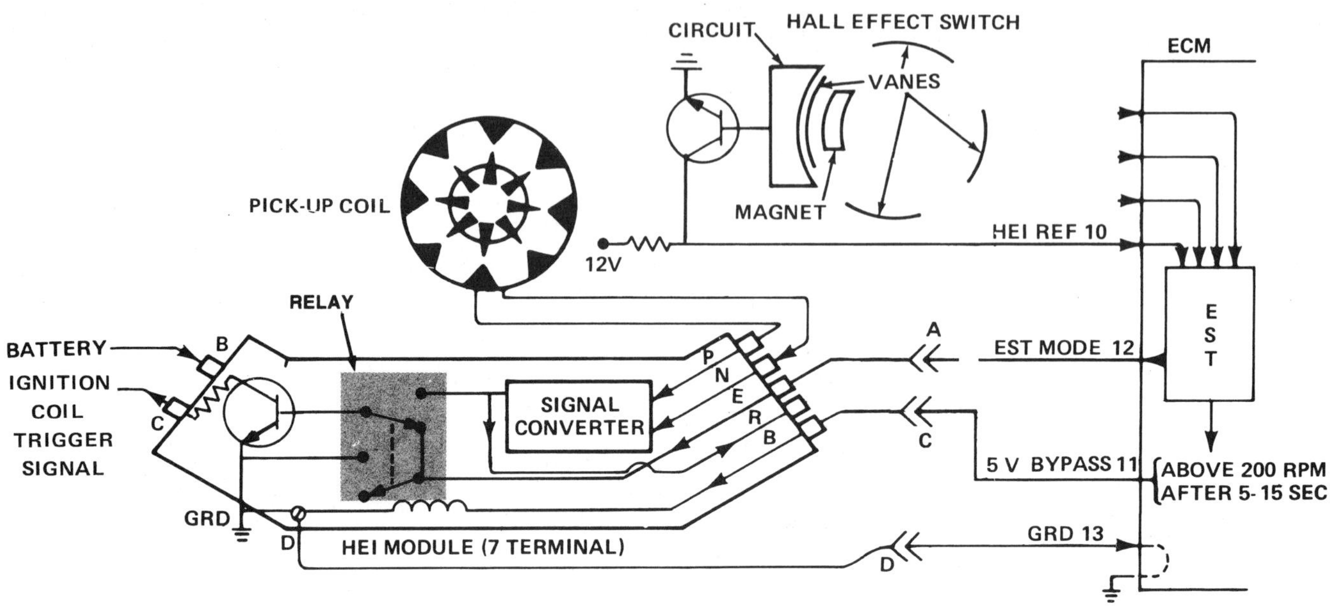

FIGURE 72
Hall Effect ignition system circuit schematic. (Courtesy of Chevrolet Motor Division. General Motors Corporation)

TRUE CASE HISTORY

Vehicle	Chevrolet Spectrum (Isuzu)
Engine	Four-cylinder, 1.5 liter
Mileage	5,100
Problem	1. Engine surge, low power, and only 17 mpg. 2. Dealer replaced the carburetor, oxygen sensor, spark plugs, coolant temperature sensor, and ECM. None of this cured the problem. 3. Dealer set the timing to 3 degrees BTDC because that was the number on the underhood emissions label.
Cure	Adjusted timing to 15 degrees BTDC, per the factory manual.

TRUE CASE HISTORY

Vehicle	Mercury Marquis equipped with an EEC III system
Engine	Eight-cylinder, 351W (5.8L)
Mileage	63,000
Problem	1. Two "dead" cylinders at all engine speeds. 2. A complete absence of spark to these two cylinders.
Cause	A broken (missing) pulse ring lobe on the vibration damper.
Comment	On EEC III systems, a crankshaft position sensor provides electrical input to the computer for engine speed measurement and control of ignition timing (Figure 73). The sensor relays voltage signals based on movement of a crankshaft pulse ring (Figure 74). With one pulse ring lobe missing, there was no electrical input to the computer for two cylinders.

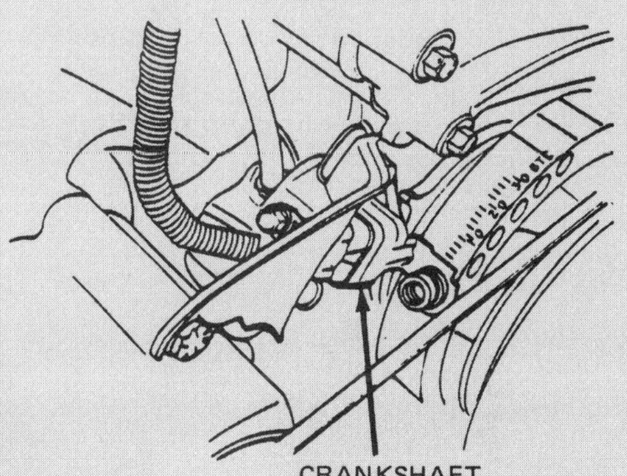

FIGURE 73
Crankshaft position sensor mounting. (Courtesy of Ford Parts and Service Division)

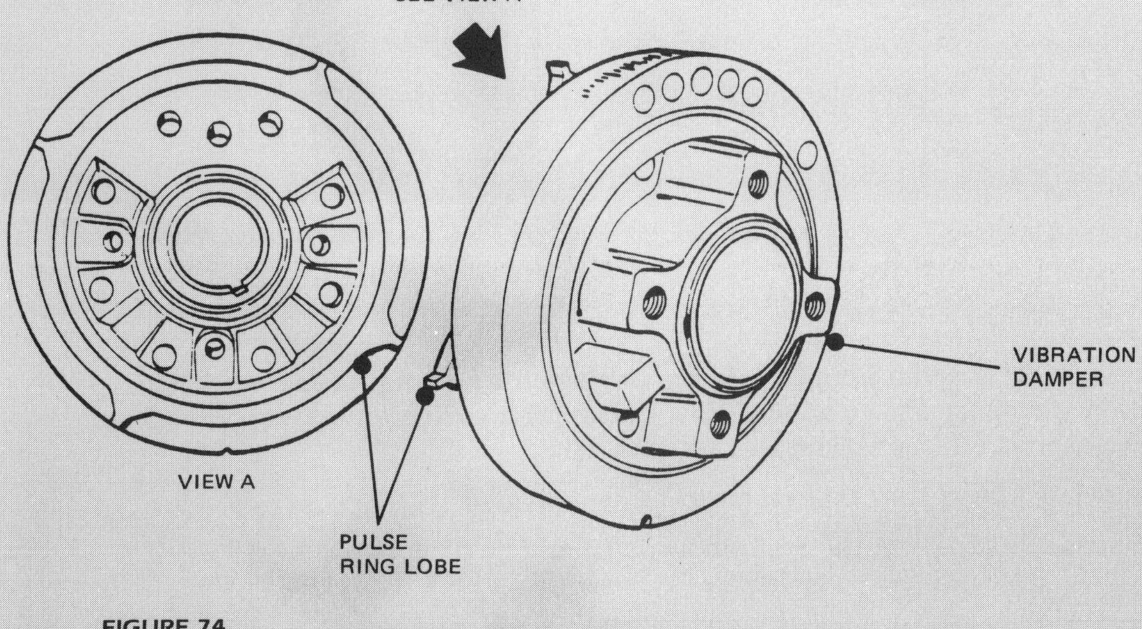

FIGURE 74
The crankshaft pulse ring assembly is part of the vibration damper. (Courtesy of Ford Parts and Service Division)

ASE STUDY QUESTIONS

1. An engine with electronic ignition will not start (fuel and carburetion are good).

   Technician A says that this could be caused by a "punctured" rotor.

   Technician B says that this could be caused by too much pickup coil air gap clearance.

   Who is right?

   a. A only **c.** Both A and B
   b. B only **d.** Neither A nor B

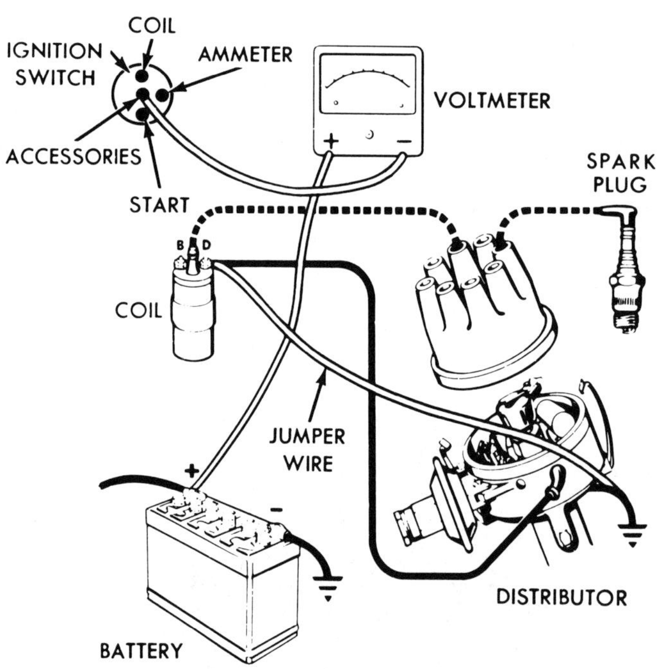

2. Voltmeter leads and a jumper wire are connected to a point-type ignition system as shown above. With the ignition switch on and all accessories off, the voltmeter reads 1.3 volts.

   Technician A says that the coil could be defective.

   Technician B says that the ignition switch could be defective.

   Who is right?

   a. A only **c.** Both A and B
   b. B only **d.** Neither A nor B

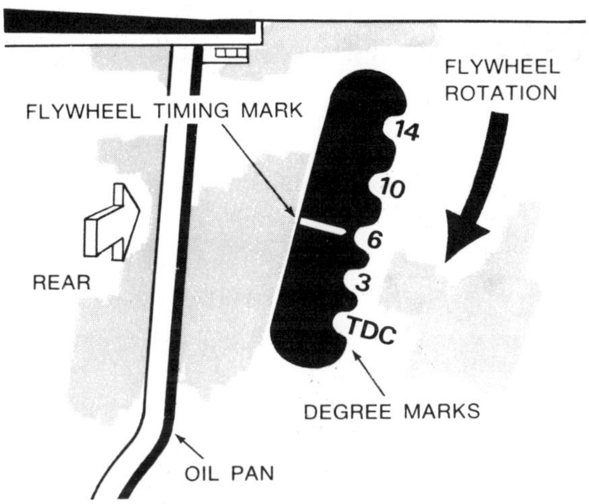

4. A minor tune-up is performed on a car. After 50 miles of driving, the engine develops a rough idle. A check shows the dwell to be way out of specification.

   Technician A says that this can be caused by failure to use cam lubricant.

   Technician B says that this can be caused by an improperly tightened point hold-down screw.

   Who is right?

   a. A only
   b. B only
   c. Both A and B
   d. Neither A nor B

3. A car performs erratically. When timing is checked with a light, the marks shown above do not return to the same spot when the rpms are increased and then dropped.

   Technician A says that this could be caused by a worn breaker plate.

   Technician B says that this could be caused by a worn weight and cam assembly.

   Who is right?

   a. A only
   b. B only
   c. Both A and B
   d. Neither A nor B

5. The spark plug shown above has chunks of aluminum imbedded between the electrodes.

   Technician A says that this can be caused by using low grade gasoline.

   Technician B says that this can be caused by preignition.

   Who is right?

   a. A only
   b. B only
   c. Both A and B
   d. Neither A nor B

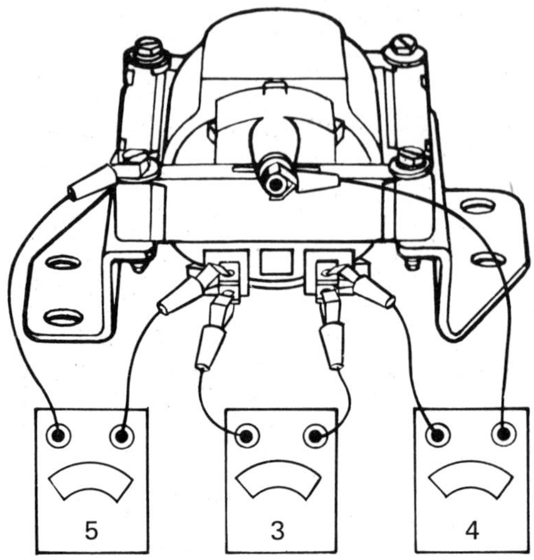

6. An ohmmeter is being used to test the ignition coil shown above.

   Technician A says that test point 3 and test point 4 should show continuity.

   Technician B says that test point 5 should show infinite resistance.

   Who is right?

   a. A only
   b. B only
   c. Both A and B
   d. Neither A nor B

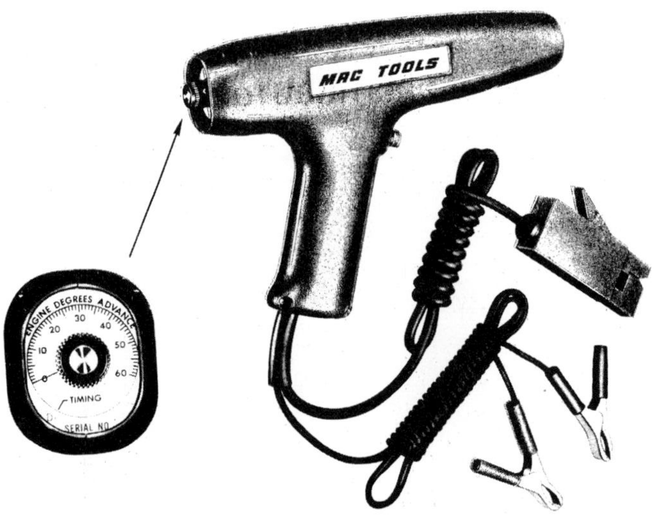

7. **Technician A** says to poke a small hole in the no. 1 spark plug wire when connecting the device shown above.

   Technician B says that the distributor vacuum hose(s) should be disconnected and plugged when using the above device.

   Who is right?

   a. A only
   b. B only
   c. Both A and B
   d. Neither A nor B

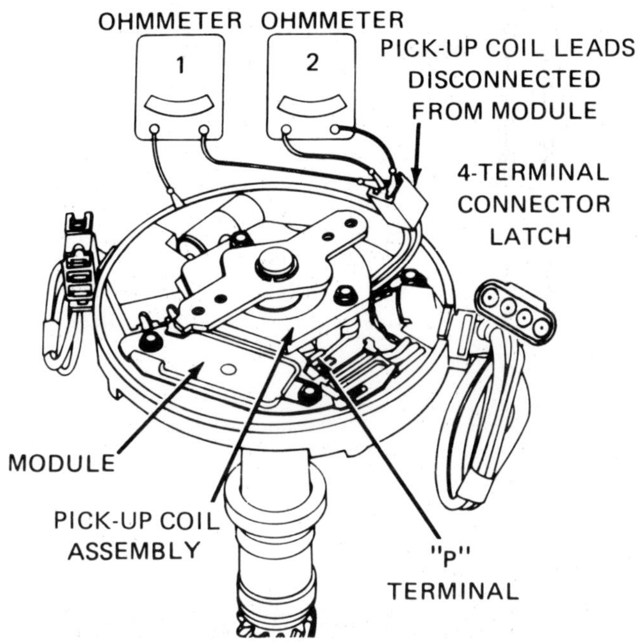

8. A pickup coil is being tested in the above picture.

   Technician A says that ohmmeter 1 should have no reading (infinite).

   Technician B says that ohmmeter 2 should have no reading (infinite).

   Who is right?

   a. A only
   b. B only
   c. Both A and B
   d. Neither A nor B

9. A car equipped with electronic ignition misses intermittently while driving. After replacing the distributor cap, rotor, and spark plug wires, the problem still exists.

   Technician A says that the cause could be a defective ignition switch.

   Technician B says that the cause could be a poor harness connection at the control unit (module).

   Who is right?

   a. A only
   b. B only
   c. Both A and B
   d. Neither A nor B

SECTION REVIEW QUESTIONS

1. What are the two basic circuits contained in the ignition system?
2. A typical electronic ignition system would include what components?
3. What is a reluctor?
4. What is the purpose of the magnetic pickup coil assembly?
5. A felt wick is located in the center of some distributor shafts, just underneath the rotor. What is its purpose?
6. When inspecting a distributor that has been removed from the engine, what should you look for?
7. An excessively worn distributor shaft bushing can cause what problems?
8. Too wide a point gap can cause what problems?
9. Too narrow a point gap can cause what problems?
10. What is the purpose of a condenser?
11. What type of feeler gauge should be used when setting electronic ignition air gap?
12. Define spark plug heat range.
13. List four conditions that could cause carbon-fouled spark plugs.
14. List four conditions that could cause oil-fouled spark plugs.
15. What is splashed fouling?
16. What is corona?
17. How tightly should gasket-type spark plugs be installed?
18. Why is spark plug wire routing important?
19. Using a timing light, how can a centrifugal advance weight check be made?
20. Using a timing light, how can a vacuum advance check be made?
21. What is a magnetic timing probe?
22. How is an applied voltage check made on a BID ignition system?
23. How is the pickup coil checked on a Chrysler electronic ignition system (not a Hall Effect system)?
24. A Ford Dura-Spark electronic ignition system would include what components?

25. A conventional HEI system would include what components?

26. What is a Hall Effect switch? How does it work?

27. Describe a C^3I system.

28. What is a waste spark?

Section F

FUEL AND EXHAUST SYSTEMS DIAGNOSIS AND REPAIR

After studying this section, and with hands-on practice on a "live" vehicle, you should be able to perform the following tasks:

- Inspect, test, and replace fuel pumps and pump controls; inspect, service, and replace fuel filters.
- Inspect, clean or replace carburetor mounting plates, fuel injection air induction system, intake manifold, and gaskets; check and adjust air–fuel ratio (mixture) and idle speed.
- Inspect and service or replace air filters and filter housings.
- Inspect, service, and replace exhaust manifold, manifold heat control valves (heat risers), exhaust pipes, mufflers, converters, resonators, tail pipes, and heat shields.
- Test the operation of positive crankcase ventilation (PCV) systems.
- Inspect, service, and replace positive crankcase ventilation (PCV) valves/metering devices.
- Inspect, service, and replace positive crankcase ventilation (PCV) filters (breather cap), tubes, and hoses.
- Inspect turbochargers for proper operation; repair as necessary.

FUEL SYSTEM

The fuel system is designed to provide an adequate supply of fuel under sufficient pressure for all operating conditions (Figure 1).

MECHANICAL FUEL PUMP

The mechanical fuel pump is located on the engine and pulls fuel from the tank to the carburetor (Figure 2). The rocker arm is usually driven by an eccentric lobe on the camshaft, or by a push rod.

Three tests (pressure, volume, and vacuum) can be performed on a mechanical fuel pump. If both fuel pump pressure and volume (flow rate) are within specifications, the pump can be considered satisfactory. However, if the fuel pressure is within specifications, but the volume is too low, a vacuum test should be performed.

If both pressure and vacuum are correct, but volume is less than specified, inspect the fuel line and fuel filter for foreign material.

Chart 1 lists additional troubleshooting information for mechanical fuel pumps.

Mechanical Fuel Pump Pressure Test

1. Disconnect the fuel line to the carburetor at the fuel pump and remove the outlet fitting.

2. Install a tee fitting in place of the outlet fitting and connect the fuel line to it.

NOTE: If there is no room to install the described fittings at the outlet of the fuel pump, disconnect the fuel line at the carburetor and perform these tests. This will allow you to test the fuel pump and the in-line filter (if one is present).

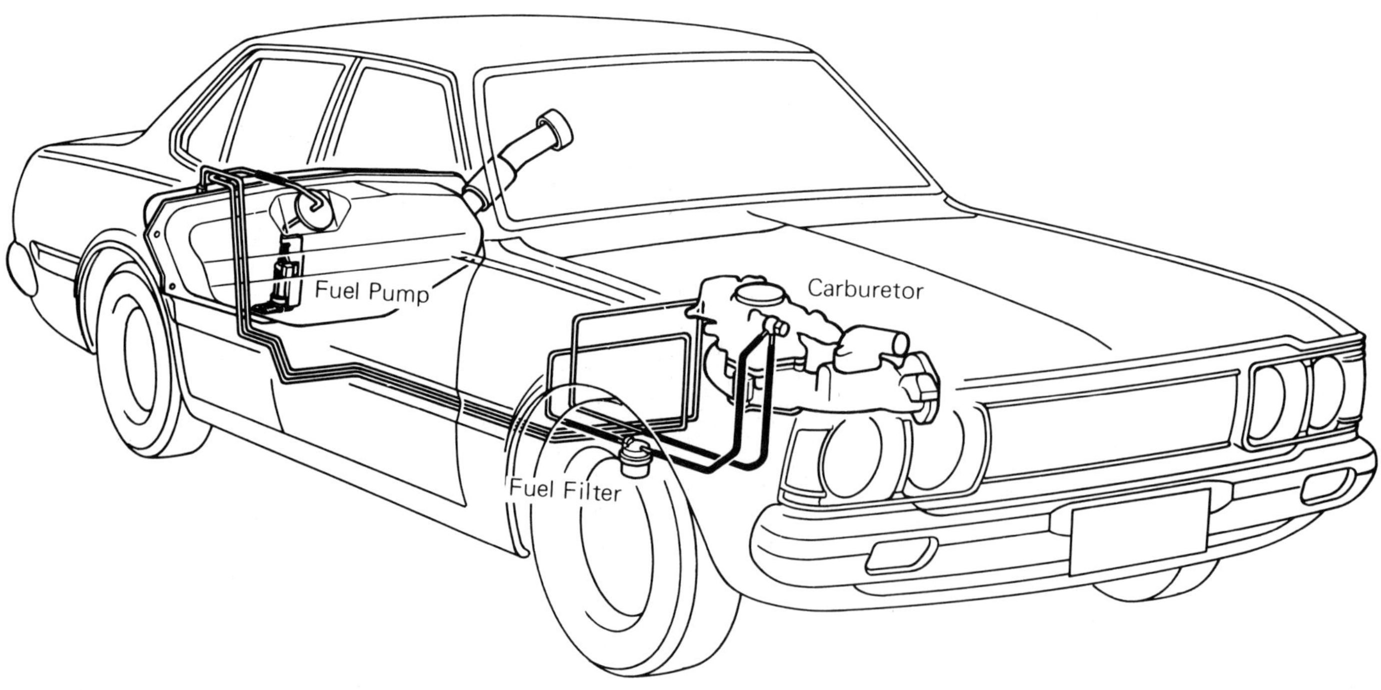

FIGURE 1
A fuel supply system.

CHART 1 TROUBLESHOOTING CHART FOR MECHANICAL FUEL PUMPS

Fuel Pump Leaks Oil

1. Defective pump-to-block gasket
2. Loose pump mounting bolts
3. Cracked or deteriorated pull rod oil seal (can allow engine oil and fuel to mix)

Fuel Pump Leaks Fuel

1. Worn, ruptured, or torn diaphragm
2. Loose inlet or outlet fittings

Fuel Pump Noise

1. Loose pump mounting bolts
2. Worn or damaged rocker arm
3. Weak or broken rocker arm spring
4. Stiff (hardened) inlet hose

Insufficient Fuel Delivery

1. Restricted vent in tank (can also cause a collapsed fuel tank)
2. Worn, ruptured, or torn diaphragm
3. Weak diaphragm spring
4. Worn push rod
5. Damaged camshaft eccentric
6. Worn or broken rocker arm
7. Dirt or restriction in fuel tank, lines, or hoses
8. Restricted fuel filter (Figure 3)
9. Improperly seating valves in fuel pump
10. Vapor lock
11. Cracked or "pinholed" inlet hose
12. Air leaking into the sediment bowl (if there is one)

High Fuel Pressure

1. Bulged diaphragm (from fuel getting between its layers)
2. "Frozen" diaphragm operating link on the rocker arm
3. Excessively strong diaphragm spring

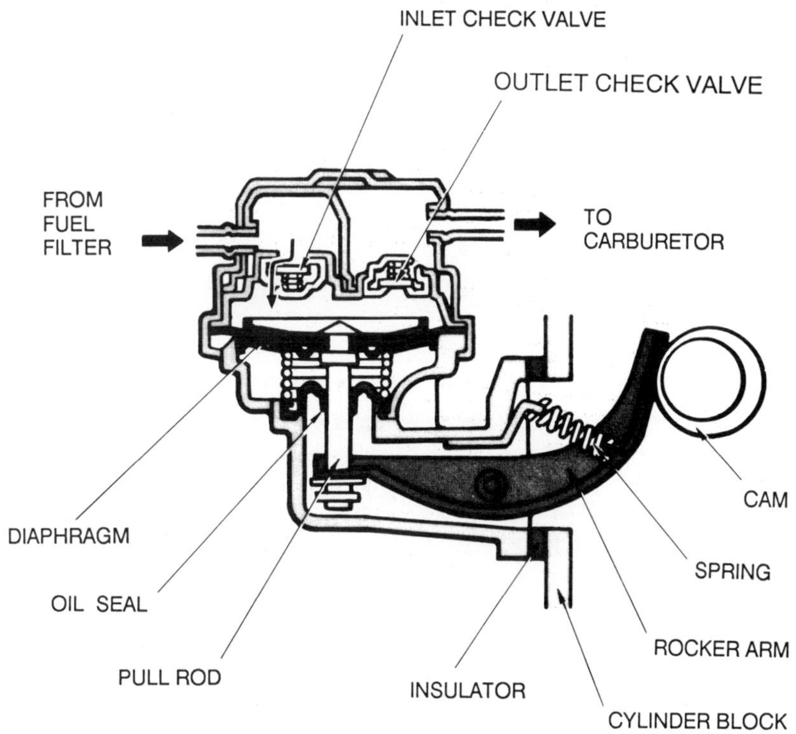

FIGURE 2
Mechanical fuel pump (typical).

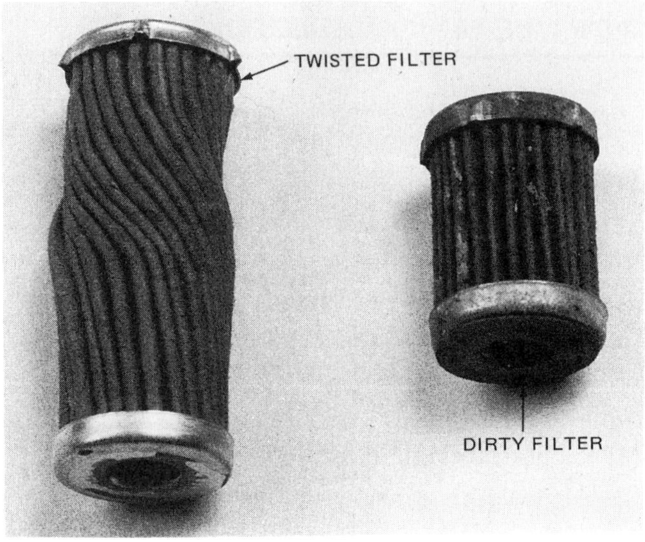

FIGURE 3
Restricted carburetor inlet filters.

3. Connect a pressure gauge to the tee fitting (Figure 4).

4. Connect a short length of fuel hose to the gauge. Position the hose so fuel can be directed into a suitable quart container.

5. Start the engine and operate at idle speed.

6. Observe the gauge. The reading should be 3 to 6 psi on the average.

Mechanical Fuel Pump Volume Test

1. Install gauge as instructed in the fuel pump pressure test.

2. Operate the engine at idle speed. One pint of fuel should be pumped within specification or in approximately 30 seconds or less.

Mechanical Fuel Pump Vacuum Test

1. Connect a vacuum gauge to the vacuum (intake) side of the fuel pump.

FIGURE 4
An example of a fuel pump pressure and volume test on an older vehicle; the test is also used on some newer vehicles. (Courtesy of Ford Parts and Service Division)

2. Run the engine at idle speed and observe the gauge. Fuel pump vacuum should not typically be less than 6″ Hg.

3. Stop the engine. The gauge should continue to indicate a vacuum for 10 seconds.

Removing Fuel Pump

1. Disconnect all lines and hoses that attach to the fuel pump.

2. Loosen, but do not remove, the two bolts that secure the pump assembly.

3. Using a remote starter switch, crank the engine until the fuel pump eccentric on the camshaft is in the position that applies the least tension on the fuel pump rocker arm.

4. Remove the two bolts and the pump assembly.

Installing Fuel Pump

1. Inspect the fuel pump push rod (if there is one) for wear or damage. If necessary, install a new one.

2. Squirt oil into the pump rocker arm cavity to lubricate the internal parts.

3. Using a new gasket, position the pump assembly and install the two bolts (Figure 5[a] and [b]).

> **SHOP TIP**
> With the fuel pump eccentric at low cam, you can use a hacksaw blade to hold the push rod up while placing the rocker arm underneath the push rod.

4. Connect lines and hoses.

5. Start engine. Check for leaks and proper operation of the pump assembly.

COMMENT: Change engine oil if you suspect the old pump has a torn or porous diaphragm. This condition can allow gasoline to leak into the crankcase.

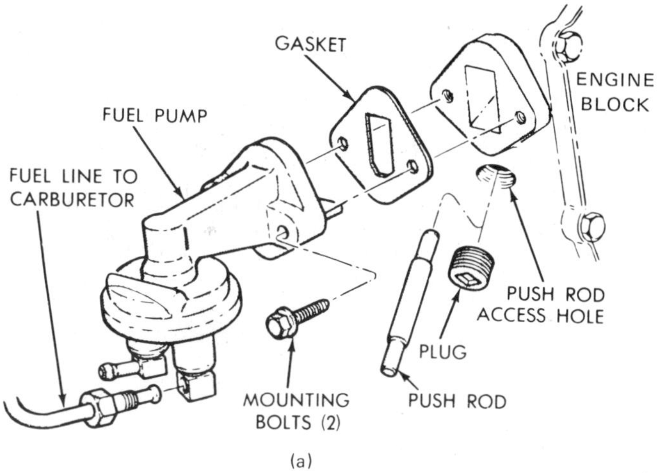

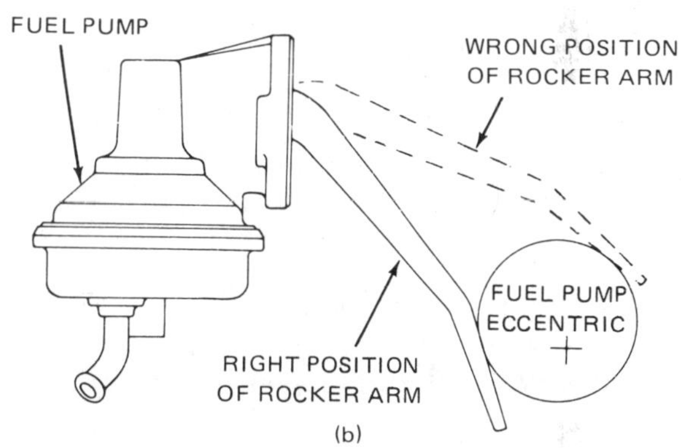

FIGURE 5
(a) This mechanical fuel pump is acutated by a push rod. (Courtesy of Chrysler Corporation) (b) Correct positioning of the fuel pump rocker arm on the camshaft eccentric. (Courtesy of Chevrolet Motor Division, General Motors Corporation)

ELECTRIC FUEL PUMP

Electric fuel pumps are designed to push fuel. On carburetor-equipped vehicles, a diaphragm or vane-type pump is generally used. The electric fuel pump is located inside the fuel tank or somewhere along the fuel line.

The roller cell positive displacement pump is most common on fuel injected vehicles. It is a "wet" type pump because fuel flows through the pump motor (Figure 6).

> **SHOP TIP**
> Every 10,000 miles, pour 3 to 4 ounces of Marvel Mystery Oil® into the fuel tank (Figure 7). This will help lubricate the fuel pump.

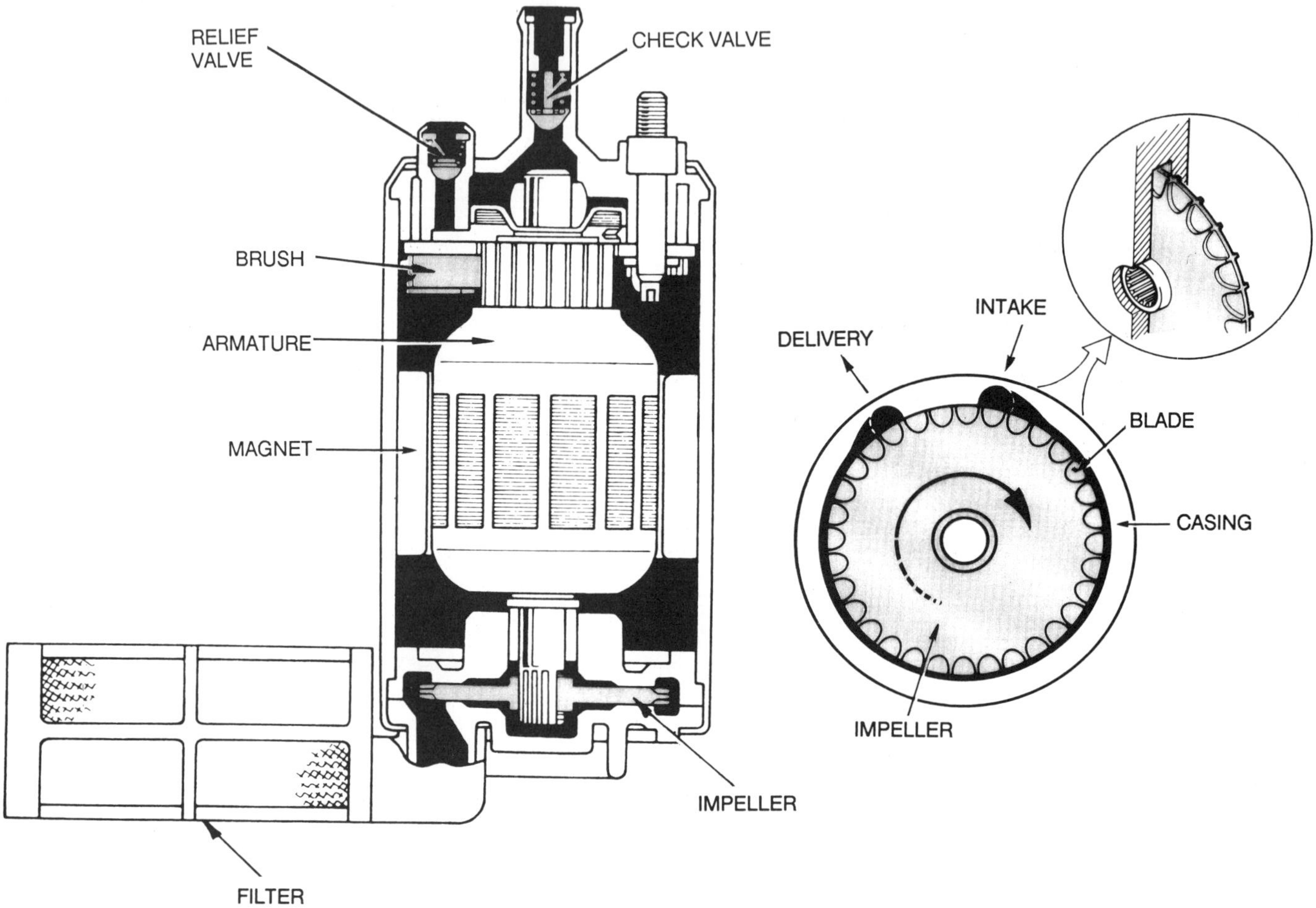

FIGURE 6
Wet-type impeller electric fuel pump.

> **SHOP TIP**
> A "frozen" electric fuel pump can sometimes be freed by tapping it. This can allow you to get the car off the street and into the shop for proper repair.

Two tests (volume and pressure) are often made on an electric fuel pump.

Electric Fuel Pump Volume Test

1. Be sure the fuel lines are not kinked, clogged, or leaking.

2. Energize the fuel pump without running the engine and measure the volume at the proper test point.

***NOTE:** If the volume is lower than specified, check for at least 11.5 volts at the fuel pump feed wire. If the voltage is acceptable, the pump is faulty.*

FIGURE 7
Marvel Mystery Oil®.

Electric Fuel Pump Pressure Test

1. Install a pressure gauge at the proper test point.

2. Operate the fuel pump without running the engine and read system pressure (Figure 8[a] and [b]).

CAUTION: Slowly bleed off (depressurize) the pressure when opening the fuel line on a fuel injection system. This can be done by "cracking" open the fuel line and catching the spray in a shop towel.

> **SHOP TIP**
> If the system pressure is lower than specified, quickly pinch and release the fuel return line. The pressure should rise to at least 100 psi (on most gasoline engines equipped with fuel injection). If it does not, there is a pump problem, or the "sock" in the fuel tank is plugged (Figure 9).

Troubleshooting the Electric Fuel Pump

Chart 2 lists some electric fuel pump troubleshooting data.

Electric Fuel Pump Replacement

Always refer to the manufacturer's shop manual before replacing an electric fuel pump; there can be major differences in procedures. For example, some in-tank pumps can be removed through an access hole in the car trunk (Figure 14), while on other cars it is necessary to remove the fuel tank.

FUEL FILTERS

A fuel system has several filters. The primary one is generally a screen or "sock" located in the fuel tank (Figure 15). The second one is either an in-line type, or a carburetor inlet type (Figures 16, 17, and 18).

On electronic fuel injection (EFI) systems, a third filter is located inside the injector. It is a fine mesh screen that can be serviced only as a complete injector assembly. Refer to page 313 (Figure 19).

> **SHOP TIP**
> When replacing an in-line EFI filter, catch the fuel in a clean pan (Figure 20). Check for the presence of metal (pump is breaking down), rust, or a blackish substance (high levels of methyl alcohol are dissolving rubber fuel hose).

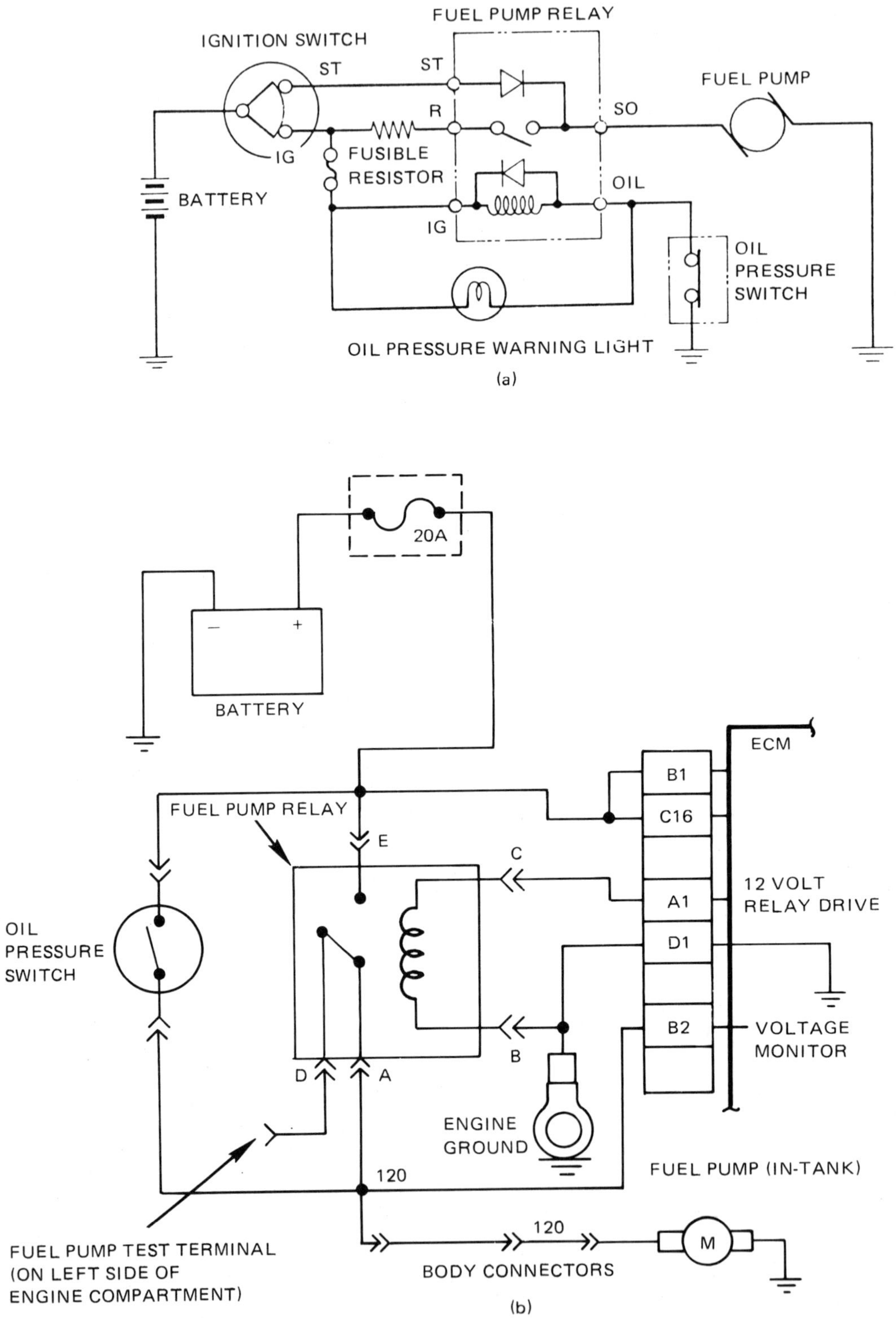

FIGURE 8
(a) Fuel pump control system circuit. This fuel pump turns while the starter is cranking and when oil pressure is present. *Do not operate the pump without the resistor in the circuit.* (Courtesy of Toyota Motor Sales, U.S.A., Inc.) (b) Fuel pump circuit. How could you operate this pump without running the engine in order to make a pressure test? (Courtesy of Chevrolet Motor Division, General Motors Corporation)

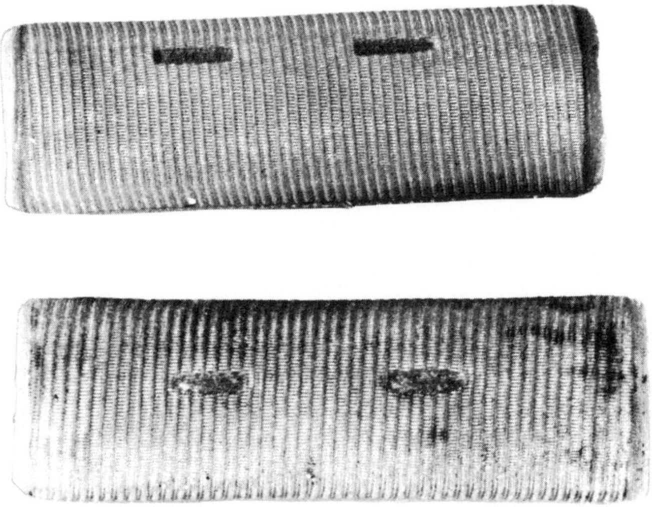

FIGURE 9
In-tank filter socks prevent contaminants from entering the fuel supply line. A clogged filter sock may prevent gas from flowing into the line. This picture shows a clogged and unclogged filter sock. Always replace the sock at the same time as replacing the fuel pump. Sometimes the contamination is on the inside of the sock and not readily visible. (Courtesy of Chrysler Corporation)

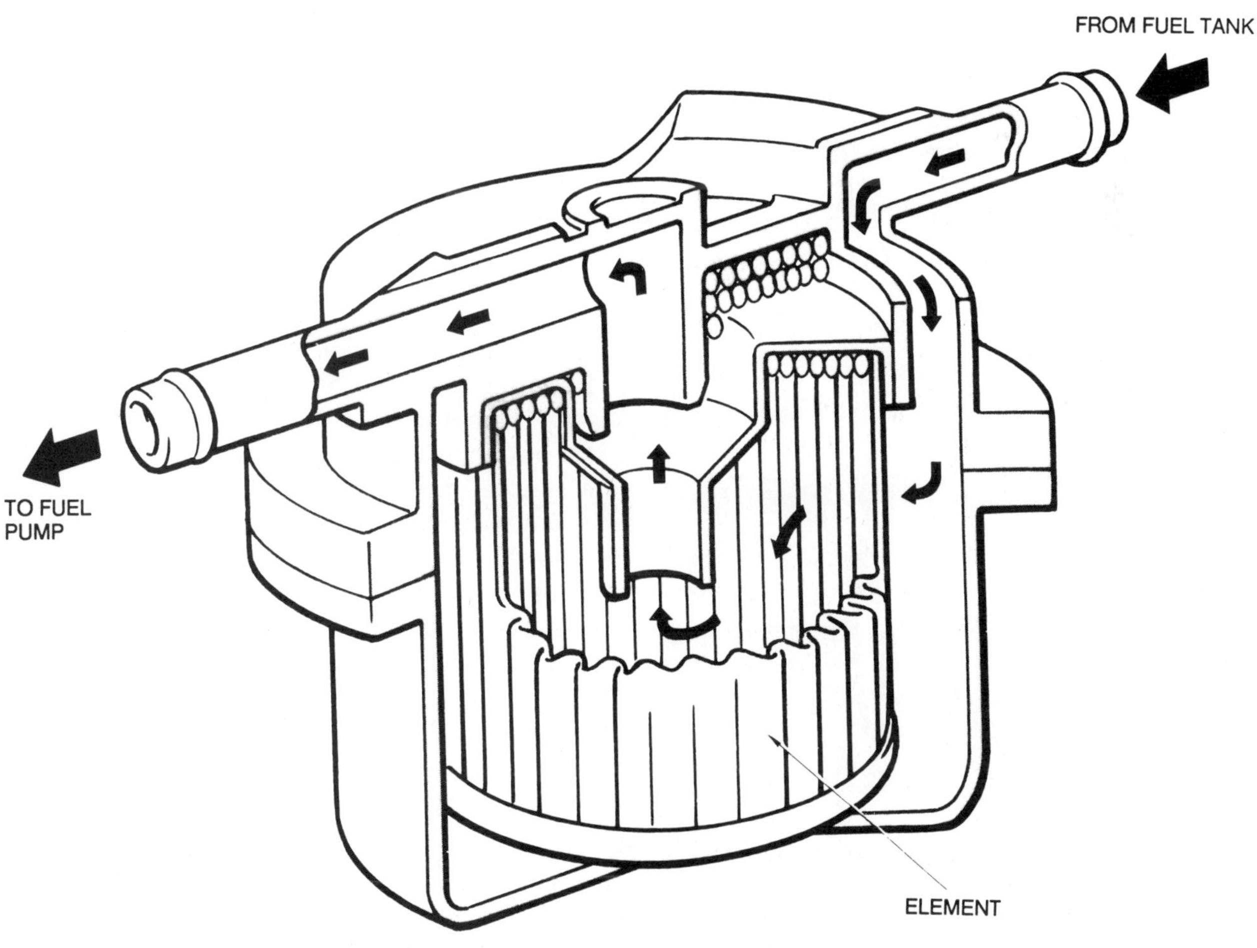

FIGURE 10
Insufficient fuel delivery is often caused by a restricted fuel filter element.

CHART 2 TROUBLESHOOTING CHART FOR ELECTRIC FUEL PUMPS

Insufficient Fuel Delivery

1. Restricted vent in tank (can also cause a collapsed fuel tank)
2. Dirt or restriction in fuel tank, lines, or hoses
3. Restricted fuel filter (Figure 10)
4. Defective fuel pump

No Fuel Delivery

1. Shorted or open electrical wiring (Figure 11)
2. Defective oil pressure switch
3. Defective fuel pump relay
4. Leaks in lines or hoses
5. Defective fuel pump
6. Open inertia switch (Figure 12)
7. Defective fuel pump switch (Figure 13)
8. Blown fuel pump fuse

CARBURETOR MOUNTING PLATES

On some engines, the exhaust gas recirculation (EGR) valve is attached to the rear of the carburetor mounting plate. Exhaust gases are extracted from a drilled passage in the exhaust crossover passage in the intake manifold (Figure 21[a] and [b]). They are then routed and metered through the EGR valve to a passage in the carburetor mounting plate and into the intake manifold.

FIGURE 11
The fuel pump feed wire on this vehicle is routed through a hard-shell connector. The connector became loose and the engine quit running while driving.

NOTE: The EGR valve will cause poor performance if it operates when it should not (idle, wide open throttle, cold starts, and warm-up). EGR can cause spark knock and driveability problems if it fails to operate when hot.

The EGR valve, mounting plate passages, and intake manifold crossover passage(s) must be cleaned at the recommended maintenance intervals. See Figure 22.

Some positive crankcase ventilation (PCV) systems have a vacuum passage that goes through a spacer located under the carburetor. Be sure this vacuum passage is not clogged.

A heated grid may be part of the carburetor mounting plate (Figure 23). This provides a source of rapid heat to the engine induction system during cold driveaway.

On some engines, an insulator is used to reduce heat flow from the engine to the carburetor (Figure 24).

FIGURE 12
In the event of a collision, the inertia switch will open and shut off the fuel pump. The switch must be reset manually to restart electric fuel pump operation. *Do not reset the switch until you are absolutely sure there are no fuel leaks.* (Courtesy of Ford Parts and Service Division)

FIGURE 13
This airflow meter (air box) has a fuel pump switch.

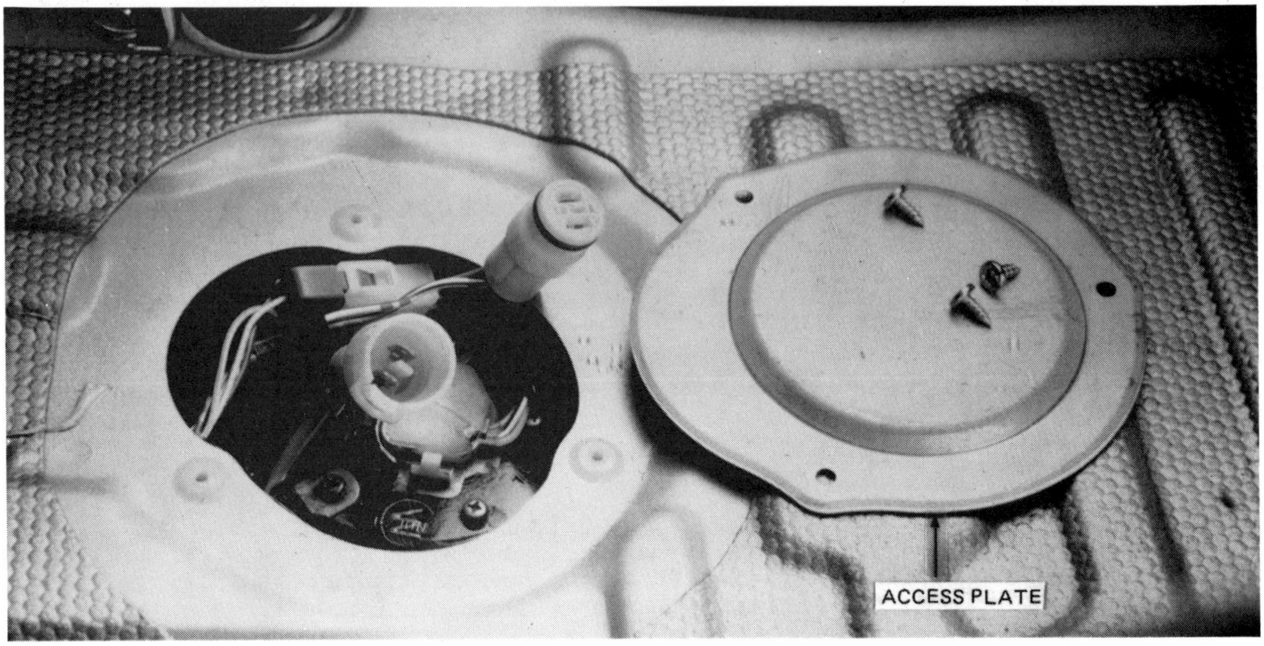

FIGURE 14
Access to the fuel pump on this vehicle is gained by removing a plate in the trunk.

FUEL INJECTION AIR INDUCTION SYSTEM

For the engine to run properly, there must be no air leaking into the air induction system between the air flow meter and the cylinder head (Figures 25 and 26).

SHOP TIP

A loose-fitting engine oil dipstick, oil filler cap, or PCV system hose can allow air to leak into the fuel injection air induction system, causing the mixture to lean out.

INTAKE MANIFOLD AND GASKETS

The intake manifold is designed to deliver an equal amount of air–fuel mixture to each of the engine's cylinders.

Intake manifold mounting surfaces and gaskets must be leakproof. No outside air must be allowed to enter past any connection into a manifold runner. The entry of air upsets the air–fuel mixture, produces a lean condition, and can cause rough idle.

Some V-8 engines are designed with the manifold serving as the lifter chamber cover. A very narrow section of gasket separates the intake manifold ports from the oil mist in the tappet chamber. An internal gasket leak at the parting face between the manifold and the cylinder head allows oil to be drawn into the intake ports (Figure 27).

Carefully check the intake manifold gasket(s) for tear damage and for evidence of water, vacuum, or oil leakage.

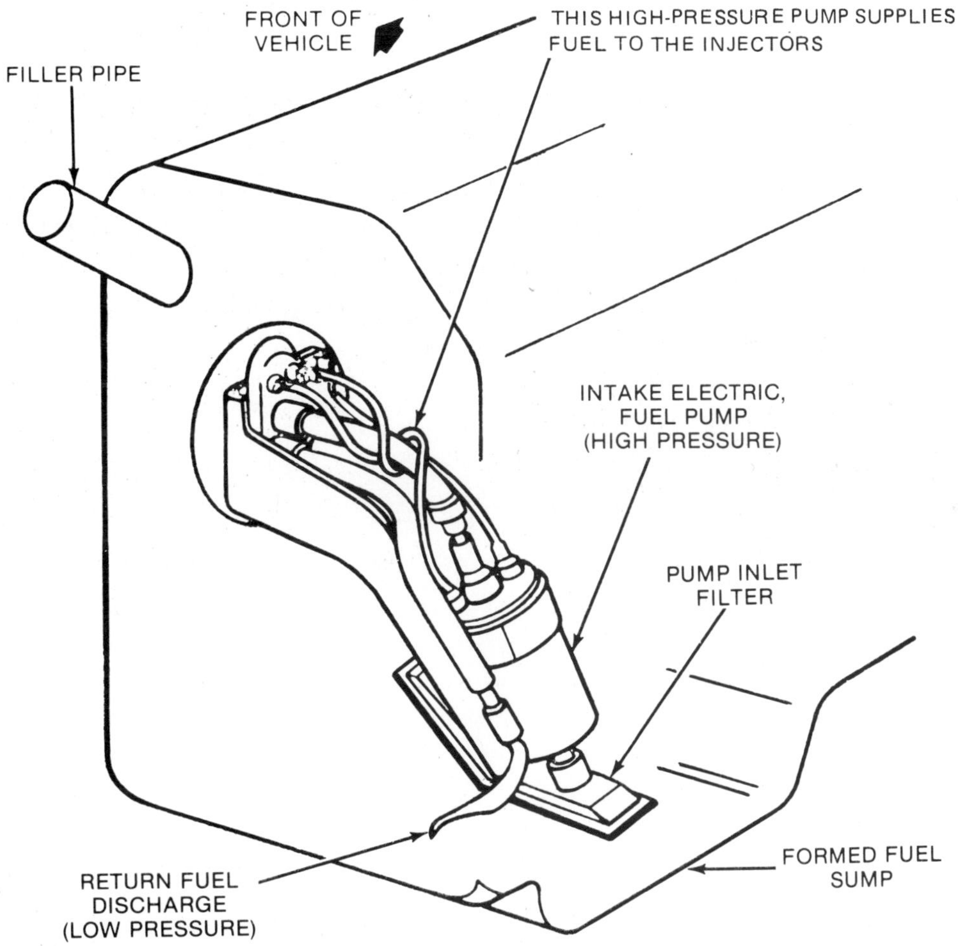

FIGURE 15
Note the location and shape of this fuel pump inlet filter. (Courtesy of Ford Parts and Service Division)

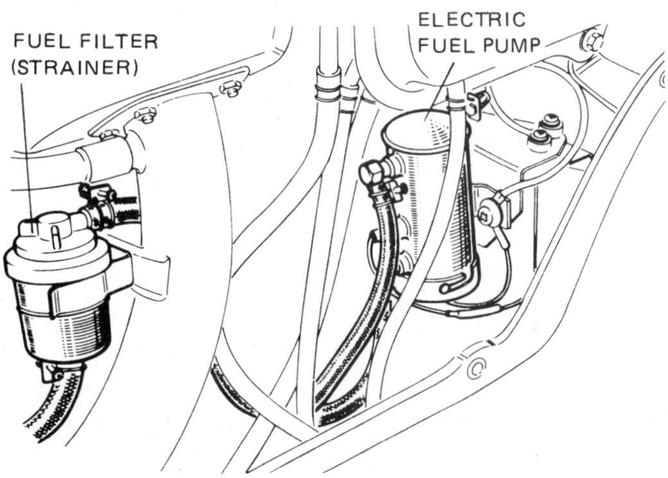

FIGURE 16
This electric fuel pump is mounted in the luggage compartment in the rear wheelhouse opening. The fuel filter is located beside the pump. (Courtesy of Nissan Motor Company, Ltd.)

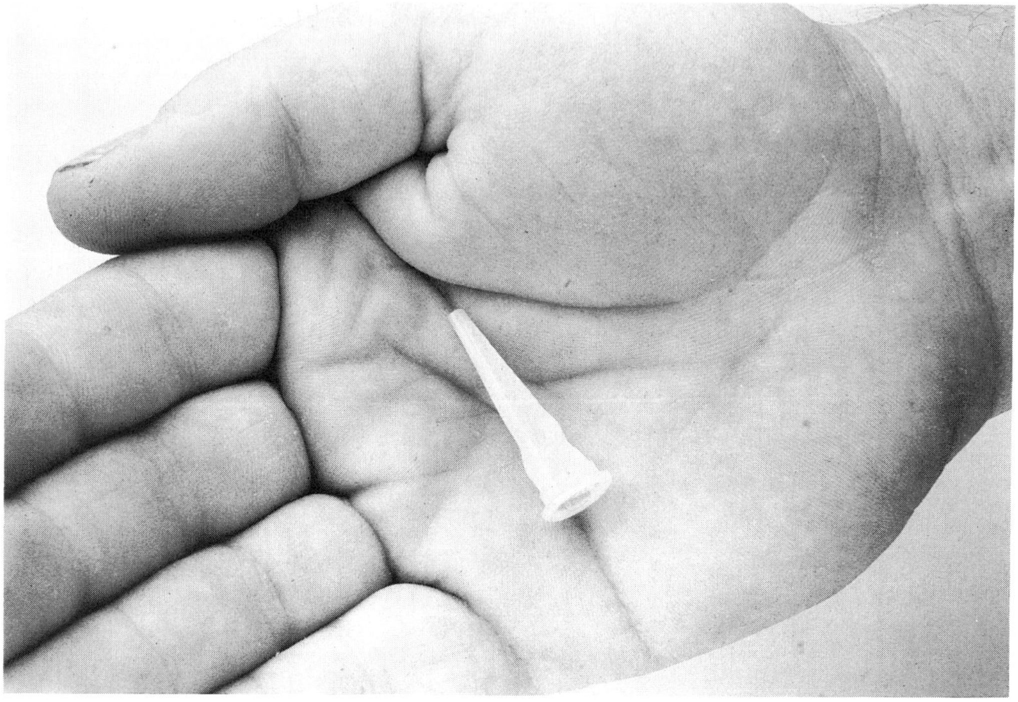

FIGURE 17
Some vehicle makers place a strainer sock in either the outlet or inlet hole of externally mounted electric fuel pumps.

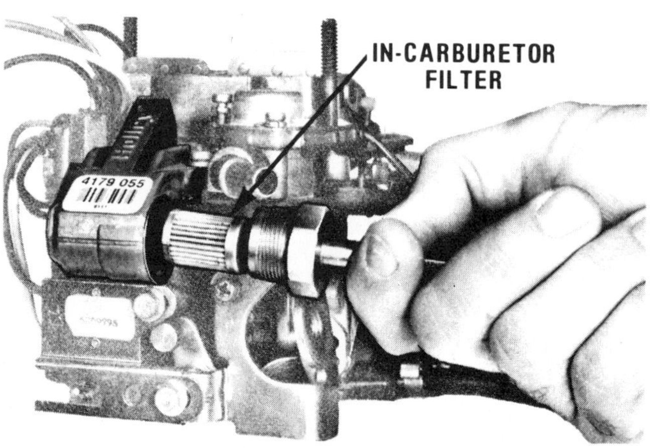

FIGURE 18
On many engines, there is a fuel filter in the carburetor. (Courtesy of Chrysler Corporation)

On some engines a sheet metal oil deflector is riveted over the outside of the intake manifold exhaust crossover passage (Figure 28). If so equipped, use a cold chisel and lift out the attaching rivets. The deflector can later be reattached by pop riveting, or by tapping the rivet holes and using machine screws.

Remove any carbon buildup and visually check the outside of the crossover passage for cracks.

Check the intake manifold mounting surfaces for warpage by using a precision straightedge and a feeler gauge (Figure 29). Mounting surfaces *must be straight within .010" and free of pits and erosion.* If they are not, the surfaces should be machined flat or the manifold should be replaced.

When installing an intake manifold, tighten the bolts to the correct torque specification and in the proper sequence (Figure 30).

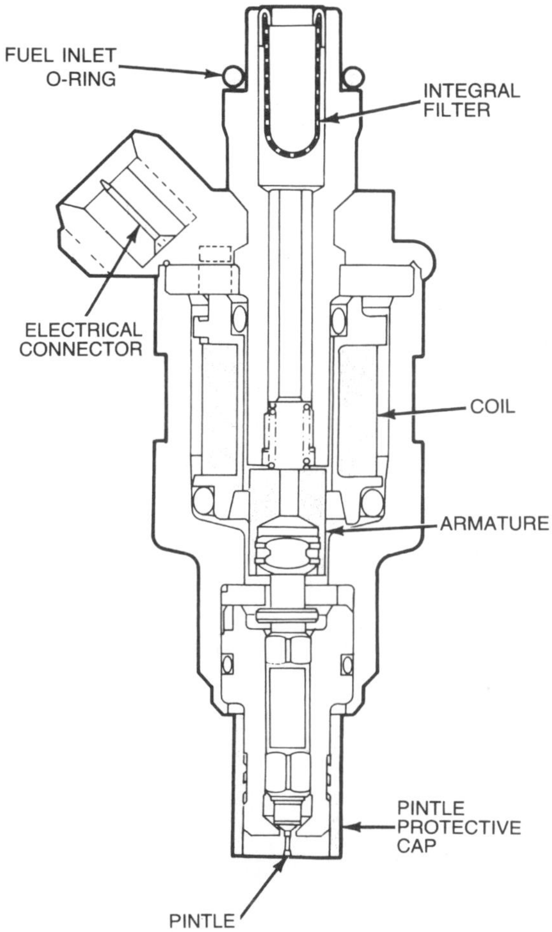

FIGURE 19
Most injectors have a screen filter inside. (Courtesy of Ford Parts and Service Division)

FIGURE 20
When replacing a fuel filter, check the contents.

CARBURETOR IDLE MIXTURE ADJUSTMENT

The volume of air–fuel mixture discharged into the air stream underneath the throttle valve(s) is controlled by the idle mixture screw(s). Turning the screw(s) clockwise reduces fuel flow to produce a leaner mixture. Turning the screw(s) counter-clockwise allows more fuel to flow for a richer mixture. See Figure 31[a].

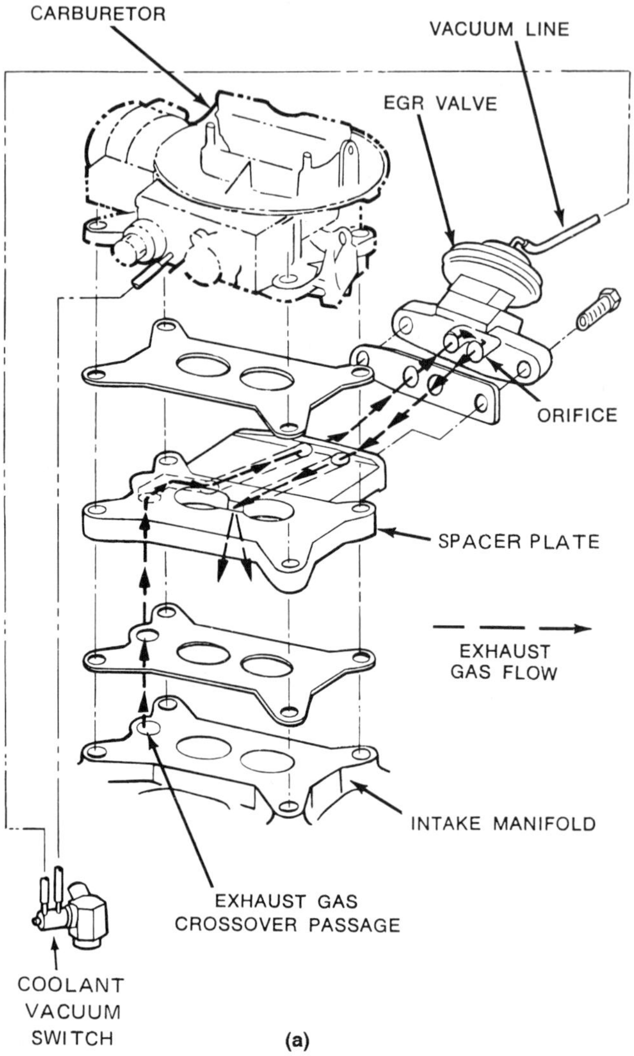

(a)

(b)

FIGURE 21

(a) This EGR system takes part of the burned exhaust gas and routes it back to the engine's intake system underneath the carburetor. (Courtesy of Ford Parts and Service Division) (b) Badly eroded carburetor spacer plate.

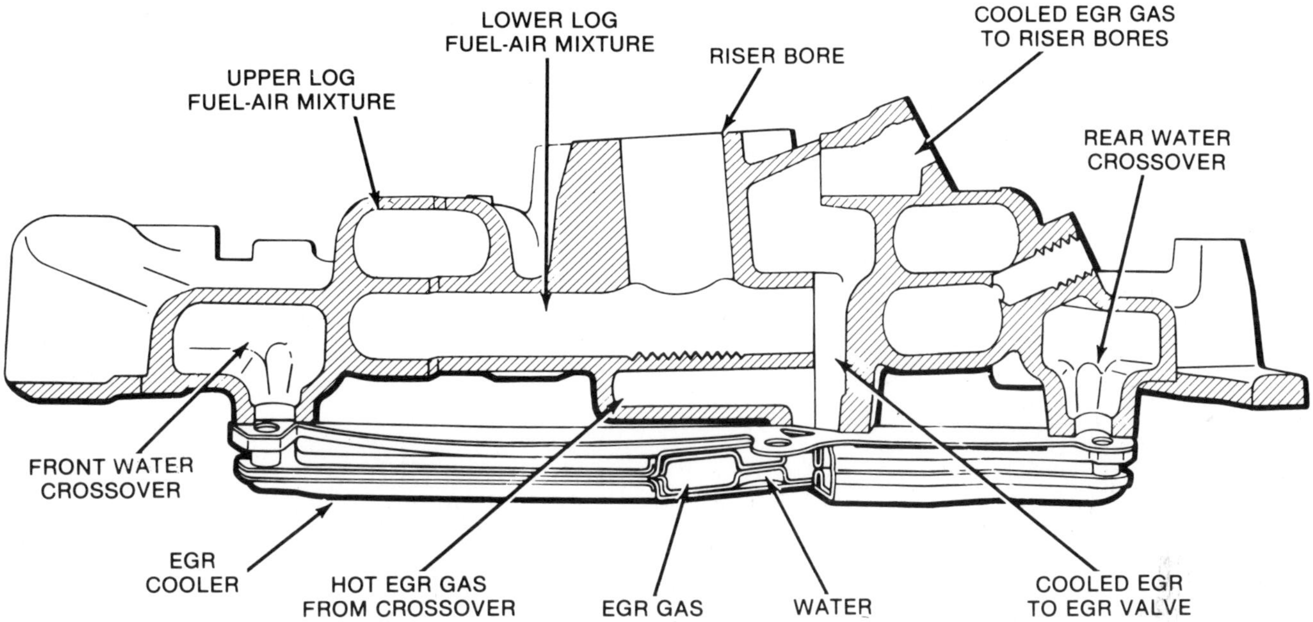

FIGURE 22
This intake manifold has an integral EGR cooler. Engine coolant enters the EGR cooler from the rear water crossover and passes forward to the front water crossover. (Courtesy of Ford Parts and Service Division)

FIGURE 23
A grid-type electrical heater may be placed between the carburetor and the intake manifold. This unit, often called an early fuel evaporation (EFE) plate, helps reduce spark plug fouling when outdoor temperatures are lower than 40°F. Emission control is also improved. (Courtesy of Ford Parts and Service Division)

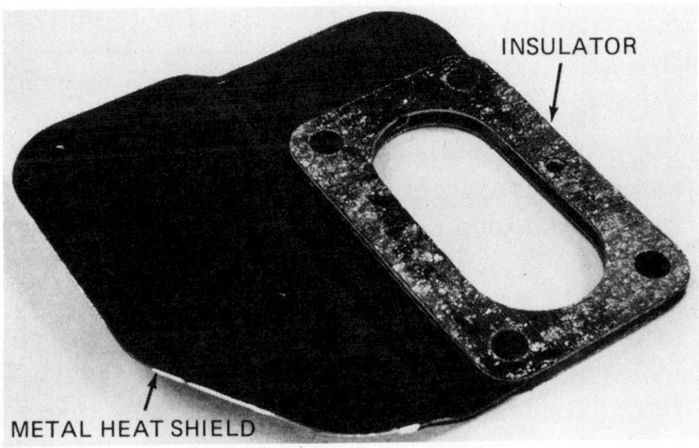

FIGURE 24
Insulator plate used between the carburetor and intake manifold. The insulator and metal shield reduce heat flow to the carburetor.

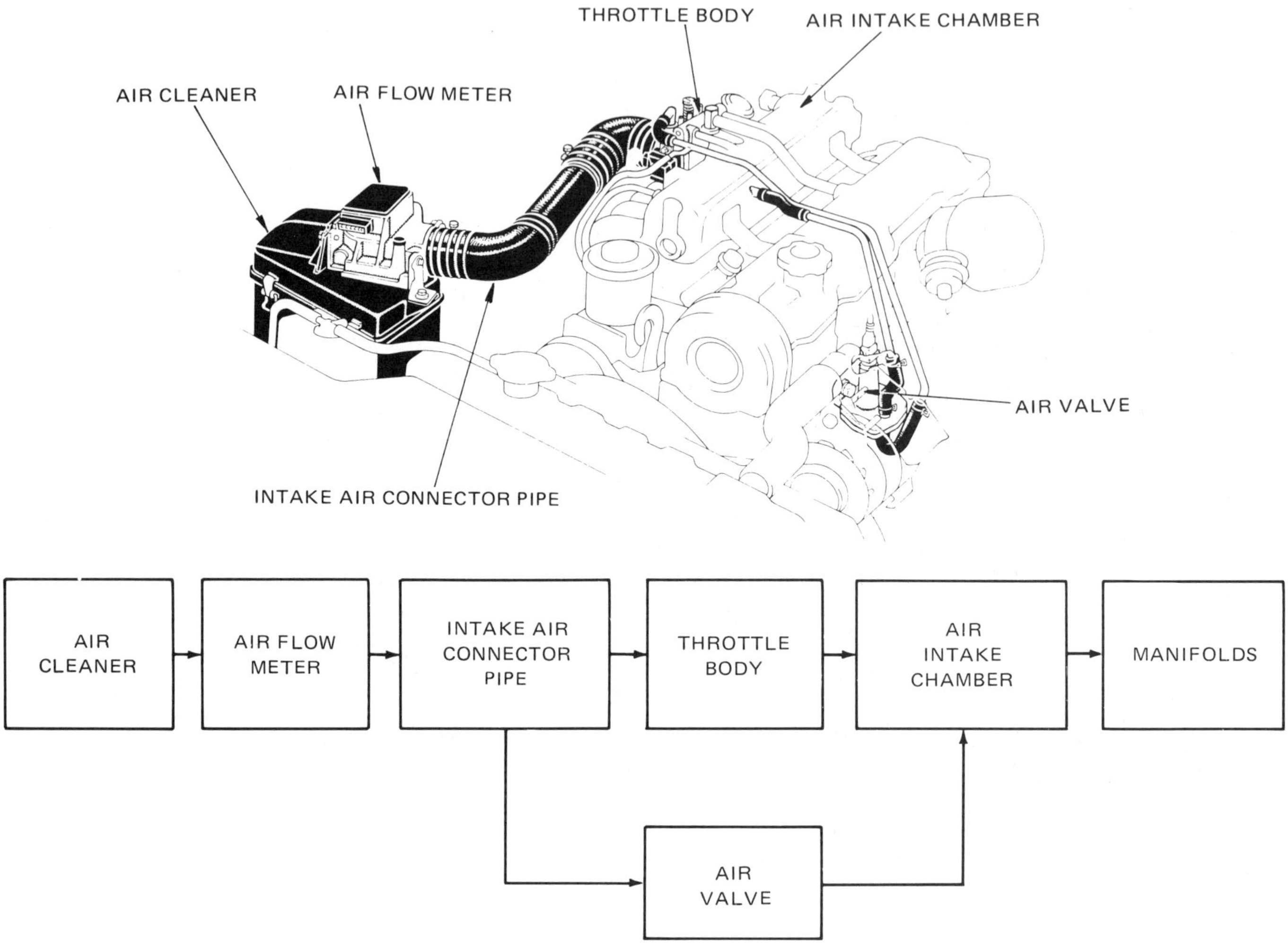

FIGURE 25
A fuel injection air induction system.

CONFIRMING CLOSED-LOOP OPERATION

When doing a tune-up on a computer-controlled vehicle, it is extremely important that you confirm closed-loop operation. The engine will run very poorly if it stays in open-loop mode! In this section we explain how to confirm closed-loop operation on any computer-controlled vehicle equipped with a mixture-controlled carburetor. Basically, closed-loop operation means that the on-board computer is accepting input from the various sensors (primarily oxygen sensor voltage) and controlling the carburetor mixture. Several conditions are required for closed-loop operation:

1. The engine coolant must be about 180°F.

2. The oxygen sensor temperature must be above 600°F.

3. The computer timer must be satisfied (15 seconds to 2 minutes).

COMMENT: *Some General Motors vehicles take only 15 seconds to go into closed-loop operation.*

4. The air pump air must be flowing downstream of the oxygen sensor to the catalytic converter.

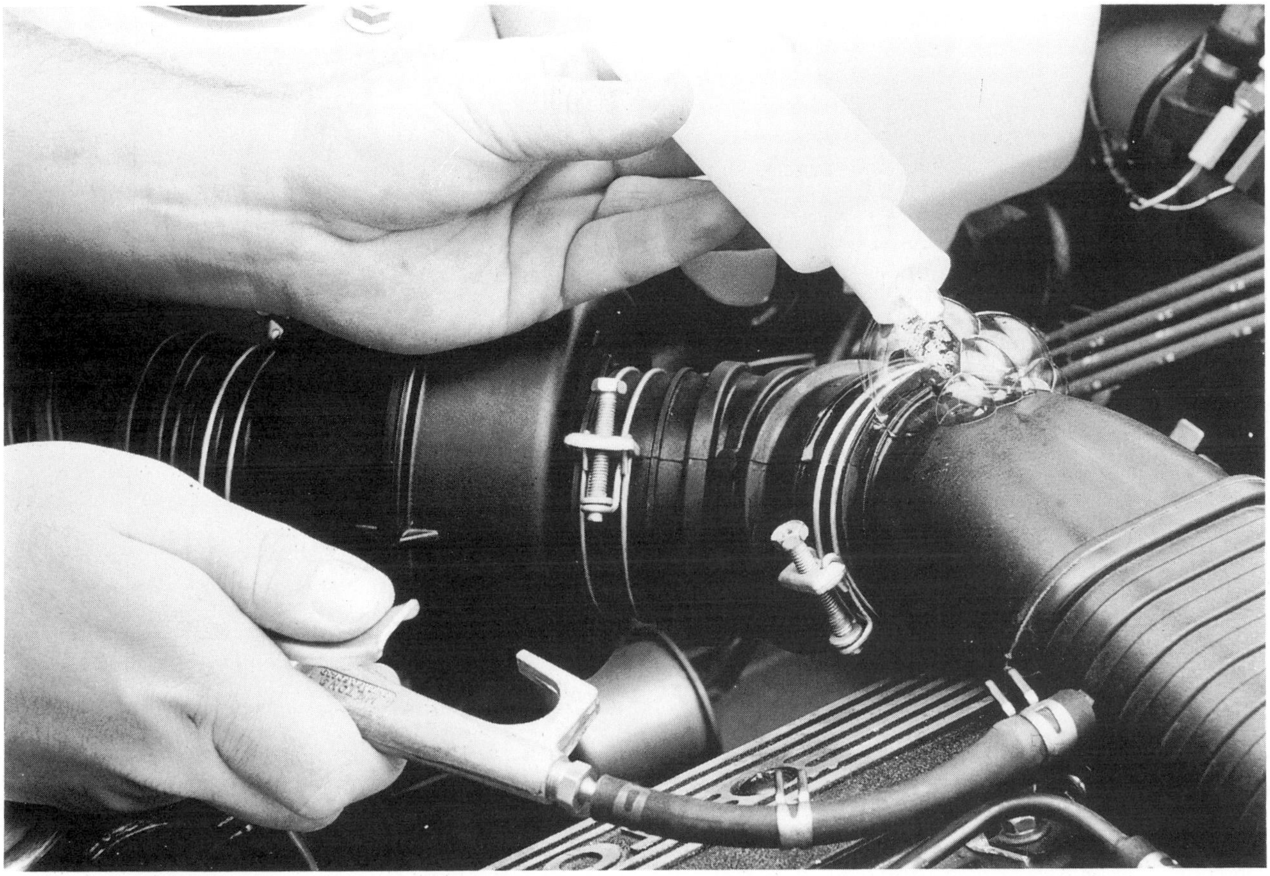

FIGURE 26
A quick test for air induction leakage can be made by using an air hose and liquid shampoo. With the engine stopped, pressurize the air induction system to about 15 psi. Look for bubbles as the shampoo is dripped on all hoses and connections.

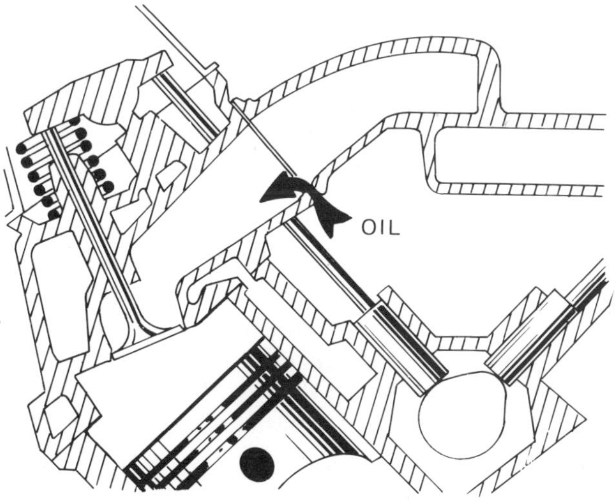

FIGURE 27
On some engines, a defective intake manifold gasket can cause oil to be pulled into the combustion chamber.

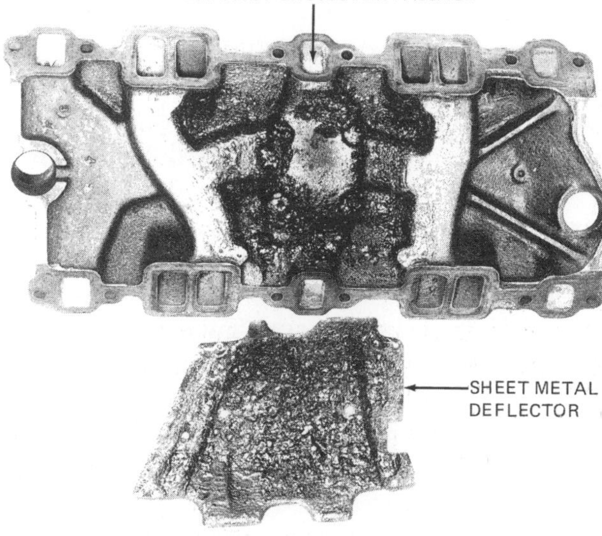

FIGURE 28
View of a Chevrolet intake manifold from below. Observe the exhaust crossover passage.

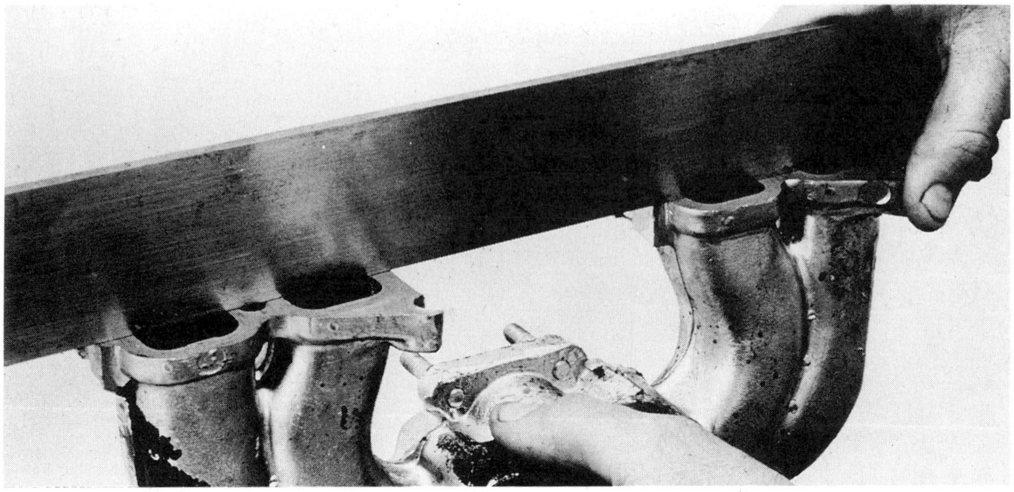

FIGURE 29
Positioning a precision straightedge for checking intake manifold warpage. If a .010″ feeler gauge can be slid between the manifold mounting surfaces and the straightedge, manifold replacement or machining is required.

NOTE: In some instances, the downstream air may be diverted to the atmosphere or air cleaner.

Observation of closed-loop operation can be confirmed by watching the oxygen sensor voltage. The procedure is as follows:

1. On a single-wire oxygen sensor, connect the red lead (+) of a high-impedance (10-MΩ) digital volt-ohmmeter (DVOM) to the oxygen sensor lead. Reconnect the oxygen sensor lead. Connect the black lead (−) to an engine ground. Figure 31[b] shows the hookup details. On a two-wire oxygen sensor, insert the DVOM leads into the oxygen sensor connector. Reconnect the oxygen sensor leads. On a three-wire oxygen sensor, connect the DVOM leads as you would a single-wire sensor. Two of the three wires are used for the heater element.

2. Run the engine until it is fully warmed up and downstream air injection is verified.

3. Operate the engine at *2000 rpm* and observe voltage. This is necessary because some systems are designed to remain in open-loop mode during idle.

4. *The oxygen sensor voltage during closed-loop operation should vary rapidly from .1 to 1.0 V.*

COMMENT: Sometimes the voltage may vary from 0 to 1.2 V. This is OK.

A high voltage (for example, .8 V) indicates that the engine combustion is rich. A low voltage (for example, .2 V) indicates lean combustion. A fixed voltage between .2 and .8 V indicates an open-loop condition.

5. Induce a lean mixture by creating a large vacuum leak. *The oxygen sensor voltage should drop rapidly to about zero.* If it does not, check for rich carburetion, a plugged PCV system, a defective carbon canister purge control system, or a defective oxygen sensor.

6. Choke the engine and watch the oxygen sensor voltage. *It should rapidly increase to .8 V or more.* If it does not, check for lean carburetion, vacuum leaks, a defective EGR system, or a defective oxygen sensor.

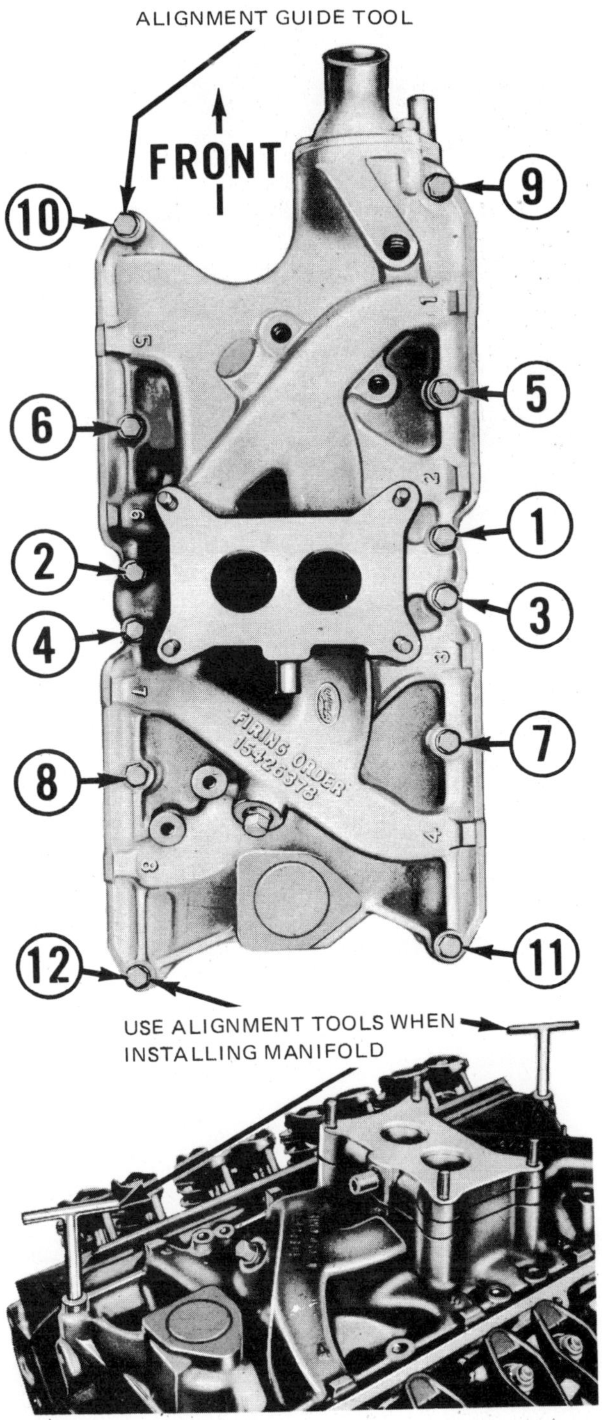

FIGURE 30
Typical intake manifold installation and bolt torque sequence. (Courtesy of Ford Parts and Service Division)

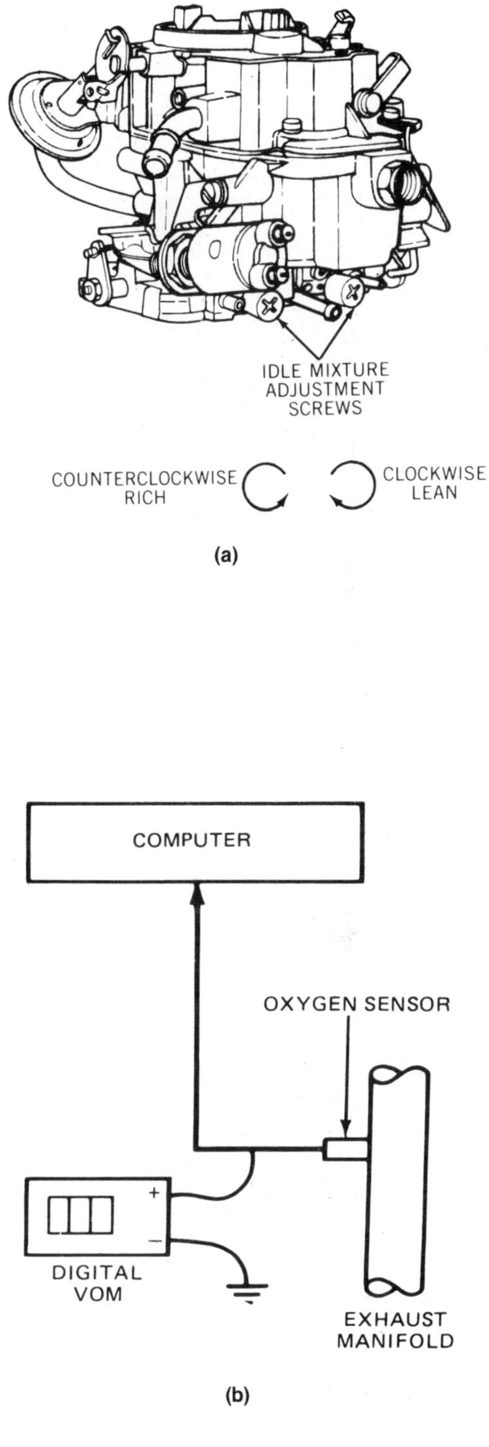

FIGURE 31
(a) Carburetor idle mixture screws. (Courtesy of Chrysler Corporation) (b) Hookup details for confirming closed loop operation.

FIGURE 32
Carburetor idle speed screw.

Idle Speed Adjustment Guidelines

1. Bring the engine up to operating temperature and make sure the fast idle cam is not holding the throttle valve(s) open. In addition, be sure the choke valve is wide open.

2. Adjust the idle speed screw (Figure 32) in "neutral" (for manual shift cars), or in "drive" (for automatic transmission cars).

3. On some cars, the idle speed is adjusted by turning a throttle solenoid plunger (Figure 33.) First, disconnect the throttle solenoid wire. Then set the low (or de-energized) idle speed by turning the low idle speed screw. Connect the throttle solenoid wire and open the throttle slightly to allow the plunger to extend. Set the high (or energized) idle speed by turning the solenoid plunger, by sliding the solenoid in a bracket, or by adjusting the solenoid bracket.

4. Some engines use a computer-controlled reversible electric motor to maintain idle speed (Figure 34). This motor keeps the idle speed constant regardless of the load imposed upon the engine. Closed-loop engine operation must be confirmed when setting this motor.

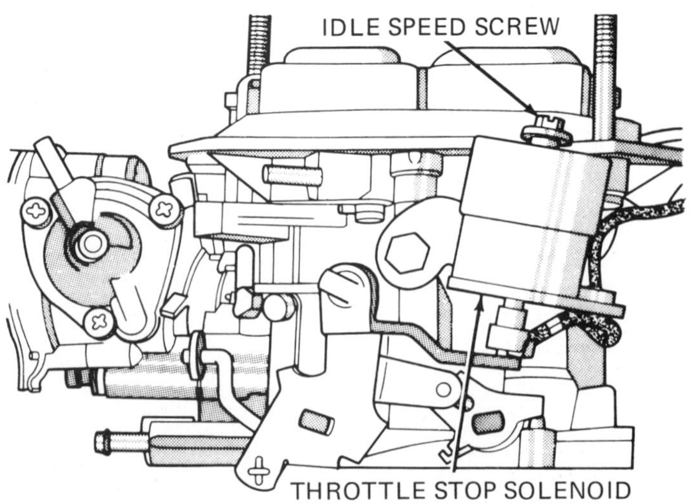

FIGURE 33
Throttle stop solenoid. (Courtesy of Chrysler Corporation)

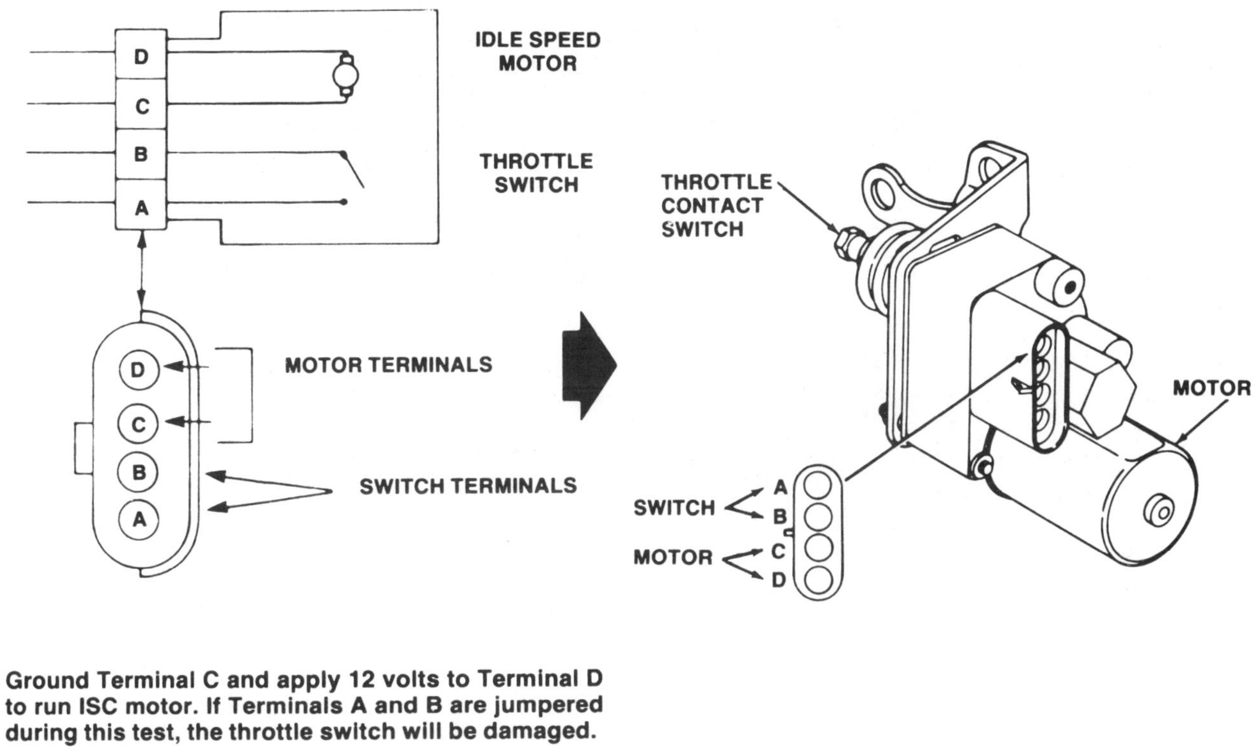

FIGURE 34
Idle speed control (ISC) motor. Electronic inputs are used to control the position of the throttle contact plunger. (Courtesy of Buick Motor Division, General Motors Corporation)

Idle Mixture Adjustment Guidelines

1. Before adjusting idle mixture, adjust dwell, timing, and idle speed.

2. When you seat an idle mixture screw, do it lightly. Overtightening can cause damage (Figure 35).

3. If specified by the vehicle manufacturer and allowed by local regulations, you may remove limiter caps or tamper-resistant plugs to adjust idle mixture (Figure 36).

4. When adjusting idle mixture using the lean-drop method, adjust the mixture screw(s) to get the smoothest idle. Readjust idle speed to specifications. Then turn the mixture screw(s) in until rpm decreases by the specified amount.

FIGURE 35
Badly damaged idle mixture screw.

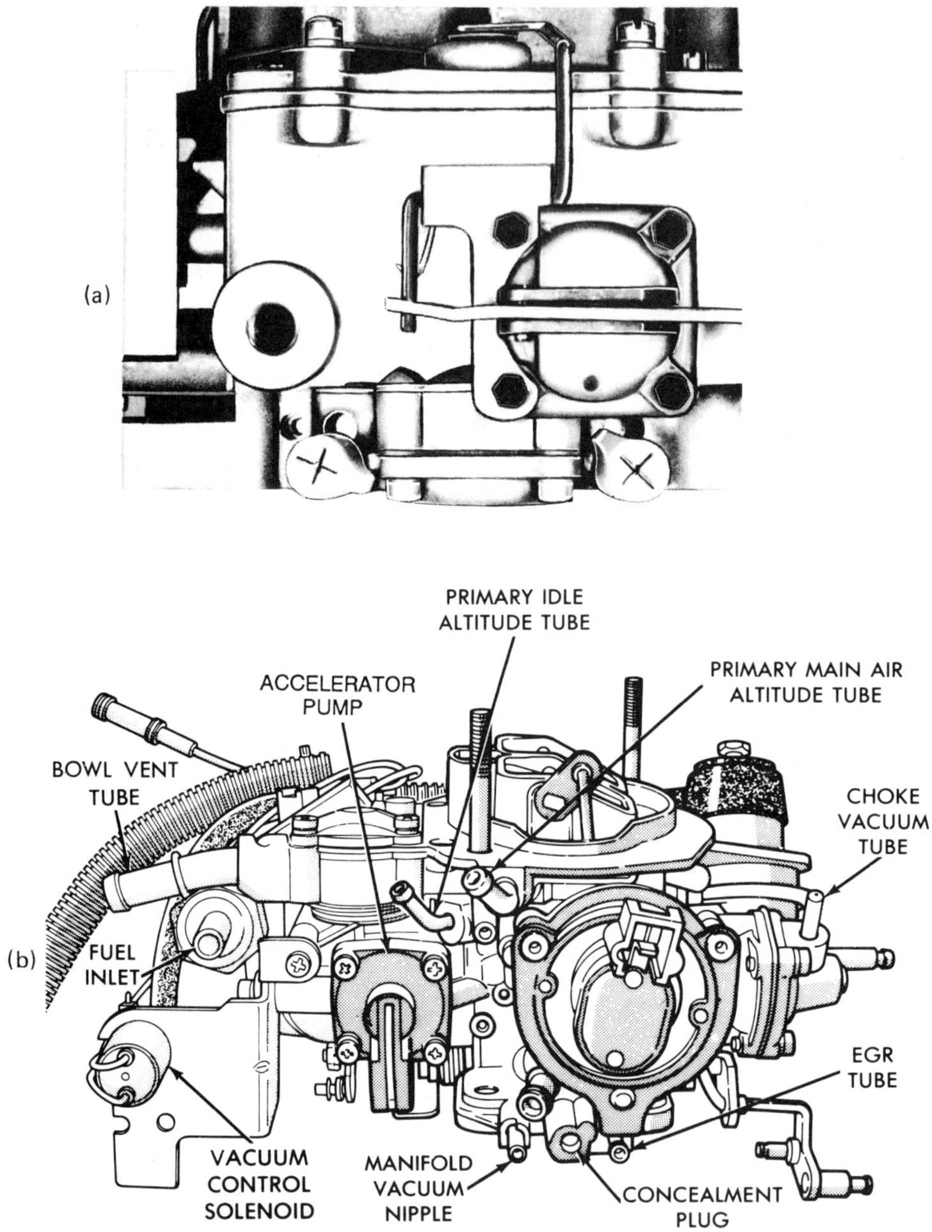

FIGURE 36
(a) External idle limiter caps are used on this carburetor. (b) On most new carburetors, tamper-resistant plugs must be removed before the idle mixture can be adjusted.

5. When adjusting idle mixture using the lean-best-idle method, adjust the mixture screw(s) to get the highest idle rpm. Then turn the mixture screw(s) in until idle rpm decreases 20 rpm. Finally, turn the screw(s) out one-half turn.

6. When adjusting idle mixture using an infrared analyzer, it may be necessary to change engine oil. Fuel dilution of the oil can increase vapors through the PCV system and cause an unstabilized reading.

7. When using the propane enrichment method to adjust idle mixture on a computer-controlled carburetor, put the engine in open-loop.

With an engine correctly timed and the carburetor throttle valve(s) firmly seated against the idle speed stop, the introduction of propane gas into the intake manifold should cause a gain in idle rpm. This is because a richer mixture is present. Failure to achieve a specified idle rpm gain will require an adjustment of the idle mixture screws to a leaner position.

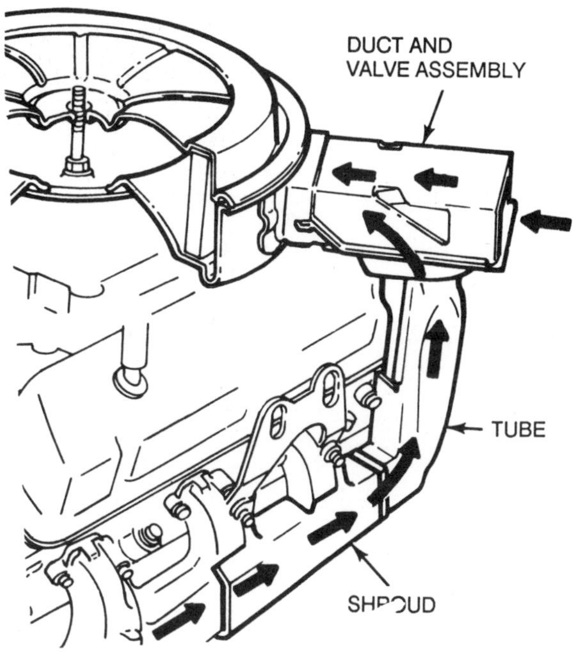

FIGURE 37
A thermostatically controlled valve in this air cleaner snorkel maintains inlet air at approximately 100°F. (Courtesy of Ford Parts and Service Division)

AIR CLEANER ASSEMBLY

The functions of the air cleaner assembly are as follows:

1. Filters dirt and abrasives from the engine intake air.

2. Acts as a flame arrestor in case of engine backfire.

3. Serves as an air silencer.

4. Is often used to control intake air temperature (Figure 37).

5. Provides filtered air to the positive crankcase ventilation (PCV) system.

Air Filter Elements

On most automobiles, the element consists of pleated paper between two sealing surfaces (Figure 38). Sometimes a synthetic foam pad is placed around the outside of the paper element to increase filter effectiveness.

Air Filter Element Inspection

1. Inspect the element and sealing surfaces for deformation or damage (Figure 39).

2. Inspect the element for excessive dirtiness or oil contamination (from blowby gases). Clean or replace if necessary.

NOTE: Some carburetor-equipped engines use a carbon-impregnated element that allows the air cleaner assembly to serve as a vapor storage area. This air filter element is very dark in appearance and should not be mistaken for being dirty.

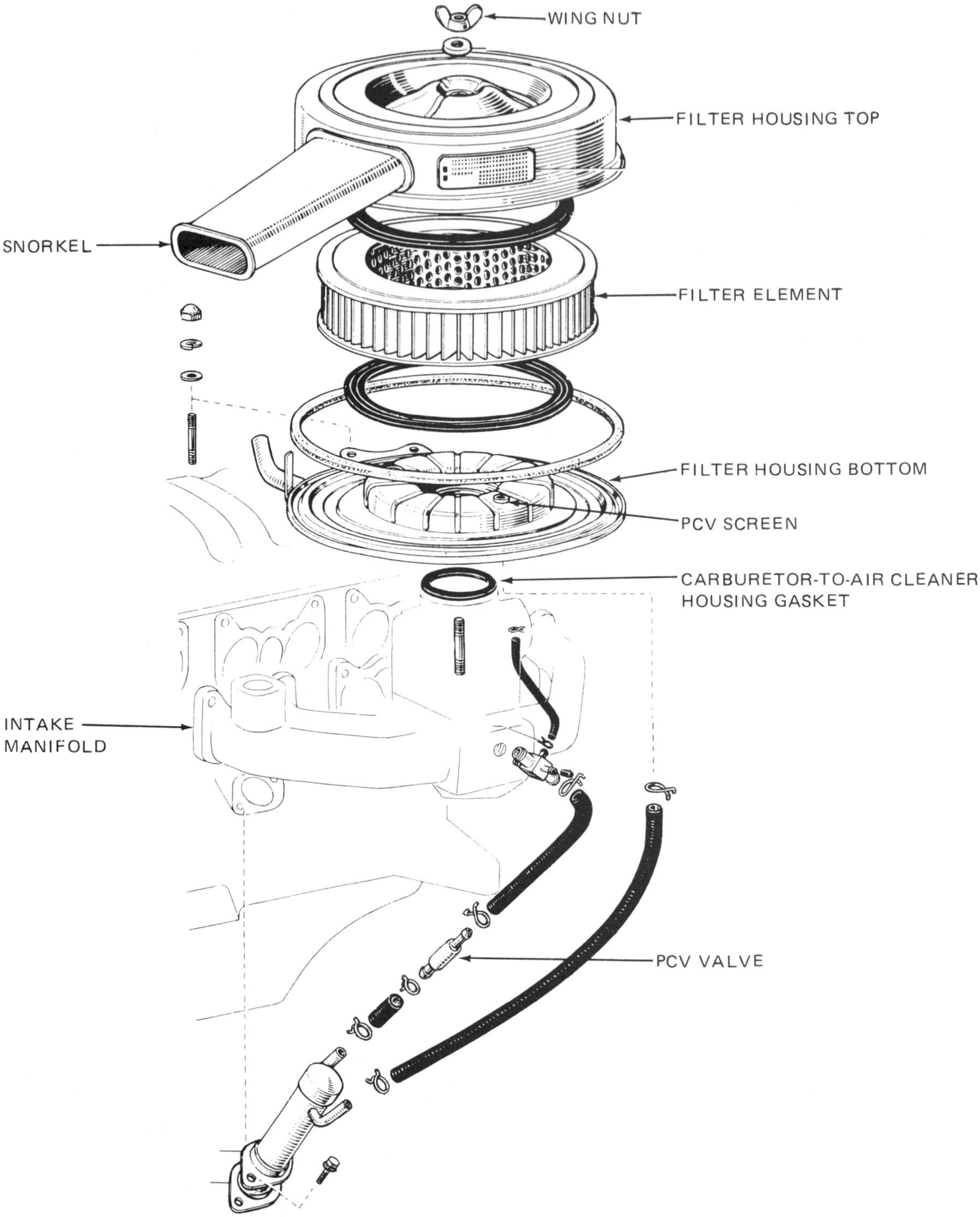

FIGURE 38
Air filter element and housing assembly. (Courtesy of American Honda Motor Co., Inc.)

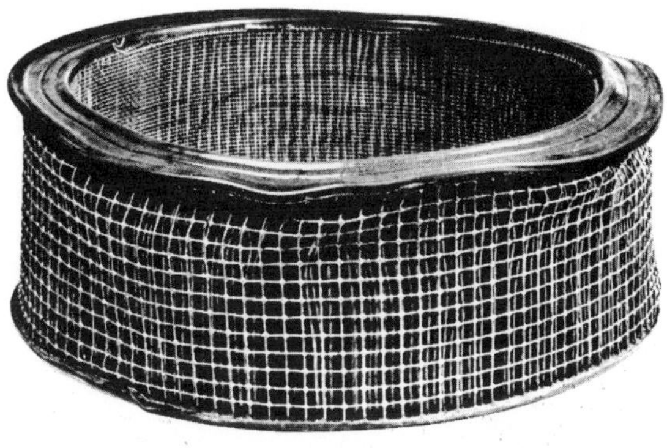

FIGURE 39
Improper cleaning ruined this air filter element.

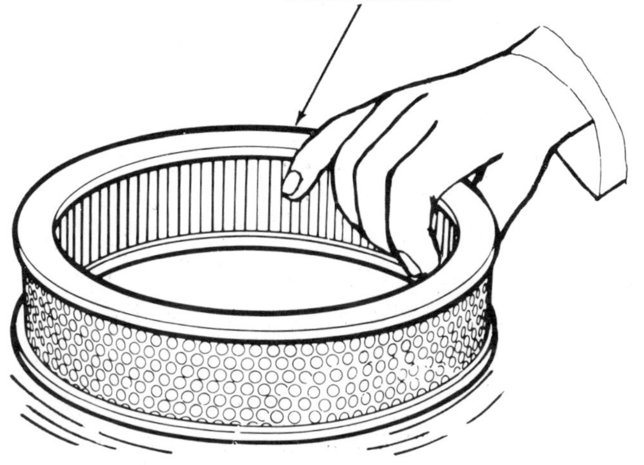

FIGURE 40
Much of the dirt in an air filter element can be dislodged by gently tapping the element on a flat surface. (Courtesy of Chrysler Corporation)

> **SHOP TIP**
> Crankcase oil collecting in the air cleaner can be caused by an external air leak into the crankcase. This causes the air flow in the PCV system to reverse itself and carry oil vapor (mist) back into the air cleaner. When the oil settles out of the vapor, it collects in the bottom of the air cleaner.

A loose oil filler cap, an improperly installed front crankshaft seal, or a dipstick not fully seated can be sources of external air leaks.

Servicing Air Filter Element

1. Remove the element.

2. Wipe the inside of the filter housing with a clean solvent-dampened rag.

3. Tap the element gently (square on the floor) to knock off the dirt (Figure 40). After tapping, low-pressure compressed air can be used to blow dirt from the inside of the element outwards (Figure 41). *Never use a close, high-pressure air blast because of the danger of rupturing the element!*

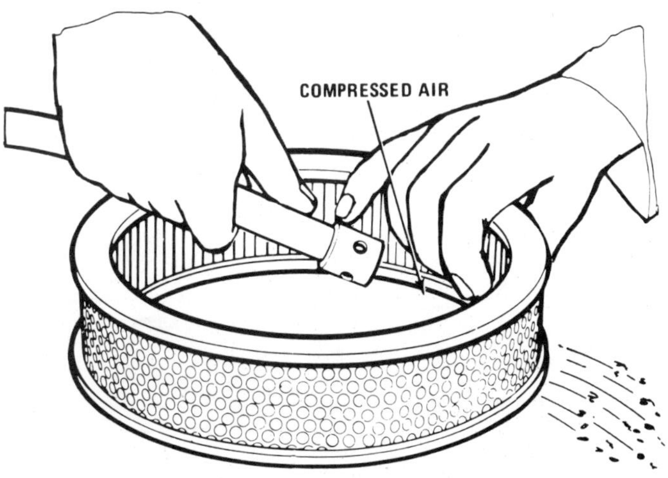

FIGURE 41
After tapping the air filter element as shown in Figure 40, low pressure air can be used to remove the remaining dirt in the element. (Courtesy of Chrysler Corporation)

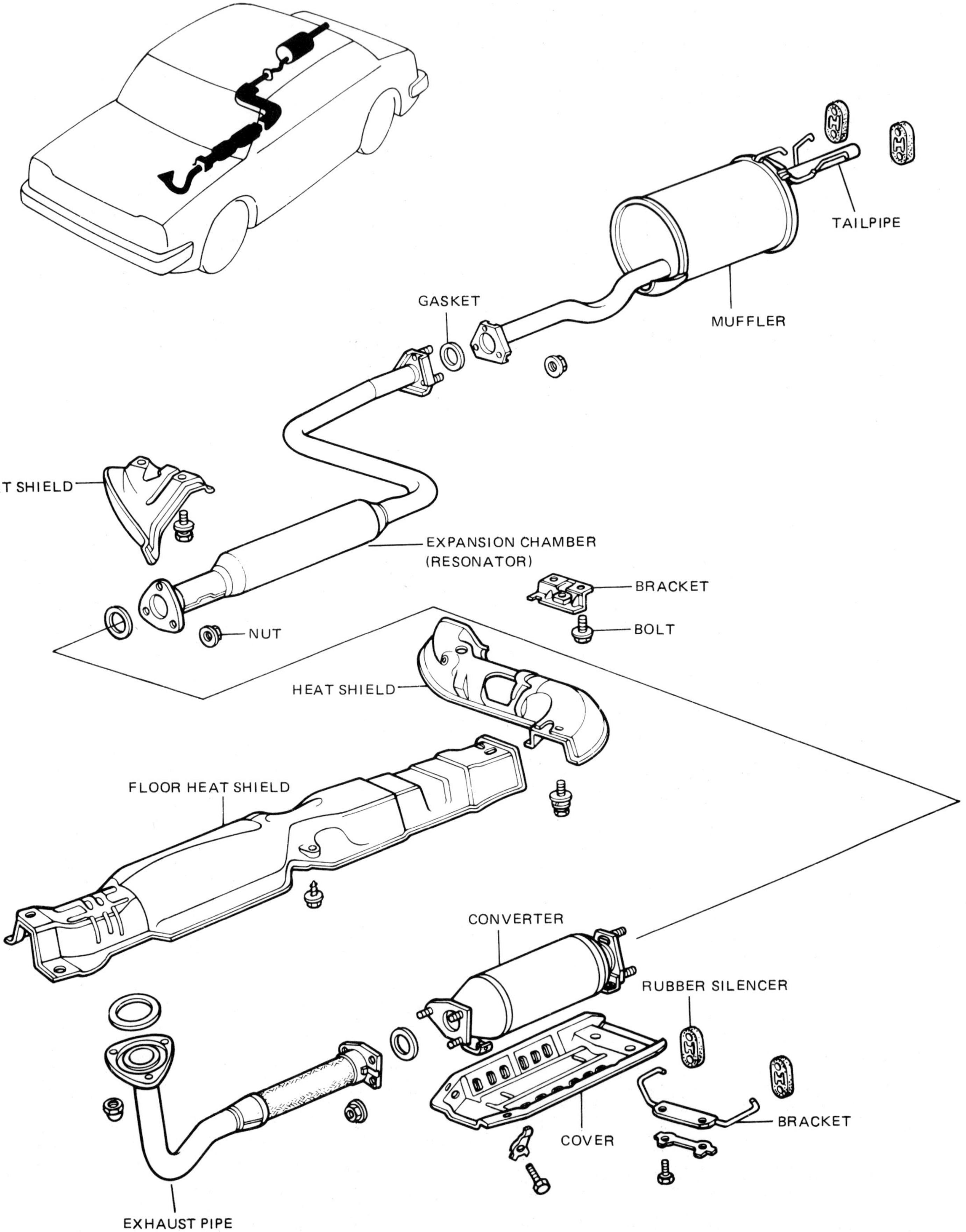

FIGURE 42
Vehicle exhaust system. (Courtesy of American Honda Motor Co., Inc.)

4. Check all air intake hoses, clamps, and connections for damage and airtight connections. Replace hoses and tighten connections as necessary.

5. Be sure to inspect the gasket between the air cleaner housing and the carburetor air horn (if so equipped). Replace if torn or misaligned.

6. Be sure the pre-heater hose to the air cleaner housing is in good condition and is correctly installed.

7. If the air cleaner has a PCV filter, it should be cleaned or replaced.

EXHAUST SYSTEM

The basic purpose of the exhaust system is to collect the high-temperature burned gases from the exhaust ports and deliver them to the atmosphere. In addition, it helps limit the noise produced. The hot gases pass from the exhaust manifold to an exhaust pipe, catalytic converter, resonator, muffler, and out the tailpipe (Figure 42). Removal of hot gases must be done with the least possible restriction or back pressure.

The exhaust system must be properly aligned to prevent stress, leakage, and noise. Refer to Chart 3 for a listing of exhaust system problems.

CHART 3 TROUBLESHOOTING CHART FOR THE EXHAUST SYSTEM

Exhaust Gas Leaks

1. Leaks at pipe joints
2. Damaged or improperly installed gaskets
3. Broken or rusted out muffler
4. Broken or rusted out pipes (Figure 43)
5. Loose, warped, or cracked exhaust manifold

Noise

1. Loose baffle in muffler
2. Blown out muffler (Figure 44)
3. Leak between exhaust manifold and cylinder head
4. Broken hanger (Figure 45)
5. Pipe-to-body contact
6. Leaks at pipe joints
7. Damaged gaskets or improperly torqued connections (Figure 46)

"Blistered" Spark Plugs

1. "Frozen" closed manifold heat control (Figure 47)
2. Air cleaner flapper valve stuck in the heat-on position (Figure 48)

Engine Slow to Warm Up

1. "Frozen" open manifold heat control valve
2. Blocked crossover passage in cylinder head or intake manifold (Figure 49)
3. Damaged or missing hot air duct

Engine Hard to Start Cold

1. Cracked or burned-through choke stovepipe (Figure 50)

Engine Lacks Power

1. Kinked or collapsed pipe (Figure 51)
2. Plugged catalytic converter

Noisy Manifold Heat Control Valve

1. Broken thermostatic spring
2. Broken, weak, or missing anti-rattle spring (Figure 52)

Cracked Intake Manifold

1. "Frozen" closed manifold heat control valve
2. Excessive carbon buildup on bottom of intake manifold exhaust crossover

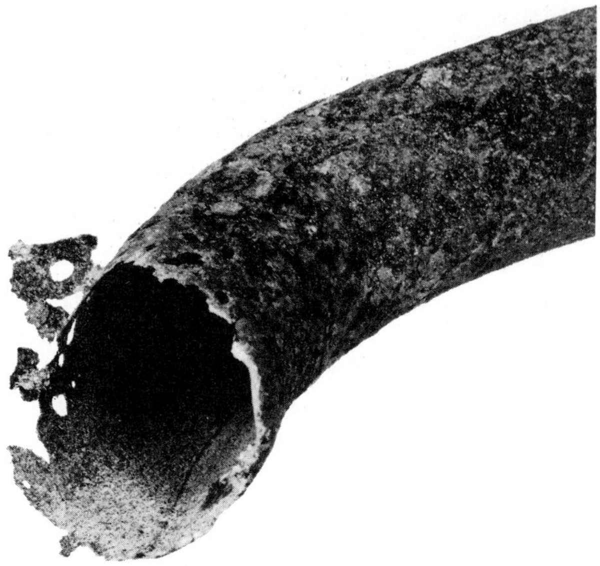

FIGURE 43
Rusted out exhaust pipe. Corrosive acids formed by the combination of lead deposits and water are usually the cause.

FIGURE 44
Repeated backfiring blew this muffler apart.

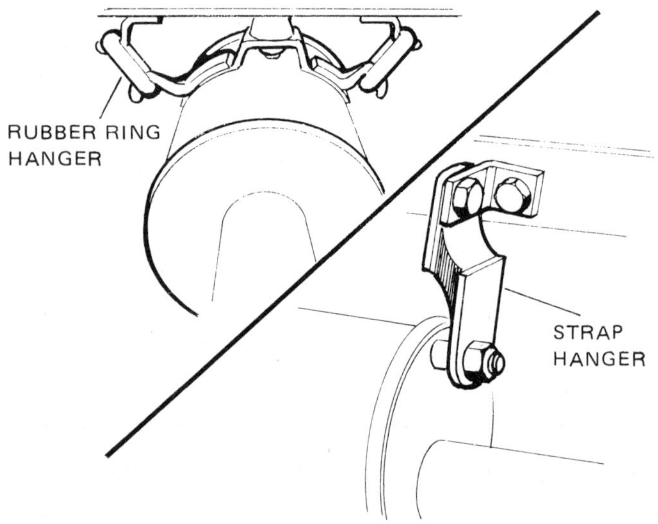

FIGURE 45
Exhaust system hangers.

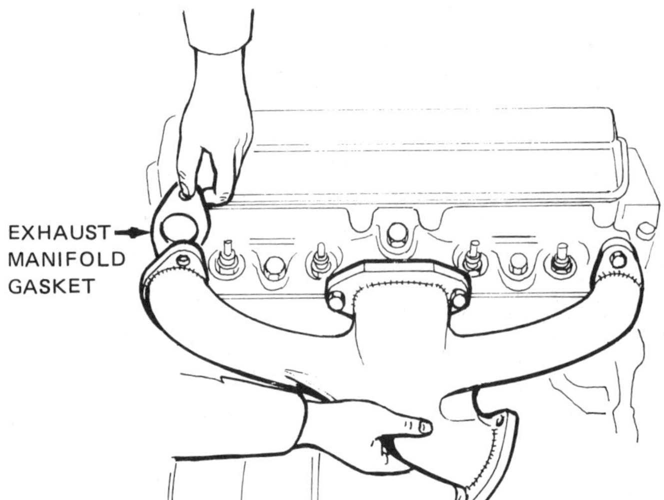

FIGURE 46
A damaged or improperly torqued exhaust manifold gasket can cause noise.

FIGURE 47
Stuck closed manifold heat control valve.

Exhaust Manifold and Gaskets

Exhaust manifolds are generally one piece, and are made of cast iron or stainless steel. Exhaust manifolds expand and contract considerably while in operation. This can cause cracking.

Many exhaust manifolds are equipped with an oxygen sensor and fittings for air injection (Figure 53).

The exhaust manifold is attached to the cylinder head with studs and nuts, or with bolts. When used, gaskets are either the embossed steel type or the perforated steel/asbestos type. Exhaust manifold gaskets can develop breaks and leak.

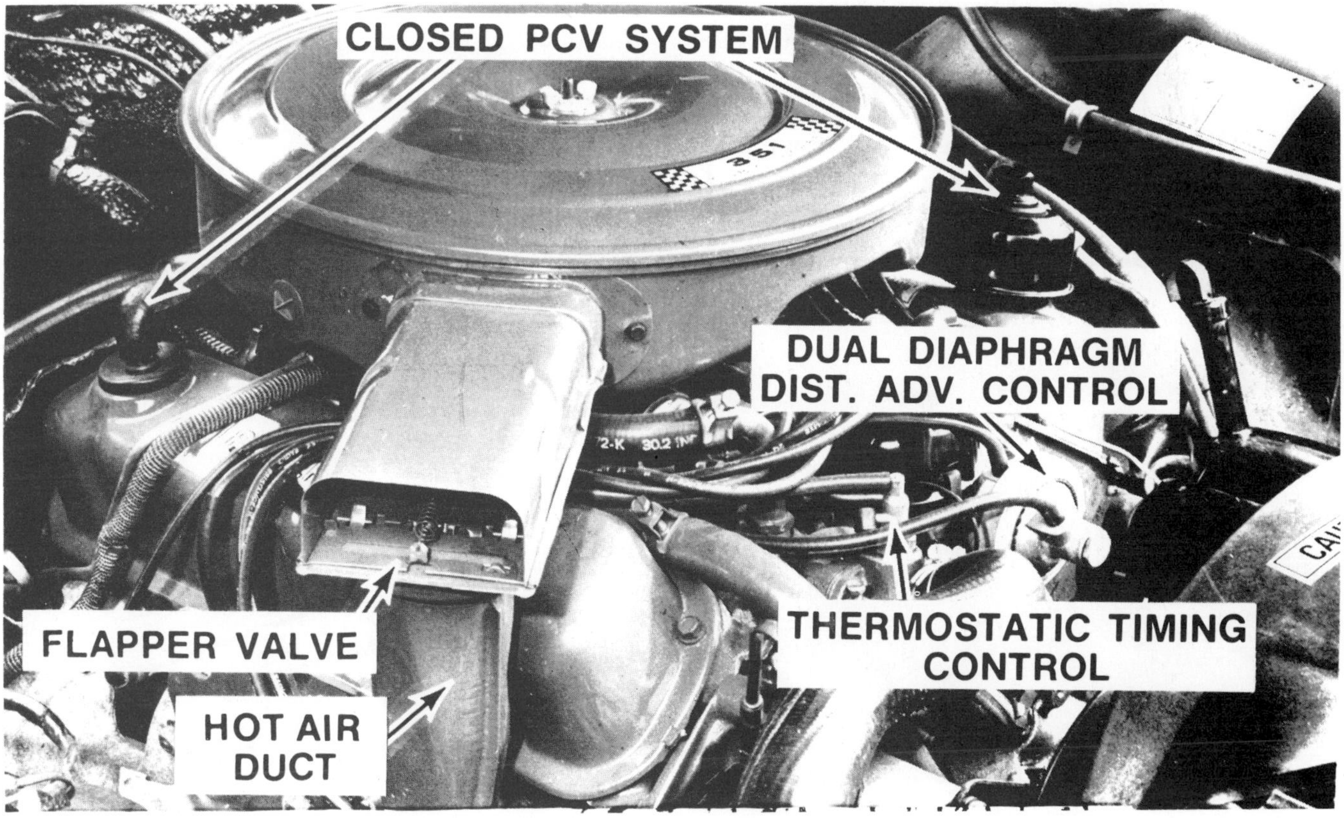

FIGURE 48
A malfunctioning flapper valve can cause driveability problems.

FIGURE 49
The center two exhaust ports of this cylinder head are siamesed to form a heat riser (crossover) passage.

SHOP TIP

When metal-to-metal exhaust manifolds or tubing-type headers are used, place a 1/8″ RTV silicone bead around exhaust port openings and tighten down the manifold or headers. This provides a very effective seal in place of any other gasket. Remember to clean metal surfaces and to wipe with solvent or lacquer thinner prior to using the silicone.

FIGURE 50
A cracked manifold stove pipe allowed exhaust gas to enter this choke assembly and damage it.

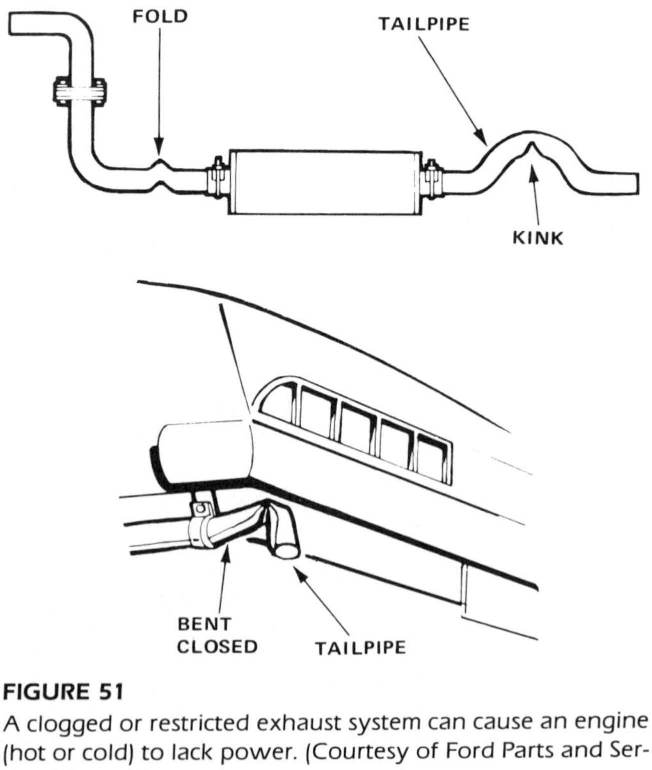

FIGURE 51
A clogged or restricted exhaust system can cause an engine (hot or cold) to lack power. (Courtesy of Ford Parts and Service Division)

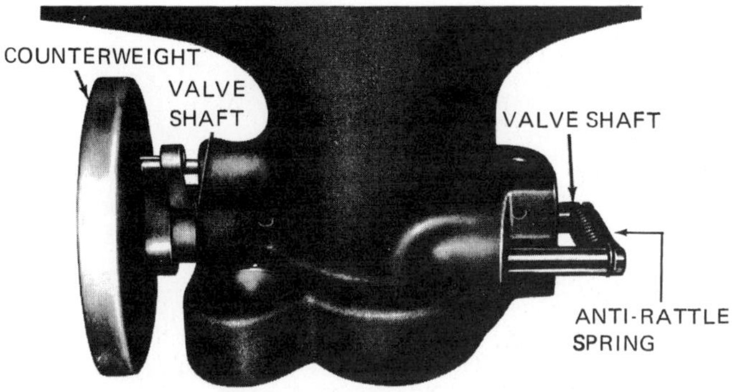

FIGURE 52
If the manifold heat control valve is noisy, check the anti-rattle spring (if so equipped). (Courtesy of Chrysler Corporation)

The exhaust manifold is connected to the exhaust pipe by a flanged connection, or by a ball joint type connection (Figure 54).

> **SHOP TIP**
> When installing an exhaust pipe flange seal (often called a donut), apply antiseize to the gasket surfaces. See Figure 55. This will allow the donut to move and align itself for a better seal when the joint is tightened.

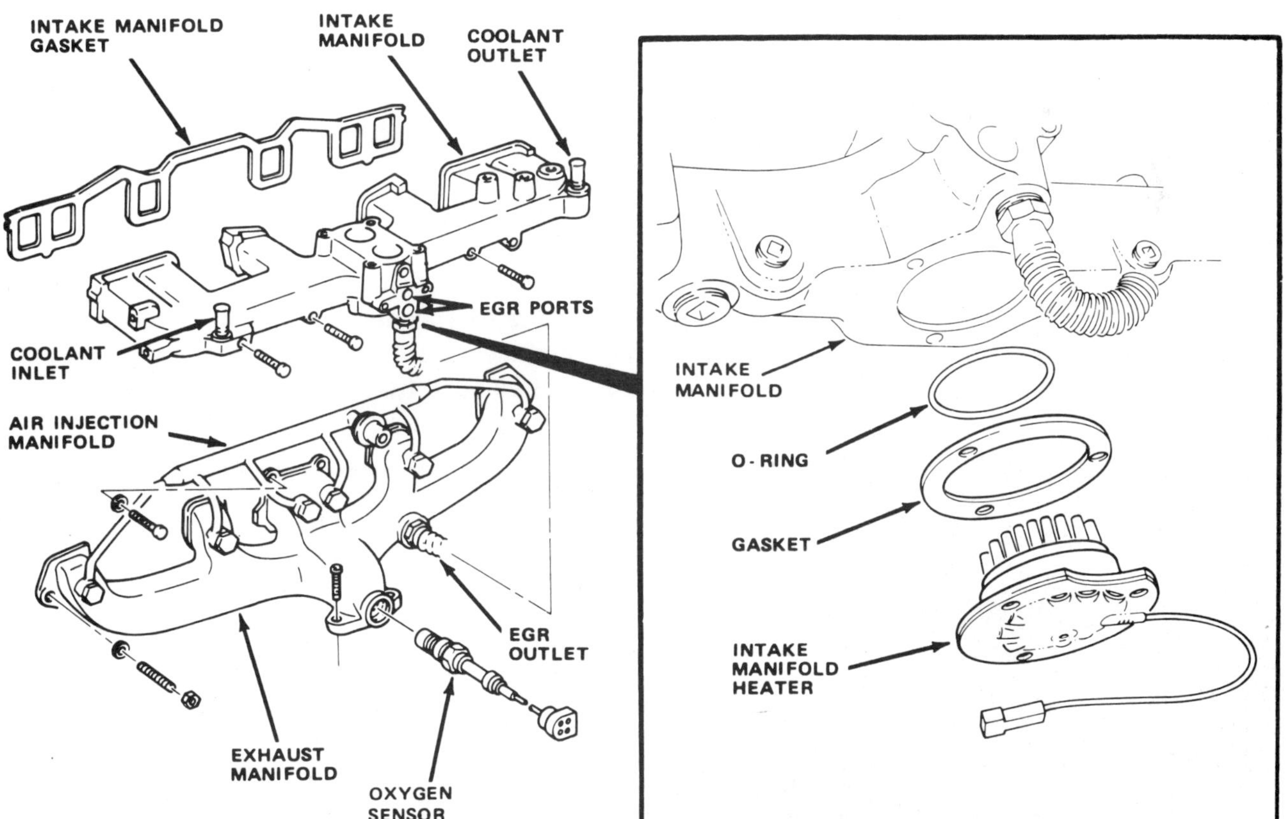

FIGURE 53
Intake and exhaust manifold assembly for an in-line six-cylinder engine. The intake manifold is electric and is coolant heated. Locate the oxygen sensor. (Courtesy of American Motors Corporation)

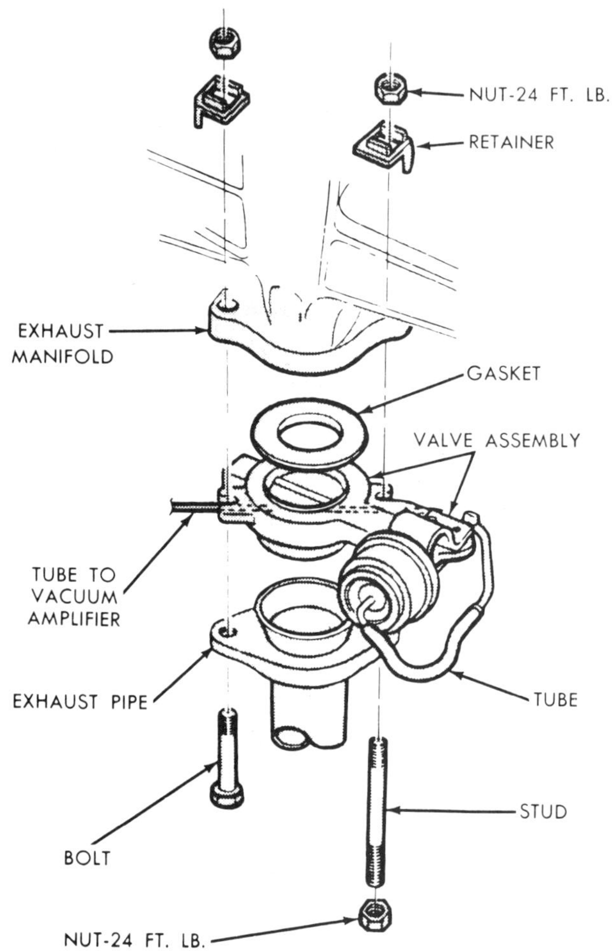

FIGURE 54
Typical flanged exhaust pipe to exhaust manifold connection. (Courtesy of Chrysler Corporation)

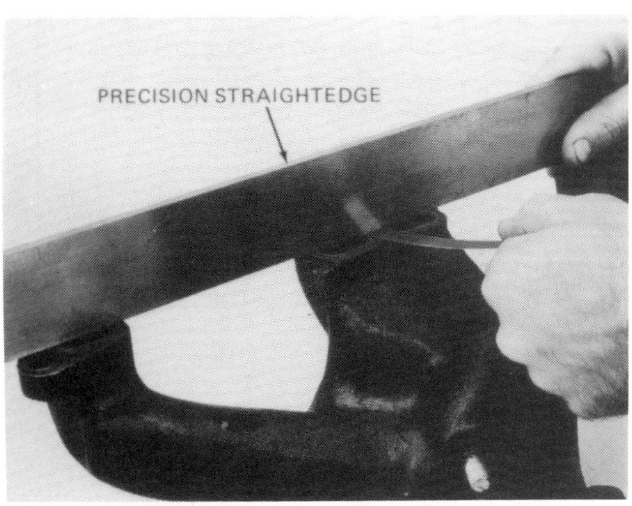

FIGURE 56
Checking an exhaust manifold for warpage.

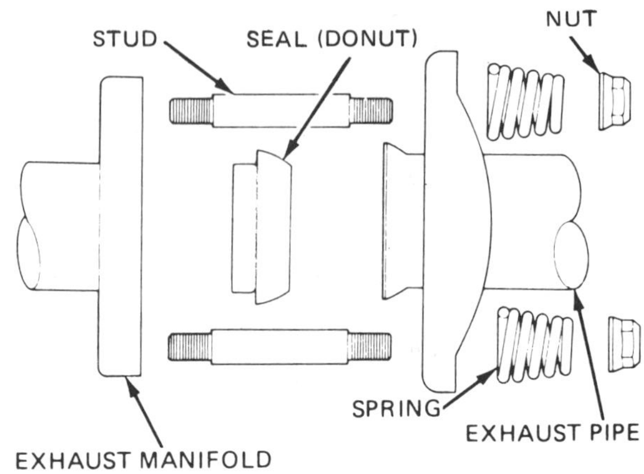

FIGURE 55
A ball joint exhaust manifold connection that uses a donut-type seal. Engine torque reaction is accommodated by this type of connection. (Courtesy of Oldsmobile Division, General Motors Corporation)

Exhaust Manifold Gasket Installation Tips

1. Before installing a new gasket, check for warping of the manifold (Figure 56). Mounting surfaces must be straight within .010″ and free of pits and erosion.

> **SHOP TIP**
> Exhaust manifold gasket life can be extended by coating the manifold-to-gasket surfaces with antiseize compound. This makes for a heat-resistant seal that allows the manifold to expand and contract.

2. When installing an exhaust manifold, do not torque at the outside ends first, since cracking may result (Figure 57). Tighten the center bolts to the maximum torque specification, and the bolts at either end to the minimum torque specification.

3. On some exhaust manifold gaskets, different material is used on each face: one face is perforated steel, and the other face is asbestos. *When installing, always place the asbestos face against the cylinder head.* This allows the exhaust manifold to slip without gouging the soft asbestos face.

FIGURE 57
Improper tightening sequence can cause an exhaust manifold to crack.

Heat Riser Valve

Many exhaust systems use a heat riser valve to direct exhaust gas to the carburetor base (Figure 58). This helps improve fuel vaporization during engine warm-up.

On V-type engines, when the heat riser valve is closed, the exhaust gas passes through a crossover passage in the intake manifold. Refer back to Figure 28.

Some vehicles use a manifold-vacuum-controlled heat riser valve (Figure 59).

Servicing Heat Riser Valve

The heat riser valve must operate freely with no sticking. If it is "frozen" or sticking, apply penetrating oil and work it back and forth until it moves freely (Figure 60). If this is not successful, replace the valve assembly.

POSITIVE CRANKCASE VENTILATION (PCV) SYSTEM

The PCV system ventilates the engine crankcase, removing harmful "blowby" gases (Figure 61[a] and [b]). If not removed, these gases can contaminate the engine oil and form sludge in the crankcase.

PCV systems are "closed" systems that circulate fresh air through the crankcase whenever the engine is running. This is accomplished by a PCV valve that allows a small amount of intake manifold vacuum to be placed on the crankcase.

PCV System Service

A PCV valve can become stuck open, or it can wear to a point where it is always operating at maximum flow. This can allow oil mist to be drawn into the intake manifold. Oil consumption, fouled spark plugs, and rough idle are the results.

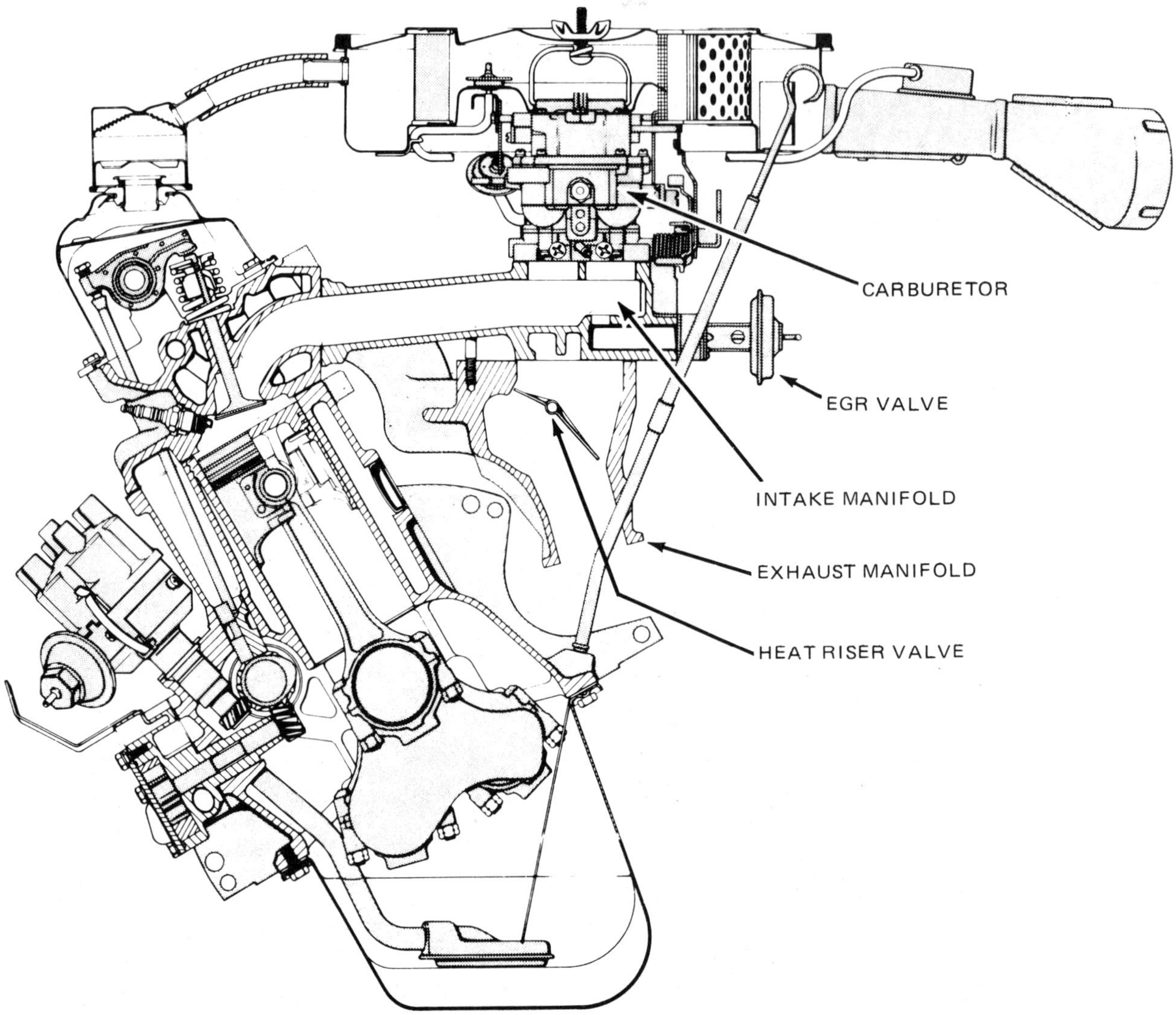

FIGURE 58
Heat riser valves are designed to direct exhaust gas to the carburetor base or bottom of the intake manifold. (Courtesy of Chrysler Corporation)

COMMENT: It is a common error to think that if a PCV valve rattles when you shake it, it is acceptable. Shaking will not indicate seat and plunger wear inside the valve.

A quick method for checking out a PCV valve is as follows:

1. Disconnect the PCV valve, leaving the vacuum signal hose attached.

2. Start the engine and read idle rpm on a tachometer.

3. Pull and disconnect the PCV valve, leaving the vacuum signal hose attached. Place your finger over the end of the PCV valve opening (Figure 62). You should feel a strong vacuum.

4. Idle rpm should normally drop no more than 80 rpm. If the rpm drops more than this, suspect a worn valve or the wrong valve.

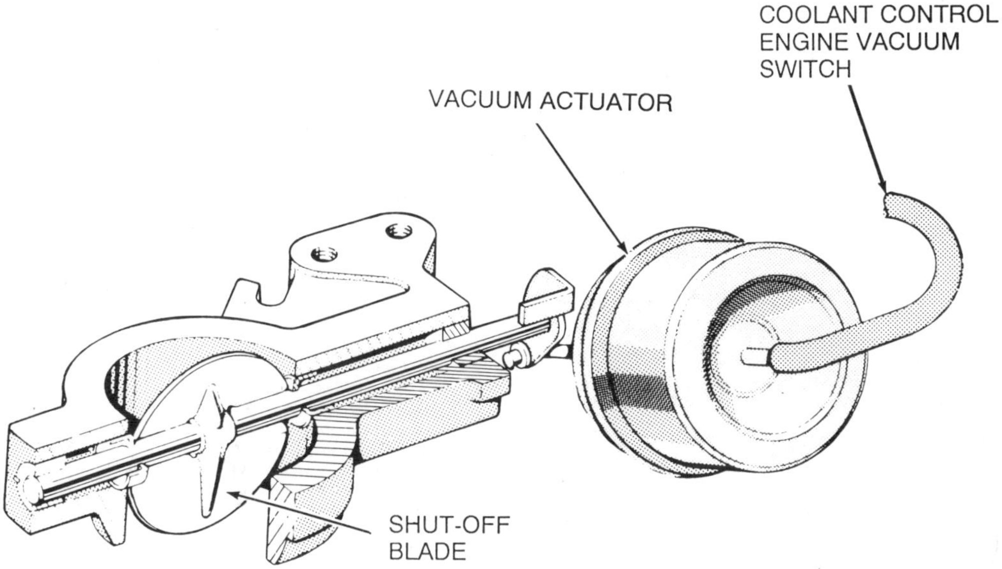

FIGURE 59
Manifold vacuum controlled heat riser valve. Such a valve can also be seen in Figure 54.

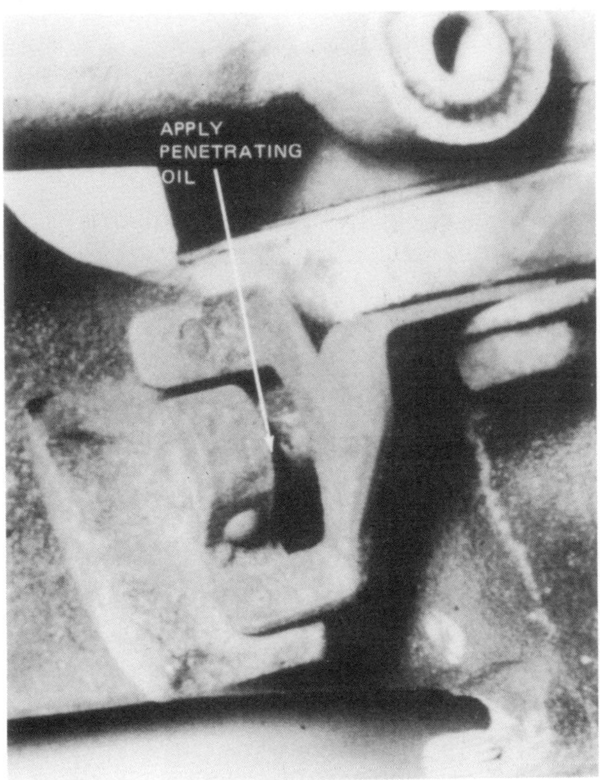

FIGURE 60
Penetrating oil applied at the proper location will often free a "frozen" or sticking heat riser valve.

A PCV system can become closed from sludge and varnish deposits. This can result in plugged oil rings, ruptured gaskets and seals, carburetor problems, rough idle, and oil inside the air cleaner assembly. You can test the PCV system for clogging with an air flow tester (Figure 63).

On some engines (Fords in particular), make sure that the vacuum passage in the carburetor spacer is not clogged. Clean all attached fittings and hoses (Figure 64). Many engines contain an oil separator filter in the PCV system (Figure 65). This must be cleaned or replaced periodically to prevent clogging.

TURBOCHARGERS

Basically, the turbocharger (often simply called a turbo) is an air compressor that is driven by engine exhaust gases. See Figure 66. The exhaust gases drive a turbine wheel, which is coupled to a compressor wheel by means of a shaft (Figure 67). As the compressor wheel turns, air is forced into the cylinders.

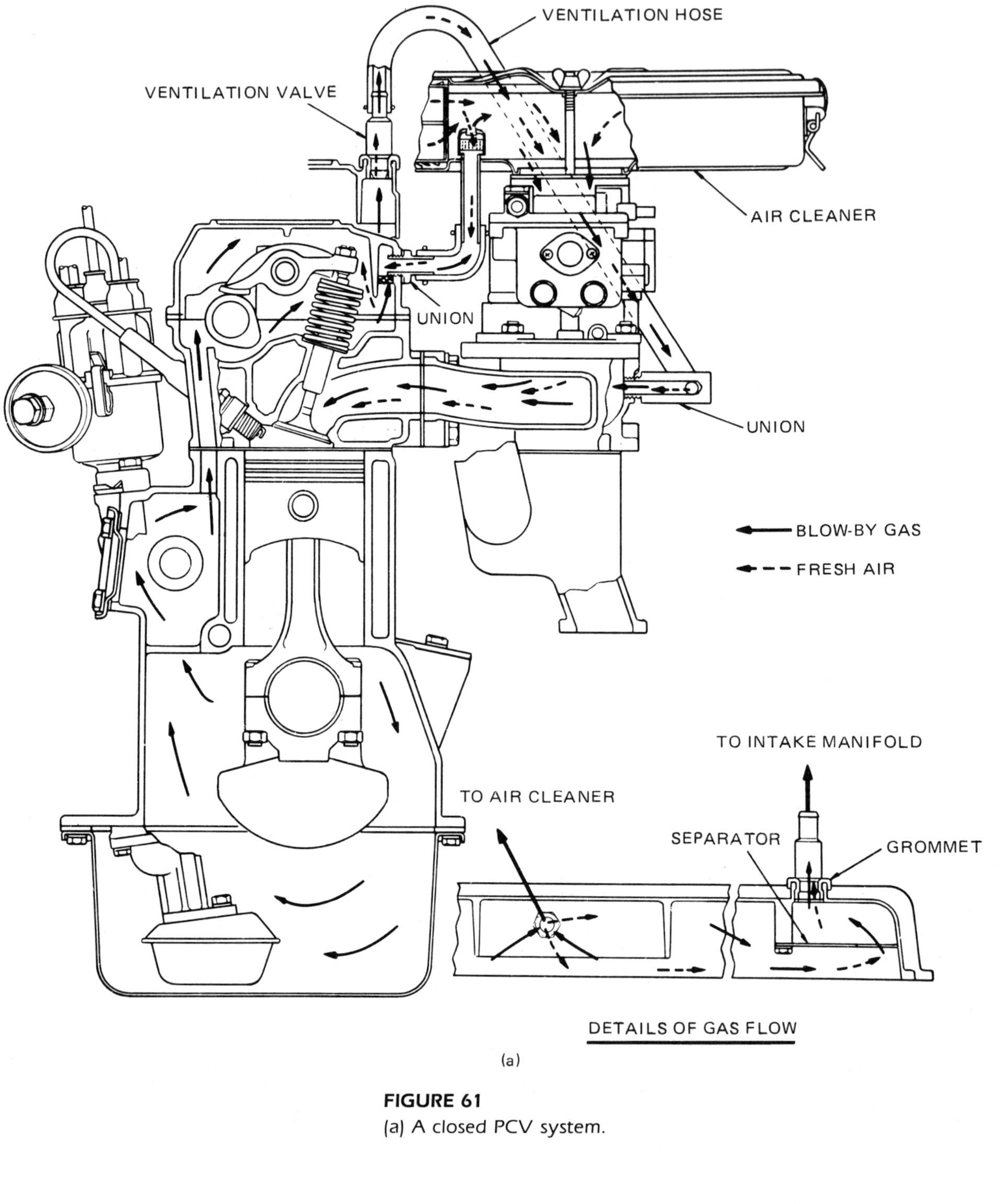

(a)

FIGURE 61
(a) A closed PCV system.

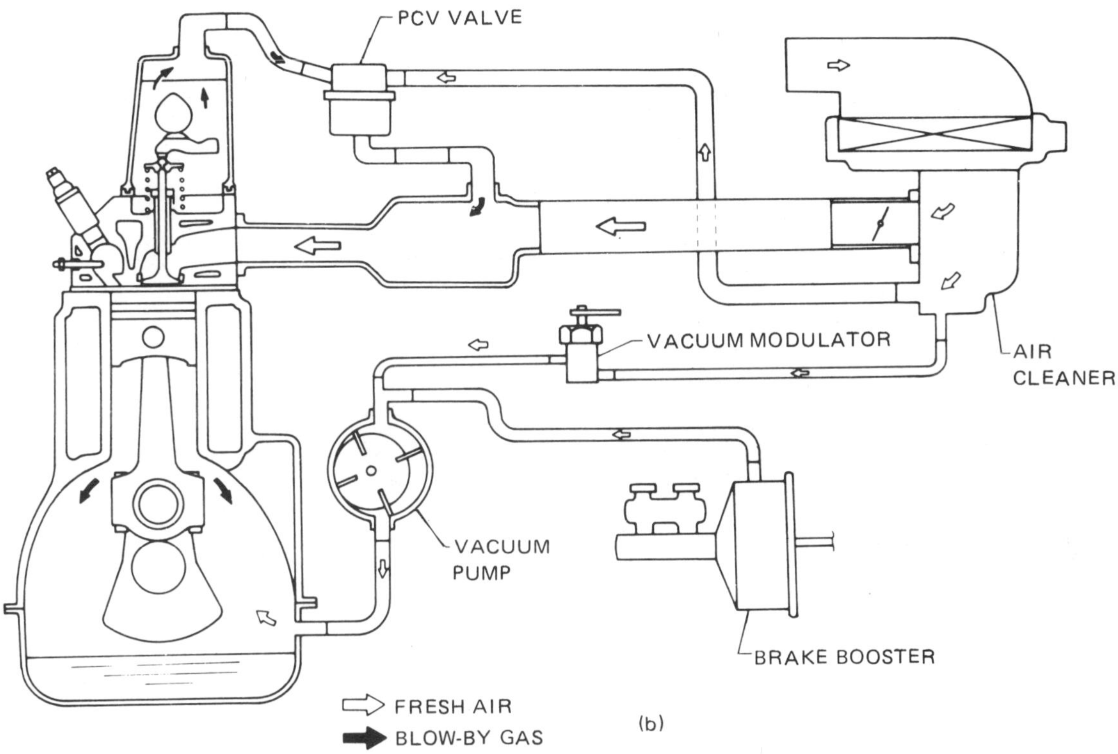

FIGURE 61 (CONTINUED)

(b) PCV system used on an automotive diesel engine. (Courtesy of Nissan Motor Company, Ltd.)

FIGURE 62

A PCV valve can be tested by placing your finger over the opening end of the valve while the engine is idling.

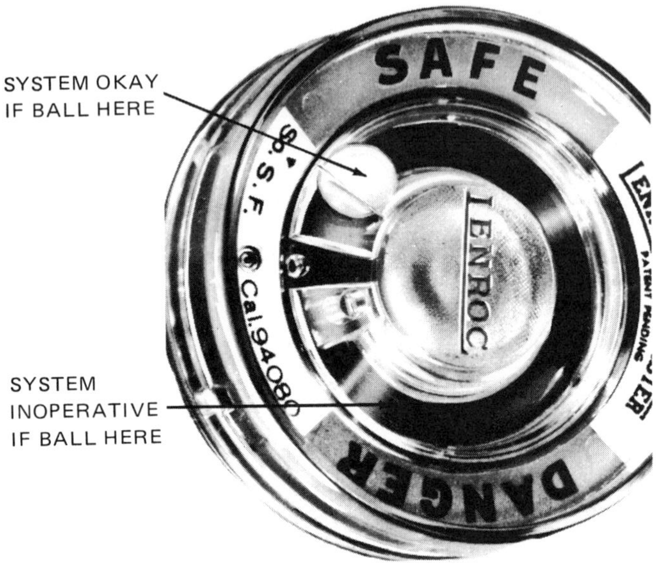

FIGURE 63
Testing a PCV system is made easy with an air flow tester.

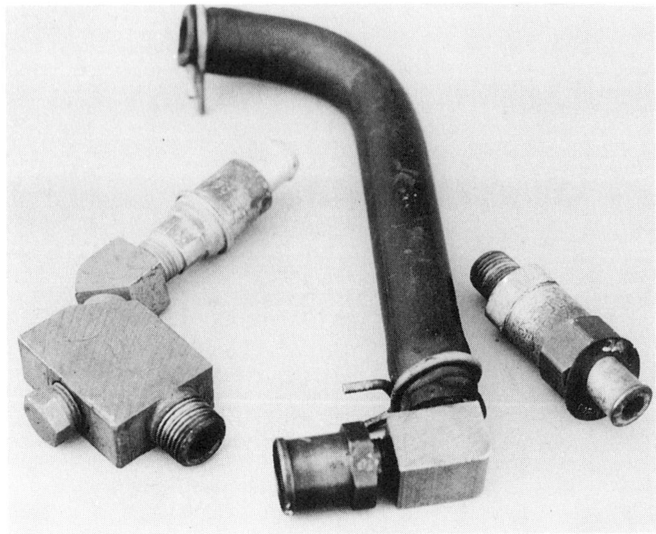

FIGURE 64
Engine roughness can be caused by a plugged PCV system. Hoses and fittings should be cleaned periodically to remove internal carbon deposits.

Pressure Control

As engine speed rises, the amount of exhaust gas entering the turbine housing also rises. This causes the speed of the turbine to rise. This, in turn, increases turbo pressure (boost pressure).

A typical control system operates as follows:

1. When boost pressure is developed, the turbo pressure switch turns on a green indicator light (Figure 68). This informs the driver that the turbo is providing boost.

2. When boost pressure reaches a certain value (the intercept point), the waste gate valve is opened by the actuator. This decreases the speed of the turbine and prevents overboost. Study Figure 69.

3. If the waste gate valve fails to open, the relief valve will open. Also, a high pressure switch turns on an amber overboost warning light.

Control of turbo pressure is illustrated in the pressure curve shown in Figure 70.

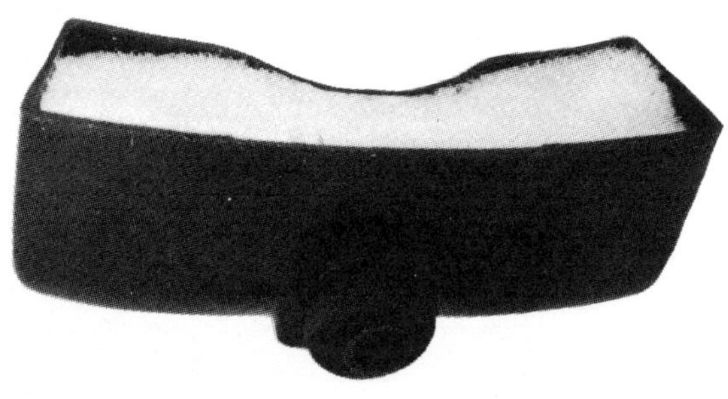

FIGURE 65
PCV system breather filter. (Courtesy of The Echlin Manufacturing Company)

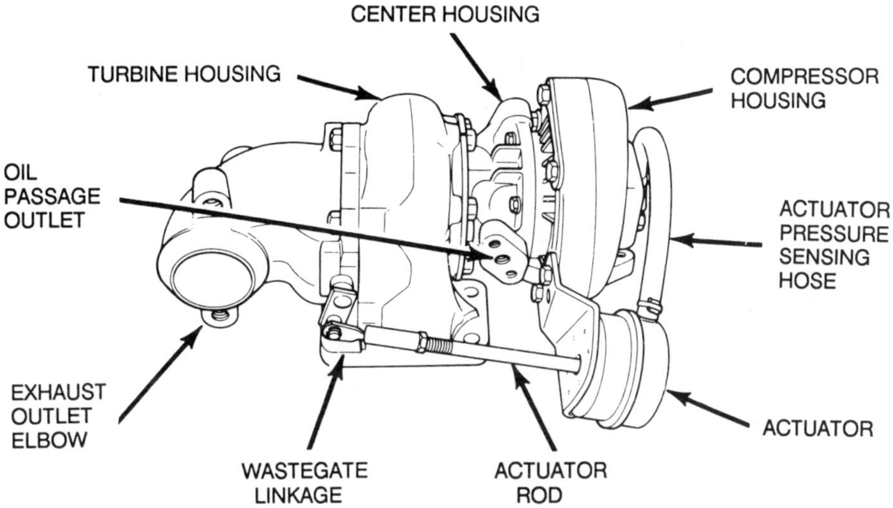

FIGURE 66
Typical turbocharger construction. (Courtesy of Ford Parts and Service Division)

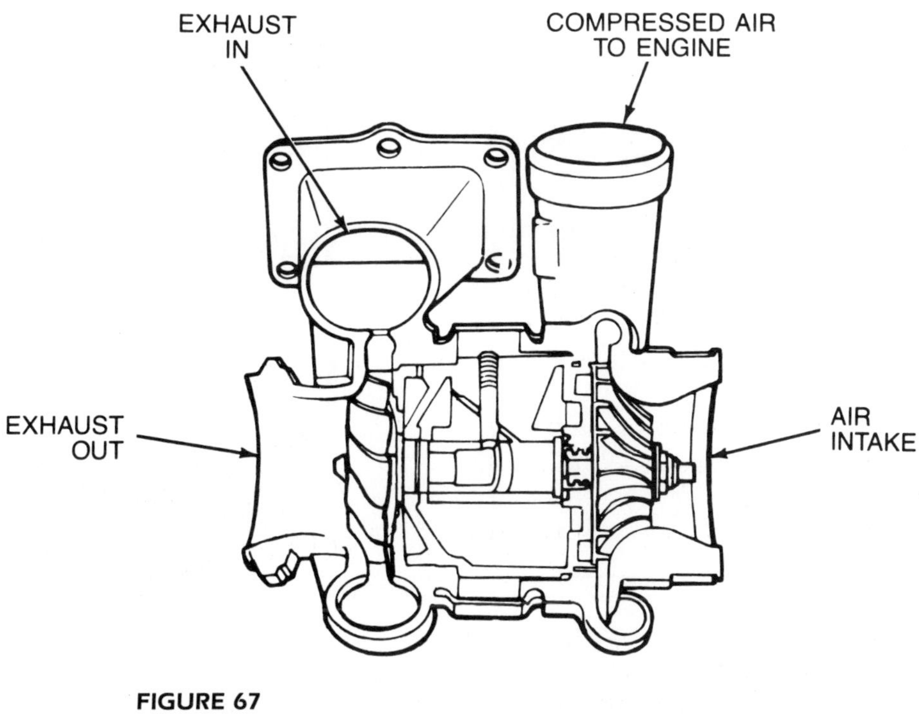

FIGURE 67
Turbocharger operation. (Courtesy of Chrysler Corporation)

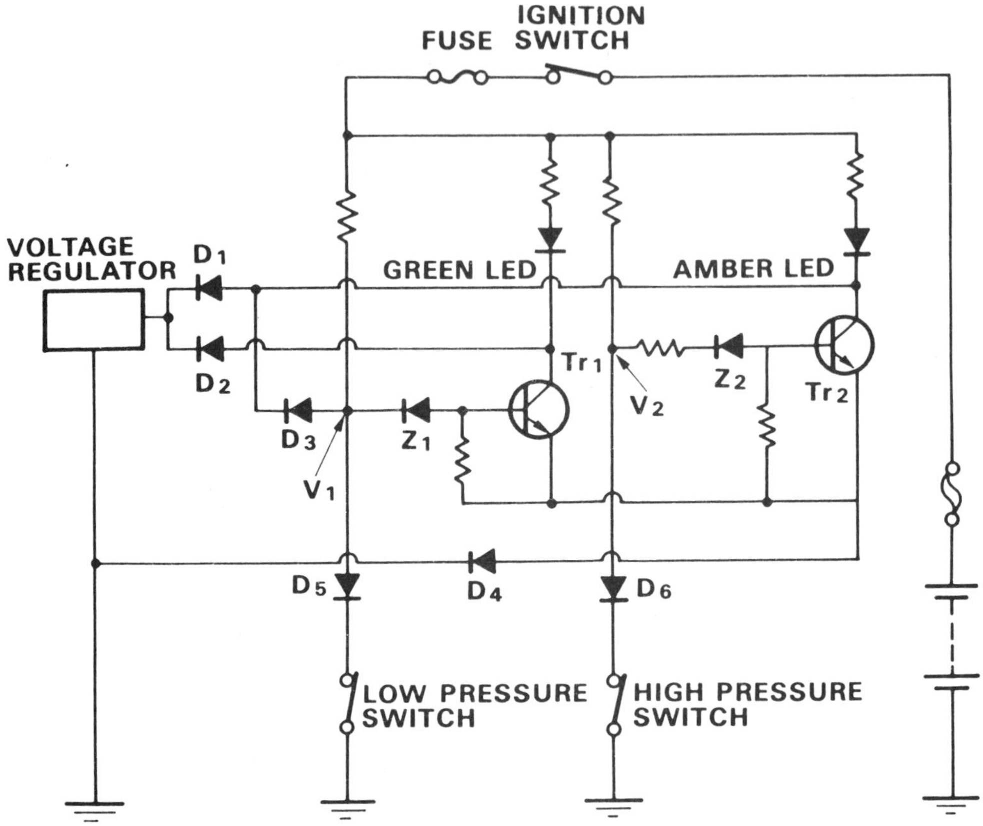

FIGURE 68
Turbocharger pressure switch circuitry. (Courtesy of Toyota Motor Sales, U.S.A., Inc.)

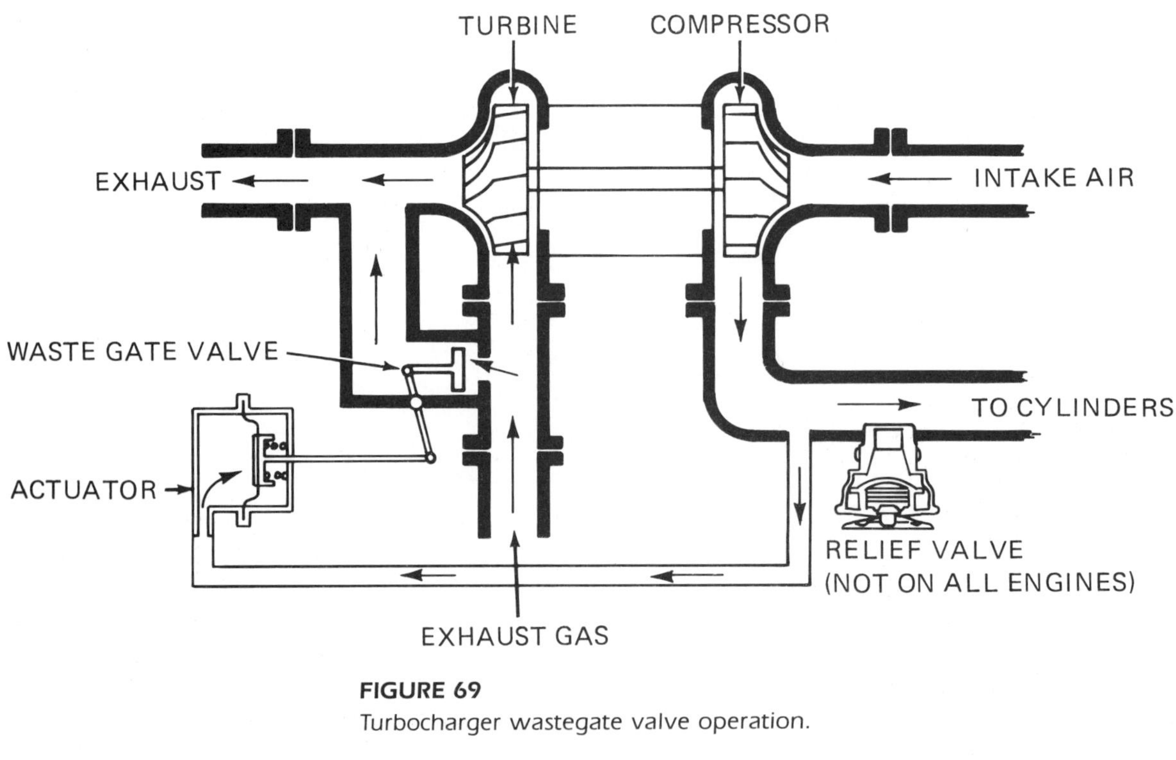

FIGURE 69
Turbocharger wastegate valve operation.

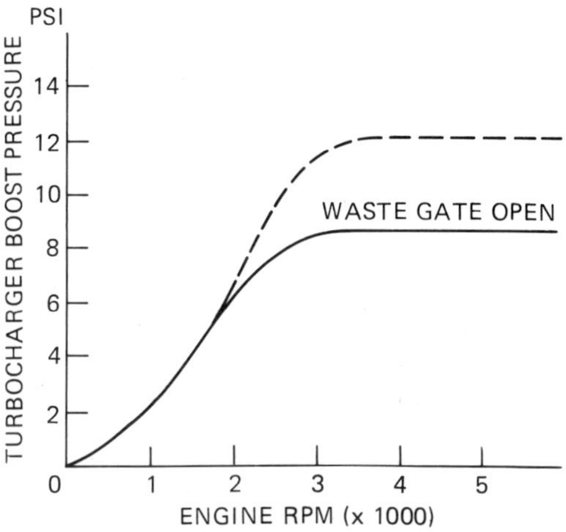

(a)

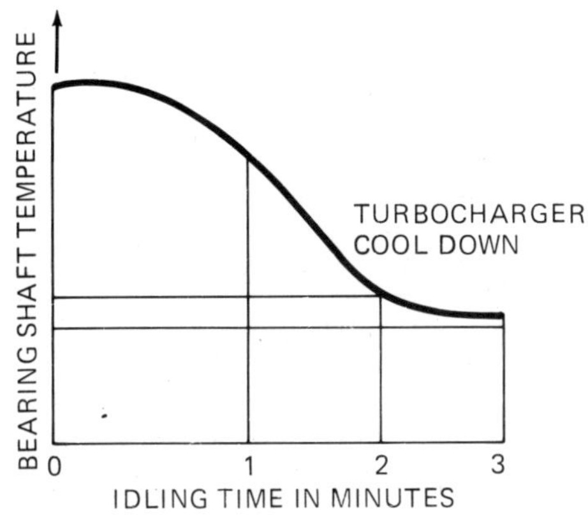

(b)

FIGURE 70
(a) Intake manifold pressure created by a turbocharger is limited to a maximum value by the wastegate. (b) Turbocharger cool down time. Critical shaft temperature is approximately 500°F.

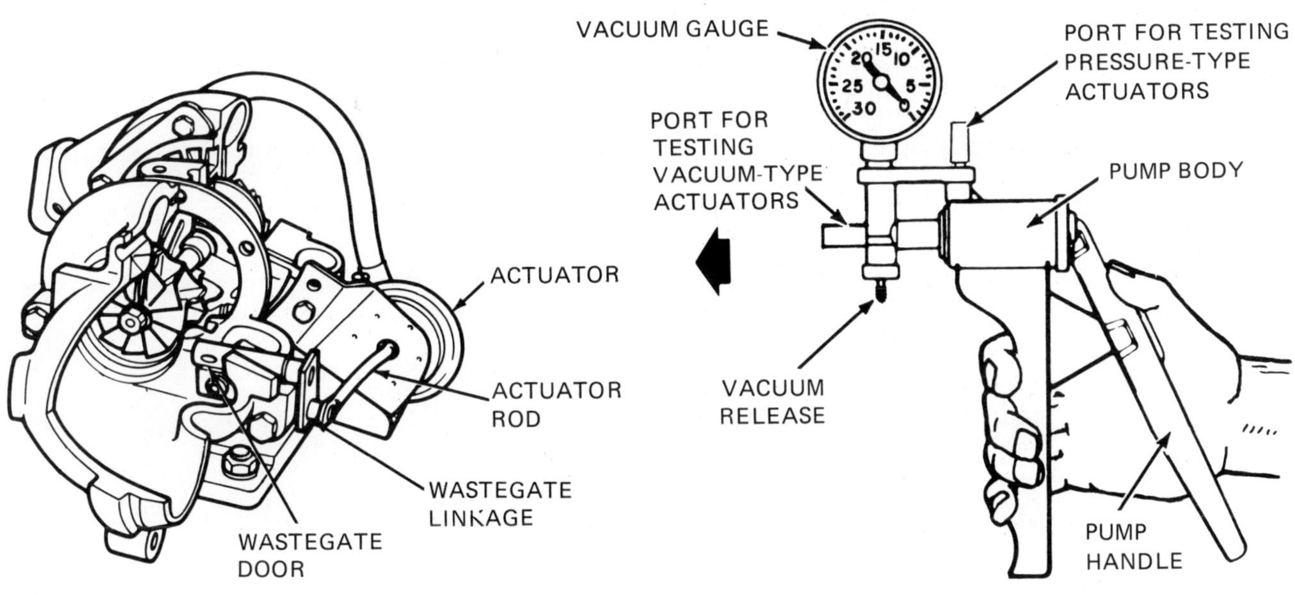

FIGURE 71
Testing the wastegate actuator. (Courtesy of Neward Enterprises Inc.)

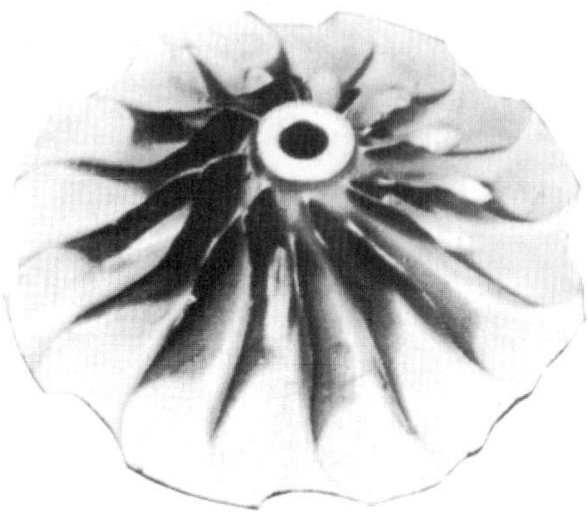

FIGURE 72
Compressor wheel damage caused by sand erosion.

FIGURE 73
Damaged (bent) compressor wheel blades.

Maintenance Precautions

A turbocharger is a precision-built part that is very durable if these simple precautions with regard to oil flow and air flow are observed.

1. When starting a turbocharged engine, let the oil pressure build up before revving the engine.

2. *Never* shut off the engine at high rpm. The turbocharger shaft will still be spinning while receiving no oil flow.

3. Always allow the engine to idle a moment before turning the ignition off (Figure 70[b]). This will help prevent oil passage choking and bearing failure.

4. Whenever installing a new or repaired turbo, prime all oil feed lines with fresh motor oil.

5. Engine oil oxidation and breakdown are accelerated under high temperature. *Therefore, the oil and filter in a turbocharged engine should be changed without fail within the stipulated time period.*

> **SHOP TIP**
> The operation of the waste gate actuator can be checked quickly by using a Mityvac® (Figure 71).

6. Periodically check the condition of all air flow hoses and clamps. Figure 72 shows an extreme example of sand erosion caused by a loose duct between the turbo inlet and the air cleaner.

FIGURE 74
Leaking shaft seals can cause excessive engine oil consumption.

System Inspection and Troubleshooting

1. Remove the air cleaner-to-turbo hose. Look inside the hose for excessive oil. This is often caused by an overfilled crankcase, or by a restricted PCV system.

NOTE: A few drops of oil in the hose is considered normal.

2. Inspect the compressor wheel for pitting or wheel damage (Figure 73). Rub marks between the compressor wheel and housing are usually caused by excessive bearing clearance. This can be verified by moving the shaft nut axially. Some movement (about .005 inch) is normal.

3. Check rotation of the compressor wheel. The wheel should turn freely with very little drag.

4. Remove the exhaust elbow and inspect the turbine wheel for visible damage. Deposits of burned oil can be caused by either a sludged center housing or by leaking seals (Figure 74).

5. Inspect the turbo oil lines for sludge buildup.

COMMENT: As a general rule, if there is visible turbo damage, the entire assembly (excluding exhaust elbow) should be replaced.

Refer to Chart 4 for additional turbocharger troubleshooting information.

CHART 4 TROUBLESHOOTING CHART FOR TURBOCHARGERS

Engine Lacks Power

1. Clogged air filter
2. Obstructed air outlet duct from compressor to intake manifold
3. Air leak in duct from compressor to intake manifold
4. Restriction in exhaust system
5. Leak in exhaust system (Figure 75)
6. Dirt caked on compressor wheel
7. Worn piston rings (excessive blowby)
8. Low boost pressure (waste gate valve leaking or incorrectly adjusted)
9. Damaged turbo (worn or bent blades)

Black Exhaust Smoke

1. Clogged air filter
2. Obstructed air intake duct to compressor
3. Air leaks in exhaust system

Excessive Engine Oil Consumption (and Blue Exhaust Smoke)

1. Obstructed oil drain line
2. Sludge or coked center housing
3. Turbo seal leakage
4. Excessive crankcase pressure

Higher Pitched Sound than Usual

1. Air leak between compressor and intake manifold
2. Air leak between engine exhaust and turbo

Rattling Noise

1. Loose exhaust pipe or outlet elbow
2. Damaged waste gate valve
3. Damaged waste gate actuator
4. Improper catalytic converter installation and torque sequence (Figure 76)

Grinding or Rubbing Sound

1. Shaft or bearing damage
2. Wheel rub against housing

Spark Knock or "Pinging"

1. Waste gate not opening properly (overboost)
2. Detonation sensor not working properly

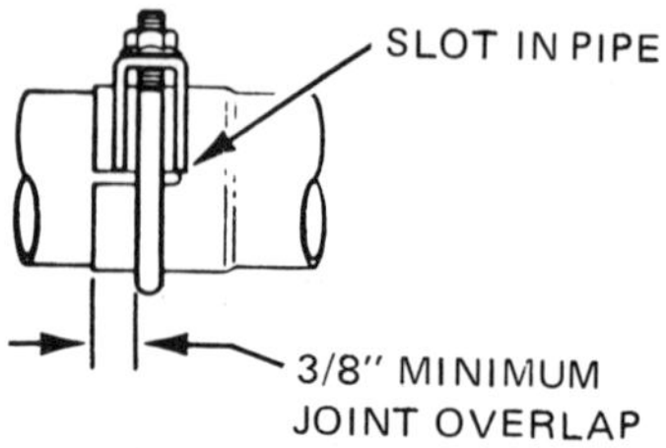

FIGURE 75
A turbocharged engine must have "tight" exhaust system clamps and joints. Otherwise, an engine power loss will result. (Courtesy of Buick Motor Division, General Motors Corporation)

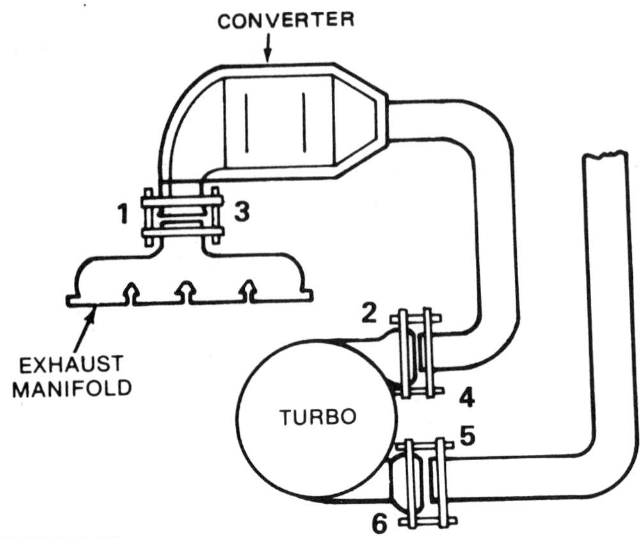

FIGURE 76
Exhaust system bolts in turbocharged engines should be torqued in correct sequence. (Courtesy of Ford Parts and Service Division)

TRUE CASE HISTORY

Vehicle	Dodge
Engine	Eight-cylinder, 318 CID
Mileage	112,000
Problem	1. Excessive oil leak from the rear main seal. 2. Seal was replaced twice without correcting the problem.
Cause	A plugged oil filler cap assembly (Figure 77). This allowed excessive crankcase pressure to develop, forcing oil through the seal.
Cure	Replaced the oil filler cap.

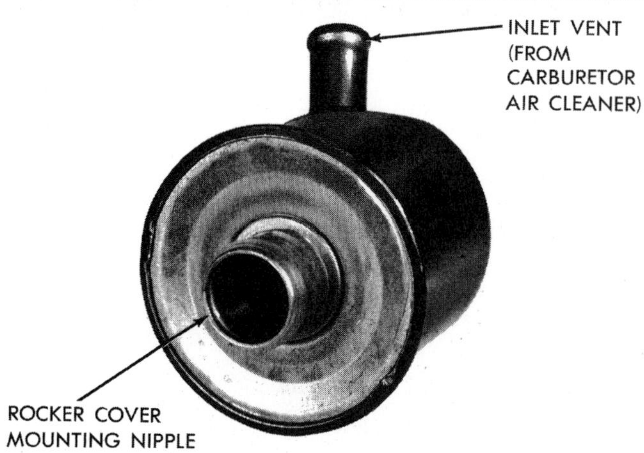

FIGURE 77
Oil filler cap.

TRUE CASE HISTORY

Vehicle	Chevrolet C-15 truck equipped with TBI
Engine	Eight-cylinder, 5.7 liter
Mileage	54,000
Problem	1. Rough idle and highway speed surge. 2. Shop replaced the PROM, ECM, IAC, and MAP sensor. None of this cured the problem.
Cause	Vacuum leak at the throttle body-to-intake manifold gasket.
Comment	The problem was diagnosed when propane was leaked near the base of the TBI assembly and the idle smoothed out.

TRUE CASE HISTORY

Vehicle	Chevrolet Camaro equipped with port fuel injection
Engine	Six-cylinder, 2.8 liter
Mileage	13,459
Problem	1. Rough idle. 2. Intermittent stall at stops (hot or cold). 3. Popping in exhaust during open-loop engine operation.
Cause	1. Heavy throttle body coking. 2. Dirty injectors. 3. TPS set at .64 volts.
Cure	1. Cleaned throttle body and injectors. 2. Set TPS to specification (.55 volts). 3. Set minimum idle speed to specification (650 rpm).

ASE STUDY QUESTIONS

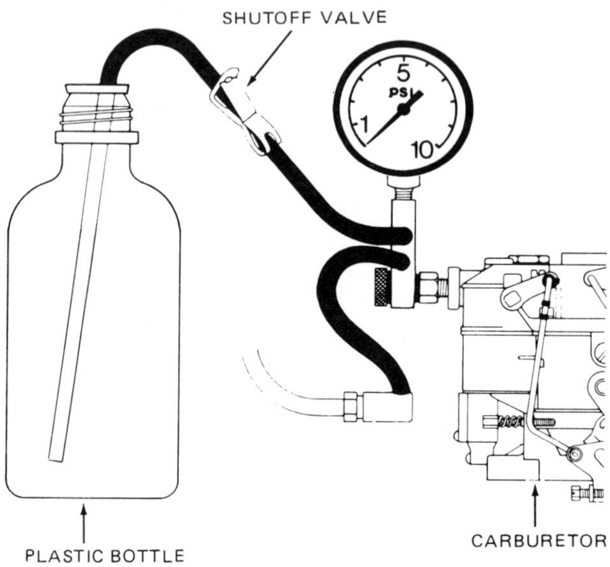

1. **Technician A** says that the setup shown above can be used to check fuel pump pressure.

   Technician B says that the setup shown above can be used to check fuel pump vacuum.

   Who is right?

   a. A only
   b. B only
   c. Both A and B
   d. Neither A nor B

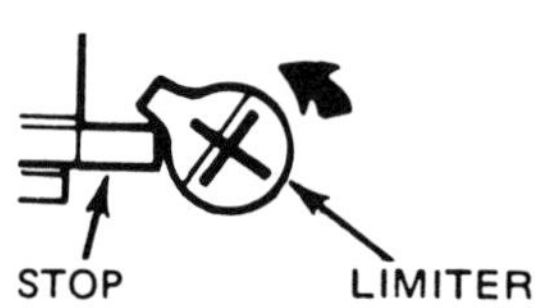

2. Refer to the above illustration.

   Technician A says that a full-rich position is shown.

   Technician B says that the limiter can be removed.

   Who is right?

   a. A only
   b. B only
   c. Both A and B
   d. Neither A nor B

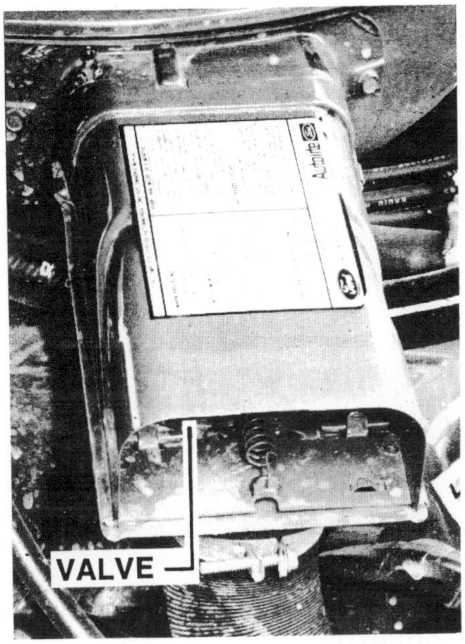

3. The flapper valve shown above is stuck open.

   Technician A says that this will allow exhaust gases to enter the intake manifold.

   Technician B says that this will prevent ignition advance during rapid acceleration.

   Who is right?

   a. A only
   b. B only
   c. Both A and B
   d. Neither A nor B

4. **Technician A** says that the manifold heat control valve (heat riser) should be lubricated periodically with 30-weight oil.

   Technician B says that the manifold heat control valve (heat riser) should be taken apart periodically and cleaned.

   Who is right?

   a. A only
   b. B only
   c. Both A and B
   d. Neither A nor B

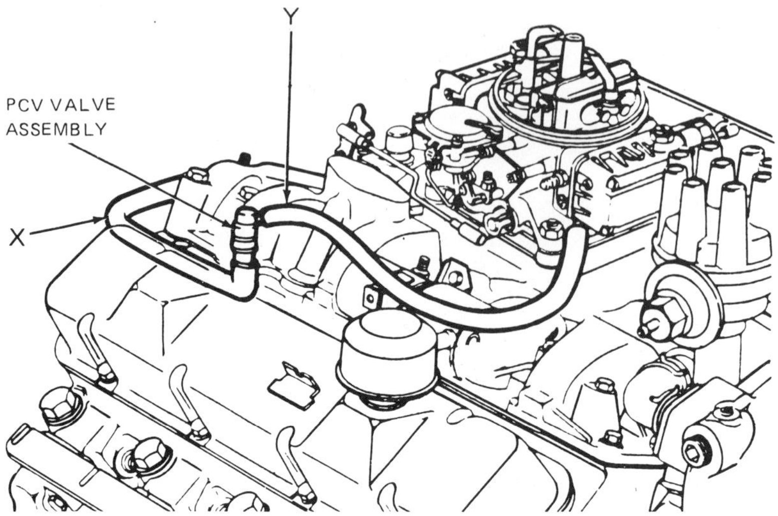

5. The engine shown above is running at idle speed. When hose X is pinched closed, there is a 50 rpm drop.

   Technician A says to replace hose X.

   Technician B says to replace hose Y.

   Who is right?

   a. A only
   b. B only
   c. Both A and B
   d. Neither A nor B

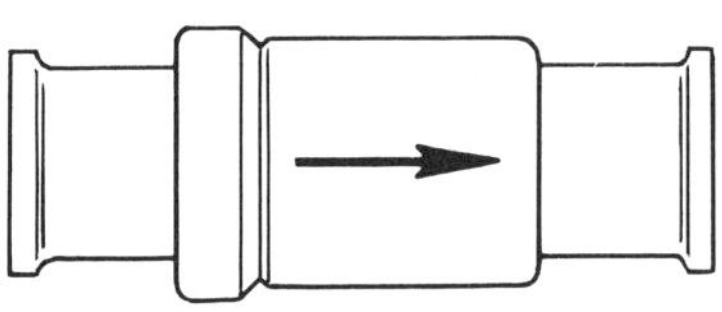

6. The parts inside the valve shown above have broken and fallen out.

   Technician A says this can cause poor or no engine idle.

   Technician B says that this can cause the oil pan seals to rupture and leak.

   Who is right?

   a. A only
   b. B only
   c. Both A and B
   d. Neither A nor B

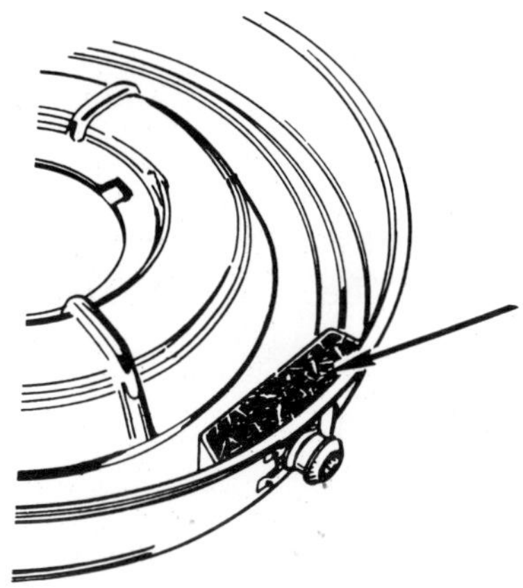

7. **Technician A** says that the filter shown above (see arrow) is part of the PCV system.

   Technician B says that excessive oil consumption can result if this filter becomes clogged.

   Who is right?

   a. A only
   b. B only
   c. Both A and B
   d. Neither A nor B

8. An engine equipped with a turbocharger has excessive oil consumption and blue exhaust smoke.

   Technician A says that this could be caused by a leaking compressor seal.

   Technician B says that this could be caused by a sludged center housing.

   Who is right?

   a. A only
   b. B only
   c. Both A and B
   d. Neither A nor B

SECTION REVIEW QUESTIONS

1. A turbocharged engine has excessive oil consumption and blue exhaust smoke. List four conditions that could be the cause.

2. An engine is slow to warm up. Name three possible exhaust system malfunctions that could be the reason.

3. What is an inertia switch?

4. What conditions can cause a mechanical fuel pump to produce high fuel pressure?

5. Name four conditions that can cause a mechanical fuel pump to make noise.

6. List six maintenance precautions regarding a turbocharger.

7. How can the operation of a waste gate actuator be checked?

8. How can a PCV valve be quickly checked?

9. Describe a "closed" PCV system.

10. What is the purpose of a heat riser valve?

11. When installing an exhaust manifold, what is the recommended procedure for tightening the bolts? Why?

12. What material can be used to provide a very effective seal when metal-to-metal exhaust manifolds or tubing-type headers are used?

13. Exhaust manifold mounting surfaces should be straight within what limits?

14. What is the basic purpose of the exhaust system?

15. What conditions can cause crankcase oil to collect in the air cleaner?

16. What are the five functions of an air cleaner assembly?

17. Explain the procedure for adjusting idle mixture using the lean-drop method.

18. Explain the procedure for adjusting idle mixture using the lean-best-idle method.

19. Explain the procedure for adjusting idle mixture using the propane enrichment method.

20. Intake manifold mounting surfaces should be straight within what limits?

21. Explain the procedure for setting idle speed on a car equipped with a throttle solenoid plunger.

22. Outline the procedure for confirming closed-loop engine operation.

23. How can a torn intake manifold gasket allow oil to be drawn into the combustion chamber(s)?

24. When replacing an in-line EFI filter, metal particles are observed in the fuel. What does this indicate?

25. What driveability problems can be caused by a malfunctioning EGR valve?

26. Where might the filters in a fuel system be located?

27. How can a "frozen" electric fuel pump sometimes be freed?

28. What periodic procedure can be performed to help lubricate an electric fuel pump?

29. What three tests can be performed on a mechanical fuel pump?

30. How might a hacksaw blade be useful when installing a mechanical fuel pump?

31. How is a mechanical fuel pump usually driven?

32. What is an EFE grid?

33. Describe a quick test for air induction leakage on an EFI system.

Section G

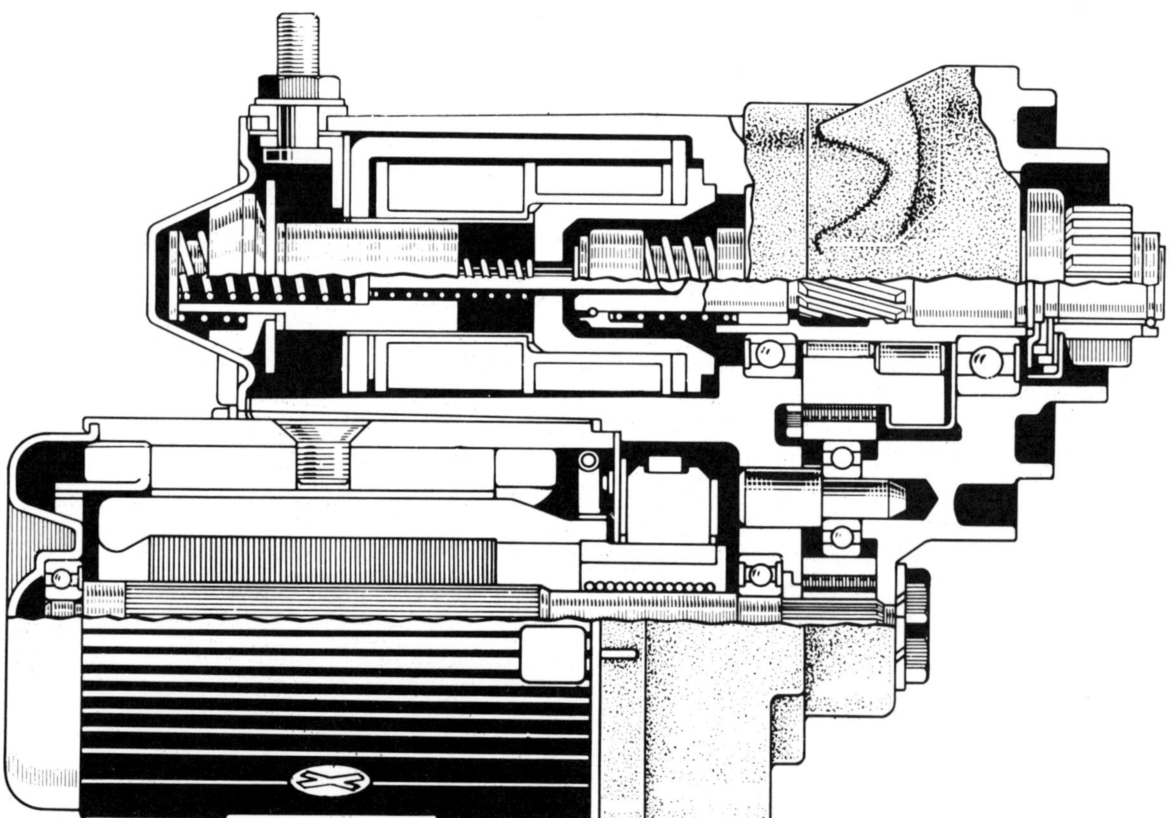

Battery and Starting System Diagnosis and Repair

After studying this section, and with hands-on practice on a "live" vehicle, you should be able to perform the following tasks:

- Inspect, test, clean, fill, or replace battery.
- Slow- and fast-charge battery.
- Jump start a vehicle with jumper cables and a booster battery or auxiliary power supply.
- Inspect, clean, repair, and replace battery cables and clamps.
- Inspect, test, and/or replace starter relays, solenoids, and circuits.
- Test, remove, and/or replace starter.

GENERAL INFORMATION

The battery is the heart of the vehicle's electrical system (Figure 1). It must be capable of delivering the necessary electricity when called for, and of accepting electricity when it is in a low state of charge (Figure 2).

If a battery is not operating up to specifications, it will not deliver the proper power to start the engine or supply other electrical circuits without the charging system operating.

Refer to Chart 1 for a listing of some battery troubles and the possible causes.

Batteries used in cars and trucks are either conventional or maintenance-free (Figures 3 and 4).

The conventional type battery must be checked regularly to ensure that the correct electrolyte level is maintained. This is done by adding water (Figure 5).

The maintenance-free battery is a sealed assembly that does not normally require the addition of make-up water to the electrolyte during its normal life. This is because the chemical composition inside the battery reduces the production of gas to an extremely small amount at normal charging voltages.

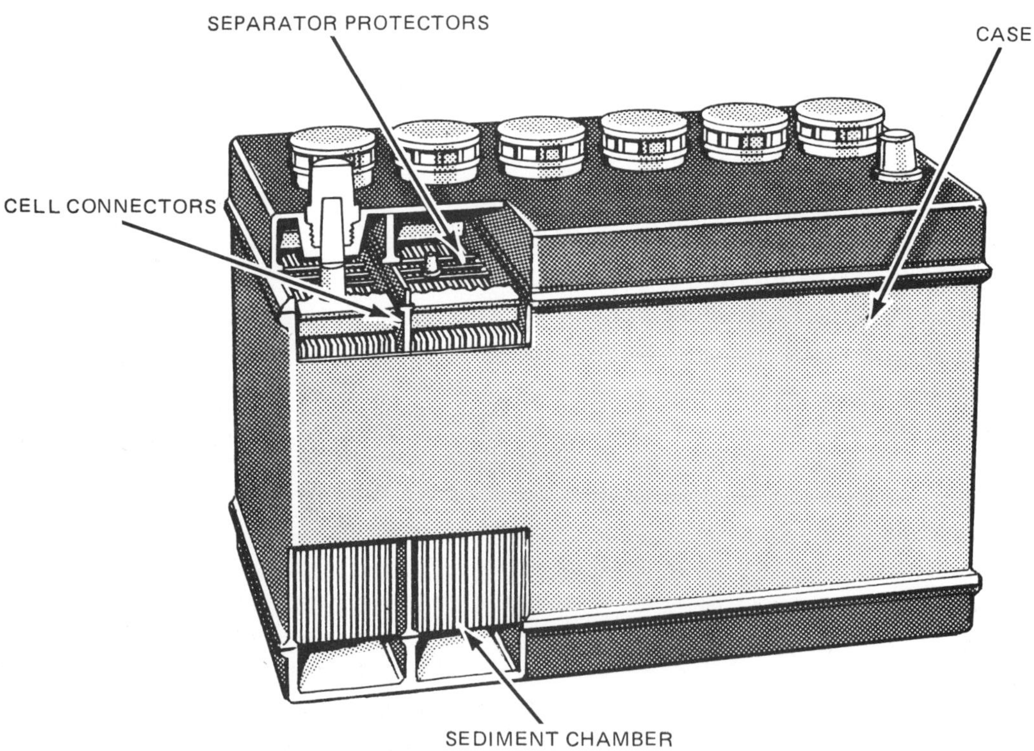

FIGURE 1
Battery construction details.

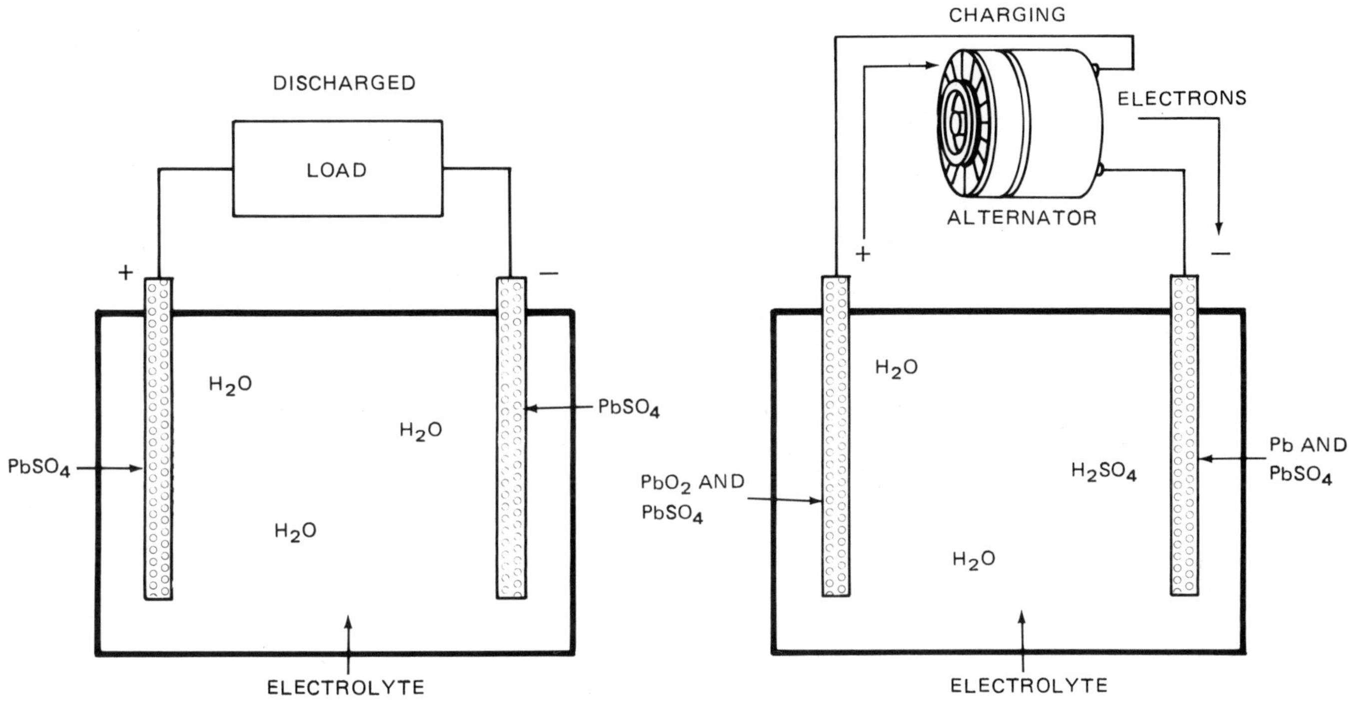

FIGURE 2
When the battery is discharged, both plates are similar and the electrolyte is very diluted. As a result, little or no power is produced. Recharging turns the battery back into dissimilar metals and returns the hydrogen and sulfate ions to the electrolyte.

CHART 1 TROUBLESHOOTING CHART FOR BATTERIES

Battery Discharge

1. Loose or broken alternator drive belt
2. Defective alternator
3. Defective voltage regulator
4. Low regulating voltage
5. Low electrolyte level
6. Corroded battery posts or connectors
7. Electrical leakage
8. Defective battery
9. Excessive electrical load
10. High circuit resistance in the charging or ground return, or in the battery connections
11. Faulty voltage regulator lead connections
12. Parasitic load on a computer vehicle can discharge battery if vehicle sits for several weeks

Battery Overcharge

1. High regulating voltage
2. Defective alternator
3. Defective voltage regulator
4. Poor or broken regulator ground connection

Battery Will Not Accept Charge

1. Sulphated battery
2. Open circuit between cells or internal defect

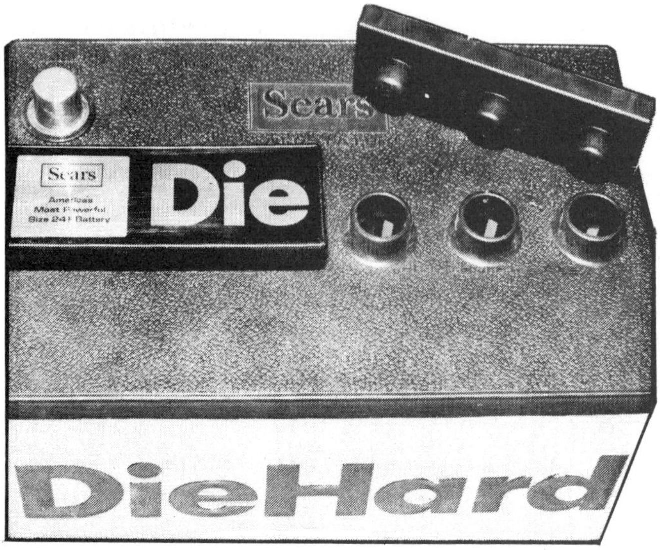

FIGURE 3
Topside view of a conventional battery.

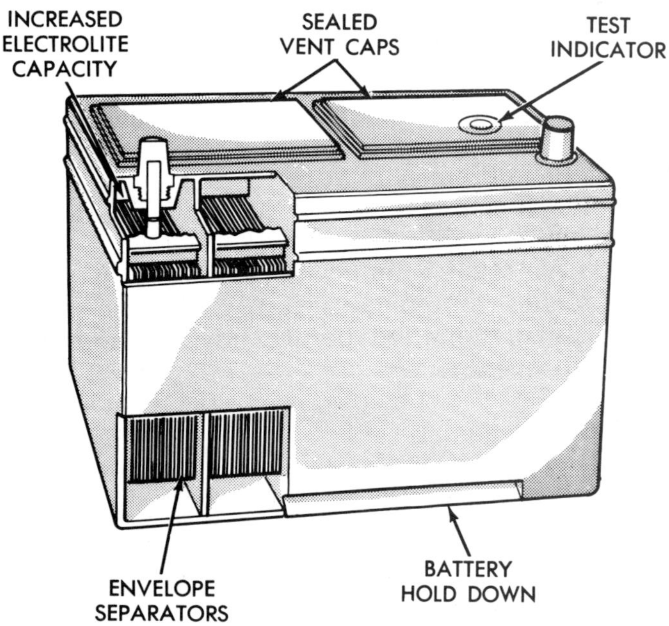

FIGURE 4
Topside view of a maintenance-free battery.

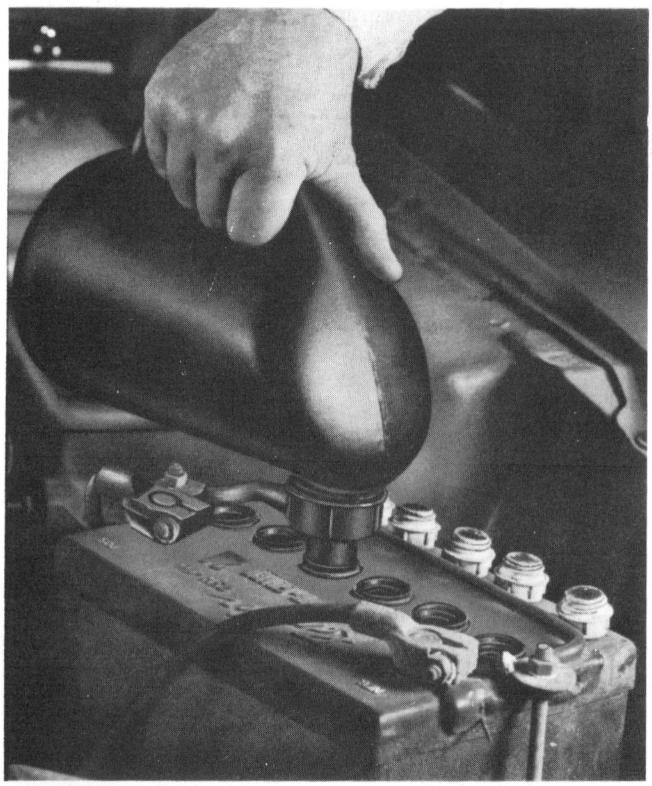

FIGURE 5
This battery filler automatically fills the battery to the correct level without overflowing. (Courtesy of Snap-On Tools Corporation)

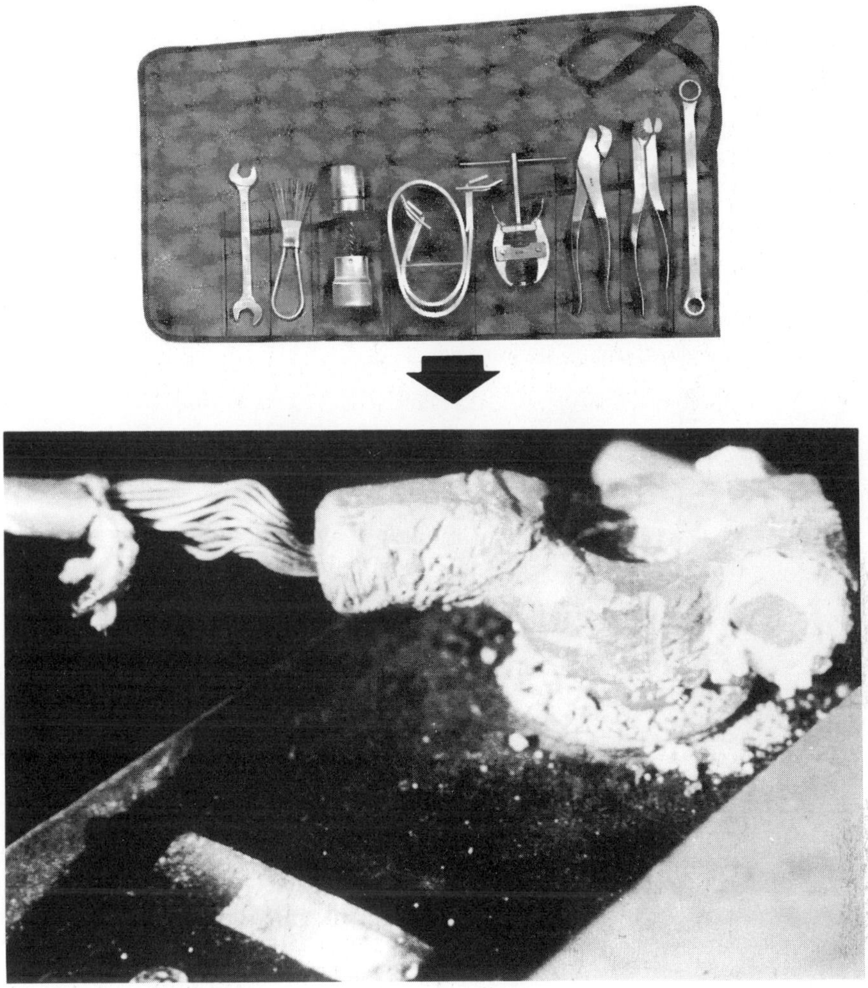

FIGURE 6
Battery tool kits are available for battery service needs.

BATTERY SERVICE

Special tools and cleaning equipment help make this job easy (Figures 6 and 7). Here are the steps to follow when cleaning a battery:

1. Be sure the ignition switch and all accessories are off.

2. Disconnect the ground cable first when removing a battery from a vehicle. When installing, connect the ground cable last. Use a battery cable puller if the cable clamp sticks to the battery post (Figure 8).

WARNING: Batteries contain sulfuric acid. Wear eye protection when servicing a battery to protect against possible splashing of the acid solution. In case of acid contact with skin or eyes, flush immediately with water for a minimum of fifteen minutes. Call a physician immediately.

3. Remove the battery hold-down clamp and the battery from the vehicle.

4. Clean the battery top with a solution of water and baking soda (Figure 9). Scrub with a stiff-bristle brush, rinse with water, and dry with a clean paper towel.

FIGURE 7
Battery spray cleaner.

5. Visually inspect the battery. There should be no cracks in the case or cover. If there are, replace the battery. The battery sides should be flat. Internal damage usually causes bulged sides. Battery posts should be secure and in good condition.

6. Inspect the battery hold-down and tray for damage caused by loss of acid from the battery (Figure 10). If acid damage is present, clean area with water and baking soda.

7. Clean battery posts and inside surfaces of cable clamps (Figure 11[a] and [b]).

> **SHOP TIP**
> Spray battery posts and connectors with WD-40. Respray every month or so for trouble-free service.

8. Install battery in vehicle.

9. Coat the cable clamps with acid-proof paint (Figure 12).

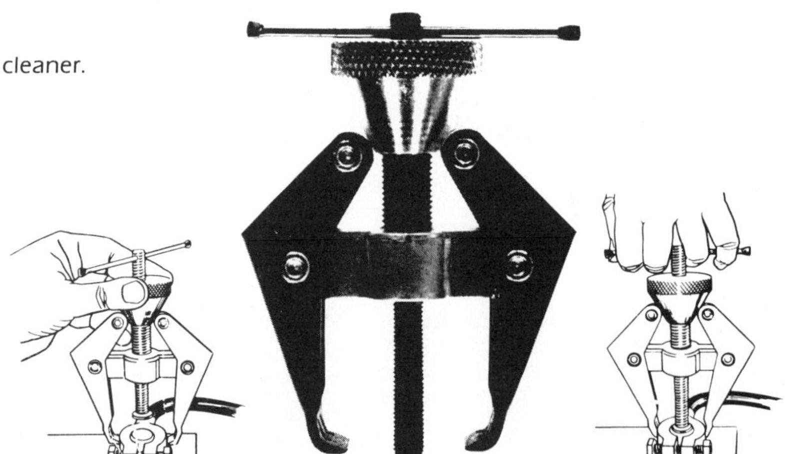

FIGURE 8
This tool makes it easy to remove cable clamps without breaking or damaging battery posts. (Courtesy of Lisle Corporation)

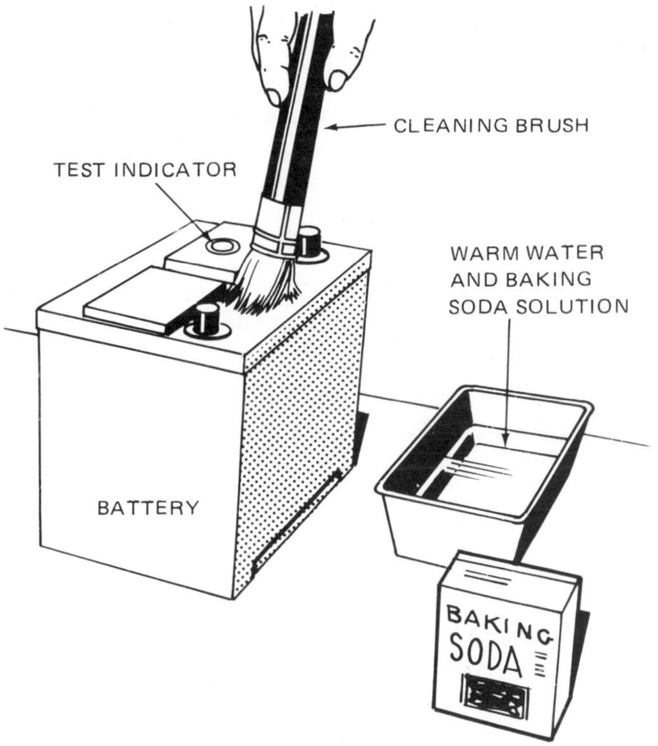

FIGURE 9
Cleaning battery top. (Courtesy of Chrysler Corporation)

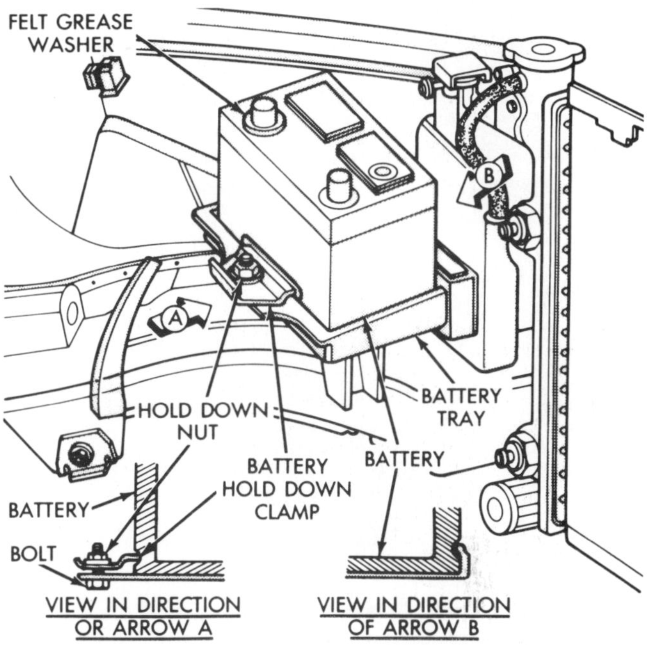

FIGURE 10
Battery hold-down and tray assembly. (Courtesy of Chrysler Corporation)

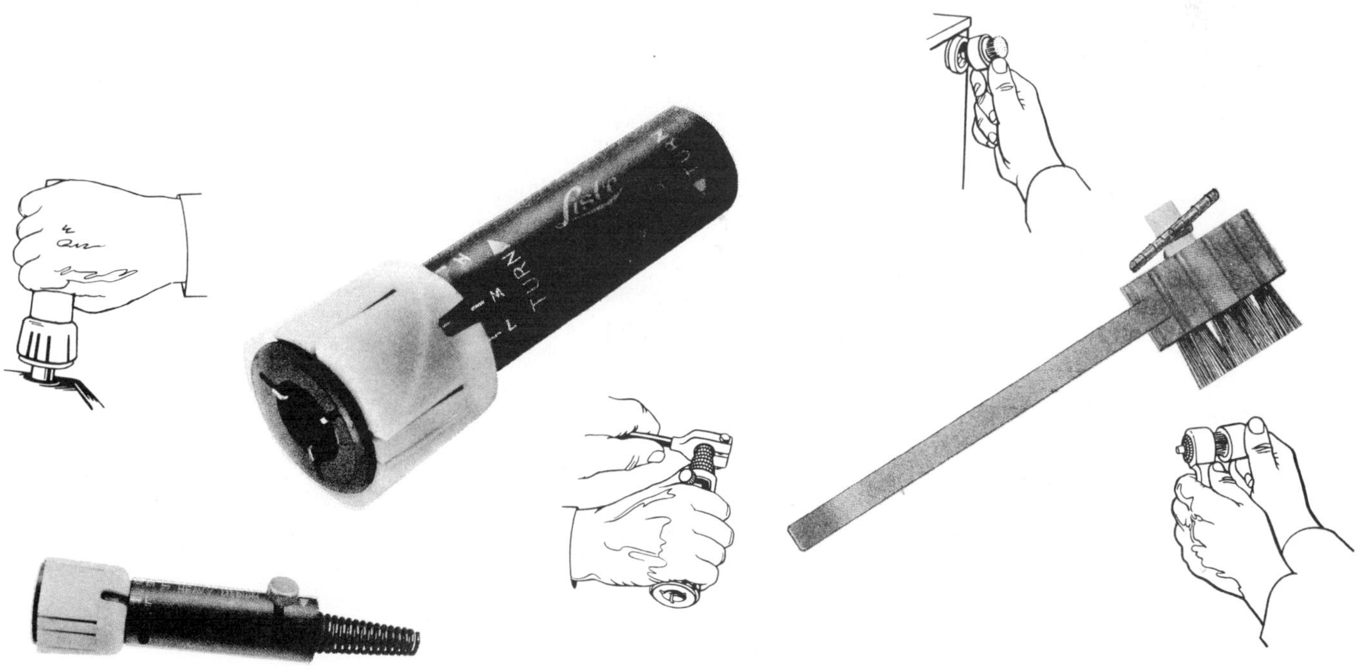

(a) (b)

FIGURE 11
(a) An adjustable battery terminal and post cleaner. (Courtesy of Lisle Corporation)
(b) Side terminal battery cleaner. (Courtesy of Lisle Corporation)

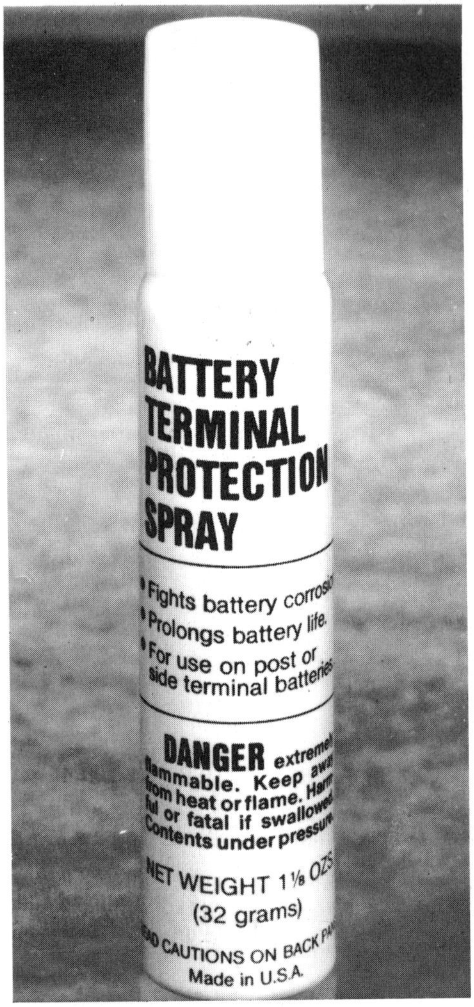

FIGURE 12
Battery terminal protection spray paint.

NOTE: Do not apply petroleum jelly or grease. This will only attract dirt and other debris. Connect cable clamps to battery posts, being sure clamps are fully seated on posts (Figure 13).

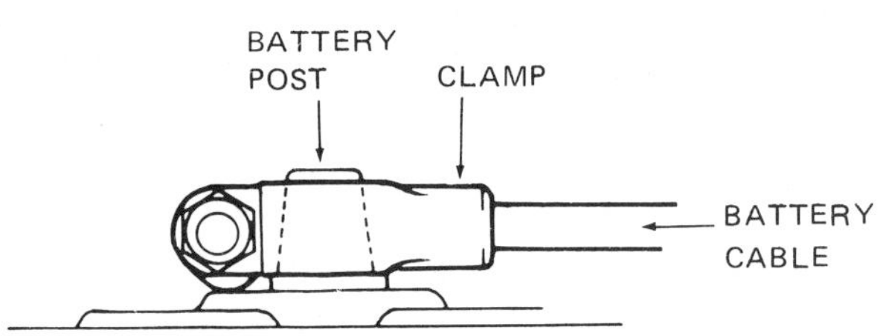

FIGURE 13
Battery cable clamps must seat fully.

> **SHOP TIP**
> A dirty battery top can cause a battery to self-discharge across the terminals. Here is a quick test. Connect the negative lead from a voltmeter to the battery negative terminal. Move the positive voltmeter lead around on the surface of the battery top. The voltmeter should read zero. If it does not, there is leakage and the battery must be cleaned.

Specific Gravity Check

The specific gravity of the electrolyte in a battery is an indication of the battery's state of charge. A battery hydrometer is used to test the electrolyte (Figure 14). Chart 2 shows various specific gravity readings and their relationship to a cell's state of charge.

When using a hydrometer to check specific gravity, be aware of the following:

1. Electrolyte temperature must be considered. In order to solve this problem, many hydrometers have a built-in thermometer and correction scale (Figure 15).

2. Never take a specific gravity reading of a cell immediately after adding water or after prolonged cranking. Let the battery stand idle for several hours, or charge the battery for 20 minutes. This will allow the water and acid to mix thoroughly.

3. Be sure the float is free and not hanging up on the side of the hydrometer when taking a reading.

4. If there are more than 50 gravity points (.050) between cells, the battery should be replaced.

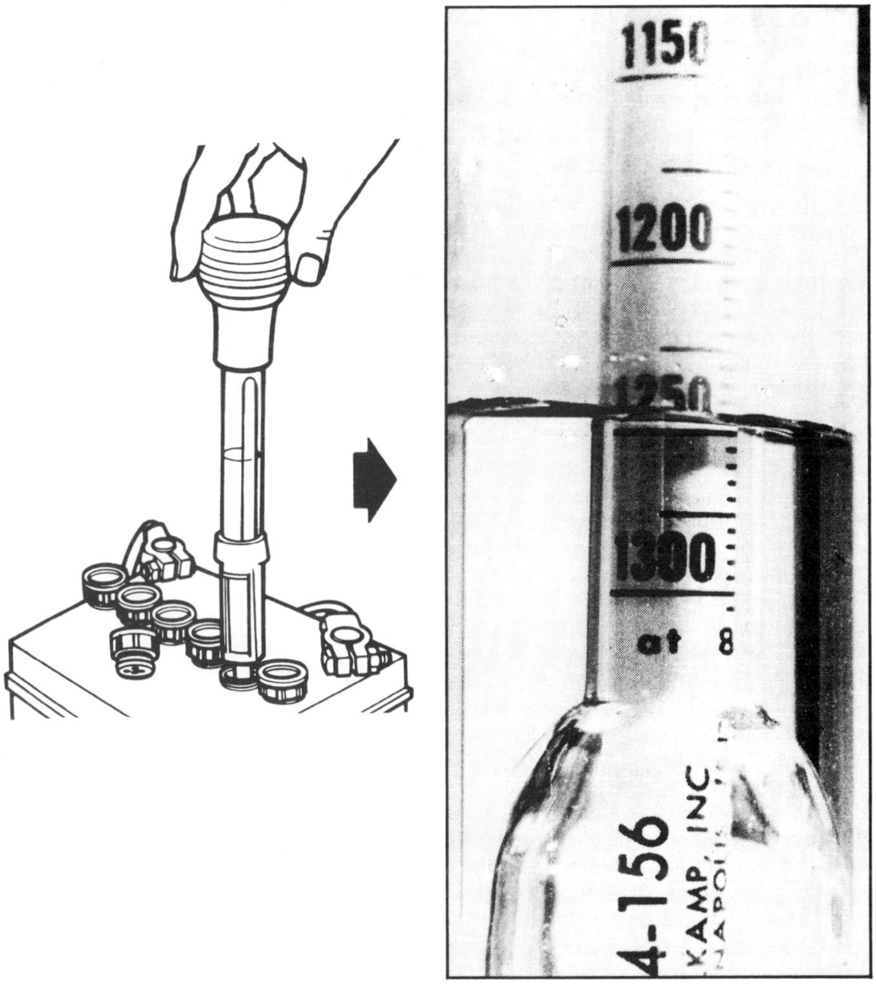

FIGURE 14
Battery hydrometer.

CHART 2 SPECIFIC GRAVITY READINGS AT 80° F

1.260–1.280 Fully charged
1.235–1.260 ¾ charged
1.205–1.235 ½ charged
1.170–1.205 ¼ charged
1.140–1.170 Barely charged
1.110–1.140 Dead

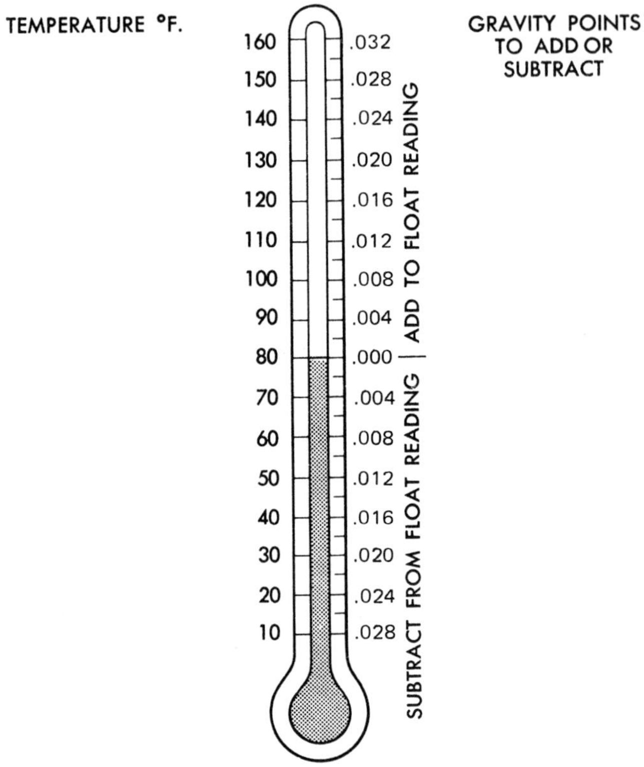

FIGURE 15
Battery hydrometer temperature correction scale.

Built-In Test Indicator (Hydrometer)

Some companies that manufacture maintenance-free batteries employ a special built-in hydrometer (Figure 16). This device allows the technician to determine the battery state of charge visually. One example of how this type of hydrometer operates is shown in Figure 17.

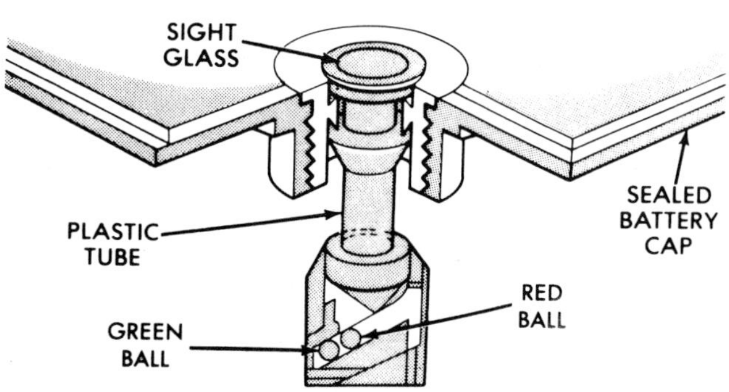

FIGURE 17
Built-in hydrometer construction details. (Courtesy of Chrysler Corporation)

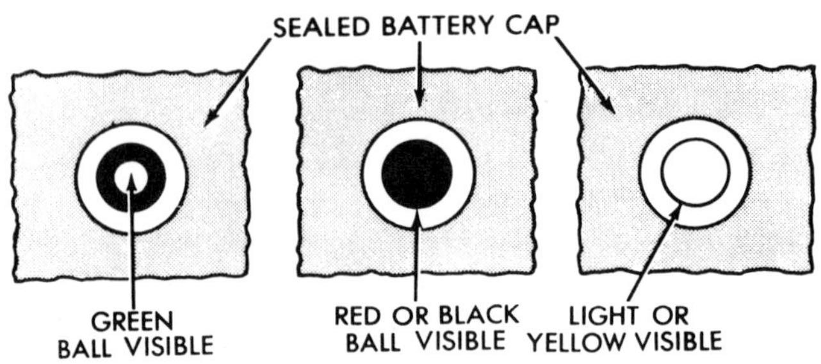

FIGURE 16
The battery may contain a visual hydrometer. When an adequate charge level exists, the green ball shows. When charging is required, red or black is visible. Yellow shows when battery replacement is required. (Courtesy of Chrysler Corporation)

NOTE: The test indicator should not be used to determine if the battery is good or bad. This can be determined by other testing procedures. Remember that this type of hydrometer indicates the condition of only one cell.

BATTERY TESTING

Refer to Chart 3 for some additional battery test procedures.

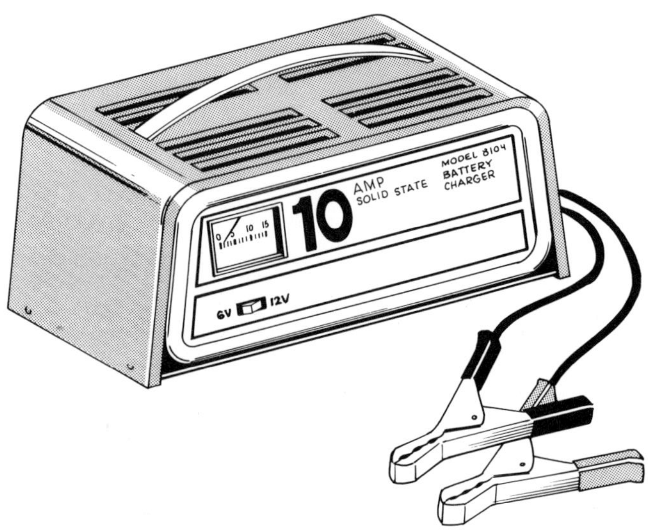

FIGURE 18
Battery "trickle" charger.

CHART 3

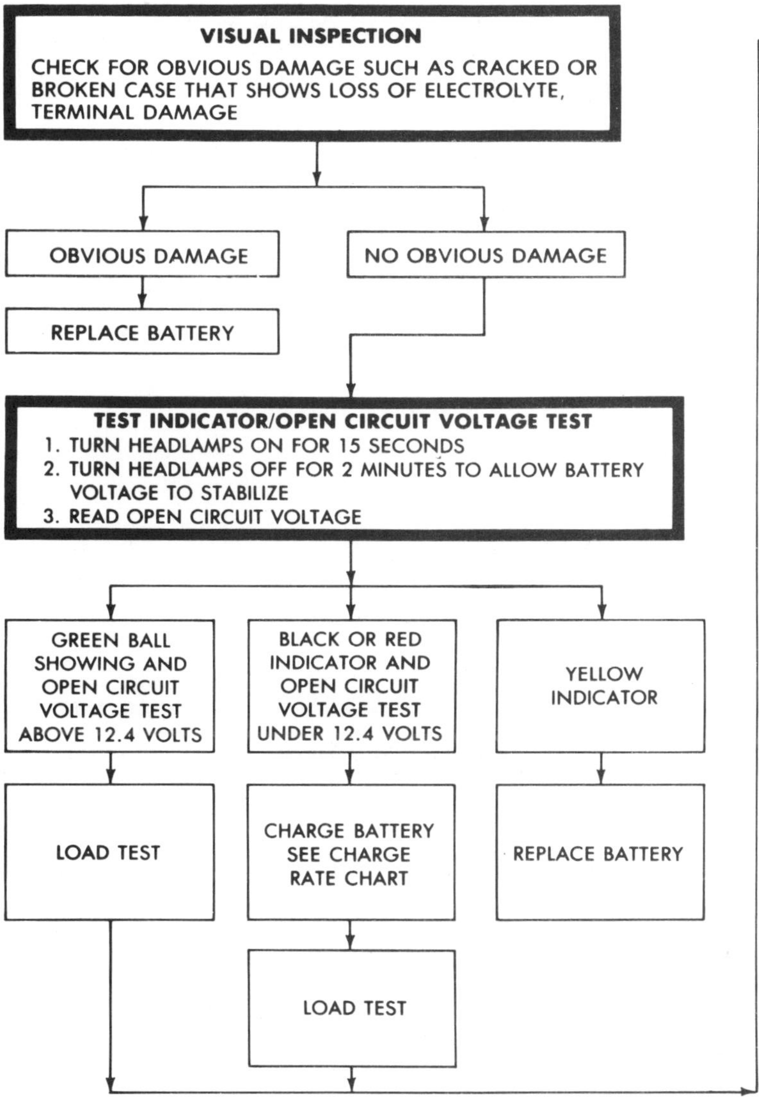

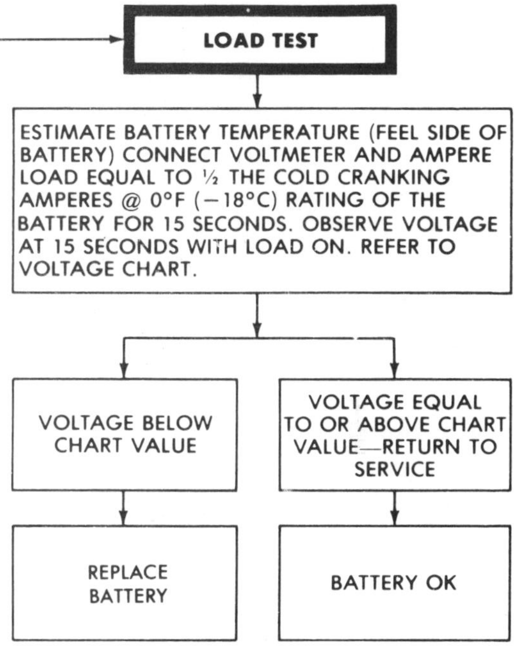

CHARGE RATE CHART

Battery	Slow Charging	Fast Charging
335 AMP	10 HRS. @ 4 AMPS 5.0 HRS. @ 10 AMPS	2.0 HRS. @ 20 AMPS 1.2 HRS. @ 30 AMPS
400 AMP	10 HRS. @ 5 AMPS 5.0 HRS. @ 10 AMPS	2.5 HRS. @ 20 AMPS 1.5 HRS. @ 30 AMPS
430 AMP	12 HRS. @ 5 AMPS 6.0 HRS. @ 10 AMPS	2.7 HRS. @ 20 AMPS 2.0 HRS. @ 30 AMPS
500 AMP	12 HRS. @ 5 AMPS 6.0 HRS. @ 10 AMPS	3 HRS. @ 20 AMPS 2 HRS. @ 30 AMPS

POSSIBLE CAUSE OF BATTERY FAILURE IN VEHICLE

1. CORRODED BATTERY POSTS OR CONNECTORS—ALWAYS CLEAN THE BATTERY POSTS OR SIDE TERMINALS AND THE BATTERY CABLE CONNECTIONS WHEN CHARGING OR REPLACING A BATTERY IN A VEHICLE
2. ELECTRICAL LEAKAGE—CHECK FOR ELECTRICAL LEAKAGE WITH ALL VEHICLE ACCESSORIES TURNED OFF BY:
   (a.) DISCONNECTING THE BATTERY GROUND CABLE AND CONNECTING A SUITABLE VOLTMETER BETWEEN DISCONNECTED BATTERY POST AND GROUND (A READING OVER 6V INDICATES AN EXCESSIVE LEAK).
   OR (b.) DISCONNECTING THE BATTERY CABLE THAT GOES TO THE STARTER AND CONNECTING AN OHMMETER BETWEEN THE CABLE AND GROUND. (A RESISTANCE OF UNDER 100 OHMS WILL DRAIN THE BATTERY)
3. VEHICLE CHARGING SYSTEM—THE VEHICLE'S CHARGING SYSTEM SHOULD ALWAYS BE CHECKED AFTER CHARGING OR REPLACING A BATTERY.

VOLTAGE CHART

ESTIMATED ELECTROLYTE TEMPERATURE	MINIMUM REQUIRED VOLTAGE UNDER 15 SEC. LOAD
70°F (21°C) & ABOVE	9.6
60°F (16°C)	9.5
50°F (10°C)	9.4
40°F (4°C)	9.3
30°F (−1°C)	9.1
20°F (−7°C)	8.9
10°F (−12°C)	8.7
0°F (−18°C)	8.5

SLOW CHARGING

If adequate time is available, a discharged battery should always be recharged by slow charging (Figure 18). The following recommendations apply when slow charging a battery:

1. If the battery is to remain in the vehicle, disconnect the negative cable. This will prevent possible damage to the electrical system during charging.

2. Slow charge rates depend upon the type and capacity of the battery. Battery manufacturers publish charts showing the suggested rate of charge. However, as a general rule, automotive batteries usually have a slow charging rate of between 4 to 5 amps.

3. The average time necessary to charge a battery by the slow charge method at normal rates is from 5 to 30 hours.

4. Watch the temperature of the electrolyte during charging. If the cell temperature reaches 125° F, or if rapid gassing occurs, lower the charging rate.

5. The battery is fully charged when there is no further rise in specific gravity over a one-hour period, or when the charging rate stabilizes.

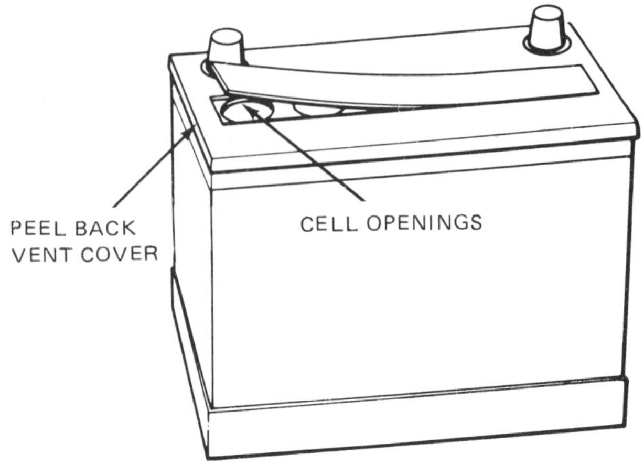

FIGURE 19
The cells of some maintenance-free batteries are covered with a plastic strip that can be peeled back.

WARNING: When a battery is being charged, an explosive gas (hydrogen) is produced. Do not create a spark or smoke near a battery on charge, or near a battery recently charged. Do not disconnect the charger clamps at the battery terminals during charge. A spark will occur when the circuit is broken. Keep the battery away from open flames. Always wear eye protection when working with batteries.

Sulphated Battery

Apply a 40-amp charge to the battery for three minutes. At the end of this time (with the charger still on), read battery terminal voltage. If it is over 15 volts, the battery is excessively sulphated (Figure 20).

Given enough time, an excessively sulphated battery can usually be reactivated by slow charging. The *rate of charge should be no more than one-half the normal slow charge rate, and the charging time should be from 60 to 100 hours.* This long charging cycle is necessary to reconvert the crystalline lead sulphate inside the battery into active material.

SHOP TIP

Specific gravity readings can be taken on most of the maintenance-free batteries. Use a small screwdriver to carefully peel off the vent cover(s). In some cases, it is necessary to first cut along the edge of the vent cover seal. See Figure 19.

FIGURE 20
Badly sulphated battery plates. (Courtesy of Delco-Remy Division, General Motors Corporation)

NOTE: When putting a battery that has been in storage for over 60 days back into service, it is recommended that it be slow charged as if it were sulphated.

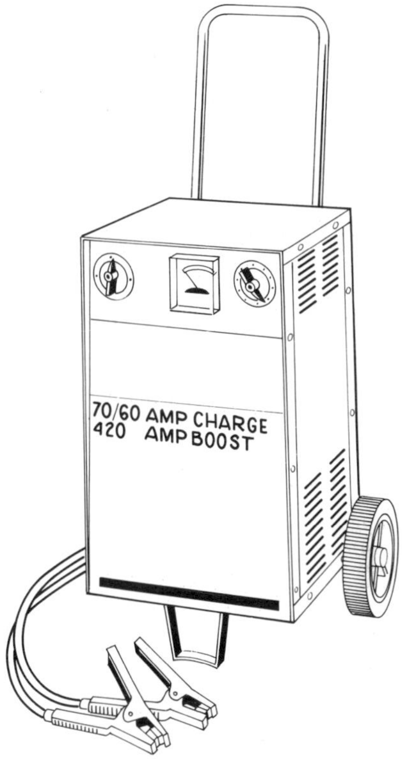

FIGURE 21
Battery fast charger.

FAST CHARGING

If adequate time for a slow charge is not available, a high-rate (fast) charge is permissible. Figure 21 shows a typical fast charger. Follow these recommendations when fast charging:

1. Do not fast charge any battery with a specific gravity of 1.225 or higher. This can cause severe battery damage due to overcharge.

2. Battery electrolyte temperature must never exceed 125° F. If this temperature is reached, or if rapid gassing is noticed, reduce the charging rate.

CAUTION: Overcharging of a maintenance-free battery can cause loss of water from the electrolyte.

3. If the battery does not show a significant change in specific gravity after one hour of fast charge, use the slow charge method.

4. A fast charger should be used only long enough to charge the battery enough to start the engine.

COMMENT: Many original equipment (OE) maintenance-free batteries can be charged only by the slow charge method.

JUMP STARTING

For safe and efficient jump starting, follow these guidelines:

HOW TO USE BOOSTER CABLES

1. Clamp the red end of the cable to the UNGROUNDED terminal—usually positive (+)—of the booster battery.

2. Connect the other red end of the cable to the weak battery terminal with the same polarity—positive (+) to positive (+) on most cars.

3. Attach the black end of the other jumper cable to the remaining grounded terminal—usually negative (-)—of the booster battery.

4. And then attach the other end to the engine block of the car being jumped—this should provide a good ground . . . . **DO NOT ATTACH THE BOOSTER CABLE TO THE GROUND TERMINAL OF THE WEAK BATTERY**. Sparks and an explosion of the battery gas could result.

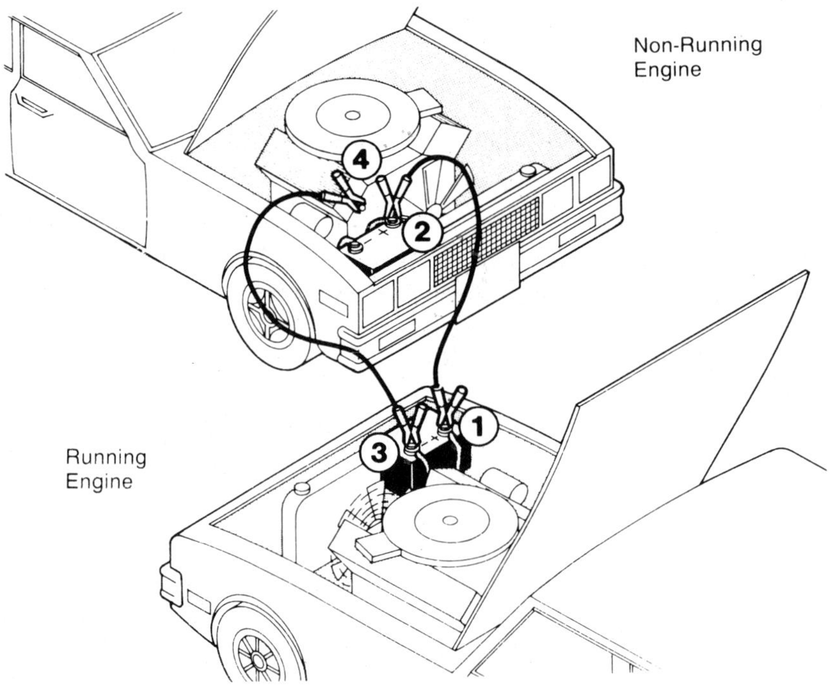

FIGURE 22
Proper connections for jump starting.

1. It is not recommended that the battery in a computer-equipped vehicle be jump started. Instead, disconnect the battery cables and charge the battery.

2. Avoid using a battery charger as a starter boost. It can cause damage to solid-state electrical components.

3. To jump start, connect a jumper cable from the positive post of the charged battery to the disabled vehicle's positive battery post. Connect one end of the other jumper cable to the negative post of the good battery. Connect the other end of this cable to the engine block on the disabled vehicle. Refer to Figure 22.

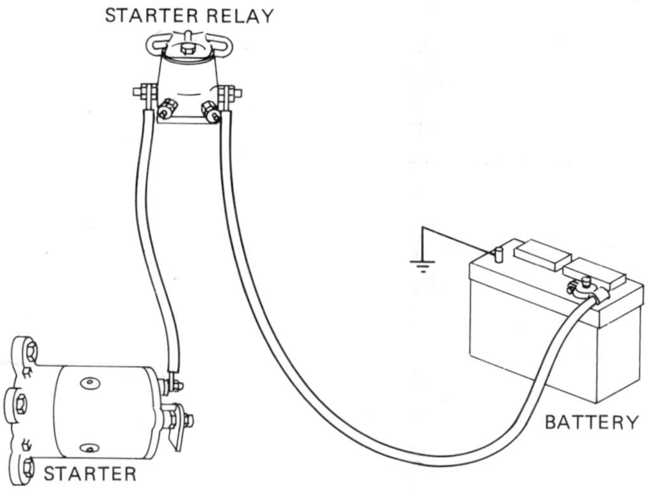

FIGURE 23
Cable connections for a vehicle battery-to-starter circuit that uses a relay.

4. Match voltages. Do not use a 24-volt battery to jump a 12-volt system.

5. When jump starting with an auxiliary power supply or battery charger, do not allow the operating voltage to exceed 16 volts. Damage to the battery, starter, alternator, or electrical system may occur.

CAUTION: When jump starting, if the wrong cable is disconnected first, a hydrogen gas explosion can occur. The negative connection on the engine block is the first to be disconnected.

BATTERY CABLES

Because of the large current draw of the starter, heavy cables are used to electrically connect the starter to the battery. In many starting systems, cables lead from the battery to the starter relay, then to the starter (Figure 23). In other systems, cables may lead directly from the battery to the starter (Figure 24). All starting system connections (at the battery, relay if it has one, and starter) must be clean and tight. Cables that are frayed or damaged should be replaced. A loose or corroded connection or a damaged cable can cause a slow or no-crank condition.

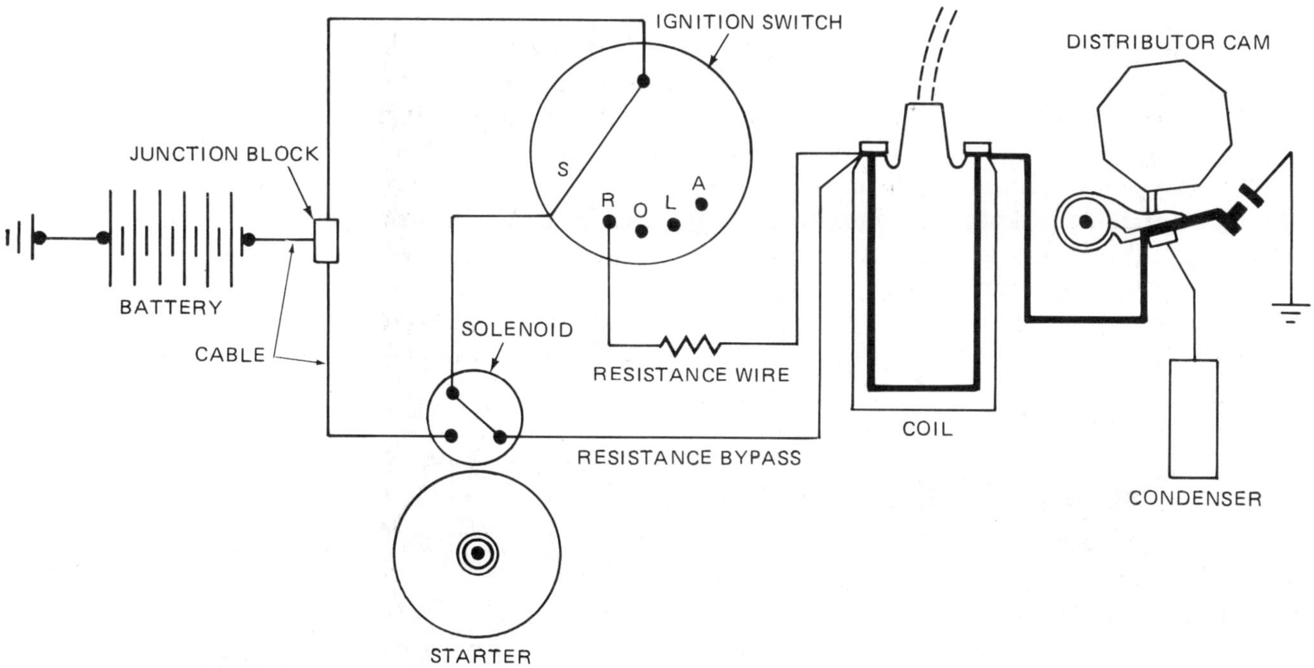

FIGURE 24
Cable connections for a vehicle battery-to-starter circuit that does not use a relay.

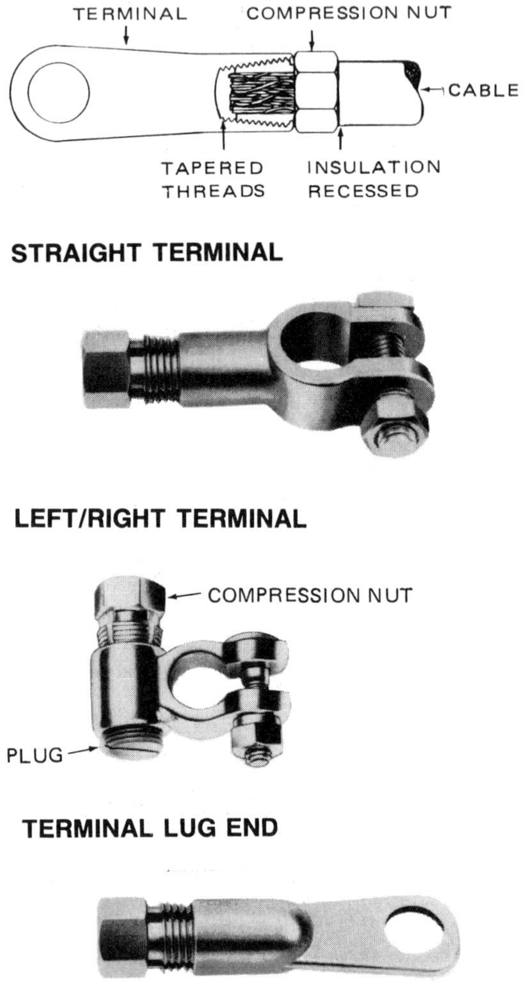

FIGURE 25
Quick-connect compression battery terminal ends. (Courtesy of Belden Corporation)

A variety of terminal styles are available to replace corroded, damaged, or broken original equipment (OE) terminal clamps, or to convert cables to side terminal ends. See Figure 25. Never install a replacement cable longer than the original (Figure 26). If the original length cable is not available, use a shorter length cable (provided that it fits). The longer the cable, the greater the resistance to current flow. As an example, a 9-inch long cable has one-third more resistance than a 6-inch long cable. This 3-inch difference can cause a starter to turn too slowly. A smaller gauge cable offers more resistance to the flow of current than a larger gauge cable. As an example, a No. 2 gauge cable has only 79% as much current carrying capacity as a No. 1 gauge cable (Figure 27).

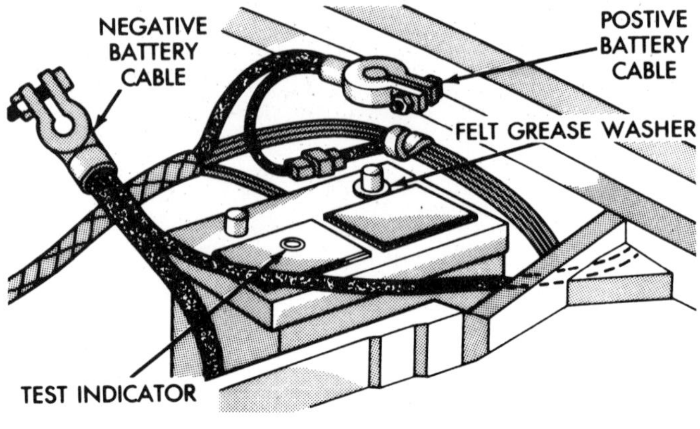

FIGURE 26
Battery cable length and diameter is important. Some negative cables have an extra small wire at one end. This wire is used on many cars as an additional ground "bond" between the battery, engine, and chassis.

STARTING SYSTEM

A starter system circuit often includes a battery, starter motor, starter relay, solenoid, ignition switch, neutral safety switch, connecting wires, and battery cables (Figures 28 and 29).

Starting systems consist of two separate circuits. The *supply circuit* supplies high current to power the starter motor, and the *control circuit* supplies low current to operate the starter relay and solenoid.

Supply Circuit

The supply circuit includes the battery, battery cables, clamps, and connectors. When checking this circuit, begin with an inspection of the battery posts and cable connections. If the posts and connections are good, proceed with an amperage draw test in order to further check the supply circuit. If amperage draw is above specified value, a faulty starter should be suspected (Chart 4). Remove the starter from the vehicle and bench test.

If amperage draw is below specified value, proceed to the supply circuit resistance test to confirm suspicions.

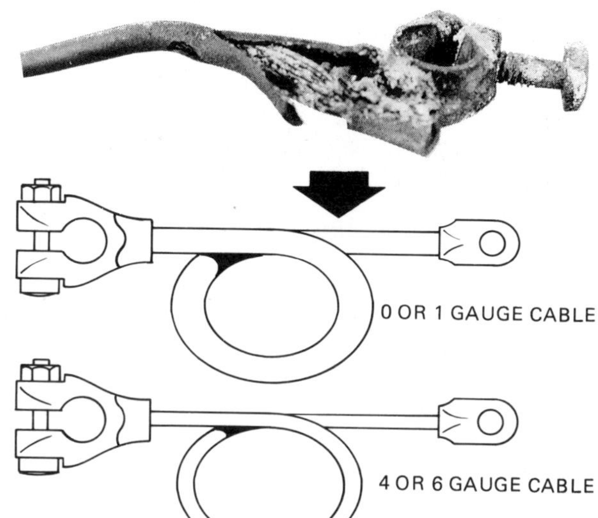

FIGURE 27
Corroded, frayed, or incorrectly sized cables cause slow cranking and hard starting.

Supply Circuit Resistance Tests

High resistance in the supply circuit wiring or connections will cause the starter to crank the engine at a slower rate than normal. The wiring and connections can be tested for high resistance (voltage drop) with a voltmeter that reads tenths of a volt. Connect the voltmeter leads across each of the connections shown in Chart 5. If any of the readings is higher than the specified limit, there is excessive resistance (voltage drop) between those connections. Clean or repair as necessary.

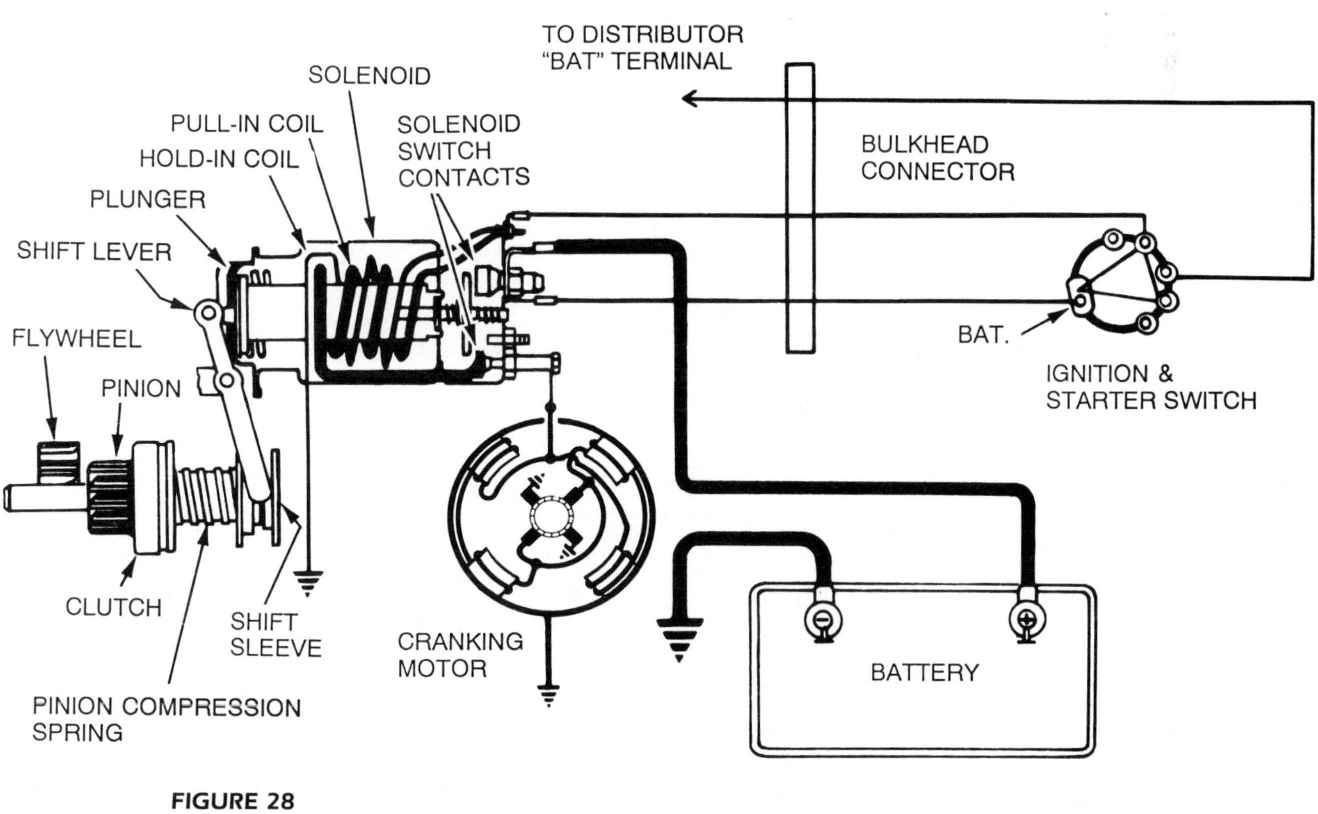

FIGURE 28
Starting system circuit. (Courtesy of Chevrolet Motor Division, General Motors Corporation)

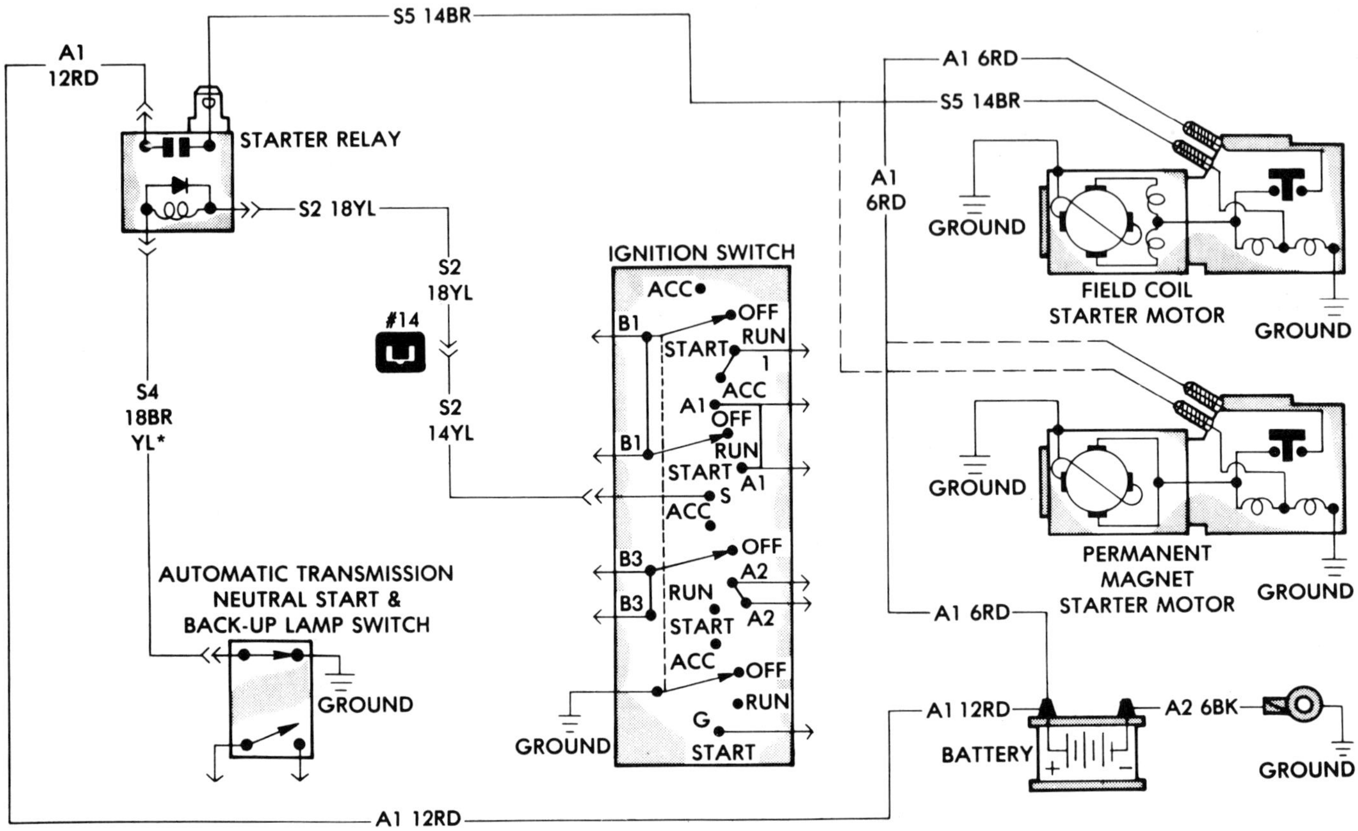

FIGURE 29
Starting system circuit. (Courtesy of Chrysler Corporation)

CHART 4

ENGINE DISPLACEMENT	12 VOLT SYSTEM MAX. CURRENT
MOST 4-6 CYLINDERS	125-175 AMPS MAX.
UNDER 300 C.I.D.	150-200 AMPS MAX.
300 C.I.D. OR OVER	175-250 AMPS MAX.

CRANKING VOLTAGE	CRANKING AMPS	PROBLEM
VOLTAGE OK	WITHIN SPECS.	SYSTEM OK
VOLTAGE OK	LOW (ENGINE CRANKS SLOWLY)	STARTER CIRCUIT (CONNECTIONS FAULTY)
VOLTAGE LOW	LOW (ENGINE CRANKS SLOWLY)	BATTERY (LOW)
VOLTAGE LOW	HIGH	STARTER MOTOR (FAULTY)

CHART 5

CIRCUIT RESISTANCE CHART			
Connection	Voltmeter Lead Connection		Voltmeter Reading
	Positive	Negative	
Positive post on battery to battery cable clamp	Positive battery post	Battery cable clamp	0 volts
Positive post on Battery to starter battery terminal	Positive battery post	Battery terminal on starter	0.2 volts
Positive post on battery to solenoid lead to field coils	Positive battery post	Solenoid lead to field coils	0.3 volts
Negative post on battery to cable clamp	Negative battery post	Battery cable clamp	0 volts
Negative post on battery to engine block	Negative battery post	Battery cable connector on engine block	0.2 volts
Starter housing to ground	Starter housing	Negative battery post	0.2 volts

Control Circuit

The starter control circuit includes the starter solenoid, starter relay, ignition switch, neutral safety switch, and all related wiring and connections.

Typical testing procedures for these components follow. Tests should be carried out in the order described.

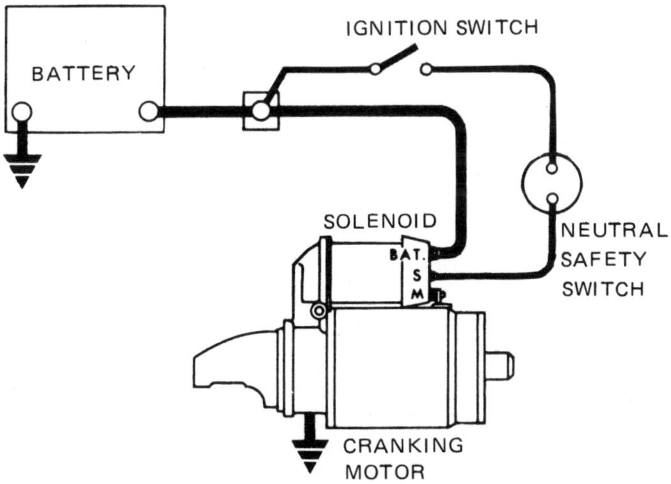

FIGURE 30
A heavy jumper wire can be used to test solenoid operation. Connect it between the BAT and S terminal on the solenoid. If the engine cranks, the solenoid tests OK. (Courtesy of The Echlin Manufacturing Company)

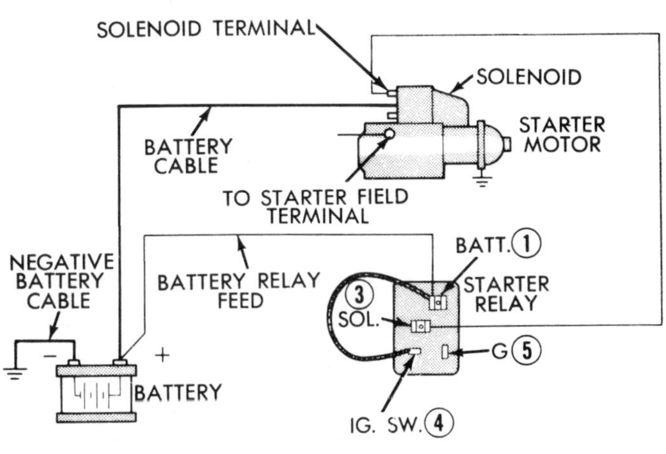

FIGURE 31
Starter relay test with one jumper wire. (Courtesy of Chrysler Corporation)

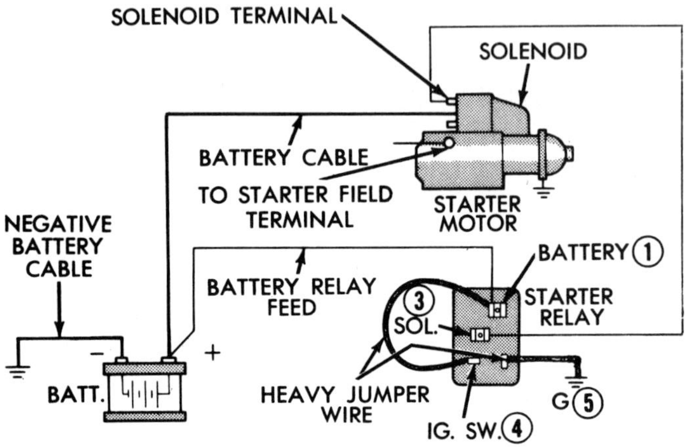

FIGURE 32
Starter relay test with two jumper wires. (Courtesy of Chrysler Corporation)

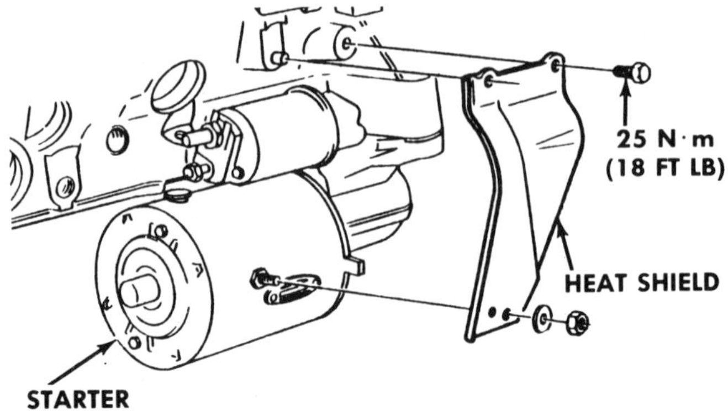

FIGURE 33
Starter heat shield. (Courtesy of Chevrolet Motor Division, General Motors Corporation)

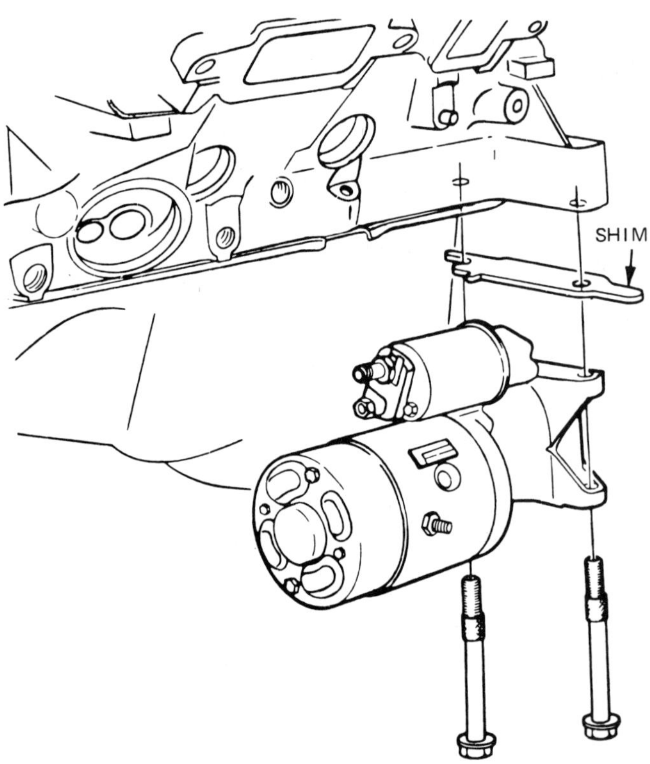

FIGURE 34
Shims are factory installed on some starters. (Courtesy of Chevrolet Motor Division, General Motors Corporation)

Starter Solenoid Test

Connect a heavy jumper wire between the battery terminal and the solenoid terminal on the starter (Figure 30). If the engine cranks, the starter solenoid is good. If the engine does not crank, or if the solenoid chatters, remove the starter for repair.

CAUTION: Before performing the starter solenoid test, disable the ignition system to prevent the engine from starting and make sure battery voltage is up.

Starter Relay Test

Jump from the battery to the ignition terminals on the starter relay (Figure 31). If the engine cranks, the starter relay is good.

If the engine does not crank, connect a second jumper wire between the relay ground terminal and a good ground (Figure 32). If the engine cranks, the starter relay is good. However, the transmission linkage or neutral safety switch could be defective or out of adjustment. If they are not, the starter relay is defective.

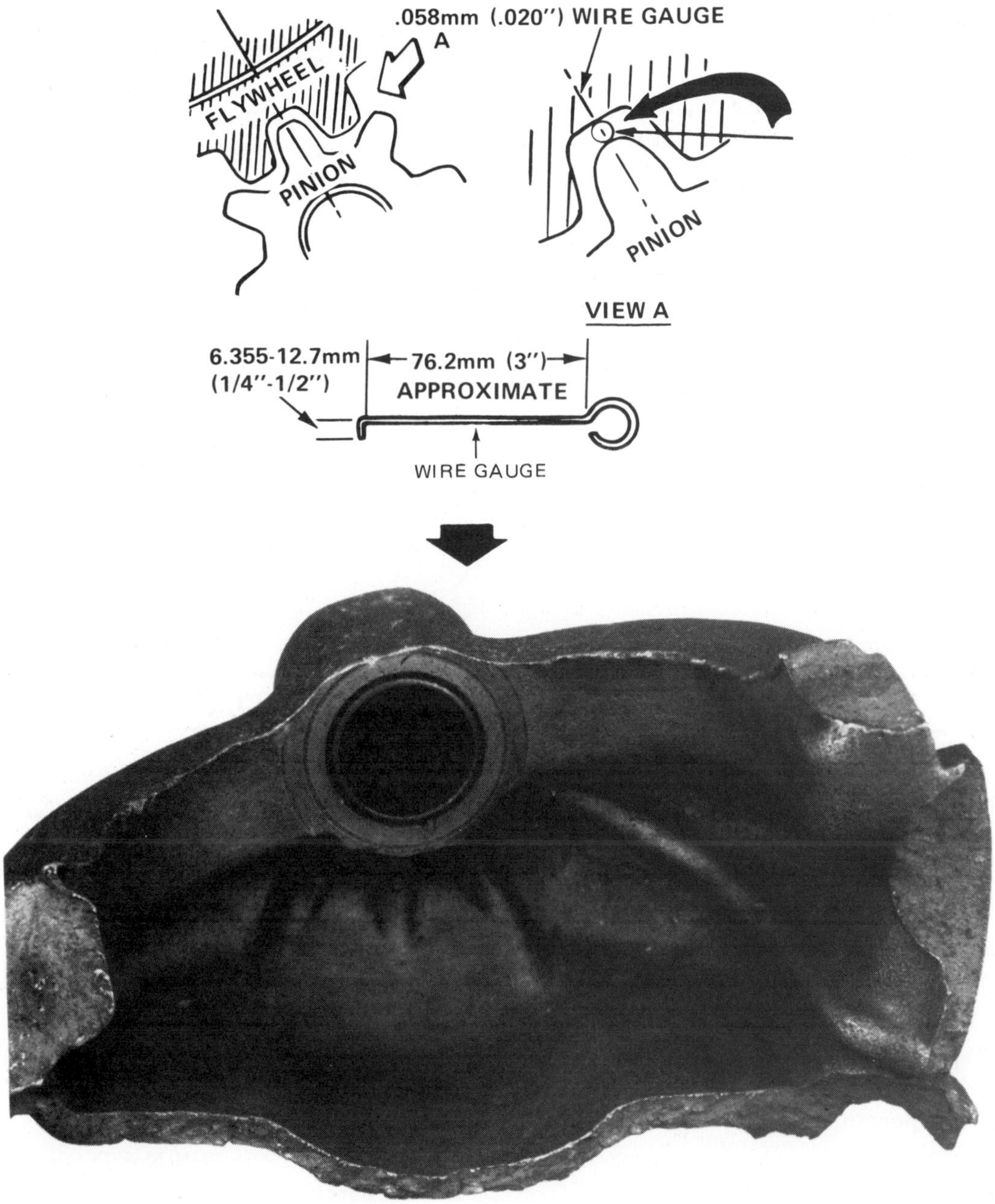

FIGURE 35
Checking starter fit. Improper fit often results in a busted starter "nose." (Courtesy of Pontiac Motor Division, General Motors Corporation)

Ignition Switch Test

If the starter solenoid and relay check out okay and the engine still does not crank, the trouble is most likely with the ignition switch or its wiring.

Check all wiring for continuity, and connections for looseness or corrosion.

STARTER INSTALLATION AND DIAGNOSIS HINTS

1. Some starters use a metal shield for protection against heat from the exhaust system (Figure 33). Do not leave this shield off when replacing the starter.

2. Many General Motors starters use factory-installed shims on the mounting surface (Figure 34). Due to variations in starter drive housings, a replacement starter may require more or fewer shims than originally used. Therefore, always check starter fit (Figure 35).

3. The information shown in Charts 6 and 7 should help in troubleshooting most starting system problems.

4. Excessive starter draw is sometimes caused by carbon buildup on the piston rings (especially when the engine is hot).

CHART 6 TROUBLESHOOTING CHART FOR STARTERS

Starter Fails to Operate
1. Weak battery
2. Loose or corroded battery cable terminals
3. Faulty ignition switch
4. Defective starter relay
5. Defective starter solenoid
6. Defective starter
7. Faulty or improperly adjusted neutral safety switch

Starter Fails to Operate and Headlights Dim
1. Weak battery
2. Loose or corroded battery cable terminals
3. Grounded field windings
4. Armature rubbing on pole shoes
5. "Spun" main or rod bearings

Starter Turns But Engine Does Not Rotate
1. Starter drive (clutch) slipping
2. Broken drive housing
3. Sheared armature shaft
4. Broken ring gear teeth
5. Dry or rusted pinion shaft due to lack of lubrication
6. Worn crankshaft thrust bearings

Starter Relay Does Not Close
1. Discharged battery
2. Faulty wiring
3. Defective starter relay
4. Faulty neutral safety switch on automatic transmission

Starter Relay Operates, But Starter Solenoid Does Not
1. Broken leads or connections inside solenoid cover
2. Open hold-in coil

Starter Operates, But Will Not Disengage when Ignition Switch Is Released
1. Faulty ignition switch
2. Ignition switch linkage "hanging up"
3. Solenoid switch contact sticking
4. Defective starter relay

Solenoid Plunger Vibrates Back and Forth When Ignition Switch Is Engaged
1. Weak battery
2. Broken leads or connections inside solenoid cover
3. Open hold-in coil
4. Loose bulkhead, ammeter, or ignition switch connections

CHART 7 THE STARTER SYSTEM TROUBLESHOOTING CHART

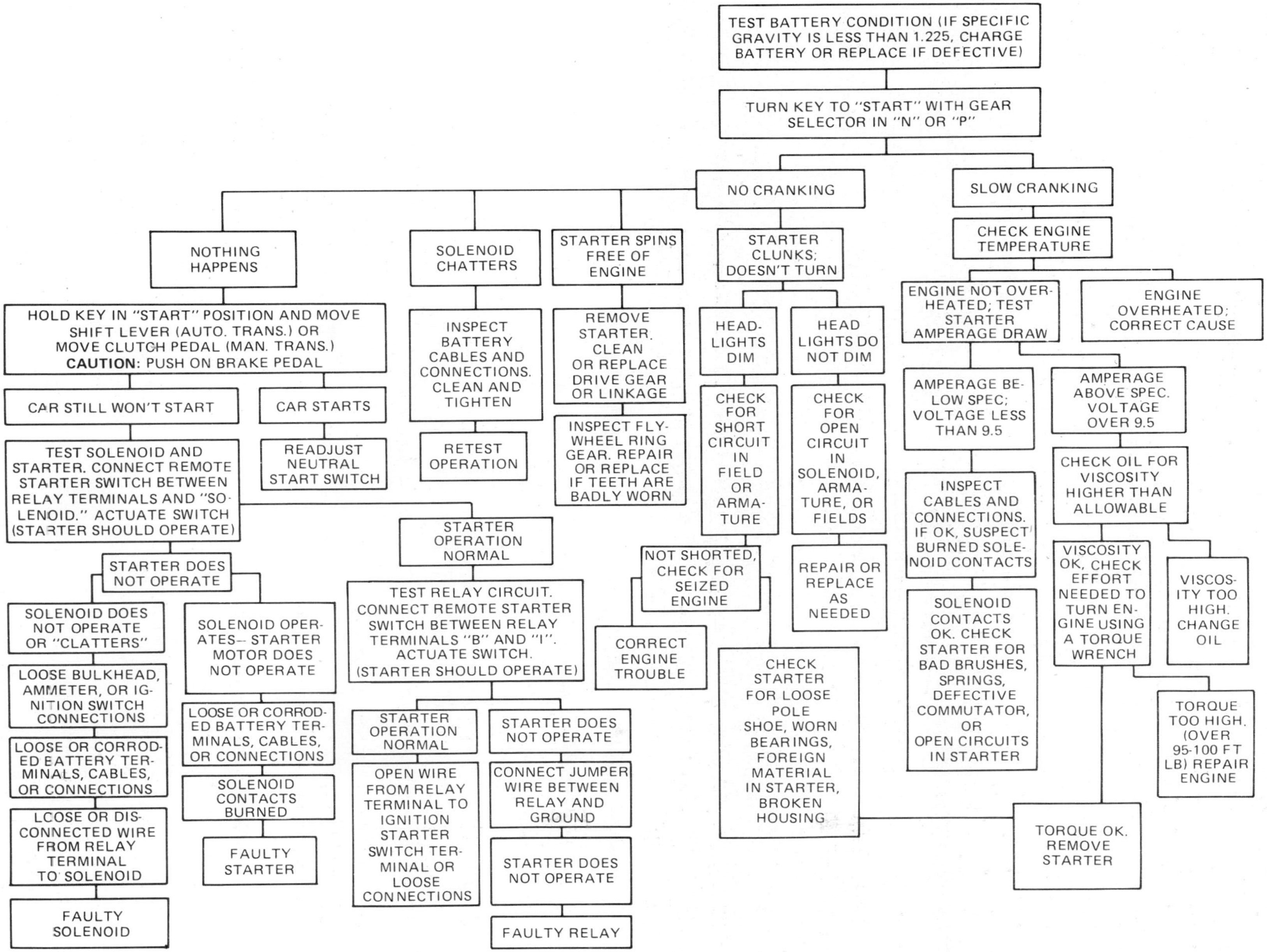

TRUE CASE HISTORY

Vehicle	Chevrolet
Engine	Six-cylinder, 250 cubic inches
Mileage	110,000
Problems	Thrown starter windings (Figure 36). This happened with two rebuilt starters on the engine.
Cause	Excessive starter overrunning after the engine started. This was a result of the ignition switch sticking and not completely releasing from the START position once the engine started. As a result, current flow was divided and kept the starter engaged while the engine was running.
Cure	Replace the ignition switch assembly.

FIGURE 36
Damage caused by excessive overrunning after the engine is started.

TRUE CASE HISTORY

Vehicle	Oldsmobile Cutlass Cierra equipped with PFI
Engine	Six-cylinder, 2.8 liter
Mileage	230. Vehicle recently purchased new.
Problem	1. Battery goes dead overnight. 2. Dealer replaced the alternator. This did not cure the problem.
Cause	Faulty fuel pump relay. The relay contacts remained closed when the engine was shut off, and continued to feed battery voltage to the fuel pump.
Correction	Replace fuel pump relay.

ASE STUDY QUESTIONS

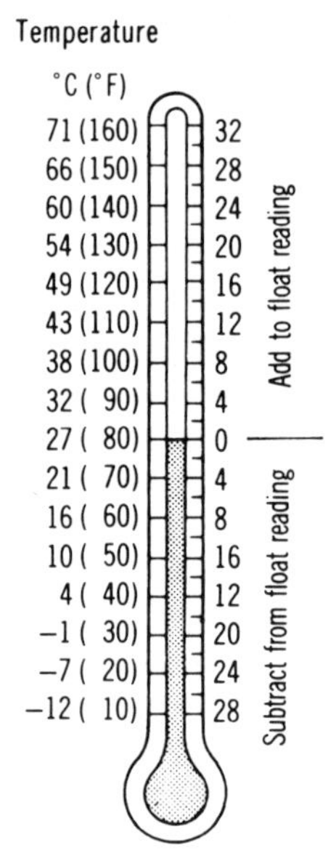

1. A battery hydrometer reading is 1.260. The electrolyte temperature is 60° F. What is the temperature-corrected reading according to the above chart?

   a. 1.320
   b. 1.284
   c. 1.268
   d. 1.252

2. A maintenance-free battery removed from a car has been in storage for six months. How shculd this battery be put back into service?

   Technician A says to dump the electrolyte, fill with new electrolyte and charge.

   Technician B says to slow charge at one-half the normal rate for 60 to 100 hours.

   Who is right?

   a. A only
   b. B only
   c. Both A and B
   d. Neither A nor B

3. A car has a discharged battery and must be jump started.

   Technician A says that the positive jumper cable should be connected to the positive battery terminal on the discharged battery.

   Technician B says that the negative jumper cable should be connected to the engine in the vehicle with the discharged battery.

   Who is right?

   a. A only
   b. B only
   c. Both A and B
   d. Neither A nor B

4. You are going to replace the battery cables in an automobile that has a negative ground system. Which of the following procedures is correct?

   a. Remove the positive cable first, and install the negative cable last.
   b. Remove the positive cable first, and intall the negative cable first.
   c. Remove the negative cable first, and install the positive cable last.
   d. Remove the negative cable first, and install the positive cable first.

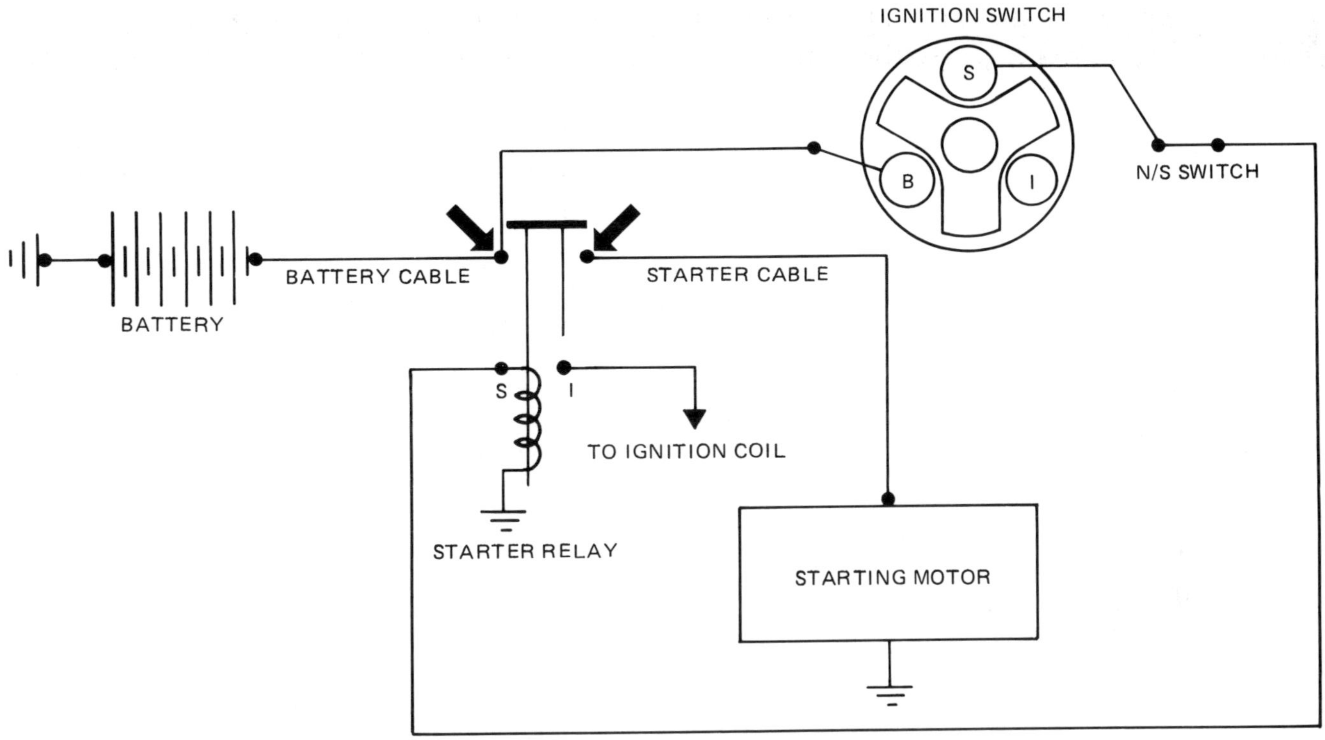

5. An engine will not crank when the ignition switch is turned to the start position. However, when the two cables shown above are disconnected at the arrows and touched together, the engine cranks.

   Technician A says that the neutral safety switch could be improperly adjusted.

   Technician B says that the ignition switch could be faulty.

   Who is right?

   a. A only
   b. B only
   c. Both A and B
   d. Neither A nor B

6. A 300-cubic-inch displacement starter draws 350 amps, and the voltage while cranking is below specifications.

   Technician A says that this could be caused by a faulty starter.

   Technician B says that this could be caused by a carbon buildup on the piston rings.

   Who is right?

   a. A only
   b. B only
   c. Both A and B
   d. Neither A nor B

SECTION REVIEW QUESTIONS

1. List three things that could cause a battery to go dead overnight.

2. Name four conditions that could cause a starter solenoid plunger to vibrate back and forth when the ignition switch is engaged.

3. A battery has a specific gravity reading of 1.205. What is its state of charge?

4. What can cause a starter to throw its windings?

5. How is a General Motors starter checked for fit?

6. The starter control circuit includes what parts?

7. The starter supply circuit includes what parts?

8. Describe how to perform a starter solenoid test.

9. A computer-equipped vehicle needs to be jump started. What is the recommended procedure?

10. Sometimes, an excessively sulphated battery can be reactivated. What is the procedure?

11. List four recommendations to follow when fast charging a battery.

12. How can you determine if a battery is excessively sulphated?

13. When a battery is being charged, what explosive gas is produced?

14. List five recommendations to follow when slow charging a battery.

15. How can you make a quick test to determine if there is leakage across a battery top?

16. If there are more than ____________ gravity points between cells, a battery should be replaced.

17. When servicing a battery, what should the cable clamps be coated with?

18. Cranking voltage should not fall below what value?

BOOK SUMMARY QUESTIONS

This test will help to further prepare you for the ASE Engine Repair Certification Examination. These questions are based on concepts presented in this book and are similar in content and style to those on the actual test.

FIGURE X

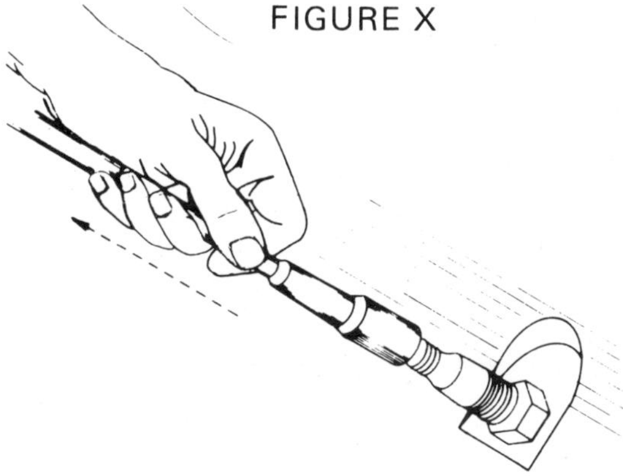

FIGURE Y

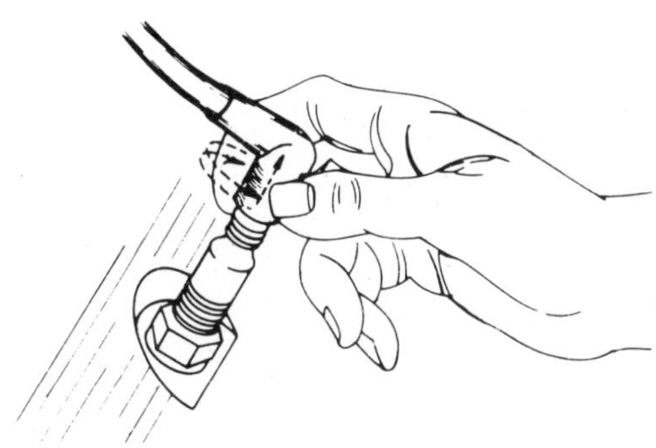

1. Which figure above shows the correct way to disconnect a spark plug cable?
   a. X only
   b. Y only
   c. Both X and Y
   d. Neither X nor Y

2. During a cylinder leakage test, the gauge reads 70% and air is heard coming from the exhaust pipe. What is the MOST LIKELY cause?
   a. Cracked piston
   b. Hole in cylinder wall
   c. Burned valve
   d. Collapsed lifter

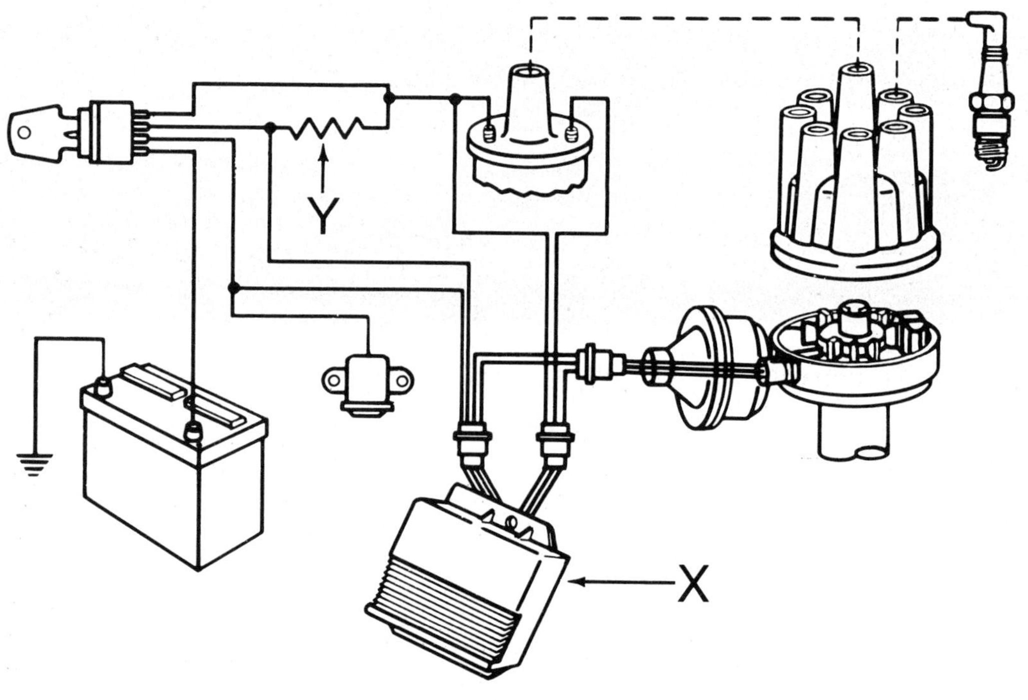

3. An engine that is equipped with the above ignition system will not start (fuel delivery is OK).

   Technician A says that part X could be defective.

   Technician B says that part Y could be defective.

   Who is right?
   a. A only
   b. B only
   c. Both A and B
   d. Neither A nor B

4. An engine equipped with the above relay is being cranked.

   Technician A says that a 12 volt test lamp should light if connected from point X to ground.

   Technician B says that a 12 volt test lamp should light if connected from point Y to ground.

   Who is right?
   a. A only
   b. B only
   c. Both A and B
   d. Neither A nor B

5. **Technician A** says that the above damage could have been caused by excessive over-running after starting.

   Technician B says that the above damage could have been caused by the driver's opening the throttle too wide during starting.

   Who is right?
   a. A only
   b. B only
   c. Both A and B
   d. Neither A nor B

6. A mechanical fuel pump makes noise when the engine is running.

   Technician A says that this could be caused by a stiff inlet hose.

   Technician B says that this could be caused by a broken rocker arm spring.

   Who is right?
   a. A only
   b. B only
   c. Both A and B
   d. Neither A nor B

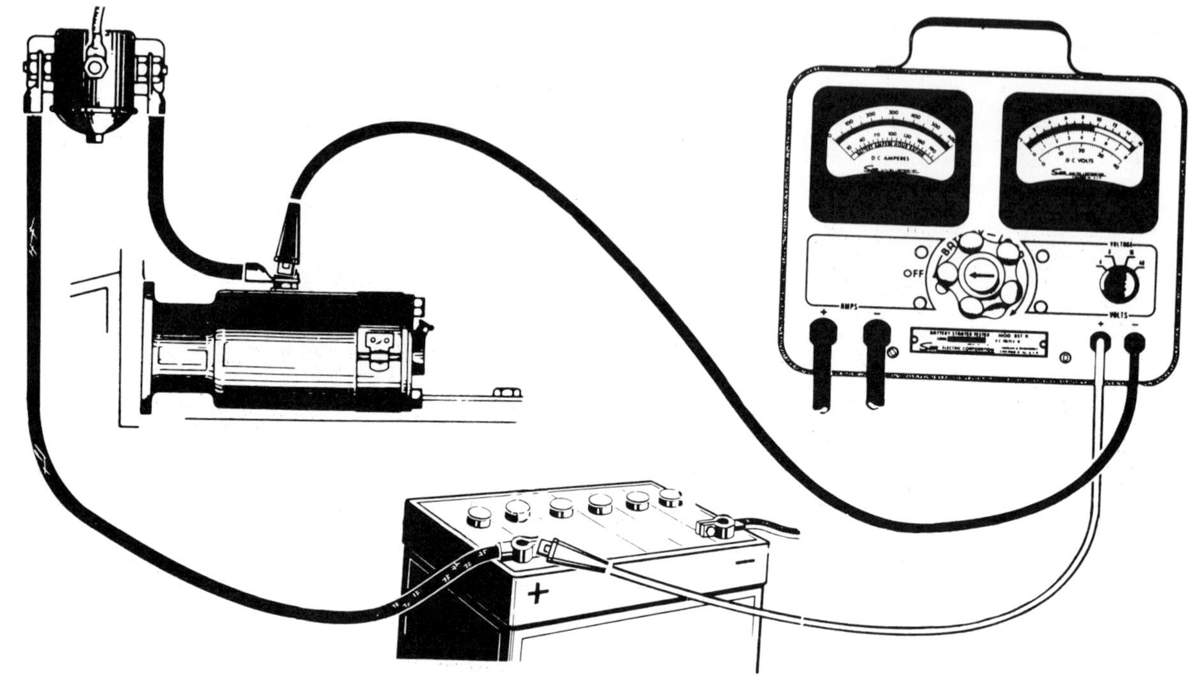

7. A voltage drop test is being made on a starting system (refer to the above illustration). While cranking, the voltmeter reads .3 volts.

   Technician A says that the negative battery cable is too small to carry the current.

   Technician B says that the starter relay has high resistance.

   Who is right?
   a. A only
   b. B only
   c. Both A and B
   d. Neither A nor B

8. **Technician A** says to span the full width of the bearing when using Plastigage.®

   Technician B says to slowly turn the crankshaft a full revolution when using Plastigage.®

   Who is right?
   a. A only
   b. B only
   c. Both A and B
   d. Neither A nor B

9. The distributor part shown above has damaged teeth.

   Technician A says that this could cause hard starting.

   Technician B says that this could cause backfiring.

   Who is right?
   a. A only
   b. B only
   c. Both A and B
   d. Neither A nor B

10. **Technician A** says that the above damage could have been caused by sharp edges in the combustion chamber.

    Technician B says that the above damage could have been caused by incandescent carbon particles.

    Who is right?
    a. A only
    b. B only
    c. Both A and B
    d. Neither A nor B

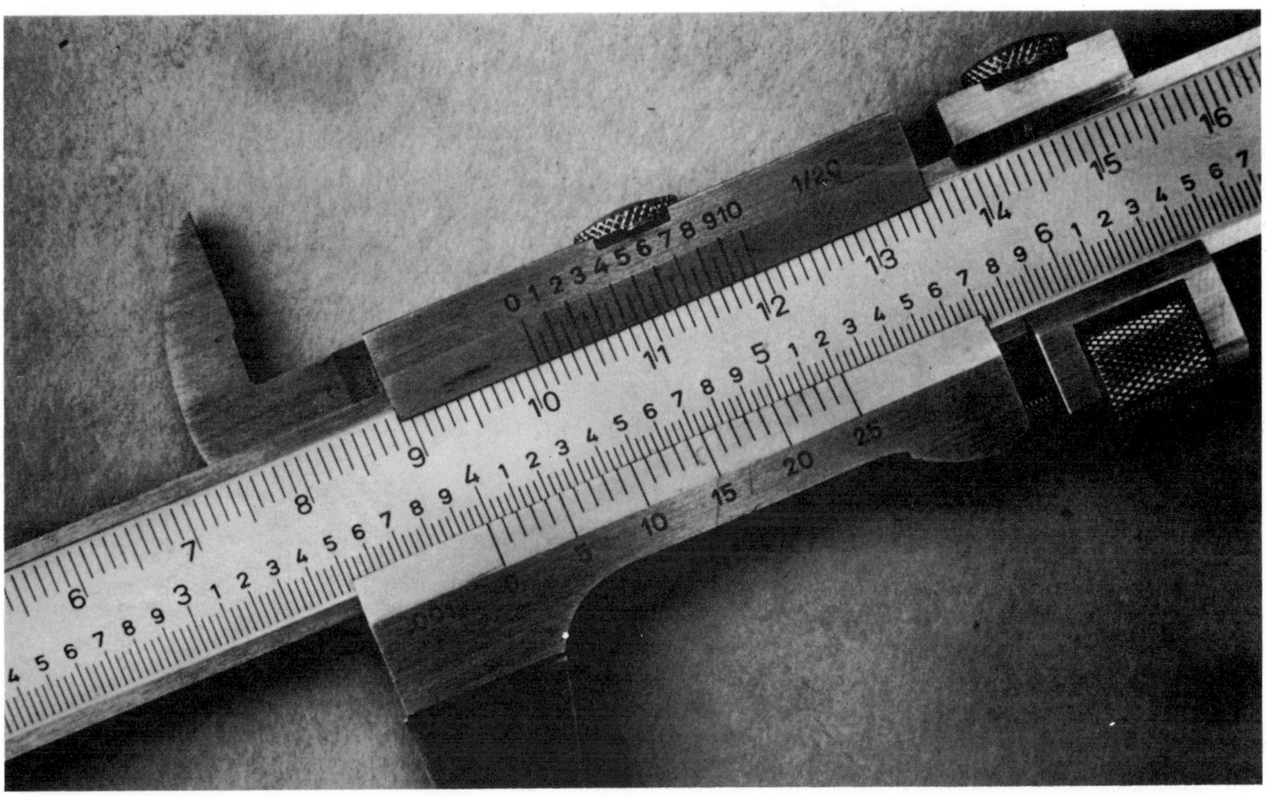

11. The above measuring tool shows a reading of
    a. 3.989″
    b. 5.225″
    c. 8.900″
    d. 10.320″

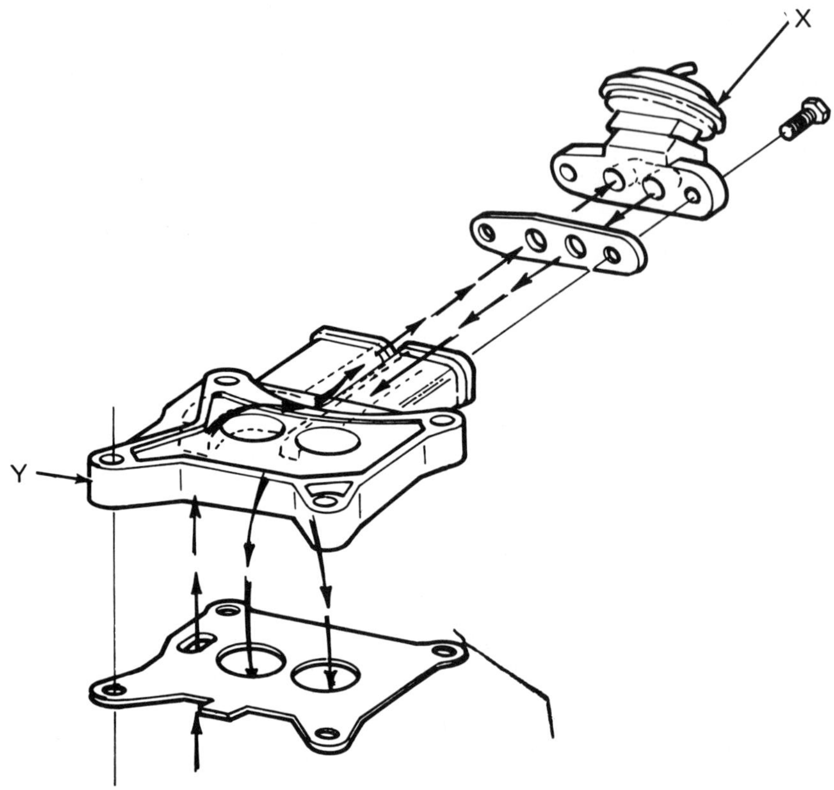

12. An engine equipped with the above system has a rough idle.

    Technician A says that part X could be defective.

    Technician B says that part Y could be defective.

    Who is right?
    a. A only
    b. B only
    c. Both A and B
    d. Neither A nor B

13. **Technician A** says that blue-gray exhaust smoke can be caused by piston rings stuck in their grooves.

    Technician B says that blue-gray exhaust smoke can be caused by plugged oil return passages in the cylinder head.

    Who is right?
    a. A only
    b. B only
    c. Both A and B
    d. Neither A nor B

14. While driving in stop-and-go city traffic, a car with an electric fan overheats.

    Technician A says that this could be caused by a stuck open thermostat.

    Technician B says that this could be caused by a bad radiator fan sensor.

    Who is right?
    a. A only
    b. B only
    c. Both A and B
    d. Neither A nor B

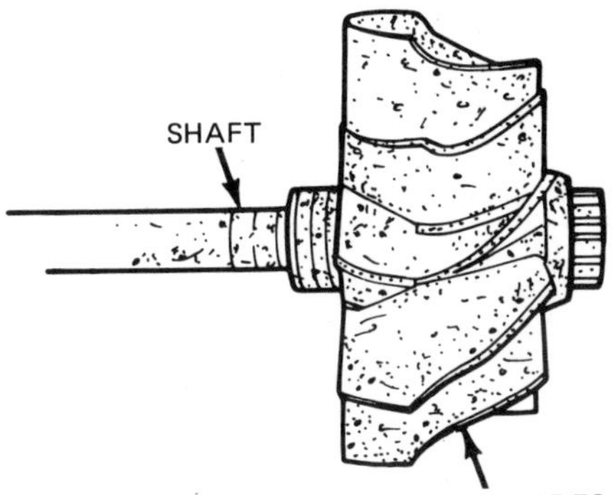

15. A turbocharger has deposits of burned oil (see the above picture) on the turbine wheel.

    Technician A says that this could be caused by a sludged center housing.

    Technician B says that this could be caused by a worn shaft assembly.

    Who is right?
    a. A only
    b. B only
    c. Both A and B
    d. Neither A nor B

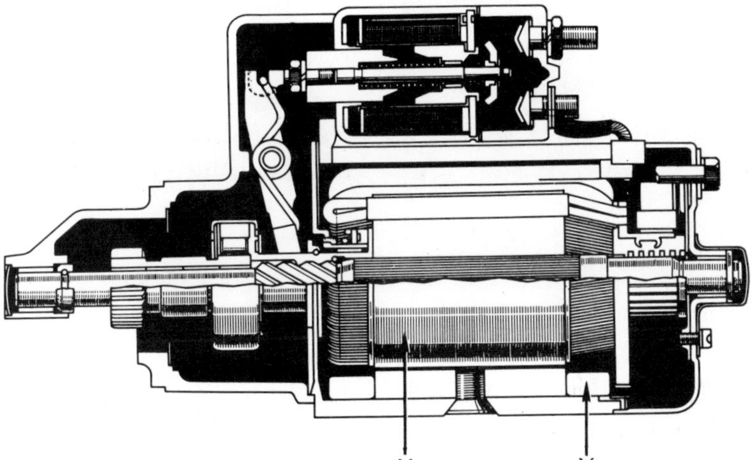

16. The above starter stays engaged after the key is released from the crank position.

    Technician A says that part X could be bad.

    Technician B says that part Y could be bad.

    Who is right?
    a. A only
    b. B only
    c. Both A and B
    d. Neither A nor B

17. Dull-gray vertical scratches are found on the piston rings during an engine teardown. This is MOST LIKELY caused by
    a. Using multi-viscosity oil
    b. Abrasive wear
    c. Over-revving
    d. Worn ring grooves

18. An engine has been a "come-back" several times with burned valves.

    Technician A says that this could be caused by blocked coolant passages.

    Technician B says that this could be caused by using the wrong torque sequence on the head bolts.

    Who is right?
    a. A only
    b. B only
    c. Both A and B
    d. Neither A nor B

19. A V-8 engine is spewing white exhaust smoke under all driving conditions.

    Technician A says that this could be caused by a cracked carburetor throttle body.

    Technician B says that this could be caused by a cracked cylinder head.

    Who is right?
    a. A only
    b. B only
    c. Both A and B
    d. Neither A nor B

21. After a recent engine overhaul the oil consumption has greatly increased.

    Technician A says that this could be caused by installing the rings into worn grooves.

    Technician B says that this could be caused by installing the rings upside down.

    Who is right?
    a. A only
    b. B only
    c. Both A and B
    d. Neither A nor B

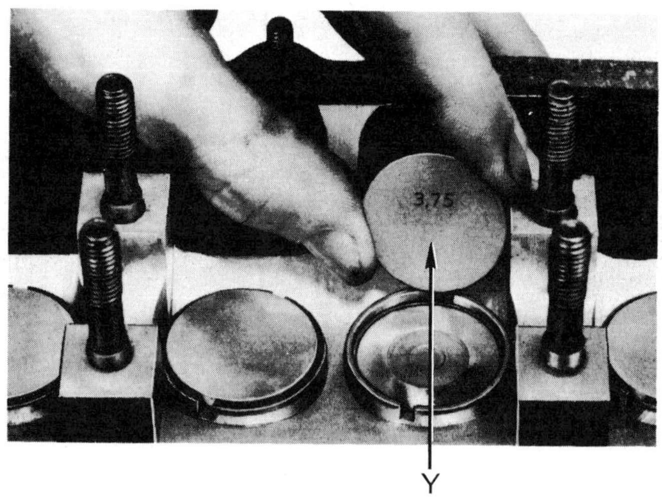

20. **Technician A** says that valve adjustment on the above engine is made by changing part X.

    Technician B says that valve adjustment on the above engine is made by changing part Y.

    Who is right?
    a. A only
    b. B only
    c. Both A and B
    d. Neither A nor B

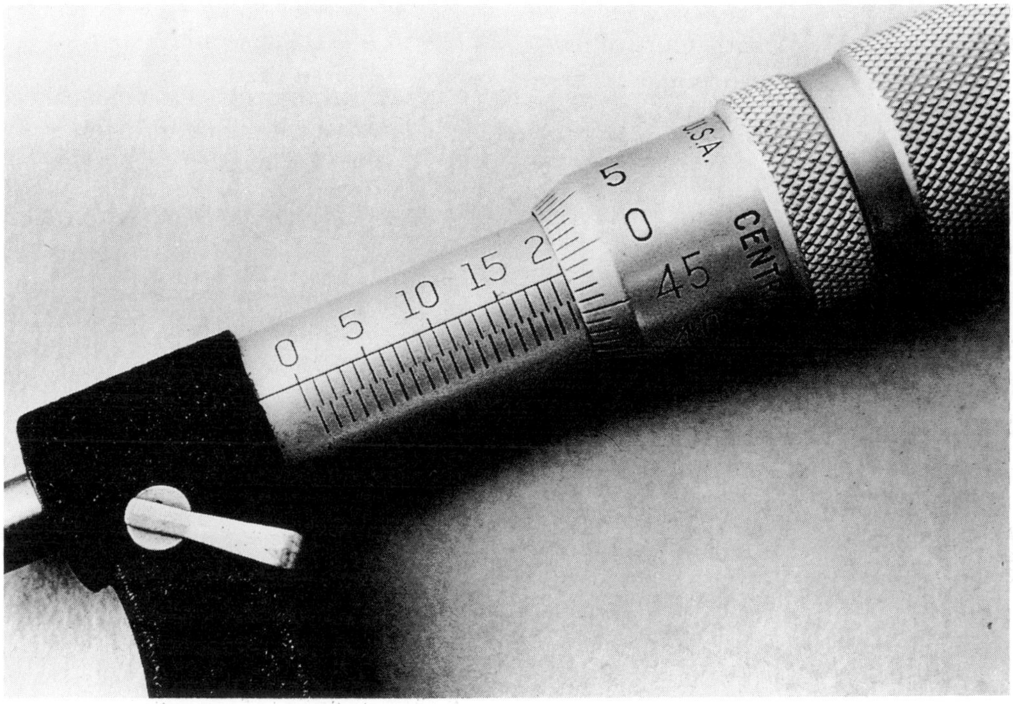

22. What is the reading on the above micrometer?
    a. 19.99 mm
    b. 20.01 mm
    c. 20.49 mm
    d. None of the above

23. The exhaust manifold heat control valve on an engine is stuck open.

    Technician A says that this could cause low cranking rpm.

    Technician B says that this could cause hesitation during cold driveaway.

    Who is right?
    a. A only
    b. B only
    c. Both A and B
    d. Neither A nor B

24. A technician removes the spark plugs from an engine. All the firing ends are coated with dry, fluffy carbon deposits.

    Technician A says that this could be caused by worn piston rings.

    Technician B says that this could be caused by a leaking carburetor base gasket.

    Who is right?
    a. A only
    b. B only
    c. Both A and B
    d. Neither A nor B

25. The rod bearing journals in an engine are .005″ out-of-round. Which of the following would be the correct procedure?
    - **a.** Grind the journals and install thicker bearings
    - **b.** Grind the journals and install thinner bearings
    - **c.** Polish the journals
    - **d.** Shim the bearings

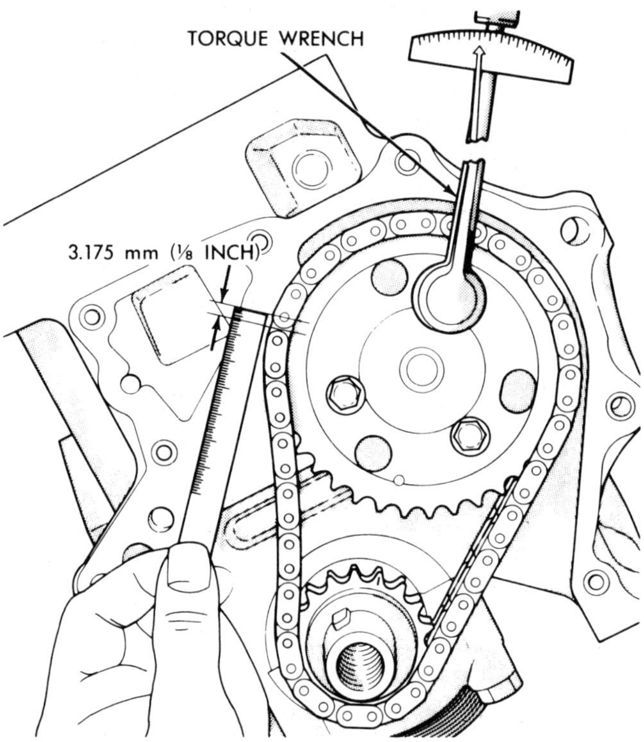

26. What procedure is being performed in the above illustration?
    - **a.** Setting cam timing
    - **b.** Locating exact TDC
    - **c.** Measuring timing chain wear and stretch
    - **d.** None of the above

27. It is necessary to add water to a battery every week.

    Technician A says that this could indicate an overcharge condition.

    Technician B says that this could indicate an incorrect voltage regulator setting.

    Who is right?
    - **a.** A only
    - **b.** B only
    - **c.** Both A and B
    - **d.** Neither A nor B

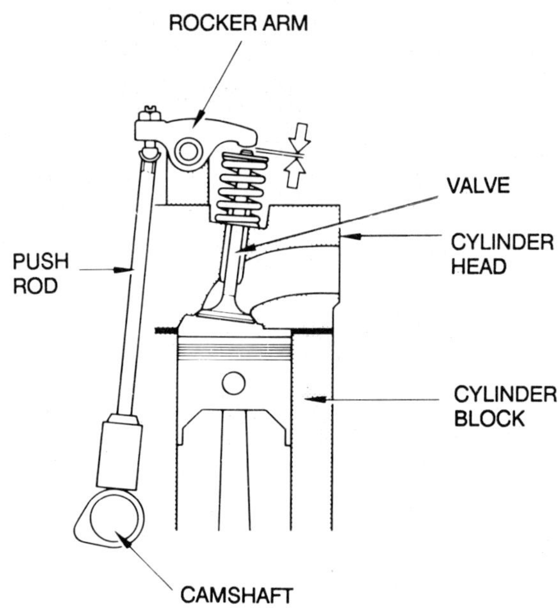

28. The above measurement (see arrows) has been set incorrectly.

    Technician A says that this can cause noise.

    Technician B says that this can cause rough idle.

    Who is right?
    - **a.** A only
    - **b.** B only
    - **c.** Both A and B
    - **d.** Neither A nor B

29. An engine runs, but the oil warning light stays on.

Technician A says that this could be caused by too much cam bearing clearance.

Technician B says that this could be caused by a ground in the warning light circuit.

Who is right?
a. A only
b. B only
c. Both A and B
d. Neither A nor B

30. A four-cylinder engine has compression readings of 135,70,130,75. Which of the following is the MOST LIKELY cause?
a. Burned valves
b. Broken valve springs
c. Incorrect valve timing
d. Missing stem seals

31. A computer controlled engine has a lean condition. Which of the following is the MOST LIKELY cause?
a. A bad PCV system
b. A bad vehicle speed sensor (VSS)
c. A bad "cat"
d. A stuck closed EGR valve

32. An engine idles rough, but runs smooth above 2500 rpm.

Technician A says that this could be caused by a burned valve.

Technician B says that this could be caused by a warped intake manifold.

Who is right?
a. A only
b. B only
c. Both A and B
d. Neither A nor B

33. These compression readings (130,135 130,140,15,135) could be caused by
a. A hole in a piston
b. A broken exhaust valve push rod
c. Both a and b
d. Neither a nor b

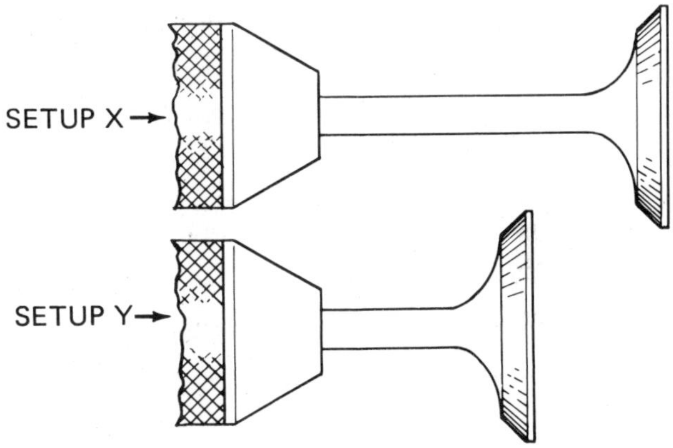

34. A valve is going to be refaced in a valve grinding machine (refer to the above picture).

Technician A says that setup X is OK.

Technician B says that setup Y is OK.

Who is right?
a. A only
b. B only
c. Both A and B
d. Neither A nor B

35. **Technician A** says that battery cable diameter affects cranking ability of the starter.

Technician B says that battery cable length affects cranking ability of the starter.

Who is right?
a. A only
b. B only
c. Both A and B
d. Neither A nor B

36. Which of the following would be the MOST LIKELY cause of an engine "pinging" complaint?
   a. Defective catalytic converter
   b. Poor fuel quality
   c. Worn air pump
   d. Bad vapor canister

37. A power balance test is being performed on a computer-controlled engine.

   Technician A says to disconnect the check engine light.

   Technician B says to disconnect the computer power supply.

   Who is right?
   a. A only
   b. B only
   c. Both A and B
   d. Neither A nor B

FIG. 1 FIG. 2 FIG. 3

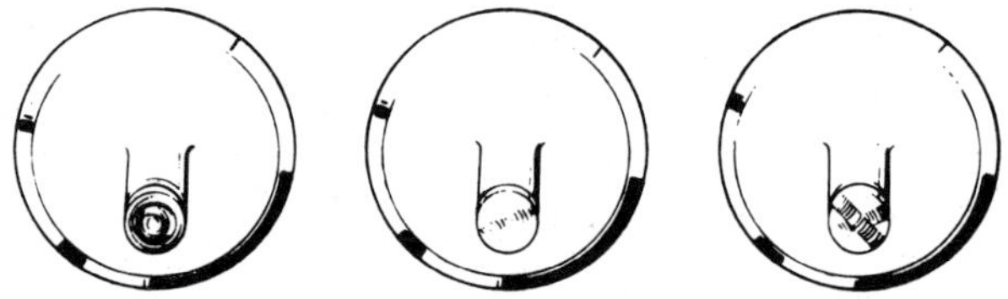

38. An engine is equipped with valve rotators. Which wear pattern shown above indicates that the rotator is functioning properly?
   a. Figure 1
   b. Figure 2
   c. Figure 3
   d. None of the above

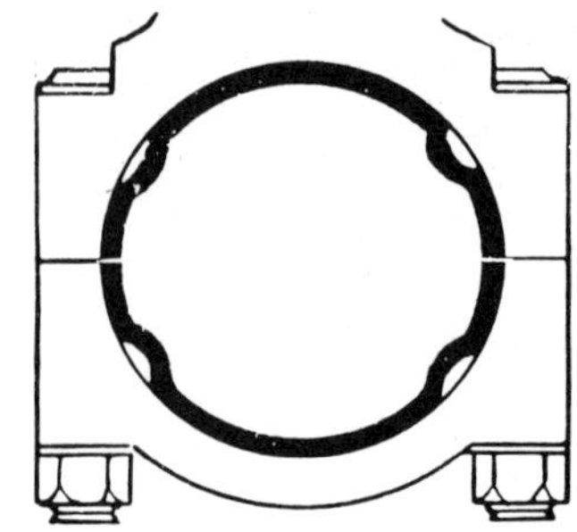

39. What could cause a connecting rod bearing assembly to buckle as pictured above?
   a. Excessive crush
   b. Over-tightening the cap
   c. Either a or b
   d. Neither a nor b

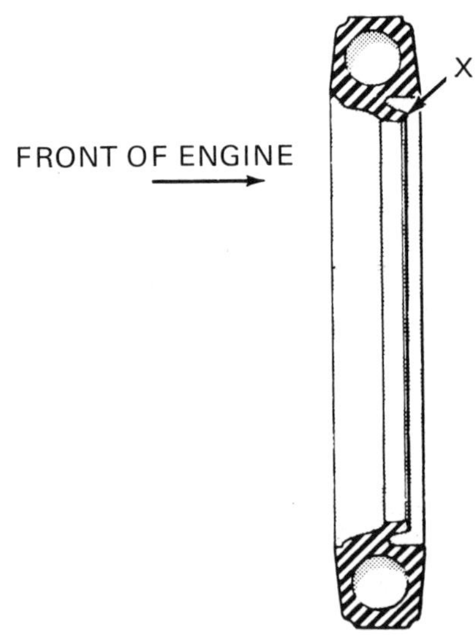

40. The above seal is being installed in an engine.

   Technician A says that lip X must face the flywheel.

   Technician B says to coat lip X with oil.

   Who is right?
   a. A only
   b. B only
   c. Both A and B
   d. Neither A nor B

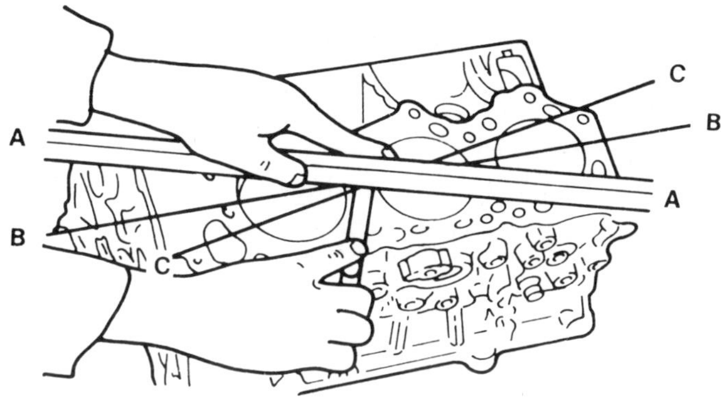

41. A .015″ (.40 mm) feeler gauge leaf can be inserted while making the above check. What should be done?
    a. Install a thicker head gasket with special sealer
    b. Deck the block
    c. Both a and b
    d. Neither a nor b

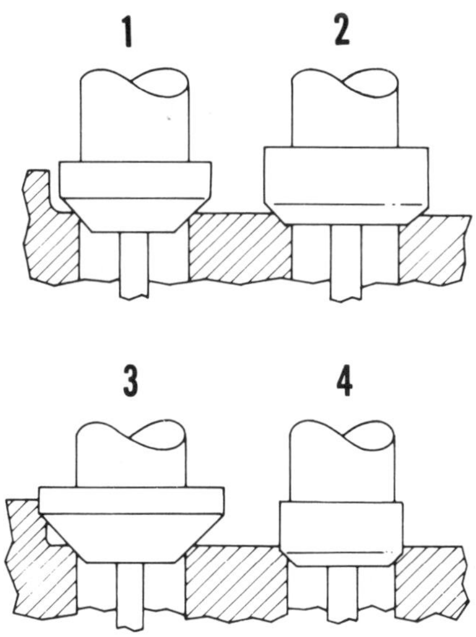

42. Which one of the above valve grinding stones fits the seat correctly and should be used?
    a. Stone 1
    b. Stone 2
    c. Stone 3
    d. Stone 4

43. An engine will not start after being rebuilt.

    Technician A says that this could be caused by incorrect valve adjustment.

    Technician B says that this could be caused by incorrect camshaft timing.

    Who is right?
    a. A only
    b. B only
    c. Both A and B
    d. Neither A nor B

44. A starter spins and makes a grinding noise, but does not crank the engine.

    Technician A says that this could be caused by a worn crankshaft thrust bearing.

    Technician B says that this could be caused by a worn flywheel ring gear.

    Who is right?
    a. A only
    b. B only
    c. Both A and B
    d. Neither A nor B

45. An engine cranks very slowly and will not start. Also, the headlights dim considerably during cranking.

    Technician A says that this could be caused by a "spun" connecting rod bearing.

    Technician B says that this could be caused by a weak battery.

    Who is right?
    a. A only
    b. B only
    c. Both A and B
    d. Neither A nor B

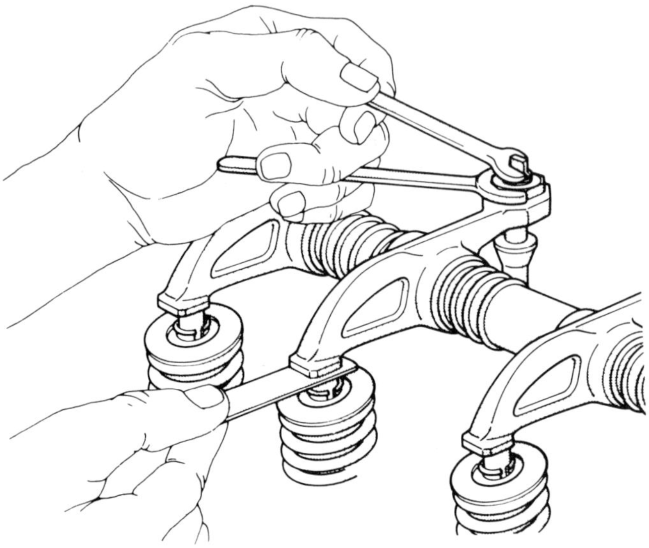

46. The above adjustment is set too wide.

    Technician A says that this will retard valve timing.

    Technician B says that this will reduce valve overlap.

    Who is right?
    a. A only
    b. B only
    c. Both A and B
    d. Neither A nor B

47. An engine runs-on (diesels) when shut off at operating temperature.

    Technician A says that this is caused by an improperly adjusted throttle stop solenoid.

    Technician B says that this can be caused by low heat range spark plugs.

    Who is right?
    a. A only
    b. B only
    c. Both A and B
    d. Neither A nor B

48. **Technician A** says that resistor-type spark plug cables reduce fouling of the spark plugs.

    Technician B says that resistor-type spark plug cables reduce the likelihood of being damaged when removing and installing.

    Who is right?
    a. A only
    b. B only
    c. Both A and B
    d. Neither A nor B

49. During a power balance test, a V-8 engine has two cylinders on opposite corners of the block that do not show any drop in engine rpm. The MOST LIKELY cause would be
    a. A bad head gasket
    b. A leaking carburetor base gasket
    c. A distributor cap cracked between two terminals
    d. Switched spark plug wires

50. A hot air intake (HAI) system is being inspected.

    Technician A says that the snorkel door supply line should be connected to ported vacuum.

    Technician B says that with the engine warm, the snorkel door should move when a wet rag cools the temperature sensor.

    Who is right?
    a. A only
    b. B only
    c. Both A and B
    d. Neither A nor B

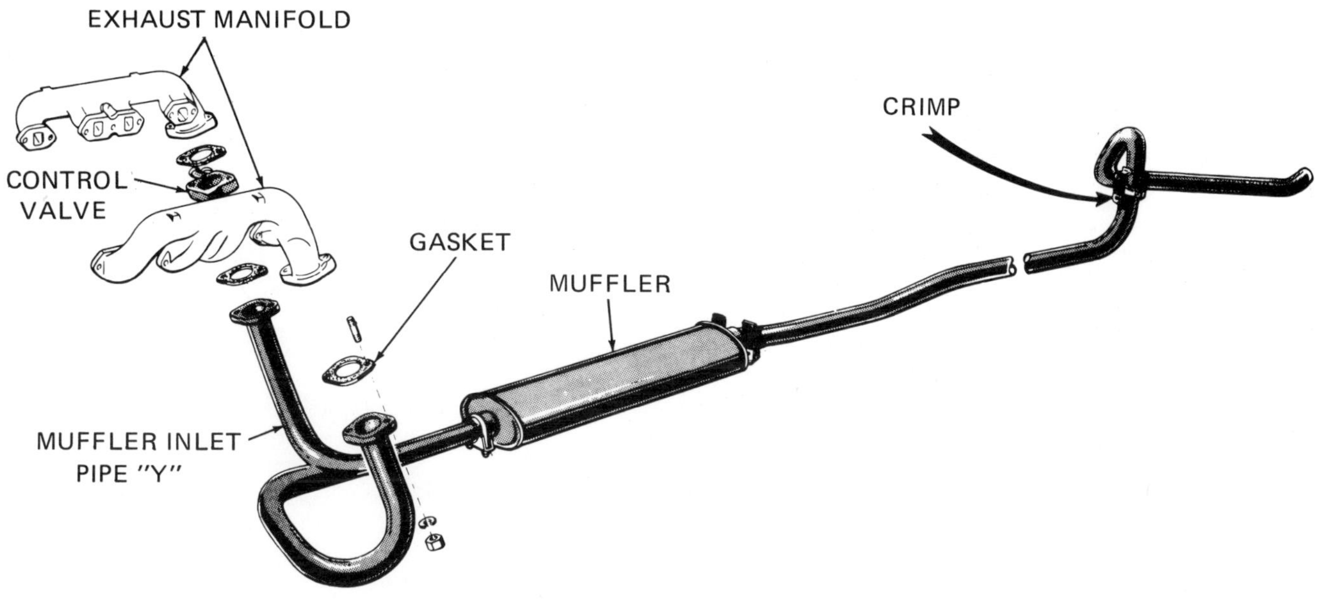

51. An exhaust system has a crimp in the pipe at the rear hanger area (see above arrow).

    Technician A says that this can cause loss of engine power.

    Technician B says that this can cause engine overheating.

    Who is right?
    a. A only
    b. B only
    c. Both A and B
    d. Neither A nor B

52. A fuel pump is being tested using a vacuum/pressure gauge at the pump inlet.

    Technician A says that this will test the condition of the diaphragm.

    Technician B says that this will test the check valves in the pump.

    Who is right?
    a. A only
    b. B only
    c. Both A and B
    d. Neither A nor B

53. A six-cylinder engine has a loose intake manifold. This would MOST LIKELY cause
    a. Rough idle
    b. Overheating
    c. Spark plug fouling
    d. Loss of power

54. A car is towed into the shop with a burst oil filter. What MOST LIKELY caused this?
    a. Using two low viscosity oil
    b. Using too high viscosity oil
    c. Oil pump pressure relief valve stuck open
    d. Oil pump pressure relief valve stuck closed

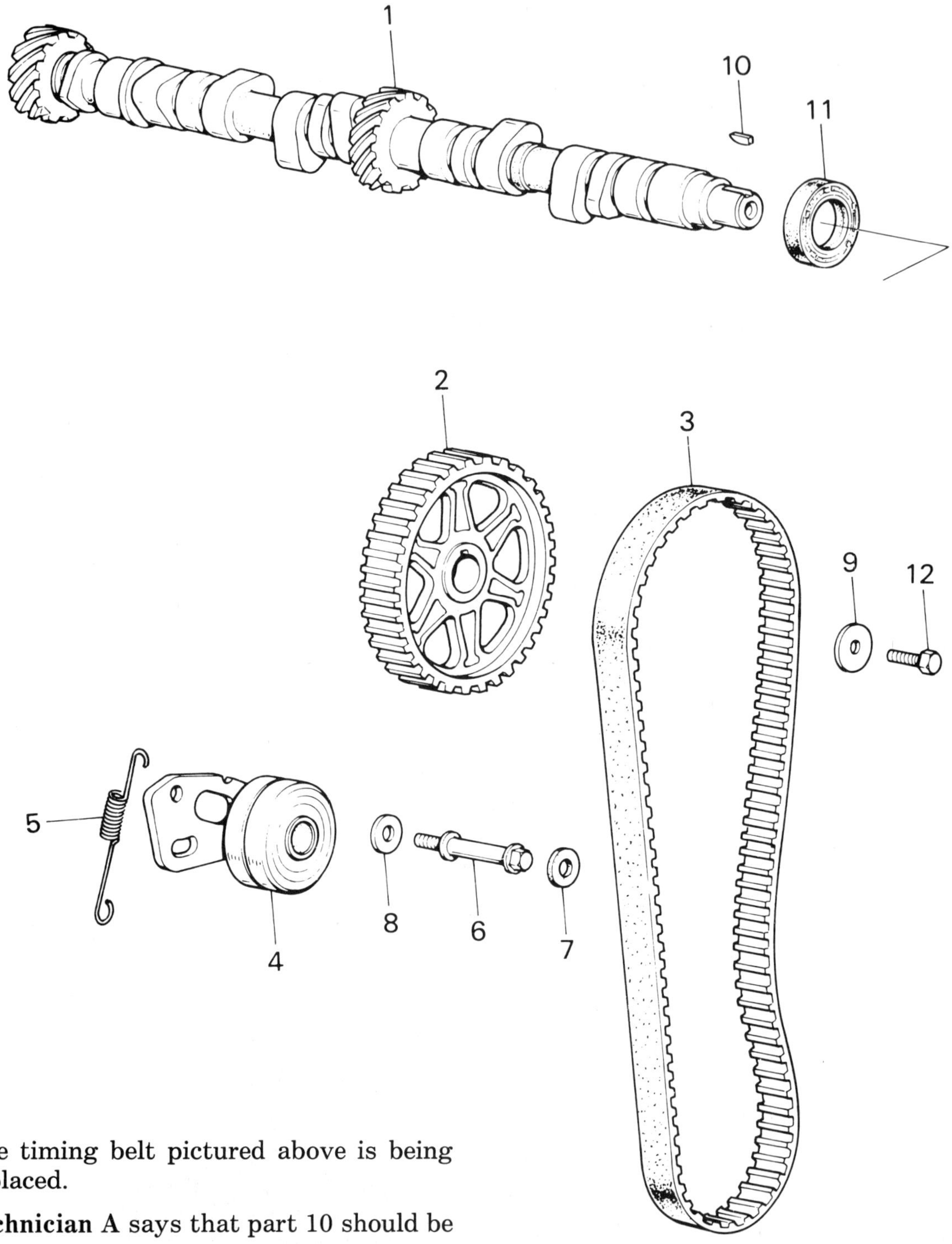

55. The timing belt pictured above is being replaced.

    Technician A says that part 10 should be new.

    Technician B says that part 4 should be coated with grease.

    Who is right?
    a. A only
    b. B only
    c. Both A and B
    d. Neither A nor B

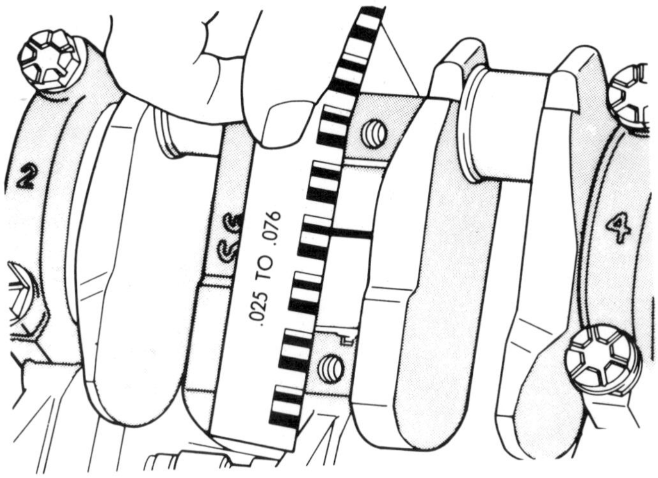

56. What is the technician doing in the above picture?
    a. Measuring rod bearing clearance
    b. Measuring main bearing clearance
    c. Determining crankshaft thrust wear
    d. Checking crankshaft journal indexing

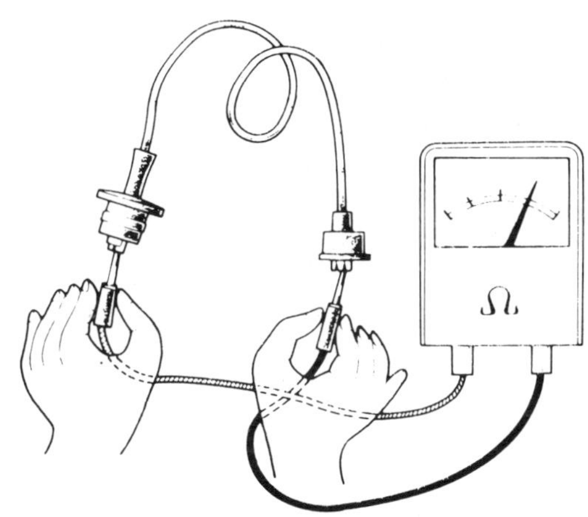

57. **Technician A** says that the above test checks resistance.

    Technician B says that the above test SHOULD NOT indicate continuity.

    Who is right?
    a. A only
    b. B only
    c. Both A and B
    d. Neither A nor B

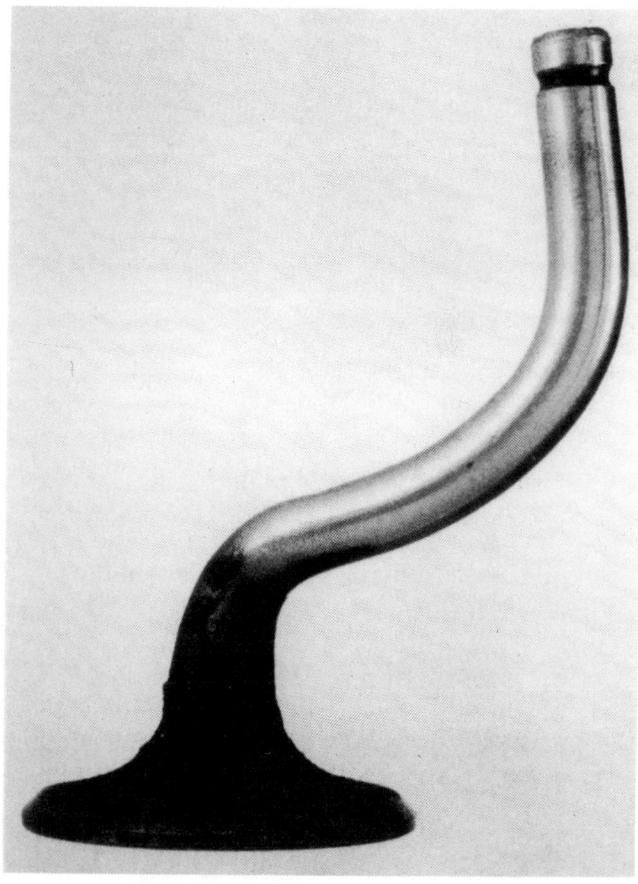

58. Which of these would be the MOST LIKELY cause of the above damage?
    a. Excessive valve clearance
    b. Worn camshaft lobes
    c. Incorrect ignition timing
    d. Broken timing belt

59. When removing the intake valves from a cylinder head, head carbon deposits under the valve heads are noticed (see the above picture). Which of these is the MOST LIKELY cause?
   a. Cracked intake manifold
   b. Malfunctioning heat riser
   c. Worn valve guides
   d. Worn camshaft lobes

60. All of the connecting rod bearing backs from an engine have a shiny, mirror-like appearance.

   Technician A says that .001″ or .002″ oversize bearings should have been installed.

   Technician B says that shim stock should have been installed behind the bearings.

   Who is right?
   a. A only
   b. B only
   c. Both A and B
   d. Neither A nor B

61. A turbocharged engine has a higher pitched sound than usual.

   Technician A says that this could be caused by an air leak between the compressor outlet and engine.

   Technician B says that this could be caused by an air leak between the engine exhaust and turbo.

   Who is right?
   a. A only
   b. B only
   c. Both A and B
   d. Neither A nor B

62. A turbocharged engine lacks power and is emitting black smoke.

   Technician A says that this could be caused by a dirty air cleaner.

   Technician B says that this could be caused by loose ducting connections.

   Who is right?
   a. A only
   b. B only
   c. Both A and B
   d. Neither A nor B

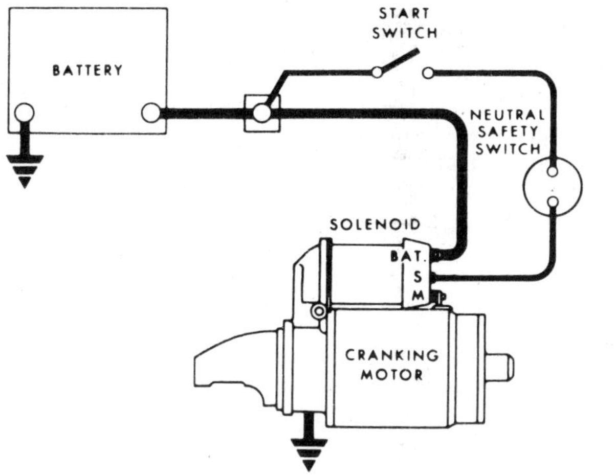

63. An engine with the above starting circuit will not crank when the start switch is closed. However, when a jumper wire is connected from the solenoid BAT terminal to the M terminal, the engine cranks.

    Technician A says that the neutral safety switch could be defective.

    Technician B says that the battery ground cable could be defective.

    Who is right?
    a. A only
    b. B only
    c. Both A and B
    d. Neither A nor B

64. Spark timing is being set. When the technician revs up the engine, the timing marks do not move. Which of these could be the reason for this?
    a. Punctured vacuum advance diaphragm
    b. Worn distributor shaft upper bushing
    c. Stuck mechanical advance weights
    d. Any of the above

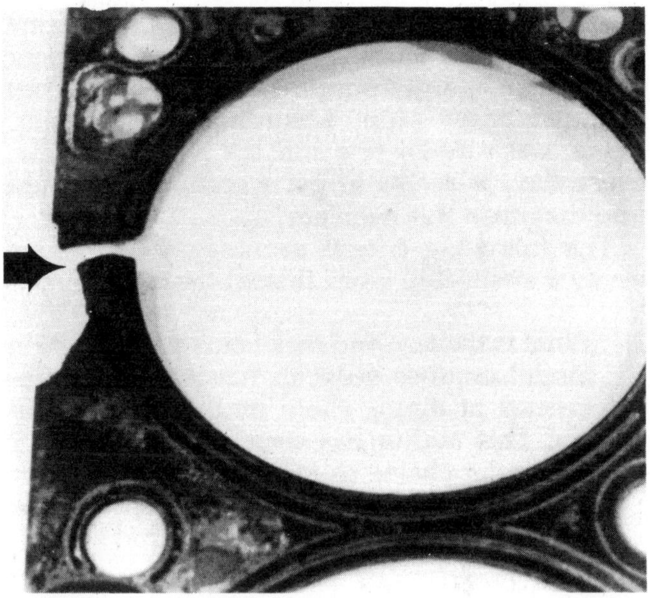

65. The above gasket has failed three times in the same place (see arrow).

    Technician A says that this could be caused by improper torque.

    Technician B says that this could be caused by incorrect tightening sequence.

    Who is right?
    a. A only
    b. B only
    c. Both A and B
    d. Neither A nor B

66. Which of these should be used to clean the cylinder bores after honing?
    a. Cleaning solvent
    b. Murphy's oil soap
    c. Isopropyl alcohol
    d. Hot water and soap

67. A FWD car with a transverse mounted engine overheats.

    Technician A says that this could be caused by a drive belt soaked with oil.

    Technician B says that this could be caused by a bad thermostatic fan clutch.

    Who is right?
    a. A only
    b. B only
    c. Both A and B
    d. Neither A nor B

68. An engine starts and idles, but will not run at higher speeds. Which of these is the MOST LIKELY cause?
    a. Bad fuel pump
    b. Worn needle
    c. Varnish on idle ports
    d. Pinched gas line

69. **Technician A** says that too much turbo boost can cause a ruptured wastegate diaphragm.

    Technician B says that too much turbo boost can cause bent compressor wheel blades.

    Who is right?
    a. A only
    b. B only
    c. Both A and B
    d. Neither A nor B

70. With the engine at idling speed, the above pointer is floating between 15 and 20. The MOST LIKELY cause would be
    a. Retarded timing
    b. Advanced timing
    c. Stuck EGR valve
    d. Too lean idle mixture

71. When installing a cylinder head, where does the tightening sequence usually start?
    a. At the ends
    b. In the middle
    c. At the bolts closest to the exhaust manifold
    d. At the bolts closest to the intake manifold

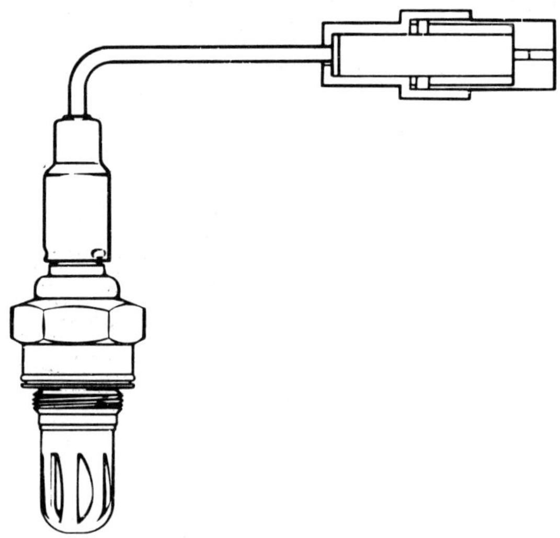

72. The above device shows a fixed voltage reading of 1.0 volt during engine idle.

    Technician A says that this indicates a lean air-fuel ratio.

    Technician B says that this indicates the absence of oxygen in the exhaust gas.

    Who is right?
    a. A only
    b. B only
    c. Both A and B
    d. Neither A nor B

73. A turbocharger is being replaced on an engine.

Technician A says to disconnect the wastegate actuator before starting the engine.

Technician B says to prime the oil feed line with fresh motor oil before starting the engine.

Who is right?
a. A only
b. B only
c. Both A and B
d. Neither A nor B

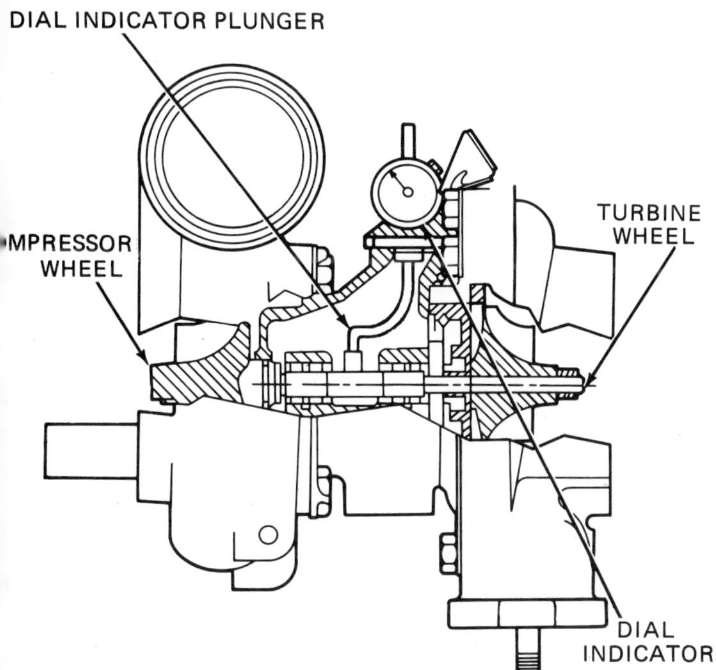

74. **Technician A** says that the above setup is measuring radial bearing clearance.

Technician B says that the above setup is measuring axial bearing clearance.

Who is right?
a. A only
b. B only
c. Both A and B
d. Neither A nor B

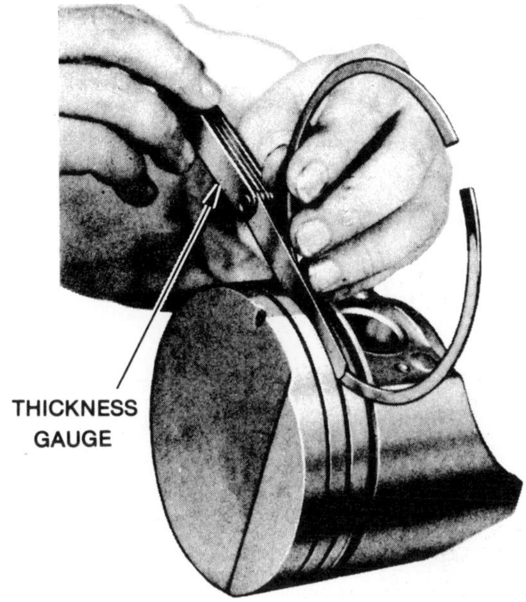

75. Refer to the above picture.

Technician A says to use a new ring when making this check.

Technician B says that a measurement of .020″ would be okay.

Who is right?
a. A only
b. B only
c. Both A and B
d. Neither A nor B

76. An engine has lifter noise.

Technician A says that this could be caused by a flat cam.

Technician B says that this could be caused by a dented oil pan.

Who is right?
a. A only
b. B only
c. Both A and B
d. Neither A nor B

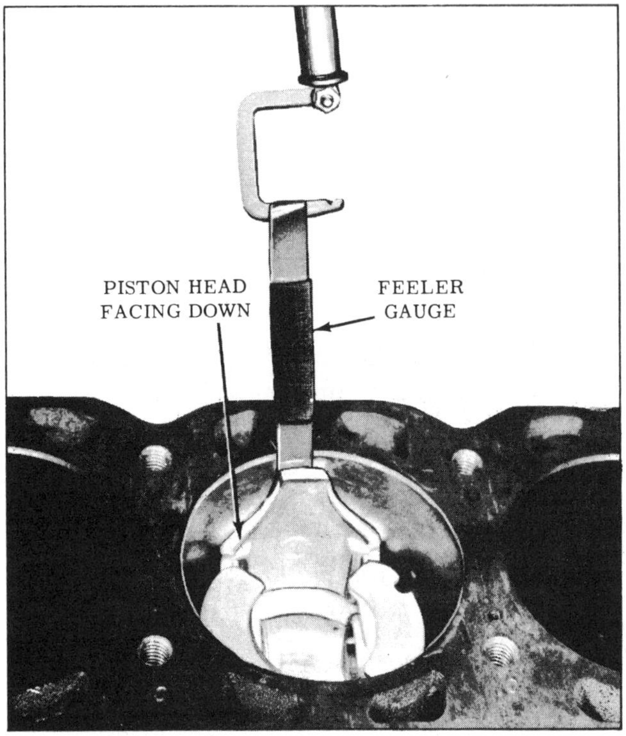

77. What is being determined in the above picture?
    a. Cylinder wall taper
    b. Piston clearance
    c. Both a and b
    d. Neither a nor b

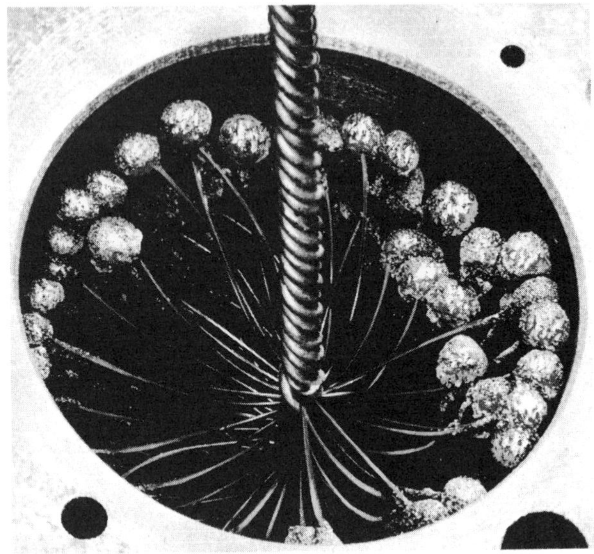

79. The above tool is used for
    a. Boring
    b. Honing
    c. Ridge reaming
    d. Cleaning

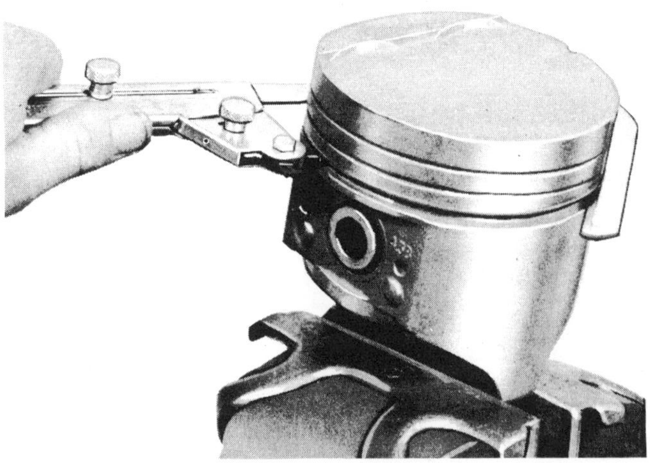

78. What is being done in the above picture?
    a. Re-sizing a worn ring groove
    b. Cleaning an oil ring groove
    c. Cleaning a compression ring groove
    d. Measuring ring groove depth

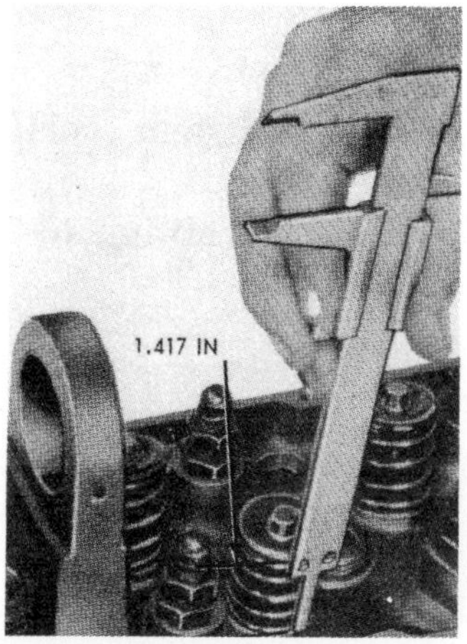

80. The above measurement is .045″ greater than the specification of 1.417″. To correct what should be done?
    a. Add shims
    b. Reduce valve stem height
    c. Replace valve spring
    d. Sink valve seat

81. You are going to replace the rod bearings in an engine. You measure the rod journal diameters at 1.948″. Standard rod journal diameter is 1.9680″ to 1.9690″. What size bearings should be installed?
    a. Undersize .010″
    b. Undersize .020″
    c. Standard
    d. Oversize .020″

82. You are overhauling a typical four-cylinder automobile engine. Which of these would be the MOST NORMAL main bearing clearance specification?
    a. .200″
    b. .020″
    c. .002″
    d. .0002″

83. When making a compression test, a low equal reading is obtained in two adjacent cylinders. This would MOST LIKELY indicate a
    a. Burned valve
    b. Blown head gasket
    c. Leaking intake manifold gasket
    d. Leaking exhaust manifold gasket

84. A piston shows the above wear pattern on the skirt. This was MOST LIKELY caused by a
    a. Twisted rod
    b. Collapsed piston
    c. Worn pin
    d. Out-of-round cylinder

85. An engine overheats only during high speed driving.

    Technician A says that this could be caused by a missing spring in the lower radiator hose.

    Technician B says that this could be caused by a bad water pump.

    Who is right?
    a. A only
    b. B only
    c. Both A and B
    d. Neither A nor B

86. A valve job has just been completed on a push rod engine that is equipped with nonadjustable hydraulic lifters. The engine will not start. A compression test is made and all the cylinders read zero. What needs to be done?
   a. Surface head(s)
   b. Install longer push rods
   c. Install thicker head gasket(s)
   d. Grind valve tips

87. **Technician A** says to replace the spring retainers when doing a valve job.

   Technician B says to replace the stem seals when doing a valve job.

   Who is right?
   a. A only
   b. B only
   c. Both A and B
   d. Neither A nor B

89. What operation is being performed in the above picture?
   a. Cleaning
   b. Knurling
   c. Peening
   d. Honing

88. **Technician A** says that the above condition can cause engine overheating.

   Technician B says that the above condition can cause water to blow out the tailpipe.

   Who is right?
   a. A only
   b. B only
   c. Both A and B
   d. Neither A nor B

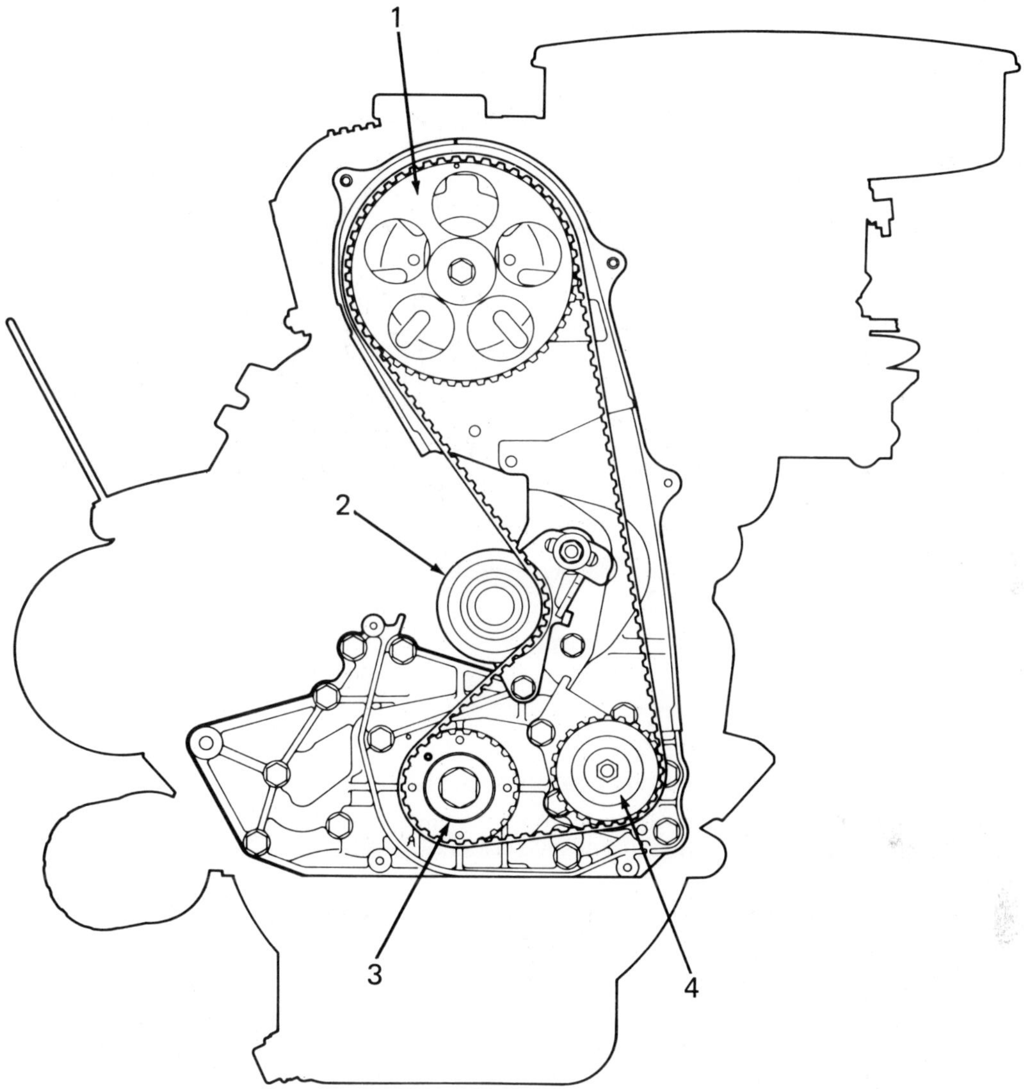

90. At which location is timing belt tension adjusted on the above engine?
    a. #1
    b. #2
    c. #3
    d. #4

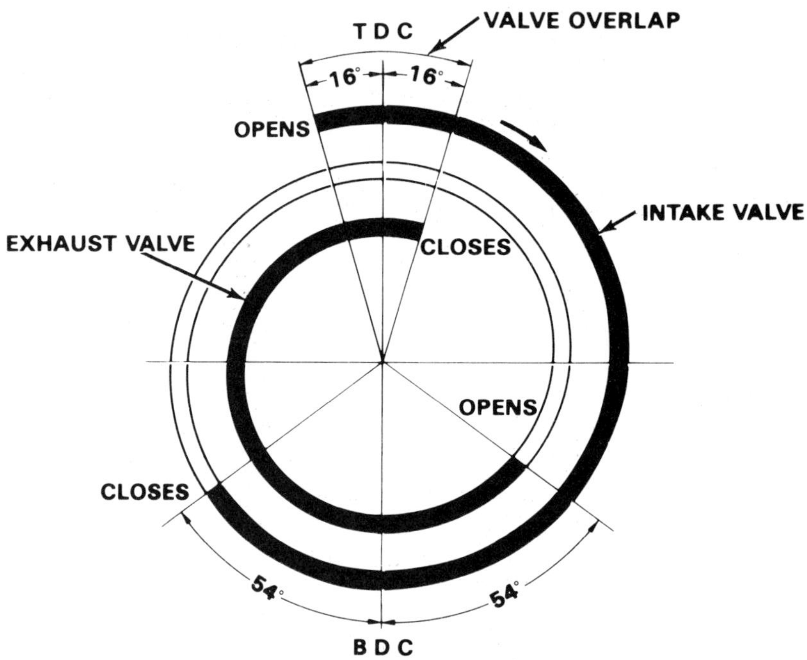

91. Refer to the above valve timing diagram. The exhaust valve is off its seat for how many degrees of crankshaft rotation?
   a. 32 deg.
   b. 108 deg.
   c. 140 deg.
   d. 250 deg.

92. Main bearing bores in a block are stretched (out-of-round).

   Technician A says that this problem can be corrected by filing the caps.

   Technician B says that this problem can be corrected by replacing the caps.

   Who is right?
   a. A only
   b. B only
   c. Both A and B
   d. Neither A nor B

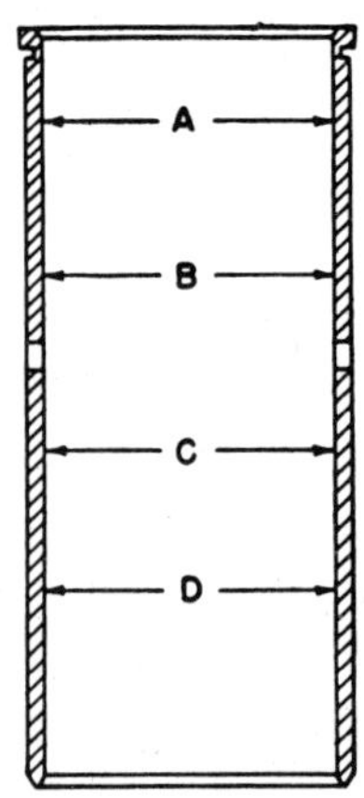

93. Refer to the cylinder cross-section picture above. The greatest wear will generally occur across point
   a. A
   b. B
   c. C
   d. D

94. A wet sleeve is fitted too loose in a cylinder block bore.

Technician A says that this can cause rapid piston ring wear.

Technician B says that this can cause cavitation.

Who is right?
a. A only
b. B only
c. Both A and B
d. Neither A nor B

95. **Technician A** says that valve burning can be caused by too thin a margin.

Technician B says that valve burning can be caused by too wide a seat.

Who is right?
a. A only
b. B only
c. Both A and B
d. Neither A nor B

96. A vehicle comes into the shop for a ring job. The cylinders measure from 4.006″ to 4.018″ directly under the ridge in the thrust direction. Original equipment (OE) bore size is 4.000″.

Technician A recommends honing the cylinders and expanding the old pistons.

Technician B recommends boring the cylinders and installing new pistons.

Who is right?
a. A only
b. B only
c. Both A and B
d. Neither A nor B

97. An engine runs at a higher than normal temperature. This could be caused by all of the following, EXCEPT
a. A plugged "cat"
b. A defective thermostat
c. A loose fan belt
d. A bad baro sensor

98. All of the compression readings on a six-cylinder engine are higher than normal.

Technician A says that this can be caused by a stretched timing belt.

Technician B says that this can be caused by excessive head milling.

Who is right?
a. A only
b. B only
c. Both A and B
d. Neither A nor B

99. An idling engine makes a hollow, muffled, bell-like sound when cold. After the engine is warm, the noise disappears.

Technician A says that this could be a valve sticking.

Technician B says that this could be piston slap.

Who is right?
a. A only
b. B only
c. Both A and B
d. Neither A nor B

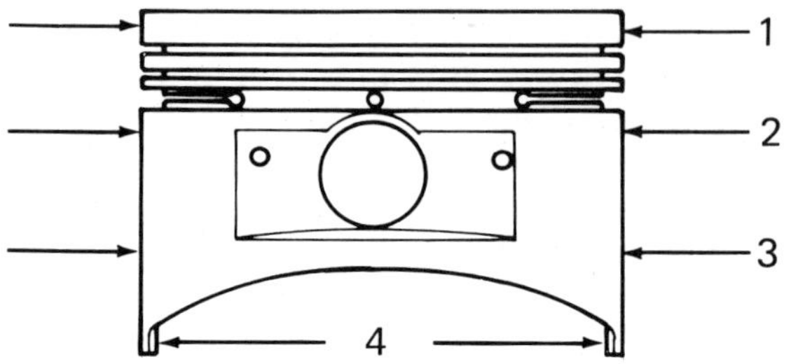

100. Refer to the above picture. Across what point is piston diameter generally measured?
   a. #1
   b. #2
   c. #3
   d. #4

101. As an engine cools off, the upper radiator hose collapses.

   Technician A says that this could be caused by a defective thermostat.

   Technician B says that this could be caused by a loose hose clamp.

   Who is right?
   a. A only
   b. B only
   c. Both A and B
   d. Neither A nor B

102. **Technician A** says that a defective head gasket can cause rust to form in the cooling system.

   Technician B says that a defective water pump can cause rust to form in the cooling system.

   Who is right?
   a. A only
   b. B only
   c. Both A and B
   d. Neither A nor B

103. What is being measured in the above picture?
   a. Tension
   b. Working height
   c. Coil clearance
   d. Squareness

104. An engine is equipped with a coolant recovery system. Coolant does not return to the radiator when the engine cools. This could be caused by a
   a. Plugged transfer hose
   b. Cracked filler neck joint
   c. Both a and b
   d. Neither a nor b

105. An engine equipped with electronic ignition will not start. This could be caused by
   a. A punctured rotor
   b. Improper pickup coil air gap
   c. Both a and b
   d. Neither a nor b

106. A fluid fan drive assembly is pictured above.

Technician A says that part X senses air temperature.

Technician B says that part X engages the drive.

Who is right?
a. A only
b. B only
c. Both A and B
d. Neither A nor B

107. A technician is adjusting hydraulic lifters by turning to a specified setting *past* zero-lash.

Technician A says that this is to position the plunger properly in the lifter body.

Technician B says that this is to set proper valve lift.

Who is right?
a. A only
b. B only
c. Both A and B
d. Neither A nor B

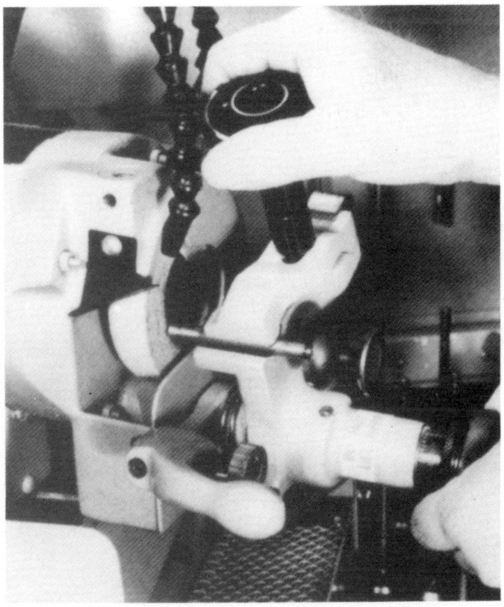

108. What is the technician doing in the above picture?
a. Putting new finish on valve end
b. Grinding tip length
c. Both a and b
d. Neither a nor b

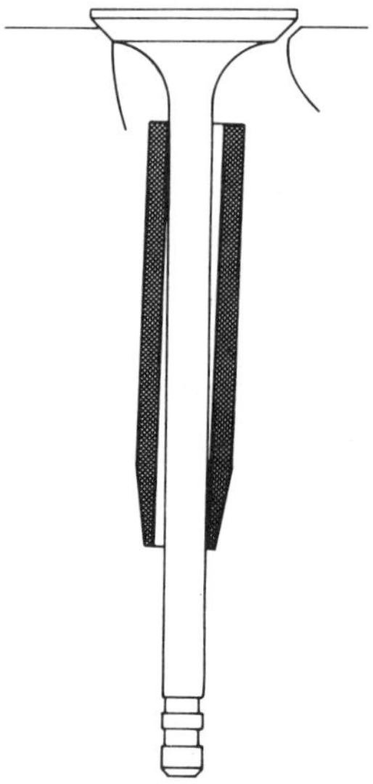

109. **Technician A** says that excessive stem-to-guide clearance is shown above.

Technician B says that valve leakage is shown above.

Who is right?
a. A only
b. B only
c. Both A and B
d. Neither A nor B

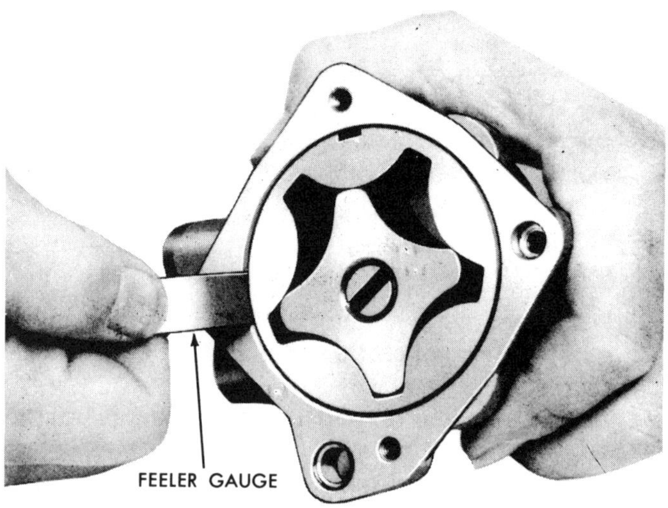

110. The above clearance measurement is excessive. What should be done?

Technician A says to add shims.

Technician B says to install oversize rotors.

Who is right?
a. A only
b. B only
c. Both A and B
d. Neither A nor B

111. **Technician A** says that some engines use head bolts that are designed to "stretch."

Technician B says that some engines use head bolts that can be used only once.

Who is right?
a. A only
b. B only
c. Both A and B
d. Neither A nor B

112. A crankshaft journal is being measured in the same direction at opposite ends. This is done when measuring for
a. Taper
b. Out-of-roundness
c. Both a and b
d. Neither a nor b

113. An engine oil light "flickers" while driving. Also, a knock can be heard when the engine is cold. This could be caused by
   a. Worn bearings
   b. Wrong viscosity oil
   c. Both a and b
   d. Neither a nor b

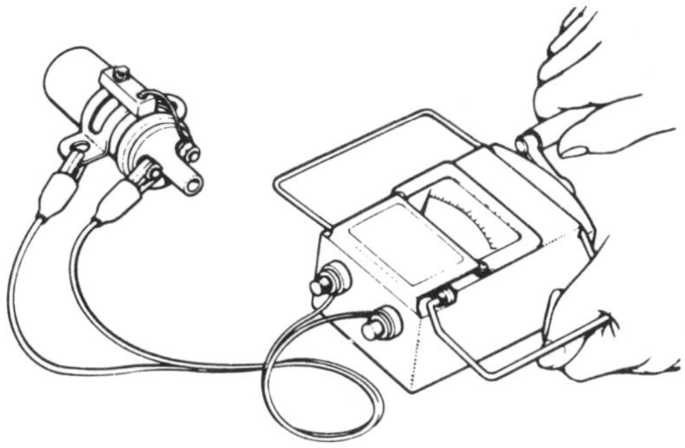

114. What is being tested in the above drawing?
   a. Primary winding resistance
   b. Secondary winding resistance
   c. Coil capacitance
   d. None of the above

115. After sitting overnight, an engine has lifter noise that goes away after running for a short period of time. The MOST LIKELY cause is
   a. Low oil pressure
   b. Oil level too low
   c. A concave lifter bottom
   d. Excessive lifter leakdown

116. An engine has excessive oil consumption. Wet and dry compression tests both are acceptable.

   Technician A says that there could be a problem in the PCV system.

   Technician B says that there could be a problem with the valve guide seals.

   Who is right?
   a. A only
   b. B only
   c. Both A and B
   d. Neither A nor B

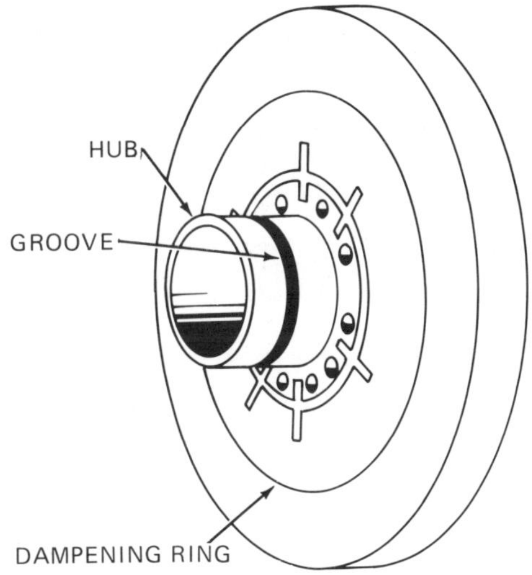

117. The above part has a groove worn into its hub.

   Technician A says that this can cause an oil leak.

   Technician B says that repair sleeves are available.

   Who is right?
   a. A only
   b. B only
   c. Both A and B
   d. Neither A nor B

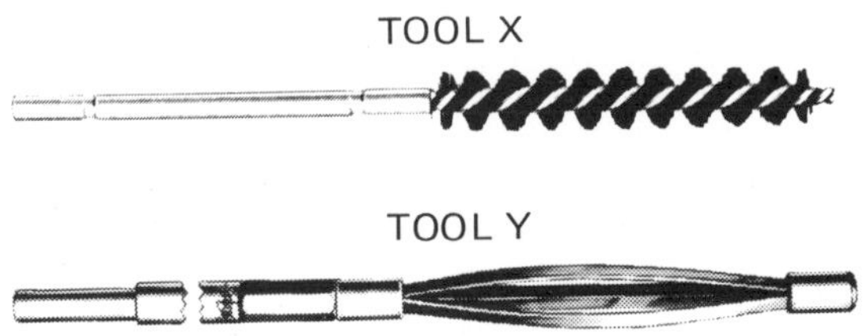

118. Refer to the above tools.

   Technician A says that tool X is used to clean valve guides.

   Technician B says that tool Y is used to ream valve guides.

   Who is right?
   a. A only
   b. B only
   c. Both A and B
   d. Neither A nor B

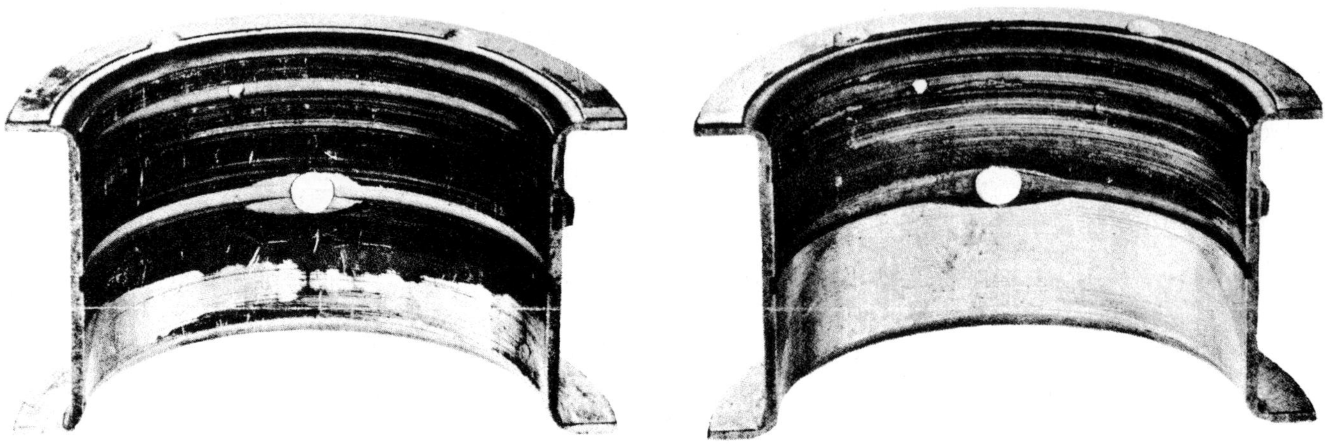

119. Which of the following is the MOST LIKELY cause of the above wear?
   a. Bent connecting rod
   b. Tapered journal
   c. Fatigue
   d. Improper installation

120. A computer controlled engine won't go into closed loop.

   Technician A says that the oxygen sensor could be coated with carbon.

   Technician B says that the thermostat could be missing.

   Who is right?
   a. A only
   b. B only
   c. Both A and B
   d. Neither A nor B

121. The electric cooling fan on a front-wheel-drive vehicle is inoperative.

   Technician A says that this could be due to a bad ground to the fan motor.

   Technician B says that this could be due to a melted wire to the fan relay.

   Who is right?
   a. A only
   b. B only
   c. Both A and B
   d. Neither a nor b

122. An engine has a knocking noise on medium to hard acceleration. Which of these would be the MOST LIKELY cause?
   a. Ignition pickup (trigger) coil shorted
   b. Vacuum advance diaphragm leaking
   c. Fuel mixture too rich
   d. EGR passage restricted

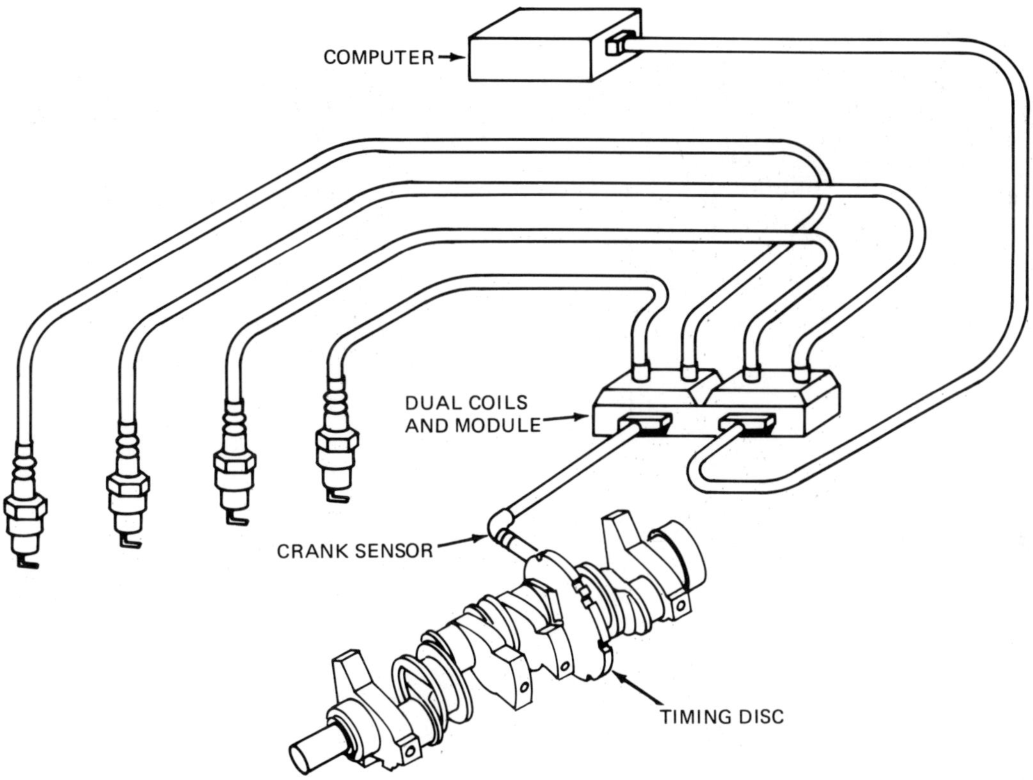

123. An engine equipped with a distributorless ignition system (DIS), as shown above, won't start.

Technician A says that this could be caused by a faulty crank sensor.

Technician B says that this could be caused by a wiring problem.

Who is right?
a. A only
b. B only
c. Both A and B
d. Neither A nor B

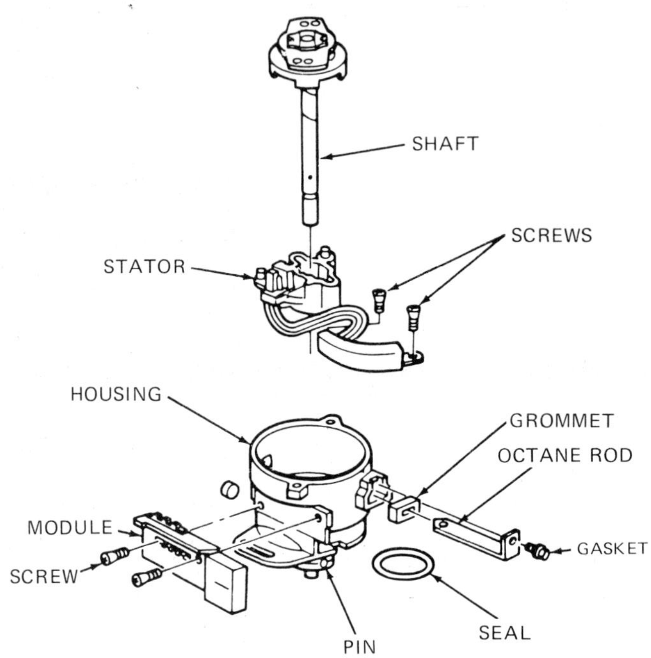

124. An engine equipped with the above distributor will crank, but there is no spark.

    Technician A says that the stator could be bad.

    Technician B says that the module could be bad.

    Who is right?
    a. A only
    b. B only
    c. Both A and B
    d. Neither A nor B

Appendix A

SOURCES OF MANUALS

Information on the availability and price of factory service manuals may be obtained by contacting the specific car manufacturer listed below:

Alfa Romeo, S.P.A.
National Headquarters
250 Sylvan Avenue
Englewood Cliffs, NJ 07632

American Honda Motor Company, Inc.
100 West Alondra Blvd.
Gardena, CA 90247

American Isuzu Motors Inc.
Izuzu Motors Ltd.
2300 Pellissier Place
Whittier, CA 90601

American Motors Corporation
14250 Plymouth Road
Detroit, MI 48232

BMW of North America, Inc.
Bayerische Motoren Werke A.G.
BMW Plaza
Montvale, NJ 07645

Buick Motor Division
General Motors Corporation
Flint, MI 48550

Cadillac Motor Car Division
General Motors Corporation
2860 Clark Avenue
Detroit, MI 48202

Chevrolet Motor Division
General Motors Corporation
General Motors Building
Detroit, MI 48202

Chrysler Corporation
P.O. Box 1919
Detroit, MI 48231

Dodge Division
Chrysler Corporation
P.O. Box 857
Detroit, MI 48231

FIAT, S.P.A.
Strada Torina 50
Orbassano, Italy 10043

Ford Motor Company
The American Road
Dearborn, MI 48121

Hyundai Motor America
7373 Hunt Avenue
P.O. Box 2669
Garden Grove, CA 92642-2669

Jaguar Rover Triumph, Inc.
600 Willow Tree Road
Leonia, NJ 07605

Lincoln-Mercury Division
Ford Motor Company
3000 Schaefer Road
Dearborn, MI 48216

Mazda Motor Corporation
Toyo Kogyo Company Ltd.
821 Brock Road South
Pickering, Ontario Canada L1W 3L6

Mercedes-Benz of North America, Inc.
Daimler-Benz A.G.
1 Mercedes Drive
Montvale, NJ 07645

Mitsubishi Motor Sales of America, Inc.
West Office Tower
10540 Talbert Avenue
Fountain Valley, CA 92728-9060

Nissan Motor Company, Ltd.
18501 S. Figueroa Street
Carson, CA 90248

Oldsmobile Division
General Motors Corporation
920 Townsend Street
Lansing, MI 48921

Peugeot Motors of America, Inc.
1 Peugeot Plaza
Lyndhurst, NJ 07071

Plymouth Division
Chrysler Corporation
P.O. Box 857
Detroit, MI 48231

Pontiac Motor Division
General Motors Corporation
1 Pontiac Plaza
Pontiac, MI 48053

Porsche-Audi
Worldwide Volkswagen Corporation
340 Rt. 59
West Nyack, NY 10994

Saab Scania of America
Saab-Scania Aktiebolag
Saab Drive
Orange, CT 06477

Subaru of America, Inc.
7040 Central Highway
Pennsauken, NJ 08109

Toyota Motor Sales, U.S.A., Inc.
19001 S. Western Avenue
Torrance, CA 90509

Volkswagen of America
Corporate Service Division
888 West Big Beaver Road
Troy, MI 48007

Volvo of America Corporation
AB Volvo
One Volvo Drive
Rockleigh, NJ 07647

SOURCES OF COMPREHENSIVE SHOP MANUALS

The following companies publish comprehensive shop manuals that contain commonly used specifications and the most frequently performed service operations.

Chilton Book Company
Chilton Way
Radnor, PA 19809

Mitchell Manuals Inc., A Cordura Company
P.O. Box 26260
San Diego, CA 92126

Motor Publications
555 West 57th Street
New York, NY 10019

Appendix B

ENGINE DISASSEMBLY AND REASSEMBLY CHECKLIST

The checklist that follows is designed to serve as a guide during engine teardown. The sequence and steps listed will apply to most engines. Completing such a checklist will help you avoid missing important steps and provide space for recording key measurements.

Make of engine ______________________________

Displacement ____________ No. of cylinders ___

Firing order ______________________________
Bore and stroke ____ x ____

Prior to engine teardown, record any compression or leakdown data in the *NOTES* column .... ☐

Disconnect battery if engine is being worked on in vehicle .... ☐

Drain engine oil and note condition .... ☐

Remove distributor cap and wires............. ☐

Remove distributor and hold-down .... ☐

Remove spark plugs and note condition... ☐

Remove oil filter assembly, any side cover(s), and fuel pump and lines .... ☐

Remove the carburetor or injector assembly .... ☐

Drain fuel from fuel lines .... ☐

NOTES:

Remove all components that are not part of the main block assembly (alternator, water pump, belts, hoses, etc.) .... ☐

Remove intake manifold .... ☐

Remove exhaust manifold(s) .... ☐

Remove valve cover(s) .... ☐

Remove rocker arm assembly .... ☐

On OHC (overhead camshaft) engines, remove camshaft timing gear(s), bearing caps, bearings, and camshaft .... ☐

Remove cylinder head(s) .... ☐

Remove valves from head(s) .... ☐

Check block and head surfaces for warpage .... ☐

Remove the crankshaft pulley .... ☐

Remove the oil pan .... ☐

Remove the timing chain cover .... ☐

Remove rod and piston assemblies .... ☐

REMARKS:

1. Use protection covers on rod bolts.
2. Position crank throw at BDC when installing or removing pistons.
3. Observe the coloration and visual condition of the rod bearings.

Amount of cylinder taper

No. 1 ______	No. 5 ______
No. 2 ______	No. 6 ______
No. 3 ______	No. 7 ______
No. 4 ______	No. 8 ______

NOTES:

Piston measurement of skirts perpendicular to pin

No. 1 ____________ No. 5 ____________

No. 2 ____________ No. 6 ____________

No. 3 ____________ No. 7 ____________

No. 4 ____________ No. 8 ____________

Piston-to-cylinder-wall clearance

No. 1 ____________ No. 5 ____________

No. 2 ____________ No. 6 ____________

No. 3 ____________ No. 7 ____________

No. 4 ____________ No. 8 ____________

Main bearing clearance

No. 1 ____________ No. 5 ____________

No. 2 ____________ No. 6 ____________

No. 3 ____________ No. 7 ____________

No. 4 ____________ No. 8 ____________

Rod bearing clearance

No. 1 ____________ No. 5 ____________

No. 2 ____________ No. 6 ____________

No. 3 ____________ No. 7 ____________

No. 4 ____________ No. 8 ____________

Crankshaft end play measurement........ ____

Check new pinlocks (if so equipped)

- Proper seating.. ☐
- Open end facing down............................ ☐

Piston rings

- Establish proper ring relationship .... ☐
- Install on pistons................................... ☐
- Ring end gaps staggered correctly .... ☐
- In the *NOTES* column, make a drawing showing ring end gap relationships .... ☐

NOTES:

Piston and rod bearing assemblies installed (use covers on rod bolts) .... ☐

Valve seat width and location satisfactory .... ☐

Cylinder head(s) and gasket(s) installed.... ☐

All other remaining parts installed .... ☐

Adjust the valve clearance (statically) in two revolutions of the crankshaft .... ☐

Verify engine oil level satisfactory .... ☐

Engine is now ready to start ... Yes ☐ No ☐

NOTES:

ENGINE BREAK-IN PROCEDURE

1. Double-check all fluid levels—water, oil, and so on.

2. Prime oil system. Use a quality detergent oil for break-in. Use *no* anti-freeze for first 500 miles.

3. Make sure distributor is installed "in-time" and all plug wires are attached in correct firing order.

4. It is *very important* that the engine starts immediately. Do not crank and crank—this could destroy the camshaft and cam bearings. If engine does not start, stop and find out why.

5. The first 15 minutes of the engine's running time is very important. *Do not idle*—keep running at 1500 rpm. Do not allow overheating.

6. After you have rechecked all fluid levels, belts, hoses, etc., please follow this 500 mile break-in procedure:
   a. Do not idle in stop and go traffic. A new engine will always run warmer than the tired one you replaced.
   b. Do not "baby" the engine, but do not abuse it. A 200 or 300 mile freeway trip would be ideal.
   c. After 500 miles, change the oil and filter, and add anti-freeze.

1981–88 FORD ESCORT 1.6L & 1.9L ENGINES CYLINDER HEAD INSTALLATION PRECAUTIONS

Escort cylinder heads are very prone to cracking and warpage! The damaged cylinder head is not the problem, it is the result of a problem. If the problem is not corrected, the replacement cylinder head will surely suffer damage too.

These cylinder heads crack or warp due to excessive heat. This heat can come from the coolant and/or combustion chamber temperature. High combustion chamber temperatures are not always indicated by high coolant temperatures. Tuning problems, such as air-fuel mixtures, spark timing, or excessive back pressure can also cause high combustion chamber temperatures.

COMPLETE AND CHECK ITEMS BEFORE INSTALLING A REPLACEMENT ESCORT CYLINDER HEAD:

☐ 1. Check the electric fan for proper operation and replace the fan switch if necessary.

☐ 2. Replace the thermostat.

☐ 3. Have radiator rodded or recored and replace questionable hoses.

☐ 4. Insure that the radiator pressure cap operates properly.

☐ 5. Use an antifreeze formulated specifically for aluminum.

☐ 6. Check water pump and belts.

☐ 7. Clean or replace air and fuel filters, crankcase vent tube and the PCV valve.

☐ 8. Check EGR valve operation.

☐ 9. This cylinder head must be installed using new head bolts. These are *torque-to-yield bolts* that must not be reused. Use antiseize on the bolt threads and underside of bolt heads; torque them to 44 ft. lbs. in small increments and in the proper sequence. Loosen them two (2) complete turns and retorque to 44 ft. lbs. in small increments and in the proper sequence. Turn all bolts in sequence 90° (1/4 turn). Turn all bolts in sequence an additional 90°. *This is the only correct way to torque Escort cylinder heads.*

☐ 10. Replace the oil and oil filter.

☐ 11. Clean the intake and exhaust manifolds thoroughly making sure that the gasket surfaces are clean and flat. Air leaks create lean conditions that elevate cylinder temperatures.

☐ 12. Check distributor operation and the advance curve. Be especially watchful of the operation of the mechanical advance weights and the vacuum advance.

☐ 13. Check fuel pump operation and pressure.

☐ 14. Check block surface for warpage.

☐ 15. Check for oil and water leaks. This engine is difficult to fill with coolant without creating air pockets. Some thermostats tend to create air pockets and prevent the system from being fully charged until coolant is circulating. Air pockets cause hot spots and damage heads. Make sure coolant is circulating and is at the proper level.

☐ 16. Check the compression in each cylinder.

☐ 17. Check the emission control systems for correct operation; don't overlook the oxygen sensor.

☐ 18. Check and note ignition timing.

☐ 19. Check and note exhaust emissions.

☐ 20. Check and note air-fuel ratio.

☐ 21. Make sure the camshaft and crankshaft are correctly timed before the rocker arms are tightened.

These instructions may seem excessive in view of the simplicity of installing other cylinder heads. **BEWARE**, the Ford Escort cylinder head is not like other cylinder heads. They are less forgiving and require all of the engine operating systems to be operating within specifications.

Appendix C

TORQUE CHARTS/ RELATED INFORMATION

U.S. STANDARD

GRADE OF BOLT		S.A.E. 1 & 2	S.A.E. 5	S.A.E. 6	S.A.E. 8		
MINIMUM TENSILE STRENGTH		64,000 P.S.I.	105,000 P.S.I.	133,000 P.S.I.	150,000 P.S.I.		
GRADE MARKING ON HEAD ➔						SOCKET OR WRENCH SIZE	
U.S. STANDARD		TORQUE (IN FOOT POUNDS)				U.S. REGULAR	
BOLT DIAMETER	DEC. EQUIV.					BOLT HEAD	NUT
1/4	.250	5	7	10	10.5	3/8	7/16
5/16	.3125	9	14	19	22	1/2	9/16
3/8	.375	15	25	34	37	9/16	5/8
7/16	.4375	24	40	55	60	5/8	3/4
1/2	.500	37	60	85	92	3/4	13/16

THE FOLLOWING RULES APPLY TO THE CHART:

1. Consult manufacturers specific recommendations when available.
2. The chart may be used directly when any of the following lubricants are used: *NEVER-SEEZ COMPOUND, MOLYKOTE, FEL-PRO C-5, GRAPHITE AND OIL OR SIMILAR MIXTURES.*
3. Increase the torque by 20% when using engine oil or chassis grease as a lubricant.
4. Reduce torque by 20% when plated bolts are used.
5. Increase torque by 20% when multiple tapered tooth lockwashers are used.

CAUTION: Tightening into aluminum usually will require less torque.

METRIC STANDARD

GRADE OF BOLT		5D	8G	10K	12K		
MINIMUM TENSILE STRENGTH		71,160 P.S.I.	113,800 P.S.I.	142,000 P.S.I.	170,674 P.S.I.		
GRADE MARKING ON HEAD →		5D	8G	10K	12K	SOCKET OR WRENCH SIZE	
METRIC		TORQUE (IN FOOT POUNDS)				METRIC	
BOLT DIAMETER	U.S. DEC. EQUIV.					BOLT HEAD	NUT
6mm	.2362	5	6	8	10	10mm	10mm
8mm	.3150	10	16	22	27	14mm	14mm
10mm	.3937	19	31	40	49	17mm	17mm
12mm	.4720	34	54	70	86	19mm	19mm
14mm	.5512	55	89	117	137	22mm	22mm

THE FOLLOWING RULES APPLY TO THE CHART:

1. Consult manufacturers specific recommendations when available.
2. The chart may be used directly when any of the following lubricants are used: *NEVER-SEEZ COMPOUND, MOLYKOTE, FEL-PRO C-5, GRAPHITE AND OIL OR SIMILAR MIXTURES.*
3. Increase the torque by 20% when using engine oil or chassis grease as a lubricant.
4. Reduce torque by 20% when plated bolts are used.
5. Increase torque by 20% when multiple tapered tooth lockwashers are used.

CAUTION: Tightening into aluminum usually will require less torque.

BOLT NOMENCLATURE

(ENGLISH) INCH SYSTEM Bolt, 1/2-13x1

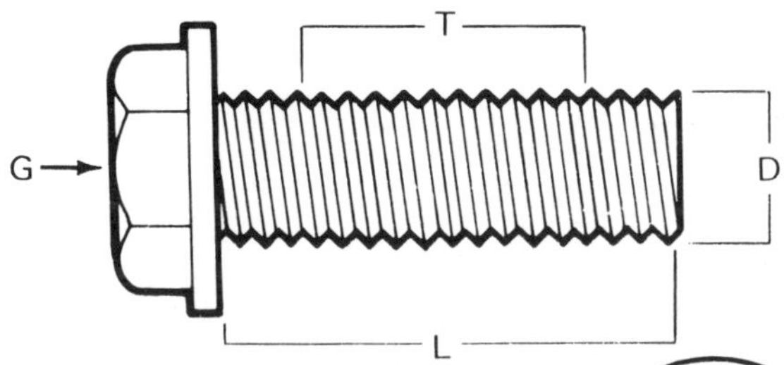

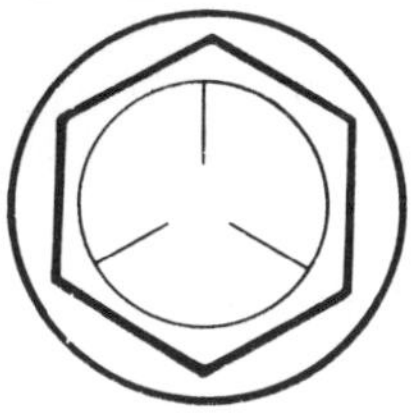

G- Grade Marking (bolt strength)
L- Length, (inches)**
T- Thread Pitch (thread/inch)
D- Nominal Diameter (inches)

METRIC SYSTEM Bolt M12-1.75x25

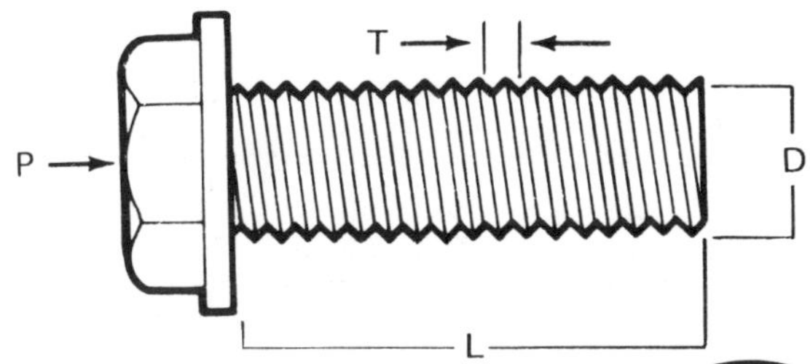

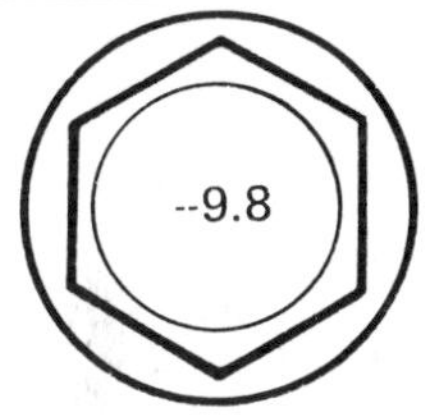

P- Property Class* (bolt strength)
L- Length (millimeters)**
T- Thread Pitch (thread width crest to crest mm)
D- Nominal Diameter (millimeters)

(ENGLISH) INCH SYSTEM

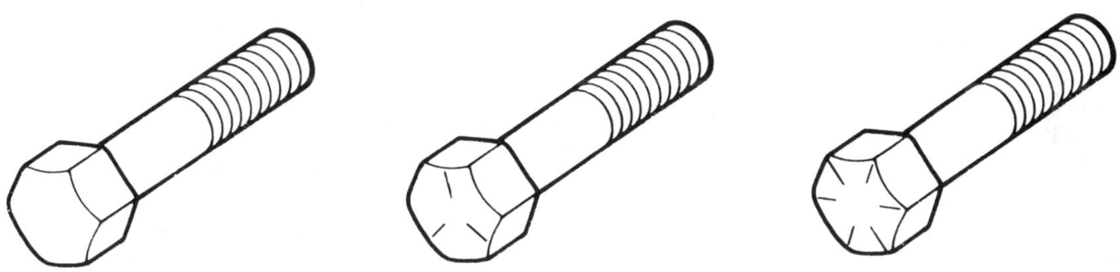

Grade 1 or 2　　Grade 5　　Grade 8

Increasing number of slashes represent increasing strength.

METRIC SYSTEM

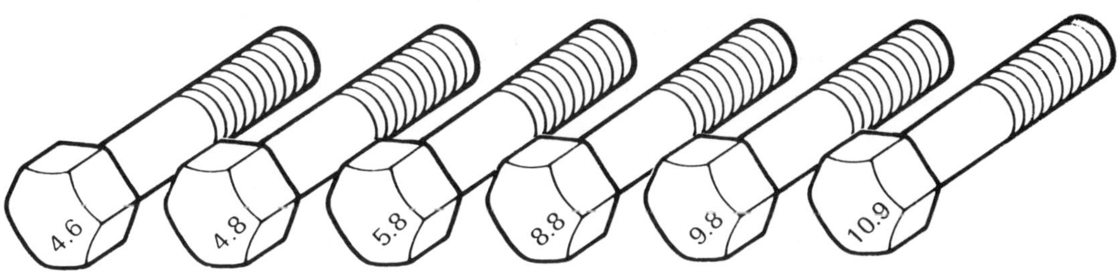

Increasing numbers represent increasing strength.

METRIC WRENCH SYSTEM

WRENCH SIZE	*7mm*	*8mm*	*10mm*	*11mm*	*13mm*	*17mm*
METRIC DIAMETER	*4mm*	*5mm*	*6mm*	*7mm*	*8mm*	*10mm*

METRIC DIAMETER	*12mm*	*14mm*	*16mm*	*20mm*
WRENCH SIZE	*19mm*	*22mm*	*24mm*	*30mm*

SOCKET SIZE COMPARISON

REGULAR SOCKET SIZES	DECIMAL EQUIVALENTS		METRIC SOCKET SIZES
1/8"	.125	.118	3mm.
3/16"	.187	.157	4mm.
1/4"	.250	.236	6mm.
5/16"	.312	.354	9mm.
3/8"	.375	.394	10mm.
7/16"	.437	.472	12mm.
1/2"	.500	.512	13mm.
9/16"	.562	.590	15mm.
5/8"	.625	.630	16mm.
11/16"	.687	.709	18mm.
3/4"	.750	.748	19mm.
13/16"	.812	.787	20mm.
7/8"	.875	.866	22mm.
15/16"	.937	.945	24mm.
1"	1.00	.984	25mm.

COMPARISON CHARTS

DIAMETER IN INCHES

Dia. in Inches	INCH SYSTEM Nominal Size	Dec. Size	Coarse TPI	Fine TPI	METRIC SYSTEM Nominal Size	Dec. Eq. (in)	Coarse Pitch mm	Coarse TPI	Fine Pitch mm	Fine TPI	Nominal Size
					M27	1.063	3	8.5	2	12.5	M27x2
1.050											
1.000	1	1.000	8	12							
.950					M24	.945	3	8.5	2	12.5	M24x2
.900											
	7/8	.875	9	14	M22	.866	2.5	10	1.5	17	M22x1.5
.850											
					M20	.787	2.5	10	1.5	17	M20x1.5
.750	3/4	.750	10	16							
					M18	.709	2.5	10	1.5	17	M18x1.5
.700											
	5/8	.625	11	18	M16	.630	2	12.5	1.5	17	M16x1.5
.600											
.550					M14	.551	2	12.5	1.5	17	M14x1.5
.500	1/2	.500	13	20							
					M12	.471	1.75	14.5	1.25	20	M12x1.25
.450											
	7/16	.437	14	20							
.400					M10	.393	1.5	17	1.25	20	M10x1.25
	3/8	.375	16	24							
.350											
	5/16	.312	18	24	M8	.315	1.25	20	1.0	25	M8x1
.300											
.250	1/4	.250	20	28							
					M6	.236	1.0	25	.75	34	M6x.75
.200					M5	.196	.8	32	.5	51	M5x.5
	10	.190	24	32							
	8	.164	32	36	M4	.157	.7	36	.35	51	M4x.5
.150											
	6	.138	32	40							
	5	.125	40	44	M3	.118	.5	51	.35	74	M3x.35
	4	.112	40	48							
.100	3	.099	48	50	M2.5	.098	.45	56	.35	74	M2.5x.35
	2	.086	56	64	M2	.079	.4	64	.25	101	M2x.25
	1	.073	64	72	M1.6	.063	.35	74	.2	127	M1.6x.2
	0	.060		80	M1.4	.055	.3	85	.2	127	M1.4x.2
.050											

DIAMETER/THREAD PITCH COMPARISON

TORQUE CONVERSION

NEWTON METRES (N•m)	POUND-FEET (LB.-FT.)
1	0.7376
2	1.5
3	2.2
4	3.0
5	3.7
6	4.4
7	5.2
8	5.9
9	6.6
10	7.4
15	11.1
20	14.8
25	18.4
30	22.1
35	25.8
40	29.5
50	36.9
60	44.3
70	51.6
80	59.0
90	66.4
100	73.8
110	81.1
120	88.5
130	95.9
140	103.3
150	110.6
160	118.0
170	125.4
180	132.8
190	140.1
200	147.5
225	166.0
250	184.4

POUND-FEET (LB.-FT.)	NEW
1	
2	
3	
4	
5	
6	
7	
8	
9	
10	
15	
20	
25	
30	
35	
40	
45	
50	
55	
60	
65	
70	
75	1
80	1
90	1
100	1
110	14
120	16
130	17
140	18
150	20
160	21
170	23
180	24

3 EASY WAYS TO

Appendix D

INCHES AND MILLIMETERS CONVERSION CHART

Many dimensions in the automotive industry are specified in millimeter (mm) notation. As a result, today's technician must think in terms of metric measurement. However, for the present and the near future, he or she will also be using inch measurements. Until a full transition to metrics takes place, there will be times when it is necessary to convert from one system to the other. The chart on the next page will serve as an aid for making metric-inch conversions.

MILLIMETERS TO DECIMAL(inches)						DECIMAL(inches) TO METRIC					
MM	DECIMAL (Inches)	MM	DECIMAL (Inches)	MM	DECIMAL (Inches)	DECIMAL (Inches)	MM	DECIMAL (Inches)	MM	DECIMAL (Inches)	MM
1	.039	41	1.614	81	3.189	.001	.02	.320	8.12	.720	18.28
2	.079	42	1.654	82	3.228	.002	.05	.330	8.38	.730	18.54
3	.118	43	1.693	83	3.268	.003	.07	.340	8.63	.740	18.79
4	.157	44	1.732	84	3.307	.004	.10	.350	8.89	.750	19.05
5	.197	45	1.772	85	3.346	.005	.12	.360	9.14	.760	19.30
6	.236	46	1.811	86	3.386	.006	.15	.370	9.39	.770	19.55
7	.276	47	1.850	87	3.425	.007	.17	.380	9.65	.780	19.81
8	.315	48	1.890	88	3.465	.008	.20	.390	9.90	.790	20.06
9	.354	49	1.929	89	3.504	.009	.22	.400	10.16	.800	20.32
10	.394	50	1.969	90	3.543	.010	.25	.410	10.41	.810	20.57
11	.433	51	2.008	91	3.583	.020	.50	.420	10.66	.820	20.82
12	.472	52	2.047	92	3.622	.030	.76	.430	10.92	.830	21.08
13	.512	53	2.087	93	3.661	.040	1.01	.440	11.17	.840	21.33
14	.551	54	2.126	94	3.701	.050	1.27	.450	11.43	.850	21.59
15	.591	55	2.165	95	3.740	.060	1.52	.460	11.68	.860	21.84
16	.630	56	2.205	96	3.780	.070	1.77	.470	11.93	.870	22.09
17	.669	57	2.244	97	3.819	.080	2.03	.480	12.19	.880	22.35
18	.709	58	2.283	98	3.858	.090	2.28	.490	12.44	.890	22.60
19	.748	59	2.323	99	3.898	.100	2.54	.500	12.70	.900	22.86
20	.787	60	2.362	100	3.937	.110	2.79	.510	12.95	.910	23.11
21	.827	61	2.402	125	4.921	.120	3.04	.520	13.20	.920	23.36
22	.866	62	2.441	150	5.906	.130	3.30	.530	13.46	.930	23.62
23	.906	63	2.480	175	6.890	.140	3.55	.540	13.71	.940	23.87
24	.945	64	2.520			.150	3.81	.550	13.97	.950	24.13
25	.984	65	2.559			.160	4.06	.560	14.22	.960	24.38
26	1.024	66	2.598			.170	4.31	.570	14.47	.970	24.63
27	1.063	67	2.638	200	7.874	.180	4.57	.580	14.73	.980	24.89
28	1.102	68	2.677	225	8.858	.190	4.82	.590	14.98	.990	25.14
29	1.142	69	2.717	250	9.843	.200	5.08	.600	15.24	1.000	25.40
30	1.181	70	2.756	275	10.827	.210	5.33	.610	15.49		
31	1.220	71	2.795			.220	5.55	.620	15.74	2.000	50.80
32	1.260	72	2.835	300	11.811	.230	5.84	,630	16.00	3.000	76.20
33	1.299	73	2.874	325	12.795	.240	6.09	.640	16.25	4.000	101.60
34	1.339	74	2.913			.250	6.35	.650	16.51	5.000	127.00
35	1.378	75	2.953			.260	6.60	.660	16.76	10.000	254.00
36	1.417	76	2.992			.270	6.85	.670	17.01		
37	1.457	77	3.031			.280	7.11	.680	17.27		
38	1.496	78	3.071			.290	7.36	.690	17.52		
39	1.535	79	3.110			.300	7.62	.700	17.78		
40	1.575	80	3.150			.310	7.87	.710	18.03		

Appendix E

MISCELLANEOUS CHARTS

FRACTIONAL/METRIC CONVERSION

Eighths	Sixteenths	Thirty-Seconds	Milli-meters	Decimal
		1/32		.03125
			1	.03937
	1/16			.0625
			2	.07874
		3/32		.09375
			3	.11811
1/8				.125
		5/32		.15625
			4	.15748
	3/16			.1875
			5	.19685
		7/32		.21875
			6	.23622
1/4				.250
			7	.27559
		9/32		.28125
	5/16			.3125
			8	.31496
		11/32		.34375
			9	.35433
3/8				.375
			10	.3937
		13/32		.40625
			11	.43307
	7/16			.4375
		15/32		.46875
			12	.47244
1/2				.500
			13	.51181

Eighths	Sixteenths	Thirty-Seconds	Milli-meters	Decimal
		17/32		.53125
			14	.55118
	9/16			.5625
			15	.59055
		19/32		.59375
5/8				.625
			16	.62992
		21/32		.65625
			17	.66929
	11/16			.6875
			18	.70866
		23/32		.71875
			19	.74803
3/4				.750
		25/32		.78125
			20	.7874
	13/16			.8125
			21	.82677
		27/32		.84375
			22	.86614
7/8				.875
			23	.90551
		29/32		.90625
	15/16			.9375
			24	.94488
		31/32		.96875
			25	.98425
1″				1.000000

ENGLISH-METRIC CONVERSION

	multiply	by	for equiv. no. of:
ACCELERATION	Foot/sec^2	0.304 8	metre/sec^2 (m/s^2)
	Inch/sec^2	0.025 4	metre/sec^2
TORQUE	Pound-inch	0.112 98	newton-metres (N•m)
	Pound-foot	1.355 8	newton-metres
POWER	horsepower	0.746	kilowatts (kw)
PRESSURE or STRESS	inches of water	0.2488	kilopascals (kPa)
	pounds/sq. in.	6.895	kilopascals (kPa)
ENERGY or WORK	BTU	1 055.	joules (J)
	foot-pound	1.355 8	joules (J)
	kilowatt-hour	3 600 000. or 3.6 x 10^6	joules (J=one W's)
LIGHT	foot candle	10.76	lumens/metre^2 (lm/m^2)
FUEL PERFORMANCE	miles/gal	0.425 1	kilometres/litre (km/l)
	gal/mile	2.352 7	litres/kilometre (l/km)
VELOCITY	miles/hour	1.609 3	kilometres/hr. (km/h)
LENGTH	inch	25.4	millimetres (mm)
	foot	0.304 8	metres (m)
	yard	0.914 4	metres (m)
	mile	1.609	kilometres (km)
AREA	inch^2	645.2	millimetres^2 (mm^2)
		6.45	centimetres^2 (cm^2)
	foot^2	0.092 9	metres^2 (m^2)
	yard^2	0.836 1	metres^2
VOLUME	inch^3	16 387.	mm^3
	inch^3	16.387	cm^3
	quart	0.016 4	litres(1)
	quart	0.946 4	litres
	gallon	3.785 4	litres
	yard^3	0.764 6	metres^3 (m^3)
MASS	pound	0.453 6	kilograms (kg)
	ton	907.18	kilograms (kg)
	ton	0.90718	tonne (t)
FORCE	kilogram	9.807	newtons (N)
	ounce	0.278 0	newtons
	pound	4.448	newtons
TEMPERATURE	degree farenheit	0.556 (°F −32)	degree Celsius (°C)

CYLINDER NUMBERING

IN-LINE ENGINES

VW AND PORSCHE

V-8 ENGINES

V-6

1984–77 GENERAL MOTORS ENGINE IDENTIFICATION

USE THIS CHART TO CORRECTLY IDENTIFY ANY ENGINE INSTALLED IN A GENERAL MOTORS PASSENGER CAR FROM 1984-77. THE ENGINE CODE IS DETERMINED BY REFERRING TO THE V.I.N. (VEHICLE IDENTIFICATION NUMBER), WHICH IS VISIBLE THRU THE LEFT-HAND SIDE OF THE WINDSHIELD.

ENGINE CODE: 1984-81: 8th VIN DIGIT
1980-77: 5th VIN DIGIT

ONCE YOU HAVE DETERMINED THE ENGINE CODE READ DOWN THIS CHART UNTIL YOU HAVE IDENTIFIED THE ENGINE.

ENGINE CODE	YEARS USED	ENGINE		MFG. DIVISION*
		C.I.	LITERS	
A	1977-84	V6-231	3.8	Buick
B	1977	L4-140	2.3	Chevrolet
B	1978-79	V8-350	5.7	Cadillac
B	1981	V8-350	5.7	Buick
B	1982-84	L4-121	2.0	Chevrolet
C	1977	V6-231	3.8	Buick
C	1978-79	V6-196	3.2	Buick
C	1982-84	L4-98	1.6	Chevrolet
D	1977-80	L6-250	4.1	Chevrolet
D*	1981-84	L4-112	1.8	Chevrolet
E	1977-79	L4-98	1.6	Chevrolet
E	1982-84	V6-181	3.0	Buick
F	1977-81	V8-260	4.3	Oldsmobile
F	1982-83	L4-151	2.5	Pontiac
G	1978	V6-231	3.8	Buick
G	1979	V8-305	5.0	Chevrolet
G	1982	L4-112	1.8	Chevrolet
H	1977	V8-350	5.7	Buick
H	1978-84	V8-305	5.0	Chevrolet
I	1977	L4-85	1.4	Chevrolet
J	1977	V8-350	5.7	Buick
J	1978	L4-98	1.6	Chevrolet
J	1979-82	V8-267	4.4	Chevrolet
J	1983-84	L4-112	1.8	Pontiac
K	1977-79	V8-403	6.6	Oldsmobile
K	1980-82	V6-229	3.8	Chevrolet
L	1977-82	V8-350	5.7	Chevrolet
L	1983-84	V6-173	2.8	Chevrolet
M	1978-79	V6-200	3.3	Chevrolet
N*	1978-84	V8-350	5.7	Oldsmobile
O	1979-80	L4-98	1.6	Chevrolet
P	1977	V8-350	5.7	Pontiac
P*	1979-80	V8-260	4.3	Oldsmobile
P	1983-84	L4-121	2.0	Chevrolet
R	1977-80	V8-350	5.7	Oldsmobile
R	1982-84	L4-151	2.5	Pontiac
R	1977	V8-350	5.7	Cadillac
S	1977-79	V8-425	7.0	Cadillac
S	1980-81	V8-265	4.3	Pontiac
S	1983	V8-305	5.0	Chevrolet
T	1977-79	V8-425	7.0	Cadillac
T	1980-81	V8-301	4.9	Pontiac
T*	1982-84	V6-263	4.3	Oldsmobile
U	1977-78	V8-305	5.0	Chevrolet
V	1977-80	L4-151	2.5	Pontiac
V*	1982-84	V6-263	4.3	Oldsmobile
W	1978-81	V8-301	4.9	Pontiac
X	1977	V8-350	5.7	Chevrolet
X	1978-80	V8-350	5.7	Buick
X	1981-84	V6-173	2.8	Chevrolet
Y	1977-79	V8-301	4.9	Pontiac
Y	1980-84	V8-307	5.0	Oldsmobile
Z	1977-79	V8-400	6.6	Pontiac
Z	1981-84	V6-173	2.8	Chevrolet
0	1982-84	L4-112	1.8	Pontiac
1	1978-79	L4-151	2.5	Pontiac
1	1982-84	V6-173	2.8	Chevrolet
2	1978-79	V6-231	3.8	Buick
2	1982-84	L4-151	2.5	Pontiac
3	1978-82	V6-231	3.8	Buick
3	1984	V6-231	3.8	Buick
4	1978-79	V8-350	5.7	Chevrolet
4	1980-84	V6-252	4.1	Buick
5	1980-84	L4-151	2.5	Pontiac
6	1980-84	V8-368	6.0	Cadillac
6	1983-84	V8-350	5.7	Chevrolet
7	1980	V6-173	2.8	Chevrolet
7	1982	V8-305	5.0	Chevrolet
8	1979-80	V8-350	5.7	Chevrolet
8	1982	V8-260	4.3	Oldsmobile
8	1983	V6-231	3.8	Buick
8	1982-84	V8-350	5.7	Chevrolet
8	1982-84	V8-252	4.1	Cadillac
9	1979	L4-151	2.5	Pontiac
9	1980-81	L4-98	1.6	Chevrolet
9	1980-84	V8-368	6.0	Cadillac
9	1983-84	V6-229	3.8	Chevrolet
9	1983-84	V8-307	5.0	Oldsmobile
9	1984	V6-231	3.8	Buick

* This column designates the GM Division which manufactures this engine. This does not indicate which division uses that respective engine.
* = Diesel Engine

IDENTIFICATION OF FORD 351 CU. IN. ENGINES

FORD HAS PRODUCED TWO DIFFERENT 351 CU. IN. ENGINES FROM 1969 THRU 1984. THIS CHART HELPS IDENTIFY WHICH VERSION A PARTICULAR ENGINE IS. CHECK AIR CLEANER COVER DECALS AND EMISSION CONTROL STICKERS UNDERHOOD.

351 C (CLEVELAND)
351 M (MODIFIED)
C–PRODUCED 1969 THRU 1974
M–PRODUCED 1975 THRU 1979
BUILT IN CLEVELAND, OHIO

351 W (WINDSOR)
W–PRODUCED 1969 THRU 1984
BUILT IN WINDSOR, ONTARIO

ITEM	351 C, M	351 W
NUMBER OF BOLTS ATTACHING VALVE COVER ON EXHAUST MANIFOLD SIDE	4	3
TYPE OF FRONT COVER	CAST AS PART OF BLOCK	BOLTED TO BLOCK
FUEL PUMP	BOLTS ONE ABOVE THE OTHER	BOLTS SIDE BY SIDE
THERMOSTAT HOUSING	BOLTS VERTICAL TO FRONT COVER	BOLTS HORIZONTAL TO INTAKE MANIFOLD

LITER TO CUBIC INCH DISPLACEMENT INTERCHANGE

All engines offered by domestic passenger car and light truck manufacturers from 1980 onward are designated by both liter and cubic inch displacement designations. Certain engines available before that year were also marketed using both measurements. The following interchange is provided to aid identification of these late-model domestic vehicle engines.

LITERS	CID	CYL.	FUEL	MANUFACTURER
1.0L	61	L3	Gas	Chevrolet
1.4L	85	L4	Gas	American Motors
1.4L	85	L4	Gas	Chevrolet
1.5L	91	L4	Gas	Chevrolet
1.5L	91	L4	Diesel	Chevrolet
1.6L	97	L4	Gas	Chevrolet
1.6L	98	L4	Gas	Chevrolet
1.6L	98	L4	Gas	Ford
1.7L	105	L4	Gas	American Motors
1.7L	105	L4	Gas	Chrysler
1.8L	110	L4	Diesel	Chevrolet
1.8L	110	L4	Gas	Ford Truck
1.8L	112	L4	Gas	Chevrolet
1.8L	112	L4	Gas	Pontiac
1.9L	116	L4	Gas	Ford
2.0L	121	L4	Gas	American Motors
2.0L	121	L4	Gas	Chevrolet
2.0L	121	L4	Gas	Ford Truck
2.0L	122	L4	Gas	Ford
2.0L	122	L4	Diesel	Ford
2.0L	122	L4	Gas	Ford Truck
2.1L	128	L4	Diesel	Jeep
2.2L	134	L4	Diesel	Ford Truck
2.2L	135	L4	Gas	Chrysler
2.3L	140	L4	Gas	Chevrolet
2.3L	140	L4	Gas	Ford
2.3L	140	L4	Diesel	Ford Truck
2.4L	149	L4	Diesel	Ford
2.5L	150	L4	Gas	American Motors
2.5L	151	L4	Gas	Pontiac
2.5L	153	L4	Gas	Chrysler
2.5L	153	L4	Gas	Ford
2.6L	156	L4	Gas	Chrysler
2.6L	159	V6	Gas	Ford
2.8L	171	V6	Gas	Ford
2.8L	171	V6	Gas	Ford Truck
2.8L	173	V6	Gas	Chevrolet
2.9L	177	V6	Gas	Ford Truck
3.0L	181	V6	Gas	Buick
3.0L	183	V6	Gas	Ford
3.2L	196	V6	Gas	Buick
3.3L	200	V6	Gas	Chevrolet
3.3L	200	L6	Gas	Ford
3.7L	225	L6	Gas	Chrysler
3.8L	229	V6	Gas	Chevrolet
3.8L	230	V6	Gas	Ford
3.8L	231	V6	Gas	Buick
3.9L	240	L6	Gas	Ford Truck
4.1L	250	L6	Gas	Chevrolet
4.1L	250	L6	Gas	Ford
4.1L	252	V6	Gas	Buick
4.1L	252	V8	Gas	Cadillac
4.2L	255	V8	Gas	Ford
4.2L	258	L6	Gas	American Motors
4.3L	260	V6	Diesel	Oldsmobile
4.3L	260	V8	Gas	Oldsmobile
4.3L	260	V8	Diesel	Oldsmobile
4.3L	262	V6	Gas	Chevrolet
4.3L	265	V8	Gas	Pontiac
4.4L	267	V8	Gas	Chevrolet
4.8L	292	L6	Gas	Chevrolet Truck
4.9L	300	L6	Gas	Ford Truck
4.9L	301	V8	Gas	Pontiac
5.0L	302	V8	Gas	Ford
5.0L	304	V8	Gas	Jeep
5.0L	305	V8	Gas	Chevrolet
5.0L	307	V8	Gas	Oldsmobile
5.2L	318	V8	Gas	Chrysler
5.4L	330	V8	Gas	Ford Truck
5.7L	350	V8	Gas	Buick
5.7L	350	V8	Gas	Cadillac
5.7L	350	V8	Gas	Chevrolet
5.7L	350	V8	Gas	Oldsmobile
5.7L	350	V8	Diesel	Oldsmobile
5.7L	350	V8	Gas	Pontiac
5.8L	351M	V8	Gas	Ford
5.8L	351W	V8	Gas	Ford
5.9L	359	V8	Gas	Ford Truck
5.9L	360	V8	Gas	Chrysler
5.9L	360	V8	Gas	Ford Truck
5.9L	360	V8	Gas	Jeep
5.9L	361	V8	Gas	Ford Truck
6.0L	366	V8	Gas	Chevrolet Truck
6.0L	368	V8	Gas	Cadillac
6.1L	370	V8	Gas	Ford Truck
6.2L	381	V8	Diesel	Chevrolet Truck
6.4L	389	V8	Gas	Ford Truck
6.4L	390	V8	Gas	Ford Truck
6.4L	391	V8	Gas	Ford Truck
6.6L	400	V8	Gas	Chevrolet Truck
6.6L	400	V8	Gas	Chrysler
6.6L	400	V8	Gas	Ford
6.6L	400	V8	Gas	Pontiac
6.6L	401	V8	Gas	Ford Truck
6.6L	403	V8	Gas	Oldsmobile
6.9L	420	V8	Diesel	Ford Truck
7.0L	425	V8	Gas	Cadillac
7.0L	427	V8	Gas	Chevrolet Truck
7.0L	428	V8	Gas	Ford Truck
7.0L	429	V8	Gas	Ford Truck
7.2L	440	V8	Gas	Chrysler
7.4L	454	V8	Gas	Chevrolet Truck
7.5L	460	V8	Gas	Ford Truck
7.8L	475	V8	Gas	Ford Truck
7.8L	477	V8	Gas	Ford Truck
8.8L	534	V8	Gas	Ford Truck

CORRECTIVE MACHINING TABLE

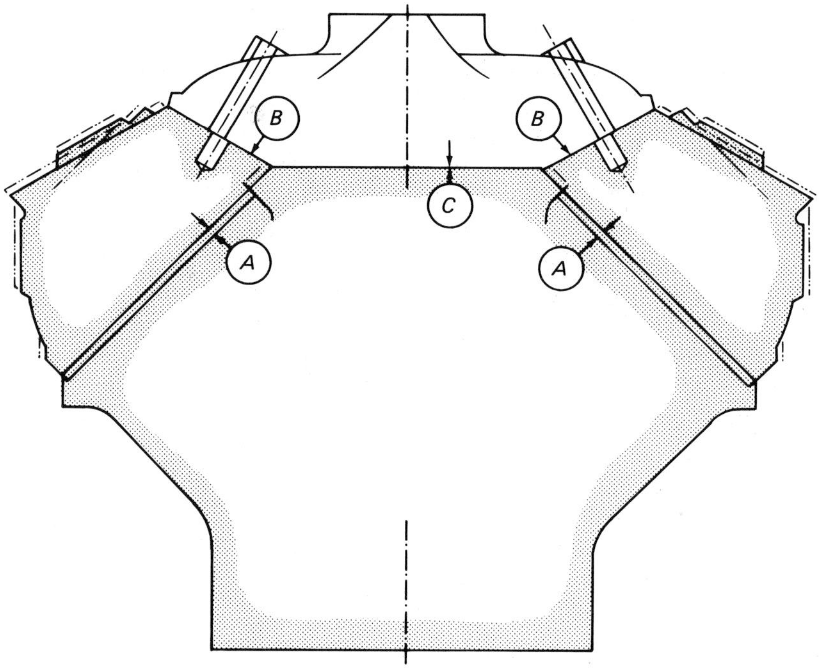

After machining the head at surface *A*, material should be machined from surface *B* for proper intake manifold alignment. Note how the bolt holes in the intake manifold do not line up after machining the head gasket surface *A*.

THE AMOUNT OF CORRECTIVE MACHINING REQUIRED VARIES WITH THE ANGLE AT WHICH THE HEAD IS MADE. HEAD ANGLE IS THE VARIANCE FROM 90°. FOR INSTANCE, A HEAD THAT HAS THE HEAD GASKET AND THE INTAKE MANIFOLD SURFACE MADE AT A RIGHT ANGLE (90°) TO EACH OTHER WOULD BE CLASSIFIED AS A 0° HEAD.

0°	Head angle (Big Block Ford) Amount removed at *A* X 1.0 = Amount to be removed at *B*
5°	Head angle Amount removed at *A* X 1.1 = Amount to be removed at *B*
10°	Head angle (Small Block Chevy and Ford) Amount removed at *A* X 1.2 = Amount to be removed at *B*
15°	Head angle (Chrysler, Y Block Ford) Amount removed at *A* X 1.4 = Amount to be removed at *B*
20°	Head angle (Oldsmobile) Amount removed at *A* X 1.7 = Amount to be removed at *B*
25°	Head angle (Cadillac, Dodge, Plymouth) Amount removed at *A* X 2.0 = Amount to be removed at *B*
30°	Head angle Amount removed at *A* X 3.0 = Amount to be removed at *B*

FASTENER TYPE IDENTIFICATION/TAP DRILL SIZES

Fastener Type Identification

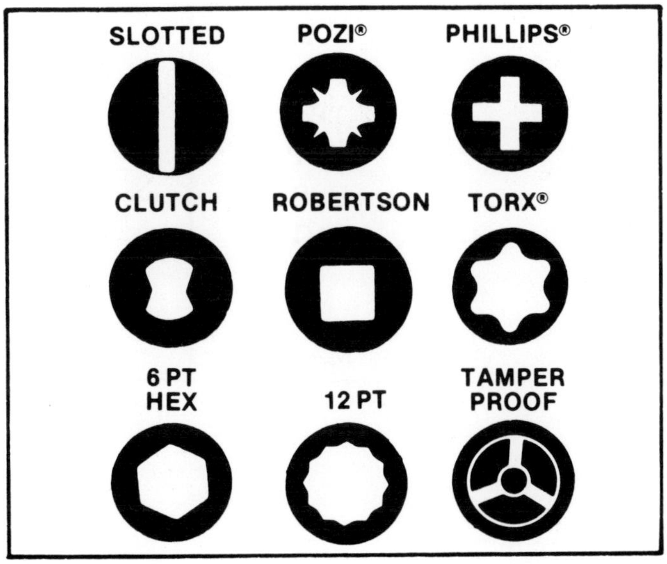

Drill Sizes for Tapping Threads

Screw Size	N. C. Pitch (Coarse)	Drill Size	Decimal Equiv.	N. F. Pitch (Fine)	Drill Size	Decimal Equiv.
No 0	. . .	. . .	. . .	80	3/64	.0469
No. 1	64	#53	.0595	72	#53	.0595
No. 2	56	#50	.0700	64	#50	.7000
No. 3	48	#47	.0785	56	#45	.0820
No. 4	40	#43	.0890	48	#42	.0935
No. 5	40	#38	.1015	44	#37	.1040
No. 6	32	#36	.1065	40	#33	.1130
No. 8	32	#29	.1360	36	#29	.1360
No. 10	24	#25	.1495	32	#21	.1590
No. 12	24	#16	.1770	28	#14	.1820
1/4	20	#7	.2010	28	#3	.2130
5/16	18	F	.2570	24	I	.2720
3/8	16	5/16	.3125	24	Q	.3320
7/16	14	U	.3680	20	25/64	.3906
1/2	13	27/64	.4219	20	29/64	.4531
9/16	12	31/64	.4844	18	33/64	.5156
5/8	11	17/32	.5312	18	37/64	.5781
3/4	10	21/32	.6562	16	11/16	.6875
7/8	9	49/64	.7656	14	13/16	.8125
1	8	7/8	.8750	14	15/16	.9375
1⅛	7	63/64	.9844			

ON-BOARD COMPUTER SYSTEM COMPONENT ABBREVIATIONS

Ford	Chrysler	G.M.	Sensor
ECA	ECU	ECM	On-board Computer
ECT	CTS	CTS	Coolant Temperature
TPS	TPS	TPS	Throttle Position
B/MAP	MAP	MAP	Load Sensor
TSS	IMS	VSS	Speed Sensor
ISC	AIS	IAC	Idle Speed Control
EEC P/R	ASD	—	Power Relay for Computer
BPS	—	BARO	Barometric Pressure
ACT	CTS	MAT	Manifold Air Temperature

FORD EEC IV (ELECTRONIC ENGINE CONTROL—4TH GENERATION)	
MAP	—Manifold Absolute Pressure Sensor (Engine load)
BARO	—Barometric Pressure Sensor (Atmospheric pressure)
TPS	—Throttle Position Sensor (Throttle position)
PIP	—Profile Ignition Pickup (RPM)
ECT	—Engine Coolant Temperature Sensor
EGO	—Exhaust Gas Oxygen Sensor (O_2 exhaust content)
BOO	—Brake On/Off Switch (Car at stop with no load)
TAB	—Thermactor Air By-pass (Air pump management)
TAD	—Thermactor Air Divert (Air pump management)
TSS	—Transmission Speed Sensor (Vehicle speed)
MAF	—Mass Air Flow (Amount of incoming air)
EVP	—Exhaust Valve Position Sensor (EGR position)
VAT	—Vane Air Temperature (Incoming air temperature)
EGRSO	—EGR Shut Off Solenoid (EGR system)
EGRC	—EGR Control solenoid (EGR vacuum control)
EGRV	—EGR Vent solenoid (EGR release vacuum control)
ISC	—Idle Speed Control (Maintains idle speed)
STO	—Self Test Output (Used for diagnostics)
TFI	—Thick Film Ignition (Type of module design)
THERMAC	—Thermostatically Modulated Air Cleaner

CHRYSLER	
CTS	—Coolant Temperature Sensor (Coolant temperature)
CTS	—Charge Temperature Sensor (Incoming air temperature)
AIS	—Automatic Idle Speed Motor (Fuel injection)
IMS	—Inferred Milage Sensor (Vehicle speed)
TPS	—Throttle Position Sensor (Throttle position)
MAP	—Manifold Absolute Pressure (Engine load)
ASD	—Automatic Shut Down Relay (Power to computer)
O_2	—Oxygen Sensor (Oxygen content in exhaust)
ESS	—Engine Speed Sensor (RPM)
ITS	—Intake Temperature Switch (Air temperature)

GENERAL MOTORS	
ECT	—Engine Coolant Temperature (Engine temperature)
ISC	—Idle Speed Control (Carbureted and DFI engines)
IAC	—Idle Air Control (Fuel injection)
VSS	—Vehicle Speed Sensor (Vehicle speed)
MAT	—Manifold Air Temperature (Fuel injection)
MAP	—Manifold Absolute Pressure (Engine load)
BARO	—Barometric Pressure (Atmospheric pressure)
TPS	—Throttle Position Sensor (Throttle position)
EFE	—Early Fuel Evaporation (Heated carburetor grid)
TK	—Throttle Kicker (Idle speed control)
DP	—Differential Pressure Sensor (Vacuum, engine load)
DIS	—Distributorless Ignition System
PROM	—Programmable Read Only Memory (Chip)
C^3I	—Computer Controlled Coil Ignition
ESC	—Electronic Spark Control

BOSCH	
ESS	—Engine Speed Sensor (RPM)
TVS	—Throttle Valve Switch (Throttle position)
ATS	—Air Temperature Sensor (Incoming air temperature)
RMS	—Reference Mark Sensor (Crank position)
ETS	—Engine Temperature Sensor (Coolant temperature)
TTS	—Thermo Time Switch (Cold start assist)
AAD	—Auxiliary-Air Device (Cold idle control)

ANSWERS

SECTION A—GENERAL ENGINE DIAGNOSIS

1. a
2. c
3. b
4. c
5. b
6. c
7. b
8. a
9. d
10. a
11. c

SECTION B—CYLINDER HEAD AND VALVE TRAIN DIAGNOSIS AND REPAIR

1. d
2. c
3. b
4. c
5. c
6. d
7. d
8. b
9. b
10. c
11. d
12. c
13. d
14. a
15. c
16. c
17. a

SECTION C—ENGINE BLOCK DIAGNOSIS AND REPAIR

1. d
2. b
3. c
4. a
5. c
6. a
7. c
8. d
9. b
10. d
11. d
12. b
13. d
14. c
15. a
16. a
17. b
18. d
19. c
20. c

SECTION D—LUBRICATION AND COOLING SYSTEMS DIAGNOSIS AND REPAIR

1. c
2. b
3. a
4. a
5. c
6. b
7. c
8. c
9. c
10. a
11. c

SECTION E—IGNITION SYSTEM DIAGNOSIS AND REPAIR

1. c
2. b
3. c
4. c
5. c
6. c
7. b
8. a
9. c

SECTION F—FUEL AND EXHAUST SYSTEMS DIAGNOSIS AND REPAIR

1. a
2. c
3. d
4. d
5. d
6. a
7. c
8. c

SECTION G—BATTERY AND STARTING SYSTEM DIAGNOSIS AND REPAIR

1. d	4. d
2. d	5. c
3. c	6. c

BOOK SUMMARY QUESTIONS

1. b	26. c	51. c	76. c	101. d
2. c	27. c	52. c	77. b	102. c
3. c	28. c	53. a	78. b	103. d
4. c	29. c	54. d	79. b	104. c
5. c	30. a	55. d	80. a	105. c
6. c	31. a	56. b	81. b	106. c
7. d	32. a	57. a	82. c	107. a
8. a	33. a	58. d	83. b	108. c
9. c	34. d	59. c	84. a	109. c
10. c	35. c	60. d	85. c	110. d
11. a	36. b	61. c	86. d	111. c
12. c	37. d	62. c	87. b	112. a
13. c	38. a	63. a	88. c	113. c
14. b	39. a	64. c	89. a	114. d
15. c	40. b	65. c	90. b	115. d
16. d	41. b	66. d	91. d	116. c
17. b	42. a	67. a	92. d	117. c
18. c	43. c	68. d	93. a	118. a
19. b	44. c	69. d	94. c	119. b
20. b	45. c	70. d	95. a	120. c
21. c	46. c	71. b	96. b	121. c
22. a	47. a	72. b	97. d	122. d
23. b	48. d	73. b	98. b	123. c
24. d	49. d	74. a	99. b	124. c
25. a	50. b	75. a	100. c	

INDEX